P9-CLH-040

LAP

DISCARD

Grzimek's
Animal Life Encyclopedia

Second Edition

● ● ● ●

Grzimek's
Animal Life Encyclopedia

Second Edition

●●●●

Volume 6
Amphibians

William E. Duellman, Advisory Editor
Neil Schlager, Editor

Joseph E. Trumpey, Chief Scientific Illustrator

Michael Hutchins, Series Editor
In association with the American Zoo and Aquarium Association

GALE®

Detroit • New York • San Diego • San Francisco • Cleveland • New Haven, Conn. • Waterville, Maine • London • Munich

THOMSON

GALE

Grzimek's Animal Life Encyclopedia, Second Edition

Volume 6: Amphibians

Produced by Schlager Group Inc.

Neil Schlager, Editor

Vanessa Torrado-Caputo, Assistant Editor

Project Editor
Melissa C. McDade

Editorial
Stacey Blachford, Deirdre Blanchfield,
Madeline Harris, Christine Jeryan, Kate
Kretschmann, Mark Springer

Permissions
Margaret Chamberlain

Imaging and Multimedia
Randy Bassett, Mary K. Grimes, Lezlie Light,
Christine O'Bryan, Barbara Yarrow, Robyn V.
Young

Product Design
Tracey Rowens, Jennifer Wahi

Manufacturing
Wendy Blurton, Dorothy Maki, Evi Seoud, Mary
Beth Trimper

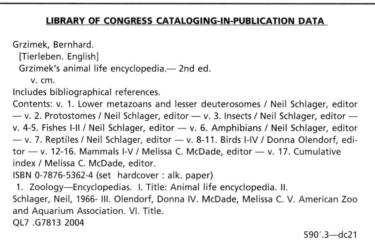

LIBRARY OF CONGRESS CATALOGING-IN-PUBLICATION DATA

Grzimek, Bernhard.
  [Tierleben. English]
  Grzimek's animal life encyclopedia.— 2nd ed.
    v. cm.
  Includes bibliographical references.
  Contents: v. 1. Lower metazoans and lesser deuterosomes / Neil Schlager, editor
  — v. 2. Protostomes / Neil Schlager, editor — v. 3. Insects / Neil Schlager, editor —
  v. 4-5. Fishes I-II / Neil Schlager, editor — v. 6. Amphibians / Neil Schlager, editor
  — v. 7. Reptiles / Neil Schlager, editor — v. 8-11. Birds I-IV / Donna Olendorf, editor — v. 12-16. Mammals I-V / Melissa C. McDade, editor — v. 17. Cumulative
  index / Melissa C. McDade, editor.
  ISBN 0-7876-5362-4 (set hardcover : alk. paper)
  1. Zoology—Encyclopedias. I. Title: Animal life encyclopedia. II.
  Schlager, Neil, 1966- III. Olendorf, Donna IV. McDade, Melissa C. V. American Zoo
  and Aquarium Association. VI. Title.
  QL7 .G7813 2004

                                                                590'.3—dc21
                                                                2002003351

Printed in Canada
10 9 8 7 6 5 4 3 2 1

Recommended citation: *Grzimek's Animal Life Encyclopedia*, 2nd edition. Volume 6, *Amphibians*, edited by Michael Hutchins, William E. Duellman, and Neil Schlager. Farmington Hills, MI: Gale Group, 2003.

Contents

· · · · ·

Foreword

Earth is teeming with life. No one knows exactly how many distinct organisms inhabit our planet, but more than 5 million different species of animals and plants could exist, ranging from microscopic algae and bacteria to gigantic elephants, redwood trees and blue whales. Yet, throughout this wonderful tapestry of living creatures, there runs a single thread: Deoxyribonucleic acid or DNA. The existence of DNA, an elegant, twisted organic molecule that is the building block of all life, is perhaps the best evidence that all living organisms on this planet share a common ancestry. Our ancient connection to the living world may drive our curiosity, and perhaps also explain our seemingly insatiable desire for information about animals and nature. Noted zoologist, E.O. Wilson, recently coined the term "biophilia" to describe this phenomenon. The term is derived from the Greek *bios* meaning "life" and *philos* meaning "love." Wilson argues that we are human because of our innate affinity to and interest in the other organisms with which we share our planet. They are, as he says, "the matrix in which the human mind originated and is permanently rooted." To put it simply and metaphorically, our love for nature flows in our blood and is deeply engrained in both our psyche and cultural traditions.

Our own personal awakenings to the natural world are as diverse as humanity itself. I spent my early childhood in rural Iowa where nature was an integral part of my life. My father and I spent many hours collecting, identifying and studying local insects, amphibians and reptiles. These experiences had a significant impact on my early intellectual and even spiritual development. One event I can recall most vividly. I had collected a cocoon in a field near my home in early spring. The large, silky capsule was attached to a stick. I brought the cocoon back to my room and placed it in a jar on top of my dresser. I remember waking one morning and, there, perched on the tip of the stick was a large moth, slowly moving its delicate, light green wings in the early morning sunlight. It took my breath away. To my inexperienced eyes, it was one of the most beautiful things I had ever seen. I knew it was a moth, but did not know which species. Upon closer examination, I noticed two moon-like markings on the wings and also noted that the wings had long "tails", much like the ubiquitous tiger swallow-tail butterflies that visited the lilac bush in our backyard. Not wanting to suffer my ignorance any longer, I reached immediately for my *Golden Guide to North*

American Insects and searched through the section on moths and butterflies. It was a luna moth! My heart was pounding with the excitement of new knowledge as I ran to share the discovery with my parents.

I consider myself very fortunate to have made a living as a professional biologist and conservationist for the past 20 years. I've traveled to over 30 countries and six continents to study and photograph wildlife or to attend related conferences and meetings. Yet, each time I encounter a new and unusual animal or habitat my heart still races with the same excitement of my youth. If this is biophilia, then I certainly possess it, and it is my hope that others will experience it too. I am therefore extremely proud to have served as the series editor for the Gale Group's rewrite of *Grzimek's Animal Life Encyclopedia*, one of the best known and widely used reference works on the animal world. *Grzimek's* is a celebration of animals, a snapshot of our current knowledge of the Earth's incredible range of biological diversity. Although many other animal encyclopedias exist, *Grzimek's Animal Life Encyclopedia* remains unparalleled in its size and in the breadth of topics and organisms it covers.

The revision of these volumes could not come at a more opportune time. In fact, there is a desperate need for a deeper understanding and appreciation of our natural world. Many species are classified as threatened or endangered, and the situation is expected to get much worse before it gets better. Species extinction has always been part of the evolutionary history of life; some organisms adapt to changing circumstances and some do not. However, the current rate of species loss is now estimated to be 1,000–10,000 times the normal "background" rate of extinction since life began on Earth some 4 billion years ago. The primary factor responsible for this decline in biological diversity is the exponential growth of human populations, combined with peoples' unsustainable appetite for natural resources, such as land, water, minerals, oil, and timber. The world's human population now exceeds 6 billion, and even though the average birth rate has begun to decline, most demographers believe that the global human population will reach 8–10 billion in the next 50 years. Much of this projected growth will occur in developing countries in Central and South America, Asia and Africa-regions that are rich in unique biological diversity.

Finding solutions to conservation challenges will not be easy in today's human-dominated world. A growing number of people live in urban settings and are becoming increasingly isolated from nature. They "hunt" in super markets and malls, live in apartments and houses, spend their time watching television and searching the World Wide Web. Children and adults must be taught to value biological diversity and the habitats that support it. Education is of prime importance now while we still have time to respond to the impending crisis. There still exist in many parts of the world large numbers of biological "hotspots"-places that are relatively unaffected by humans and which still contain a rich store of their original animal and plant life. These living repositories, along with selected populations of animals and plants held in professionally managed zoos, aquariums and botanical gardens, could provide the basis for restoring the planet's biological wealth and ecological health. This encyclopedia and the collective knowledge it represents can assist in educating people about animals and their ecological and cultural significance. Perhaps it will also assist others in making deeper connections to nature and spreading biophilia. Information on the conservation status, threats and efforts to preserve various species have been integrated into this revision. We have also included information on the cultural significance of animals, including their roles in art and religion.

It was over 30 years ago that Dr. Bernhard Grzimek, then director of the Frankfurt Zoo in Frankfurt, Germany, edited the first edition of *Grzimek's Animal Life Encyclopedia*. Dr. Grzimek was among the world's best known zoo directors and conservationists. He was a prolific author, publishing nine books. Among his contributions were: *Serengeti Shall Not Die*, *Rhinos Belong to Everybody* and *He and I and the Elephants*. Dr. Grzimek's career was remarkable. He was one of the first modern zoo or aquarium directors to understand the importance of zoo involvement in *in situ* conservation, that is, of their role in preserving wildlife in nature. During his tenure, Frankfurt Zoo became one of the leading western advocates and supporters of wildlife conservation in East Africa. Dr. Grzimek served as a Trustee of the National Parks Board of Uganda and Tanzania and assisted in the development of several protected areas. The film he made with his son Michael, *Serengeti Shall Not Die*, won the 1959 Oscar for best documentary.

Professor Grzimek has recently been criticized by some for his failure to consider the human element in wildlife conservation. He once wrote: "A national park must remain a primordial wilderness to be effective. No men, not even native ones, should live inside its borders." Such ideas, although considered politically incorrect by many, may in retrospect actually prove to be true. Human populations throughout Africa continue to grow exponentially, forcing wildlife into small islands of natural habitat surrounded by a sea of humanity. The illegal commercial bushmeat trade-the hunting of endangered wild animals for large scale human consumption-is pushing many species, including our closest relatives, the gorillas, bonobos and chimpanzees, to the brink of extinction. The trade is driven by widespread poverty and lack of economic alternatives. In order for some species to survive it will be necessary, as Grzimek suggested, to establish and enforce a system of protected areas where wildlife can roam free from exploitation of any kind.

While it is clear that modern conservation must take the needs of both wildlife and people into consideration, what will the quality of human life be if the collective impact of short-term economic decisions is allowed to drive wildlife populations into irreversible extinction? Many rural populations living in areas of high biodiversity are dependent on wild animals as their major source of protein. In addition, wildlife tourism is the primary source of foreign currency in many developing countries and is critical to their financial and social stability. When this source of protein and income is gone, what will become of the local people? The loss of species is not only a conservation disaster; it also has the potential to be a human tragedy of immense proportions. Protected areas, such as national parks, and regulated hunting in areas outside of parks are the only solutions. What critics do not realize is that the fate of wildlife and people in developing countries is closely intertwined. Forests and savannas emptied of wildlife will result in hungry, desperate people, and will, in the long-term lead to extreme poverty and social instability. Dr. Grzimek's early contributions to conservation should be recognized, not only as benefiting wildlife, but as benefiting local people as well.

Dr. Grzimek's hope in publishing his *Animal Life Encyclopedia* was that it would "...disseminate knowledge of the animals and love for them", so that future generations would "...have an opportunity to live together with the great diversity of these magnificent creatures." As stated above, our goals in producing this updated and revised edition are similar. However, our challenges in producing this encyclopedia were more formidable. The volume of knowledge to be summarized is certainly much greater in the twenty-first century than it was in the 1970's and 80's. Scientists, both professional and amateur, have learned and published a great deal about the animal kingdom in the past three decades, and our understanding of biological and ecological theory has also progressed. Perhaps our greatest hurdle in producing this revision was to include the new information, while at the same time retaining some of the characteristics that have made *Grzimek's Animal Life Encyclopedia* so popular. We have therefore strived to retain the series' narrative style, while giving the information more organizational structure. Unlike the original *Grzimek's*, this updated version organizes information under specific topic areas, such as reproduction, behavior, ecology and so forth. In addition, the basic organizational structure is generally consistent from one volume to the next, regardless of the animal groups covered. This should make it easier for users to locate information more quickly and efficiently. Like the original Grzimek's, we have done our best to avoid any overly technical language that would make the work difficult to understand by non-biologists. When certain technical expressions were necessary, we have included explanations or clarifications.

Considering the vast array of knowledge that such a work represents, it would be impossible for any one zoologist to have completed these volumes. We have therefore sought specialists from various disciplines to write the sections with which they are most familiar. As with the original *Grzimek's*, we have engaged the best scholars available to serve as topic editors, writers, and consultants. There were some complaints about inaccuracies in the original English version that may have been due to mistakes or misinterpretation during the complicated translation process. However, unlike the original *Grzimek's*, which was translated from German, this revision has been completely re-written by English-speaking scientists. This work was truly a cooperative endeavor, and I thank all of those dedicated individuals who have written, edited, consulted, drawn, photographed, or contributed to its production in any way. The names of the topic editors, authors, and illustrators are presented in the list of contributors in each individual volume.

The overall structure of this reference work is based on the classification of animals into naturally related groups, a discipline known as taxonomy or biosystematics. Taxonomy is the science through which various organisms are discovered, identified, described, named, classified and catalogued. It should be noted that in preparing this volume we adopted what might be termed a conservative approach, relying primarily on traditional animal classification schemes. Taxonomy has always been a volatile field, with frequent arguments over the naming of or evolutionary relationships between various organisms. The advent of DNA fingerprinting and other advanced biochemical techniques has revolutionized the field and, not unexpectedly, has produced both advances and confusion. In producing these volumes, we have consulted with specialists to obtain the most up-to-date information possible, but knowing that new findings may result in changes at any time. When scientific controversy over the classification of a particular animal or group of animals existed, we did our best to point this out in the text.

Readers should note that it was impossible to include as much detail on some animal groups as was provided on others. For example, the marine and freshwater fish, with vast numbers of orders, families, and species, did not receive as detailed a treatment as did the birds and mammals. Due to practical and financial considerations, the publishers could provide only so much space for each animal group. In such cases, it was impossible to provide more than a broad overview and to feature a few selected examples for the purposes of illustration. To help compensate, we have provided a few key bibliographic references in each section to aid those interested in learning more. This is a common limitation in all reference works, but *Grzimek's Encyclopedia of Animal Life* is still the most comprehensive work of its kind.

I am indebted to the Gale Group, Inc. and Senior Editor Donna Olendorf for selecting me as Series Editor for this project. It was an honor to follow in the footsteps of Dr. Grzimek and to play a key role in the revision that still bears his name. *Grzimek's Animal Life Encyclopedia* is being published by the Gale Group, Inc. in affiliation with my employer, the American Zoo and Aquarium Association (AZA), and I would like to thank AZA Executive Director, Sydney J. Butler; AZA Past-President Ted Beattie (John G. Shedd Aquarium, Chicago, IL); and current AZA President, John Lewis (John Ball Zoological Garden, Grand Rapids, MI), for approving my participation. I would also like to thank AZA Conservation and Science Department Program Assistant, Michael Souza, for his assistance during the project. The AZA is a professional membership association, representing 205 accredited zoological parks and aquariums in North America. As Director/William Conway Chair, AZA Department of Conservation and Science, I feel that I am a philosophical descendant of Dr. Grzimek, whose many works I have collected and read. The zoo and aquarium profession has come a long way since the 1970s, due, in part, to innovative thinkers such as Dr. Grzimek. I hope this latest revision of his work will continue his extraordinary legacy.

Silver Spring, Maryland, 2001
Michael Hutchins
Series Editor

How to use this book

Grzimek's Animal Life Encyclopedia is an internationally prominent scientific reference compilation, first published in German in the late 1960s, under the editorship of zoologist Bernhard Grzimek (1909–1987). In a cooperative effort between Gale and the American Zoo and Aquarium Association, the series has been completely revised and updated for the first time in over 30 years. Gale expanded the series from 13 to 17 volumes, commissioned new color paintings, and updated the information so as to make the set easier to use. The order of revisions is:

Volumes 8–11: Birds I–IV
Volume 6: Amphibians
Volume 7: Reptiles
Volumes 4–5: Fishes I–II
Volumes 12–16: Mammals I–V
Volume 3: Insects
Volume 2: Protostomes
Volume 1: Lower Metazoans and Lesser Deuterostomes
Volume 17: Cumulative Index

Organized by taxonomy

The overall structure of this reference work is based on the classification of animals into naturally related groups, a discipline known as taxonomy—the science in which various organisms are discovered, identified, described, named, classified, and catalogued. Starting with the simplest life forms, the lower metazoans and lesser deuterostomes, in volume 1, the series progresses through the more advanced classes of classes, culminating with the mammals in volumes 12–16. Volume 17 is a stand-alone cumulative index.

Organization of chapters within each volume reinforces the taxonomic hierarchy. In the case of the volume on Amphibians, introductory chapters describe general characteristics of the class Amphibia, followed by taxonomic chapters dedicated to order and family. Species accounts appear at the end of family chapters. To help the reader grasp the scientific arrangement, each type of taxonomic chapter has a distinctive color and symbol:

▲ = Family Chapter (yellow background)

● = Order Chapter (blue background)

As chapters narrow in focus, they become more tightly formatted. Introductory chapters have a loose structure, reminiscent of the first edition. Although not strictly formatted, chapters on orders are carefully structured to cover basic information about the group. Chapters on families are the most tightly structured, following a prescribed format of standard rubrics that make information easy to find. These chapters typically include:

Thumbnail introduction
   Common name
   Scientific name
   Class
   Order
   Suborder
   Family
   Thumbnail description
   Size
   Number of genera, species
   Habitat
   Conservation status
Main chapter
   Evolution and systematics
   Physical characteristics
   Distribution
   Habitat
   Behavior
   Feeding ecology and diet
   Reproductive biology
   Conservation status
   Significance to humans
Species accounts
   Common name
   Scientific name
   Subfamily
   Taxonomy
   Other common names
   Physical characteristics
   Distribution
   Habitat
   Behavior
   Feeding ecology and diet
   Reproductive biology
   Conservation status
   Significance to humans

Color graphics enhance understanding

Grzimek's features approximately 3,500 color photos, including 120 in the Amphibians volume; 3,500 total color maps, including over 150 in the Amphibians volume; and approximately 5,500 total color illustrations, including more than 300 in the Amphibians volume. Each featured species of animal is accompanied by both a distribution map and an illustration.

All maps in *Grzimek's* were created specifically for the project by XNR Productions. Distribution information was provided by expert contributors and, if necessary, further researched at the University of Michigan Zoological Museum library. Maps are intended to show broad distribution, not definitive ranges.

All the color illustrations in *Grzimek's* were created specifically for the project by Michigan Science Art. Expert contributors recommended the species to be illustrated and provided feedback to the artists, who supplemented this information with authoritative references and animal skins from University of Michigan Zoological Museum library. In addition to illustrations of species, *Grzimek's* features drawings that illustrate characteristic traits and behaviors.

About the contributors

All of the chapters were written by herpetologists who are specialists on specific subjects and/or families. Topic editor William E. Duellman reviewed the completed chapters to insure consistency and accuracy.

Standards employed

In preparing the volume on Amphibians, the editors relied primarily on the taxonomic structure outlined in *Herpetology: An Introductory Biology of Amphibians and Reptiles*, 2nd edition, edited by George R. Zug, Laurie J. Vitt, and Janalee P. Caldwell (2001). Systematics is a dynamic discipline in that new species are being discovered continuously, and new techniques (e.g., DNA sequencing) frequently result in changes in the hypothesized evolutionary relationships among various organisms. Consequently, controversy often exists regarding classification of a particular animal or group of animals; such differences are mentioned in the text.

Grzimek's has been designed with ready reference in mind, and the editors have standardized information wherever feasible. For **Conservation Status**, *Grzimek's* follows the IUCN Red List system, developed by its Species Survival Commission. The Red List provides the world's most comprehensive inventory of the global conservation status of plants and animals. Using a set of criteria to evaluate extinction risk, the IUCN recognizes the following categories: Extinct, Extinct in the Wild, Critically Endangered, Endangered, Vulnerable, Conservation Dependent, Near Threatened, Least Concern, and Data Deficient. For a complete explanation of each category, visit the IUCN web page at <http://www.iucn.org/themes/ssc/redlists/categor.htm>.

In addition to IUCN ratings, chapters may contain other conservation information, such as a species' inclusion on one of three Convention on International Trade in Endangered Species (CITES) appendices. Adopted in 1975, CITES is a global treaty whose focus is the protection of plant and animal species from unregulated international trade.

In the species accounts throughout the volume, the editors have attempted to provide common names not only in English but also in French, German, and Spanish. Unlike for birds, there is no official list of common names for amphibians of the world, but for species in North America an official list does exist: *Scientific and Standard English Names of Amphibians and Reptiles of North America, North of Mexico, with Comments Regarding Confidence in our Understanding*, edited by Brian I. Crother (2000). A consensus of acceptable common names in English, French, German, Portuguese, and Spanish for European species exists in the *Atlas of Amphibians and Reptiles in Europe*, edited by Jean-Pierre Gasc, et al. (1997). Two books purportedly contain common names of amphibians worldwide, but these are names mostly coined by the authors and do not necessarily reflect what the species are called in their native countries. The first of these books, *Dictionary of Animal Names in Five Languages. Amphibians and Reptiles*, by Natalia B. Anajeva, et al. (1988), contains names in Latin, Russian, English, German, and French. The second is *A Complete Guide to Scientific Names of Reptiles and Amphibians of the World*, by Norman Frank and Erica Ramus (1995); for those species for which no commonly accepted common name exists, the name proposed in this book has been used in the volume on Amphibians.

Grzimek's provides the following standard information on lineage in the **Taxonomy** rubric of each species account: [First described as] *Ophryophryne microstoma* [by] Boulenger, [in] 1903, [based on a specimen from] Tonkin, Vietnam. The person's name and date refer to earliest identification of a species, although the species name may have changed since first identification. However, the entity of amphibian is the same.

Anatomical illustrations

While the encyclopedia attempts to minimize scientific jargon, readers will encounter numerous technical terms related to anatomy and physiology throughout the volume. To assist readers in placing physiological terms in their proper context, we have created a number of detailed anatomical drawings. These can be found on pages 16 to 26 in the "Structure and function" chapter. Readers are urged to make heavy use of these drawings. In addition, terms are defined in the **Glossary** at the back of the book.

Appendices and index

In addition to the main text and the aforementioned *Glossary,* the volume contains numerous other elements. *For further reading* directs readers to additional sources of information about amphibians. Valuable contact information for *Organizations* is also included in an appendix. An exhaustive *Amphibians species list* records all known species of amphibians as of November 2002, based on information in Amphibian Species of the World <http://research.amnh.org/herpetology/amphibia/> and organized according to *Herpetology,* 2nd edition, by Zug, Vitt, and Caldwell; further information was obtained from AmphibiaWeb <http://www.amphibiaweb.org>. And a full-color *Geologic time scale* helps readers understand prehistoric time periods. Additionally, the volume contains a *Subject index.*

Acknowledgements

Gale would like to thank several individuals for their important contributions to the volume. Dr. William E. Duellman, advisory editor for the Amphibians volume, oversaw all phases of the volume, including creation of the topic list, chapter review, and compilation of the appendices. Neil Schlager, project manager for the Amphibians volume, coordinated the writing and editing of the text. Dr. Michael Hutchins, chief consulting editor for the series, and Michael Souza, program assistant of conservation and science at the American Zoo and Aquarium Association, provided valuable input and research support.

Library advisors

James Bobick
Head, Science & Technology Department
Carnegie Library of Pittsburgh
Pittsburgh, Pennsylvania

Linda L. Coates
Associate Director of Libraries
Zoological Society of San Diego Library
San Diego, California

Lloyd Davidson, PhD
Life Sciences bibliographer and head, Access Services
Seeley G. Mudd Library for Science and Engineering
Evanston, Illinois

Thane Johnson
Librarian
Oklahoma City Zoo
Oklahoma City, Oklahoma

Charles Jones
Library Media Specialist
Plymouth Salem High School
Plymouth, Michigan

Ken Kister
Reviewer/General Reference teacher
Tampa, Florida

Richard Nagler
Reference Librarian
Oakland Community College
Southfield Campus
Southfield, Michigan

Roland Person
Librarian, Science Division
Morris Library
Southern Illinois University
Carbondale, Illinois

Contributing writers

Amphibians

Kraig Adler, PhD
Cornell University
Ithaca, New York

Ronald Altig, PhD
Mississippi State University
Mississippi State, Mississippi

Janalee P. Caldwell, PhD
University of Oklahoma
Norman, Oklahoma

David Cannatella, PhD
Section of Integrative Biology and
    Texas Memorial Museum
University of Texas
Austin, Texas

Alan Channing, PhD
University of the Western Cape
South Africa

Martha Lynn Crump, PhD
Northern Arizona University
Flagstaff, Arizona

Margaret Davies, PhD
University of Adelaide
South Australia, Australia

Alain Dubois, Docteur d'Etat
Museum National d'Histoire Na-
    turelle
Paris, France

William E. Duellman, PhD
Natural History Museum and Biodi-
    versity Research Center
University of Kansas
Lawrence, Kansas

Harold A. Dundee, PhD
Tulane University Museum of Natural
    History
Belle Chasse, Louisiana

Jinzhong Fu, PhD
University of Guelph
Guelph, Ontario
Canada

Frank Glaw, PhD
Zoologische Staatssammlung
Munich, Germany

David M. Green, PhD
Redpath Museum
McGill University
Montreal, Quebec
Canada

Tim R. Halliday, PhD
Open University
United Kingdom

Amy Lathrop, MA
Royal Ontario Museum
Toronto, Ontario
Canada

Michel Laurin, PhD
Centre National de Recherche Scien-
    tifique, UMR 8570
Paris, France

Anne M. Maglia, PhD
University of Missouri-Rolla
Rolla, Missouri

Max A. Nickerson, PhD
University of Florida
Gainesville, Florida

Ronald A. Nussbaum, PhD
Museum of Zoology
University of Michigan
Ann Arbor, Michigan

José P. Pombal, Jr., PhD
Museu Nacional
Universidade Federal do Rio de Janeiro
Rio de Janeiro, Brazil

Arne Schiøtz, DSc
Zoologisk Museum
Copenhagen, Denmark

Stanley K. Sessions, PhD
Hartwick College
Oneonta, New York

H. Bradley Shaffer, PhD
University of California
Davis, California

Kyle B. Summers, PhD
East Carolina University
Greenville, North Carolina

Linda Trueb, PhD
Natural History Museum and Biodi-
    versity Research Center
University of Kansas
Lawrence, Kansas

David B. Wake, PhD
Museum of Vertebrate Zoology
University of California
Berkeley, California

Marvalee H. Wake, PhD
University of California
Berkeley, California

Richard J. Wassersug, PhD
Dalhousie University
Halifax, Nova Scotia
Canada

Kentwood D. Wells, PhD
Department of Ecology & Evolution-
    ary Biology
University of Connecticut
Storrs, Connecticut

Erik R. Wild, PhD
University of Wisconsin-Stevens Point
Stevens Point, Wisconsin

Jeffery Wilkinson, PhD
California Academy of Sciences
San Francisco, California

Richard G. Zweifel, PhD
American Museum of Natural History
New York, New York

Contributing illustrators

Drawings by Michigan Science Art

Joseph E. Trumpey, Director, AB, MFA
Science Illustration, School of Art and Design, University
of Michigan

Wendy Baker, ADN, BFA

Brian Cressman, BFA, MFA

Emily S. Damstra, BFA, MFA

Maggie Dongvillo, BFA

Barbara Duperron, BFA, MFA

Dan Erickson, BA, MS

Patricia Ferrer, AB, BFA, MFA

Gillian Harris, BA

Jonathan Higgins, BFA, MFA

Amanda Humphrey, BFA

Jacqueline Mahannah, BFA, MFA

John Megahan, BA, BS, MS

Michelle L. Meneghini, BFA, MFA

Bruce D. Worden, BFA

*Thanks are due to the University of Michigan, Museum of Zoology,
which provided specimens that served as models for the images.*

Maps by XNR Productions

Paul Exner, Chief cartographer
XNR Productions, Madison, WI

Tanya Buckingham

Jon Daugherity

Laura Exner

Andy Grosvold

Cory Johnson

Paula Robbins

· · · · ·

Topic overviews

What is an amphibian?

Early evolution and fossil history

Structure and function

Reproduction

Larvae

Behavior

Amphibians and humans

Conservation

What is an amphibian?

Almost everyone recognizes a fish, a bird, or a mammal, even a reptile. But what about an amphibian? Most people recognize frogs and toads as amphibians, but these animals are not the only Amphibia, a class of vertebrates (back-boned animals). There are three living groups of amphibians. The most generalized are salamanders, order Caudata (= with tail), having a cylindrical body, long tail, distinct head and neck, and usually well-developed limbs of approximately equal length. Most salamanders are terrestrial, but some are aquatic, a few are burrowers, and some others are arboreal. Frogs, order Anura (= without tail), have a robust body continuous with the head, no tail, and long hind limbs. Most frogs are terrestrial or arboreal, but many are aquatic, and a few are burrowers. The third group contains the caecilians, order Gymnophiona, also called Apoda (= without foot). These limbless amphibians superficially resemble earthworms and have blunt heads and tails, and their elongate bodies are encircled by grooves (annuli). A few caecilians are aquatic, but most burrow in soil in tropical regions of the world.

Defining characteristics

In some ways amphibians are intermediate between the fully aquatic fishes and the terrestrial amniotes (reptiles, birds, and mammals), but they are not simply transitional in their morphology, life history, ecology, and behavior. During their nearly 350 million years of evolution, amphibians have undergone a remarkable adaptive radiation, and the living groups exhibit a greater diversity of life history than any other group of vertebrates.

Basically, amphibians can be defined as quadrupedal vertebrates (four-legged, or tetrapods) with a skull having two occipital condyles (articulating surfaces with the first element of the vertebral column). The attachment of the pelvic girdle to the vertebral column incorporates only one sacral vertebra. In anurans (frogs and toads), the postsacral vertebrae are fused into a rodlike structure, the urostyle (coccyx), and a tail is absent. Caecilians and some salamanders lack limbs and girdles, whereas in anurans the hind limbs are elongated and modified for jumping. The skin is glandular and contains both mucous and poison glands but lacks external structures such as scales, feathers, or hair, characteristic of other groups of tetrapods. The heart has three chambers, two atria and one ventricle,

which may be partially divided. The aortic arches are symmetrical. Typically, amphibians have two lungs, but the lungs may be reduced or absent in some salamanders, and the left lung is proportionately small in most caecilians (as it is in snakes). Some features are unique to amphibians, all of which have teeth that consist of a pedicel and a crown, and specialized papillae for sound reception in the inner ear. Amphibians are ectotherms (cold-blooded). They are unable to regulate their body temperatures physiologically, as do birds and mammals; therefore, their body temperatures approximate those of the immediate environment, especially the substrate.

The life histories of amphibians are highly diverse. The classic amphibian life history of aquatic eggs and larvae is only one of many modes of reproduction, which include direct development of terrestrial eggs (no aquatic larval stage) and live birth. The eggs of amphibians lack a shell and the embryonic membranes (e.g., amnion, allantois, and chorion) of reptiles, birds, and mammals. Instead, amphibian eggs are protected only by mucoid capsules that are highly permeable; thus, amphibian eggs must develop in moist situations.

Phylogenetic relationships and classification

The living groups of amphibians are most closely allied with diverse fossils, the basal tetrapod vertebrates commonly placed in the class Amphibia. The phylogenetic relationships among these groups of fossils is equivocal. Based on morphological and molecular evidence, salamanders and anurans form a monophyletic group (i.e., have a common ancestor) and together are referred to as batrachians. Batrachians and caecilians form another monophyletic group, the lissamphibians.

Classification reflects biologists' knowledge of the relationships of groups of organisms. Consequently, as new characteristics, both morphological and molecular, as well as behavioral and developmental, are discovered and analyzed, the classification changes. New evidence may reveal that a group of species or genera that were once believed to be members of one family are actually more closely related to another group or are not related to the family with which they formerly were associated. For example, salamanders in the families Dicamptodontidae and Rhyacotritonidae formerly were placed in the Ambystomatidae. Likewise, African treefrogs now recognized as the family Hyperoliidae formerly were in the Rhacophoridae, and frogs formerly recognized as the family Pseudidae are now assigned to a subfamily of Hylidae.

Systematics (the study of evolution and classification of organisms) is a dynamic field, and the relationships of many groups are still being unraveled. Depending on which kinds of evidence are used, the results may differ and different classifications may be proposed. The relationships of some groups of living amphibians have not been resolved with a high level of confidence. For example, a group of frogs endemic to Madagascar has been recognized as a family, Mantellidae, a subfamily of Ranidae, and a subfamily of Rhacophoridae (adopted herein). The classification used in this volume is order Gymnophiona (caecilians) with five families, order Caudata (salamanders) with 10 families, and order Anura (frogs and toads) with 28 families.

Historical biogeography

The distributions of the families of amphibians reflect the history of Earth, especially from the time of the breakup of the supercontinent Pangaea, beginning about 190 million years ago. The early fragmentation resulted in two major land masses: Laurasia, consisting of what is now North America, Europe, and most of Asia; and Gondwana, which included what are now South America, Africa, Madagascar, the Indian subcontinent, Australia, New Zealand, and Antarctica. Prototypic lissamphibians apparently were rather widely distributed in Pangaea before the continental fragmentation.

Although a fossil caecilian is known from the Jurassic of North America, these amphibians now all live in regions that were part of Gondwana. Two families are restricted to the Indian subcontinent (one in adjacent southeastern Asia), one family is endemic to Africa, and another to South America.

Salamanders evolved in Laurasia. One family is restricted to Asia, and four families are shared by Eurasia and North America, where five families are endemic. Only one lineage (Plethodontidae) has dispersed from North America to South America.

The biogeography of anurans is somewhat more complicated. One early lineage containing the living Ascaphidae in North America and Leiopelmatidae in New Zealand has been allied with fossils from the Jurassic of South America, thereby

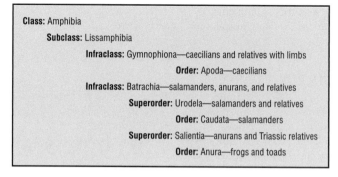

Class: Amphibia
  Subclass: Lissamphibia
    Infraclass: Gymnophiona—caecilians and relatives with limbs
      Order: Apoda—caecilians
    Infraclass: Batrachia—salamanders, anurans, and relatives
      Superorder: Urodela—salamanders and relatives
        Order: Caudata—salamanders
      Superorder: Salientia—anurans and Triassic relatives
        Order: Anura—frogs and toads

The hierarchical classification of living amphibians and their close relatives. (Illustration by Argosy. Courtesy of Gale.)

indicating that this lineage had diversified prior to the breakup of Pangaea. The fossil records and present distributions of other lineages of archaeobatrachians (primitive frogs) are in Laurasian continents: Bombinatoridae, Discoglossidae, Megophryidae, Pelodytidae, and the fossil Paleobatrachidae in Eurasia; Pelobatidae in Eurasia and North America; and Rhinophrynidae in North America. However, the historical biogeography of most anurans (neobatrachians or advanced frogs) is associated with Gondwana, the fragmentation of which into the existing continents played a major role in the differentiation of many lineages of anurans. Many lineages are restricted to one continent: six families in South America, three in Africa, two in Australia, and one each in Madagascar and the Seychelles. Others are shared with two or more Gondwanan continents: one (Pipidae) in Africa and South America; one (Hylidae) in South America and Australia (also via dispersal into North America and Eurasia); one (Hyperoliidae) in Africa, Madagascar, and the Seychelles; and another (Rhacophoridae) in those three regions plus the Indian subcontinent and adjacent southeastern Asia. Microhylidae is present on all Gondwanan land masses except the Seychelles, and it has dispersed into southeastern Asia and southern North America. True frogs (Ranidae) occur throughout the world, though only in northern Australia and northern South Amer-

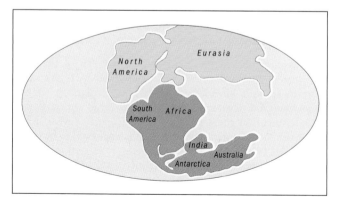

Configuration of the continents in the Early Cretaceous (130 million years ago).

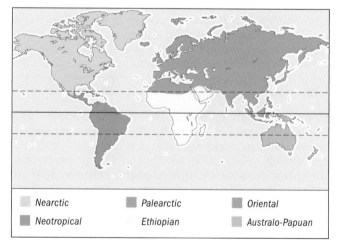

| ☐ Nearctic | ■ Palearctic | ■ Oriental |
| ■ Neotropical | Ethiopian | ■ Australo-Papuan |

Numbers of families/genera/species of amphibians in biotic regions of the world. Nearctic: 17/40/243; Neotropical: 19/185/2782; Palearctic: 15/34/192; Ethiopian: 13/95/770; Oriental: 12/75/825; Australo-Papuan: 6/58/450.

ica on these continents, and toads (Bufonidae) occur on all continents, except Australia (one species introduced).

Regional diversity

As a group, amphibians are distributed throughout the world, except for polar regions, most oceanic islands, and some desert regions. However, the patterns of distribution differ among the three living groups of amphibians. Anurans occur throughout the world but are most diverse in the tropics; salamanders are most diverse in the northern continents; and caecilians are restricted to the tropics.

Globally, except for the Arctic and Antarctic regions (which are not inhabited by amphibians), six biogeographic regions are recognized. The largest of these, the Palearctic (Europe and temperate Asia) has the fewest species of amphibians (192), followed by the Nearctic (temperate North America) with 243 species. Historically, these regions are part of the former Laurasia and have the greatest diversity of salamanders, especially in the Nearctic. In contrast, the amphibian faunas of the southern continents consist mainly of anurans. The Australo-Papuan region (Australia, New Zealand, New Guinea, and associated islands) has 450 species of anurans, but no salamanders or caecilians. The Ethiopian or Afrotropical region (sub-Saharan Africa and Madagascar) has 770 species, of which 29 are caecilians. The Oriental region (tropical and subtropical southeastern Asia, India, and associated islands harbors 825 species, of which 29 are salamanders and 44 are caecilians. By far the greatest amphibian diversity is in the Neotropical region (South America, tropical Mesoamerica, and the West Indies) with 82 species of caecilians, about 200 species of salamanders, and more than 2,500 species of anurans.

Although caecilians are pantropical, they are absent in Madagascar. Ichthyophiidae and Uraeotyphlidae are endemic to the Oriental region, Scolecomorphidae to the Ethiopian region, and Rhinatrematidae to the Neotropical region. The large family Caeciliidae is most diverse in the Neotropical region (14 genera and 73 species) and is present in Africa (6 genera and 17 species), Oriental region (2 genera and 4 species), and in the Seychelles Islands in the Indian Ocean (3 genera and 7 species).

Most salamanders live in the Northern Hemisphere; they are absent in the Australo-Papuan and Ethiopian regions. At the family level, the greatest diversity is in the Nearctic region, where all families (except Hynobiidae) occur, and five families (Ambystomatidae, Amphiumidae, Dicamptodontidae, Rhyacotritonidae, and Sirenidae) are endemic. Cryptobranchidae and Proteidae are represented by one genus each in the Nearctic and Palearctic regions. Salamandridae is the most widespread family of salamanders with nine genera in the Palearctic region, and two genera in the Nearctic region. Hynobiidae is the only family of salamanders restricted to the Palearctic region. By far, the largest family of salamanders is Plethodontidae with 25 genera in the Nearctic; one of these (*Hydromantes*) is shared with Europe. However, the greatest diversity of plethodontids is in tropical America, especially Central America and southern Mexico, where 12 genera with about 200 species occur; two of these genera also occur in South America, one as far south as Bolivia.

Foothill yellow-legged frog (*Rana boylii*) transforming from a tadpole in Mendocino County, California. (Photo by Dan Suzio/Photo Researchers, Inc. Reproduced by permission.)

Only four of the 28 families of anurans occur in both the Old and New Worlds. Bufonidae is global in its distribution, except for Australia, New Zealand, and Madagascar. Ranidae has a similar pattern, but also occurs in Madagascar and in northern Australia. Microhylidae has a few representatives in the Nearctic and Palearctic regions and is highly diverse on the southern continents, including Madagascar and New Guinea, but not in New Zealand. Hylidae is most diverse in the Neotropical region and secondarily in the Australo-Papuan region. Two genera are endemic to North America, and only a few species of *Hyla* inhabit the Oriental and Palearctic regions.

With the exception of Pelobatidae in the Nearctic and Palearctic regions, all other families of anurans are restricted to the New World or the Old World, and only a few of these are in the Northern Hemisphere. Ascaphidae is endemic to the Nearctic region, Megophryidae to the Oriental region, Discoglossidae and Pelodytidae to the Palearctic region, and Bombinatoridae in the Palearctic and Oriental regions. The greatest diversity is in the southern land masses. Leiopelmatidae is endemic to New Zealand, and Limnodynastidae and

The wide-mouthed frog (*Amietia vertebralis*) is found at high altitudes—over 10,000 ft (3,000 m)—in South Africa. There is an umbraculum in each eye that serves as a sun-shade to protect against UV light. (Photo by Alan Channing. Reproduced by permission.)

Amphibians in the ecosystem

Although amphibians are generally restricted to moist environments, such as humid forests, marshes, ponds, and streams, many species venture far from free-standing water and inhabit trees, rocky cliffs, and soil under the surface of the ground. In such diverse habitats, amphibians feed on a great variety of smaller organisms, principally invertebrates, of which insects are the most common in the diets of anurans and salamanders. However, their diets also include earthworms (especially in caecilians), small snails, spiders, and other small invertebrates. Body size plays an important role in prey selection. Some aquatic salamanders feed on tadpoles, and a few larger aquatic salamanders feed on fishes; the eel-like aquatic amphiumas feed almost exclusively on crayfish. Many species of frogs are less than 1 in (25 mm) in head-body length, and their diets are restricted to small insects and spiders. In tropical forests, many of these small frogs specialize on ants and termites, both of which are abundant. Large frogs with wide gapes tend to eat larger prey, which may include other frogs, lizards and small snakes, birds, and mammals. Tadpoles feed primarily on decaying vegetation, algae, and plankton in ponds and streams.

The marsupial frog (*Gastrotheca riobambae*) does not produce and lay clutches of eggs like most frogs, but has its own method of reproduction. The male frog fertilizes the eggs externally and then places them in a pouch on the back of the female frog. The female carries the eggs until they reach tadpole stage. She then deposits them in a pool. This female has eggs in her pouch. (Photo from Natural History Museum, University of Kansas. Reproduced by permission.)

The dietary habits of amphibians are important in the ecosystem because as adults they consume vast quantities of insects and thus help to maintain a balance in the ecosystem. Areas where local anurans have been eliminated have witnessed large population increases in some kinds of insects, and mountain streams that once were relatively free of algae can become choked with algae when algal-feeding tadpoles disappear.

Because of their abundance and relative ease of capture, amphibians are included in the diets of a great variety of animals, especially many small mammals, birds, and many kinds of snakes. Wading birds feast on tadpoles and metamorphosing frogs in shallow ponds. A few snakes specialize on salamanders, and many kinds of snakes in the tropics feed almost exclusively on frogs. Small salamanders and frogs also fall prey to spiders. Even subterranean caecilians cannot escape predation by some snakes, especially coral snakes of the genus *Micrurus*.

In summary, amphibians are a significant part of the food web in most terrestrial ecosystems on the planet. In the late 1980s, biologists realized that populations of amphibians were declining in many parts of the world. Gradual, and especially precipitous, declines result not only in the potential loss of species of amphibians, but have a significant impact on the populations of their prey and those of their predators and animals farther up the food chain. The long-term effects of these declines have yet to be determined.

Myobatrachidae are endemic to the Australo-Papuan region. The Ethiopian region has six endemic families of anurans: Arthroleptidae, Heleophrynidae, Hemisotidae, Hyperoliidae (Africa, Madagascar, and Seychelles), Scaphiophryinidae (Madagascar only), and Sooglossidae (Seychelles only). The greatest diversity of Rhacophoridae is in the Oriental region, but the family also is diverse in Madagascar and has one genus with three species in Africa.

The Neotropical region has the world's greatest diversity of anurans. In addition to many genera and species of Bufonidae, Hylidae, and Microhylidae, there are seven endemic families: Allophrynidae, Brachycephalidae Centrolenidae, Dendrobatidae, Leptodactylidae, Rhinodermatidae, and Rhinophrynidae. Four of these (Allophrynidae, Brachycephalidae, Rhinodermatidae, and Rhinophrynidae) contain a total of only eight species, but Centrolenidae and Dendrobatidae have a total of more than 300 species, and Leptodactylidae contains more than 1,000 species, of which *Eleutherodactylus* is the most speciose and widespread. Other families in the Neotropical region are Pipidae (shared with Africa) and Ranidae (shared with much of the world).

Resources

Books

Duellman, William E., ed. *Patterns of Distribution of Amphibians.* Baltimore: Johns Hopkins University Press, 1999.

Duellman, William E., and Linda Trueb. *Biology of Amphibians.* Baltimore: Johns Hopkins University Press, 1994.

Zug, George R., Laurie J. Vitt, and Janalee P. Caldwell. *Herpetology.* 2nd ed. San Diego: Academic Press, 2001.

William E. Duellman, PhD

• • • • •

Early evolution and fossil history

The appearance of limbed vertebrates—the stegocephalians

The origin of amphibians and amniotes (reptiles, birds, and mammals) must be sought among stegocephalians (i.e., four-limbed vertebrates with digits), which appeared about 370 million years ago (mya) in the Devonian. The group of bony vertebrates from which tetrapods (the group that includes extant stegocephalians) evolved is known informally as osteichthyans, and includes two groups of vertebrates—the actinopterygians and sarcopterygians. Actinopterygians are ray-finned fishes, the group containing nearly all bony fish that are familiar today. Sarcopterygians, or "lobe-finned" fishes, contains three living groups—coelacanths, lungfishes, and tetrapods. Our closest known relatives that retained paired fins are panderichthyids. The ancestor of panderichthyids and tetrapods lacked the dorsal and anal fins typical of other lobe-finned fishes, and, unlike them, its skull was not divided into anterior and posterior parts by an intracranial joint. In addition, the cranium contained a new dorsal roofing bone, the frontal.

Panderichthyids (*Panderichthys* and *Elpistostege*) were large, lobe-finned fishes about 39 in (1 m) long from the Late Devonian of Europe and North America. Unlike its osteolepiform relatives (e.g., *Eusthenopteron*), panderichthyids had a massive, flattened head and body, and a long, rather pointed snout with one external narial opening near the margin of the jaw on each side of the skull—a feature they shared with early stegocephalians. Basal ray-finned fishes (e.g., *Polypterus* and the bowfin *Amia*), lungfishes, and tetrapods possess lungs as a means of breathing; therefore, it is presumed that all early osteichthyans possessed lungs. In the lineage leading to modern ray-finned fishes, the lung was modified into a swim bladder (an organ of buoyancy), but in the lobe-finned fishes, lungs were retained. The challenge facing the early aquatic ancestor of tetrapods was to develop a mechanism for ventilating the lungs in a terrestrial environment.

The earliest known stegocephalians are moderate-sized animals: *Ichthyostega* (61 in; 1.5 m long) and *Acanthostega* (19.7 in; 0.5 m long) from Greenland, and *Tulerpeton* (ca. 26.4 in; 0.67 m long) from Russia. Based on the remains of other animals associated with these stegocephalians and features of their anatomy, it seems likely that they were aquatic. *Acanthostega* and *Ichthyostega* probably lived in freshwater or brackish environments, whereas *Tulerpeton* inhabited a marine environment. For much of the twentieth century, stegocephalians were thought to have originated in freshwater, but an increasing number of early amphibians have been found in coastal, presumably brackish and saltwater environments. Some classical Permo-Carboniferous (Garnett, Hamilton, and Robinson, Kansas, in the United States) and Devonian (Miguasha, Quebec, Canada) fossiliferous localities that previously were interpreted as freshwater environments, now are known to have been coastal, lagoonal, deltaic, or estuarine environments.

Early stegocephalians had many features associated with an aquatic lifestyle. All retained a lateral-line system, a series of sensory receptors in the skin that sense mechanical disturbances in the water and that are typical of fishes, and larval and aquatic amphibians today. *Acanthostega* seems to have retained functional internal gills in addition to lungs. Both *Acanthostega* and *Ichthyostega* had finned tails resembling that of *Panderichthys* and probably similarly used to provide propulsive force for swimming. The backbones of these stegocephalians were poorly developed and probably not capable of supporting the weight of the animal on land. *Ichthyostega* had a substantial rib cage. The overlapping ribs would have provided protection for internal organs, but would not have been sufficiently flexible to facilitate ventilation of the lungs in a terrestrial environment. The limbs were short and stout, and the forelimbs markedly larger than the hind limbs. The limbs seem to have been capable of only a restricted range of movement, and are thought to have been positioned more to the side of the body than beneath it. The hands and feet were paddlelike, having six to eight digits, depending on the species. Thus, it seems likely that in these Devonian vertebrates, the limbs may have been used to walk on a submerged substrate, perhaps in intertidal areas or in obstructed environments such as mangrove swamps. The locomotor system of these animals may be thought of as preadapted for a terrestrial lifestyle. The presence of limbs, digits, pectoral and pelvic girdles, and a rib cage that were useful in aquatic habitats provided the morphological features that their descendants could elaborate for successful exploitation of terrestrial habitats.

Archaic amphibians and other early stegocephalians

There are three groups of early stegocephalians: stem-tetrapods, amphibians, and reptiliomorphs. Stem-tetrapods include all stegocephalians that appeared before the diver-

Eocaecilia fossil and artist's concept of a living specimen. (Illustration by Brian Cressman)

gence between amphibians and reptiliomorphs, a divergence that ultimately resulted in the appearance of the lissamphibians and amniotes (reptiles, birds, and mammals), respectively. Stem-tetrapods include all known Devonian stegocephalians and perhaps many Carboniferous and Permian ones, as well. There is considerable uncertainty about the relationships of many of these taxa, because the phylogeny that was long accepted by most paleontologists was challenged in the late 1990s. Nevertheless, it is clear that a major evolutionary radiation of stegocephalians took place at the end of the Devonian and at the beginning of the Carboniferous, and that all of the main lineages (including amphibians and reptiliomorphs) existed by the Lower Carboniferous. However, reptiliomorphs may not be represented in the fossil record before the Upper Carboniferous.

Diversity of post-Devonian stegocephalians

Baphetids (formerly known as loxommatids) include five genera of seemingly aquatic stegocephalians from the Mississippian and Pennsylvanian (340–305 mya) of Europe and North America. These fossils are known primarily from skulls, which typically are broad and flat with a strange keyhole-shaped orbit. Baphetids may be allied with stem-tetrapods.

The three genera of crocodile-like colosteids, which are approximately contemporaneous with the baphetids, are important, because they once were considered to be closely related to the temnospondyls. However, the skulls of these fossils from Australia and North America lack a squamosal notch, a feature that characterizes temnospondyls, and its ab-

Triadobatrachus fossil and artist's concept of a living specimen. (Illustration by Brian Cressman)

sence in colosteids and the presence of a lateral-line organ suggest an aquatic existence.

Temnospondyls are a large group of more than 150 described genera; they extend from the Mississippian to the Lower Cretaceous, a span of about 200 million years. Most of the early temnospondyls and all of the Mesozoic representatives were aquatic, but others were amphibious, and some are thought to have been terrestrial. Among the latter are the dissorophoids, which are noteworthy because they have been argued to be closely related to living amphibians. Temnospondyls ranged in size from less than 12 in (30 cm) to more than 9.8 ft (3 m) in length. Many, if not all, dissorophoids had aquatic larvae with external gills. Some larval and/or paedomorphic (i.e., having larval or juvenile features

maintained in sexually mature adults) dissorophids had bicuspid (and possibly pedicellate) teeth, as do lissamphibians; hence, dissorophids are thought by some scientists to be closely related to lissamphibians. Other paleontologists consider temnospondyls to be stem-tetrapods. Temnospondyls had a large opening in the palate (interpterygoid vacuity) that may have been involved in a buccal pump mechanism that is similar to that used by all lissamphibians to ventilate their lungs. The stapes (middle ear bone) of most temnospondyls is more slender and oriented more laterally than that of earlier stegocephalians. Because of this, some researchers think these animals possessed a tympanum (eardrum) in the otic notch (or squamosal embayment) of the skull. If this hypothesis is correct, then the absence of a tympanum in caecilians and salamanders must be secondary (i.e., having resulted from

Early fossils of: 1. *Ichthyostega*, 2. *Panderichthys*, and 3. *Eusthenopteron*. (Illustration by Emily Damstra)

an evolutionary loss). The vertebrae of early temnospondyls retained the rhachitome pattern that is primitive for stegocephalians. Rachitomous vertebrae had a large, ventral, crescentic intercentrum and small, paired, dorsal pleurocentra that supported the neural arch; such vertebrae are poorly suited to a terrestrial lifestyle. In Mesozoic temnospondyls, the vertebral column was consolidated to form the stereospondylous pattern, characterized by a large intercentrum, along with a small pleurocentrum that sometimes was cartilaginous. The functional reason for this strengthened vertebral column is unclear, because most, if not all, Mesozoic temnospondyls were strictly aquatic.

Embolomeres ranged in length from 12 in (30 cm) up to 9.8 ft (3 m) and were mostly aquatic and amphibious predators. The fossil record of embolomeres extends from the Upper Carboniferous into the Triassic, and they are known from Europe, North America, and Russia. Their vertebrae are composed of cylindrical intercentra and pleurocentra. Because they have a massive stapes, it is thought that embolomeres lacked a tympanum, but the stapes may have conducted low-frequency ground-borne and water-borne sounds. Embolomeres are important because earlier they were thought to be related to amniotes. However, now they are considered by many scientists to represent another group of stem-tetrapods.

Seymouriamorphs are another group of stem-tetrapods that was formerly thought to be closely related to amniotes. This relatively small (12 genera), but widespread, group is known from the Permian of North America, Europe, and Asia, and is represented by larvae bearing external gills and adults (ca. 3 ft [90 cm] long). The absence of lateral-line organs and gills in the adults suggests that they were terrestrial. Adult seymouriamorphs had a long, slender stapes, which suggests that they might have had a tympanum, and a rib architecture that suggests the capacity for costal ventilation of the lungs.

Diadectomorphs are represented by eight genera in the Upper Carboniferous and Lower Permian. Once considered to be amniotes, they now are thought to be closely related to them. These animals attained lengths of 6.5 ft (2 m). In addition to some carnivorous or piscivorous forms, diadectomorphs include some of the earliest herbivorous stegocephalians, the diadectids.

An assemblage of small amphibians (most less than 12 in [30 cm] long) comprising five groups (aïstopods, nectrideans, "microsaurs," adelogyrinids, and lysorophids) forms an evolutionary grade informally known as "lepospondyls." These animals are known from the Lower Carboniferous to the Upper Permian. Some seem to have been strictly aquatic (e.g., many nectrideans, adelogyrinids), whereas others (e.g., the "microsaurs" *Pantylus* and *Tuditanus*) apparently were amphibious or terrestrial. Most lepospondylous amphibians either lacked or had only a small otic notch or squamosal embayment (e.g., aelogyrinids); thus, they must have lacked a tympanum. If these amphibians include the closest known relatives of lissamphibians, then the ancestor of caecilians and salamanders probably lacked a tympanum, and the tympanum of anurans may have appeared only in the Triassic. The name of the group derives from the structure of their vertebrae, which are dominated by a large, cylindrical pleurocentrum that fused to the neural arch early in development, as it does in lissamphibians. In some lepospondyl amphibians, a small crescentic intercentrum remains, but in others it is lost, as it is in most lissamphibians.

Lissamphibia

The oldest known lissamphibians, a group consisting of the caecilians, salamanders, anurans, and their fossil allies, date from the Triassic, some 250 mya. The fossil record of this group is extremely scanty. Indeed, in the Triassic, which lasted about 37 million years, only two species of lissamphibians are known and both are closely related to anurans. Not all paleontologists agree that Lissamphibia is a natural group, but most neontologists consider it to be monophyletic. Regardless of which phylogenetic arrangement one prefers, there is a gap of several tens of million years between the sister group (i.e., dissorophoids or lysorophids) and the earliest known lissamphibian. Beginning in the Jurassic (206 mya) fossil lissamphibians become more common; however, the best-represented groups are those that inhabited an aquatic environment.

Salientia (anurans and *Triadobatrachus*)

Triadobatrachus, the oldest fossil lissamphibian, is from the Lower Triassic of Madagascar. It is known from a single specimen, a largely complete and articulated skeleton (4.2 in [10.6 cm] long), which reveals it to be closely related to anurans. It shares many cranial features with frogs and toads, but it differs from them by having a longer trunk, a less specialized pelvic girdle, shorter limbs, and a short tail. Another, slightly younger fossil, *Czatkobatrachus*, from the Lower Triassic of Poland, is based on a few, isolated bones; thus, little can be said about its affinities or phylogenetic position. Beginning in the Lower Jurassic, there are fossil representatives of each of the modern orders. However, the record for caecilians and the general quality of fossil salamanders are exceedingly poor in contrast to that for anurans.

Triadobatrachus is considered to be closely allied to anurans (frogs and toads). Together, these animals compose a taxonomic group known as Salientia. The earliest known anurans, *Prosalirus* and *Vieraella*, are from the Lower Jurassic of Arizona (United States), and Argentina, respectively, and approximately contemporaneous with *Eocaecilia*, the stem-caecilian. Each of these moderate-sized (2 in [50 mm] and 1.25 in [30 mm] in snout-vent length, respectively) frogs differs from the larger *Triadobatrachus* in having a shorter trunk, urostyle, lacking a tail, and possessing long hind limbs

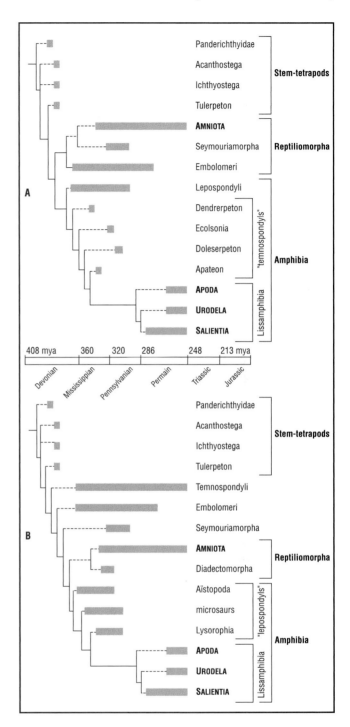

A and B: Possible phylogenetic trees for amphibians. (Illustration by Argosy. Courtesy of Gale.)

in which the ankles are modified to form an extra limb segment. Unlike *Triadobatrachus*, these frogs clearly were capable of saltatorial (i.e., jumping, hopping) locomotion typical of extant anurans.

One genus, *Eodiscoglossus*, is represented by Middle Jurassic remains from Great Britain. This rather large frog (3.25

Artist's concept of a living specimen of *Prosalirus bitis*. (Illustration by Brian Cressman)

in [80 mm] in snout-vent length) is remarkably similar to living discoglossids, especially species of *Discoglossus*.

By the Late Jurassic, the fossil record of anurans is much more diverse taxonomically and geographically. *Eobatrachus* and *Comobatrachus* are known from Wyoming (United States), and *Enneabatrachus* from Wyoming and Utah (United States); all are of uncertain affinities. *Notobatrachus*, from several localities in Patagonia in Argentina, is one of the most important Middle–Late Jurassic finds, because it is represented by whole, articulated skeletons and numerous individuals, including juveniles and subadults. This frog was large, reaching a snout-vent length of about 5.7 in (14.5 cm). It has several primitive features, including free ribs on some of the vertebrae, a poorly developed sacrum, a relatively short

pelvic girdle, and stout, relatively short hind limbs. This frog is thought to be ancestral to all living anurans. Two other taxa, *Callobatrachus* and *Mesophryne*, were described from Jurassic/Cretaceous fossil beds of China; the former is allied with discoglossids, but the affinities of the latter are unknown. The oldest fossil pipoid is *Rhadinosteus* from the Late Jurassic of Utah.

The Cretaceous produced a proliferation of anurans. Among the more significant Lower Cretaceous finds are the discoglossid *Eodiscoglossus* from Spain, and a variety of pipoid frogs—*Thoraciliacus*, *Cordicephalus*, and *Shomronella* (larvae) from Israel. Middle/Upper Cretaceous anurans include gobiatids from Mongolia and Uzbekistan, pipoid frogs from Argentina (*Saltenia* and *Avitabatrachus*) and Niger (*Pachybatrachus*), and lepto-

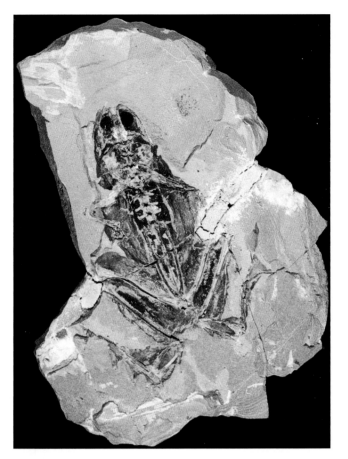

Fossilized frog skeleton embedded in rock. Frogs and toads first appeared 190–160 million years ago, in the early Jurassic period. (Photo by Volker Steger/Science Photo Library/Photo Researchers, Inc. Reproduced by permission.)

dactylids (*Baurubatrachus* and *Estesius*) from Brazil and Bolivia, respectively.

The Tertiary record of anurans is too extensive to recount here. Suffice it to say that most major families are represented by fossil remains from Europe, Africa, Asia, and North America.

Gymnophionans (caecilians and *Eocaecilia*)

Extant apodans or caecilians comprise a peculiar group of limbless, snake-like amphibians that are terrestrial or aquatic and specialized for burrowing; they possess a tentacle on each side of the head beneath the reduced eye. Most scientists consider caecilians to be the most basal of the lissamphibians. The Lower Jurassic (about 204 mya) fossil *Eocaecilia* from Arizona is thought to be a stem-caecilian. It differs from extant members of the group in having small, well-ossified limbs and girdle elements that are absent in living caecilians. Although the eyes were larger, and the skull contained more bones than living species, the lower jaw and the jaw-closing mechanism seem to resemble those of living representatives. The only other fossil remains associated with gymnophionans are vertebrae from the Upper Cretaceous of the Sudan, Early Paleocene of Bolivia, and Late Paleocene of Brazil. These fossils provide little useful information except that the modern families of caecilians had begun to differentiate by the late Mesozoic.

Urodeles (salamanders and related fossils)

Fossil urodeles are known from Middle Jurassic–Lower Cretaceous (about 180–127 mya) lake and lagoon deposits of Europe, the Upper Cretaceous–Eocene (about 90–35 mya) flood-plain deposits of North America, and the Eocene–Miocene (about 35–5 mya) brown-coals of Europe. In addition, there are remains from the Cretaceous of Bolivia, the Sudan, Niger, and Israel. *Laccotriton*, *Sinerpeton*, *Jeholotriton*, and *Liaoxitriton* are salamanders from the Upper Jurassic/Lower Cretaceous of China. With three exceptions mentioned below, all these remains clearly are caudate, i.e., belonging to salamanders. The affinities of Albanerpetontidae from Middle Jurassic–Miocene deposits of northern continents are equivocal. This group may represent a peculiar, early offshoot of the salamander lineage, or a separate lissamphibian group, the origin of which preceded the phylogenetic divergence between caudates and anurans. The phylogenetic status of *Ramonellus* from the Lower Cretaceous of Israel is unknown. A third taxon, *Triassurus*, was described from the Triassic of Uzbekistan. If the latter poorly preserved remains prove to be those of a urodele, then the group would have originated in the Triassic or earlier.

The earliest fossil remains of salamanders that can be definitely identified as caudates are the Jurassic remains from China, the karaurid salamanders, *Karaurus* and *Kokartus*, from Kazakhstan and Kirghizstan, and *Marmorerpeton* from England. The extant groups of salamanders are represented by a surprising number of fossils. (1) Sirenidae: Upper Cretaceous and Paleocene of North America; Cretaceous remains of several taxa from southern continents may also be related to sirenids; (2) Hynobiidae: Upper Pliocene of Kazakhstan; (3) Cryptobranchidae: Paleocene of Mongolia and Russia, Upper Oligocene–Pliocene of Europe, Upper Paleocene–Upper Miocene of North America, with Pleiocene–Pleistocene occurrences of *Cryptobranchus* in North America; (5) Salamandroids Ambystomatidae, Amphiumidae, Dicamptodontidae, Plethodontidae, Proteidae, Rhyacontritonidae, and Salamandridae): Stem-salamandroids from Lower Cretaceous of Europe; Amphiumidae from Paleocene–Pleistocene of North America; Dicamptotontidae from Paleocene of Canada; Salamandridae from Upper Paleocene and onwards of Europe and Upper Miocene of East Asia, and Neogene of North America. In addition, there are three groups of fossil caudates of uncertain affinities: batrachosauroids from Mid-Cretaceous–Lower Pliocene of North America and Europe; scapherpetontids from Upper Cretaceous, Paleocene, and Eocene of North America, and possibly Cretaceous of Asia; and dicamptodontid-like salamanders from Paleocene–Miocene of Europe.

Resources

Books

Duellman, William E., and Linda Trueb. *Biology of Amphibians.* New York: McGraw-Hill Book Co., 1986.

Estes, Richard. *Handbuch der Paläoherpetologie*, Vol. 2. Stuttgart: Gustav Fischer Verlag, 1988.

Heatwole, Harold, and Robert L. Carroll, eds. *Amphibian Biology.* Vol. 4, *Palaeontology. The Evolutionary History of Amphibians.* Chipping Norton, Australia: Surrey Beatty & Sons Pty. Limited, 2000.

Sanchíz, Borja. *Encyclopedia of Paleoherpetology.* Part 4, *Salientia.* Munich: Verlag Dr. Friedrich Pfeil, 1998.

Schultze, Hans-Peter, and Linda Trueb, eds. *Origins of the Higher Groups of Tetrapods. Controversy and Consensus.* Ithaca, NY: Comstock Publishing Associates, Cornell University Press, 1991.

Periodicals

Báez, Ana Maria, Linda Trueb, and Jorge O. Calvo. "The Earliest Known Pipoid Frog from South America: A New Genus from the Middle Cretaceous of Argentina." *Journal of Vertebrate Paleontology* 20 (2000): 490–500.

Gao, Ke-Qin, and Yuan Wang. "Mesozoic Anurans from Liaoning Province, China, and Phylogenetic Relationships of Archaeobatrachian Anuran Clades." *Journal of Vertebrate Paleontology* 21 (2001): 460–476.

Laurin, Michel. "A Reevaluation of the Origin of Pentadactyly." *Evolution* 52 (1998): 1476–1482.

———. "The Importance of Global Parsimony and Historical Bias in Understanding Tetrapod Evolution. Part I— Systematics, Middle Ear Evolution, and Jaw Suspension." *Annales des Sciences Naturelles, Zoologie* 13ème Série, 19 (1998): 1–42.

Laurin, Michel, Marc Girondot, and Armand de Ricqlès. "Early Tetrapod Evolution." *Trends in Ecology and Evolution* 15 (2000): 118–123.

Laurin, Michel, and Rodrigo Soler-Gijon. "The Oldest Stegocephalian from the Iberian Peninsula: Evidence that Temnospondyls Were Euryhaline." *Comptes Rendus de l'Académie des Sciences de Paris, Sciences de la vie/Life Sciences* 324 (2001): 495–501.

Linda Trueb, PhD
Michel Laurin, PhD

Structure and function

The three living orders of lissamphibians represent only a small fraction of the amphibian diversity reflected by the fossil record, which contains examples of lineages that flourished and diversified over long periods of time as well as short-lived, less successful evolutionary experiments. It is assumed that the orders Gymnophiona (caecilians), Caudata (salamanders), and Anura (frogs and toads) arose from a common ancestor in the Triassic. Thus, each of these amphibian lineages has evolved along a unique trajectory for some 300 million years to produce the three distinct groups of organisms recognized today. Superficially, it would seem that frogs and toads, the snakelike caecilians, and salamanders have few anatomical traits in common, but shared features of their integument, musculoskeletal system, internal organs, endocrine system, and sensory structures indicate their common ancestry.

Morphological features common to all lissamphibians

Integument

Skin is the interface between the organism and its environment. As a water-permeable covering, the skin functions as an organ of osmoregulation and respiration; it also supports internal structures. All lissamphibians have mucous and granular (poison or serous) glands distributed over the body. The mucous glands secrete mucopolysaccharides (mucus) that keep the skin moist and facilitate oxygen transport. Moisture is critical to respiration—as much as 90% of the animal's oxygen needs are met by passive transport of oxygen into the skin and its capillary vessels, rather than by ventilation through the lungs. Granular glands secrete a variety of substances (e.g., peptides and alkaloids) that commonly are noxious and sometimes highly toxic, and therefore are important defense mechanisms.

The colors and patterns of lissamphibians are determined by pigments produced by chromatophore cells in the skin. Stimuli such as changes in hormone, light, and temperature levels cause changes in the amounts and distributions of pigments in chromatophores. Thus, the colors of an organism can vary from day to night, seasonally, and throughout the life of the animal.

The skin also is the site of the lateral-line system, which is present in all lissamphibian larvae and in adults that are aquatic. This sensory-receptor system consists of a series of sense organs distributed in the skin of the head and along the body. These are of two types: mechanoreceptors and electroreceptors. Neuromasts are lateral-line organs that are responsive to mechanical stimuli (e.g., water currents). Ampullary organs (electroreceptors) occur only in larval caecilians and aquatic salamanders.

Musculoskeletal system

The basic architecture of a vertebrate is its musculoskeletal system—the framework of bones and the muscles associated with them that are covered by the skin and that, internally, enclose and support the viscera and sensory organs. In lissamphibians, as in other vertebrates, the musculoskeletal system can be divided into three architectural units: the head and associated structures, the trunk or backbone, and the girdles and appendages.

The head or cranium contains the brain and the primary sense organs (eyes, olfactory organs, ears, equilibrium organs) in a cartilage and bone housing—the skull. The upper and lower jaws are also part of the cranium. If the jaws bear teeth, the teeth are of a type unique to lissamphibians among living vertebrates. Each tooth is composed of a bicuspid crown that sits atop a base pedicel; as a tooth is lost, it is replaced by another that has formed on the inside (lingual margin) of the jaw adjacent to the older tooth. The lower jaw or mandible and its associated musculature form the floor of the mouth. Seated in the musculature of the throat is a complex skeletal assemblage known as the hyobranchial apparatus. Muscles associated with the hyobranchium, mandible, and cranium form the mechanical systems for securing food (opening and closing jaws, the tongue and its movement) and breathing. Mechanical ventilation of the lungs is accomplished, in part, by the buccal pump mechanism. When the muscles in the floor of the mouth contract, the volume of the buccal chamber is reduced and air is forced out through the open nostrils. When these muscles are relaxed, the floor of the mouth drops and the volume of the buccal chamber increases; this creates a vacuum that pulls air into the mouth through the nares. Then the nares are closed, the muscles are contracted, and air is forced from the mouth into the lungs.

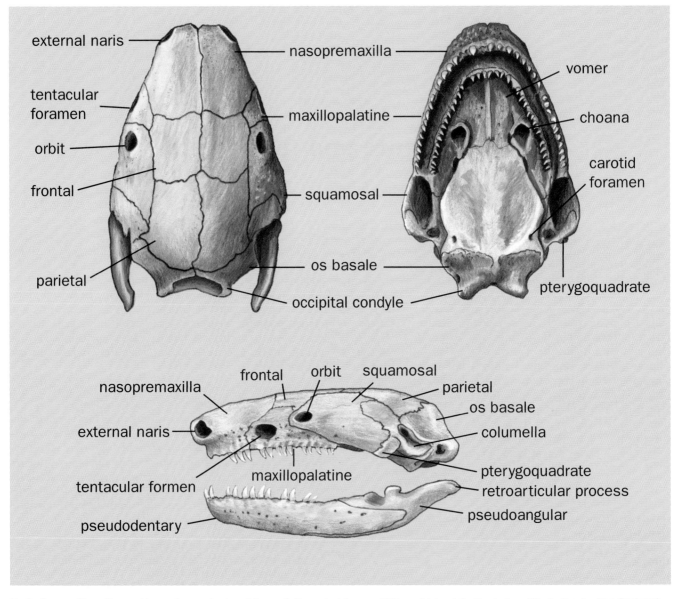

Skull of a caecilian, *Dermophis mexicanus*, in dorsal (upper left), ventral (upper right), and lateral (bottom) view. (Illustration by Dan Erickson)

The vertebral column (backbone) provides axial support for the head, appendicular skeleton (pectoral girdle and forelimbs anteriorly and pelvic girdle and hind limbs posteriorly), viscera, and tail if it is present. The column is composed of a series of bony elements, the vertebrae, each of which can be visualized as being composed of a spool that bears a bony arch on its top and a rib on each side. The ends of each spool abut one another to form a segmented, flexible column of bone. Collectively, the bony arches on top of each spool form a longitudinal canal that houses the spinal cord. The ribs (if present) are short and extend laterally from the vertebrae and their associated muscles to form a sling that supports the internal organs. The first vertebra behind the head (atlas) is specialized to support the skull; it bears a pair of hemispherical depressions (cotyles) into which the pair of rounded condyles at the end of the skull fits. This paired articular arrangement

in lissamphibians eliminates their ability to move their heads from side to side. In contrast, reptiles, birds, and mammals have a single ball-and-socket arrangement that allows the head to be moved up and down, as well as from side to side. In those lissamphibians with limbs, one of the posterior vertebrae is modified into a sacrum—an enlarged trunk vertebra with elaborate lateral processes that support the pelvic girdle. Tails are composed of caudal vertebrae that lie behind (posterior) the sacrum and lack ribs.

If present, limbs are suspended from the axial column by girdles and complex muscular connections. The pectoral girdle consists of broad blades (suprascapulae) located behind the head on either side of the back. The lower part of each blade is connected to a bone (scapula) that bears a fossa (cavity) in which the head of the upper bone (humerus) of the arm articulates. The medial (chest region) parts of the pectoral girdle consist of highly variable systems of bracing cartilages and bones. The forelimb consists of an upper arm (humerus), forearm (radius and ulna), and hand with four or fewer digits. The pelvic girdle consists of three pairs of bones. The largest and most anterior bones are the ilia, each of which bears an elongated process that articulates with the sacrum. The posterior parts of the ilia, along with the ventral paired pubes and posterior paired ischia, form a thick vertical plate with a concavity (acetabulum) on each side which receives the head of the upper limb bone, the femur. The hind limb consists of the femur, the tibia and fibula, and a foot with five or fewer digits.

Visceral anatomy

The visceral anatomy includes the circulatory system and lungs and the digestive and urogenital systems. The circulatory system consists of the blood, heart, and blood and lymph vessels that transport oxygen and metabolic products through the body. All lissamphibians (except the salamanders *Siren* and *Necturus* with four-chambered hearts) have three-chambered hearts composed of two atria and one ventricle, from which blood is routed to the head, body, skin, or lungs. Blood returning from the head, body, and skin enters the left atrium. The rest of the vascular system is composed of arteries carrying blood away from the heart and veins that route blood back to the heart. The lungs are an important adjunct to the circulatory system, because it is here that carbon dioxide is released from the blood and oxygen is acquired. Lungs are present as paired structures in all lissamphibians except plethodontid salamanders and two genera of salamandrids (*Chioglossa* and *Salamandrina*). In most, they are relatively simple structures; as little as 10% of the necessary oxygen is exchanged across the lung surfaces, with the remaining 90% being exchanged across the organism's skin. In aquatic lissamphibians, the lungs seem to be more important as hydrostatic (buoyancy), rather than respiratory, organs. The lymphatic system is composed of a third series of vessels that collect substances that seep through the walls of capillaries or are not picked up by the capillaries (e.g., fats from intestinal capillaries). Fluid movement through the lymph vessels is directed to the venous system by a series of lymph hearts (valves that restrict lymph flow in one direction). The spleen is a large aggregation of lymph tissue located on the left side adjacent to the intestine below the stomach; in lissamphibians, this is a major site of production of red blood cells and recovery of remnants of worn-out blood cells.

The mouth (buccal cavity) is an important part of the lissamphibian digestive system. The mouth has many different kinds of glands; the most notable are the intermaxillary glands, which produce a sticky secretion that is deposited on the tongue and helps to entrap prey. Food (e.g., insects) usually is crushed partially in the mouth so that the process of digestion can begin. The pharynx is an expanded chamber behind the mouth; it also bears many glands that produce mucus to help move food into the thin-walled esophagus, a short tube that connects the pharynx to the stomach. The stomach usually lies to the left of the midline; posteriorly, it is separated from the intestines,

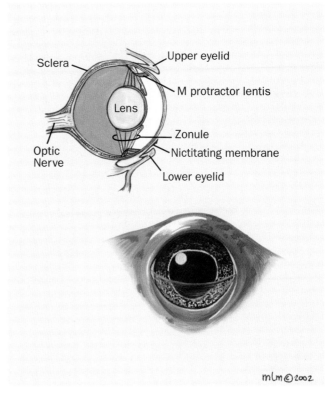

Eye anatomy of amphibians. The upper eyelid is immoveable, but the lower eyelid is moveable—the upper part of which (the nictitating membrane) is usually transparent. (Illustration by Michelle Meneghini)

where nutrients are absorbed, by the pyloric sphincter. The terminal part of the digestive tract is the cloaca, which opens to the outside by means of the vent. Two important glandular organs are associated with the digestive system—the liver and the pancreas. The liver removes toxic substances from the blood and delivers them to the small intestine via the gall bladder and bile duct. The pancreas is an exocrine and endocrine gland that lies between the small intestine and the stomach; it produces pancreatic juice, which contains an enzyme (trypsin) that is delivered to the small intestine to break down proteins.

The reproductive and excretory systems are closely allied with one another to form the urogenital system, which includes the kidneys, gonads, urogenital ducts, urinary bladder, cloaca, and, in lissamphibians, fat bodies. The pair of kidneys flanks the dorsal aorta from which each kidney receives numerous arteries that branch to form clusters of capillaries (glomeruli). Each capillary cluster is encased by an expanded end of a kidney tubule known as Bowman's capsule. This is the primary site of filtration of metabolic by-products from the blood, which if retained would upset the physiological balance of the organism. The collective wastes (urine) are conducted through the kidney tubules to the Wolffian duct to the cloaca. The urinary bladder is a pouchlike outgrowth of the cloaca where urine can be safely stored, rather than being voided constantly as it is formed. Water conservation is critical to many lissamphibians; therefore, they store urine in

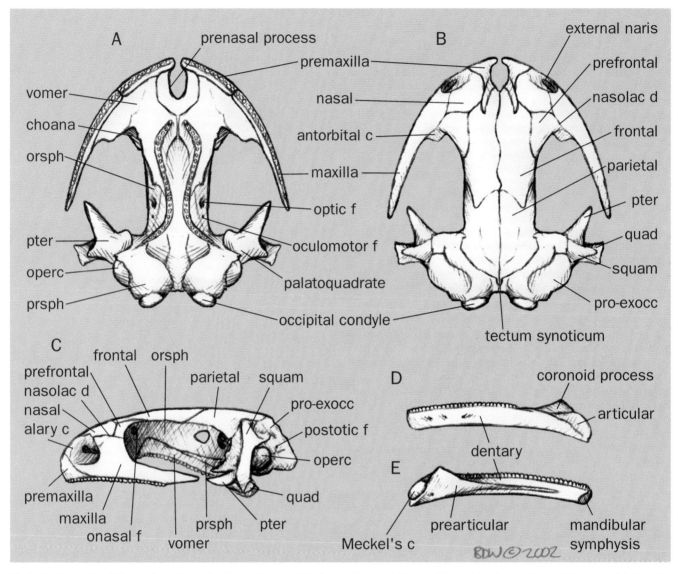

Skull of salamander, *Salamandra salamandra*. Views: A=ventral; B=dorsal; C=lateral; D=mandible in lateral view; E=mandible in medial view. Abbreviations: orsph=orbitosphenoid; pter=pterygoid; operc=operculum; prsph=parasphenoid; c=cartilage; f=foramen; nasolac d=nasolacrimal duct; quad=quadrate; pro-exocc=prootic-exoccipital; onasal=orbitonasal; squam=squamosal. (Illustration by Bruce Worden)

the bladder so that it does not create osmotic pressure that would draw water from the tissues of the animal.

The paired gonads are closely associated with the kidneys and are the site of gametes—sperm from the testes of males and eggs from the ovaries of females. Each testis has a membranous attachment to the kidney; this membrane supports the ductules for sperm transport. The sperm pass through the kidney to the Wolffian duct and then to the cloaca. Each ovary is suspended by a membrane from the middle side of the kidney. As eggs are released from the ovary into the body cavity, they are moved forward by ciliary action of the coelomic epithelium toward the lung, where the opening of the oviduct is located. The oviduct lies parallel to, and at the side of, the kidney. Ciliated epithelium and smooth muscles of the oviduct wall move eggs from the opening of the oviduct to the cloaca,

where they are extruded through the vent. All lissamphibians have fat bodies associated with the gonads; the fat bodies are thought to provide nutrients for the gonads and are largest just before hibernation and smallest following breeding. The common receptacle for the intestine, Wolffian ducts, oviducts, and the bladder is the cloaca. The opening of the cloaca to the exterior, the vent, is controlled by a muscular sphincter.

Endocrine system

Endocrine glands produce complex chemical substances (hormones) that, in combination with the activity of the nervous system, regulate and coordinate the activities of various organs. These glands include the pituitary, pineal body, thyroid, parathyroids, ultimobranchial bodies, thymus, pancreatic islets, adrenals, and gonads.

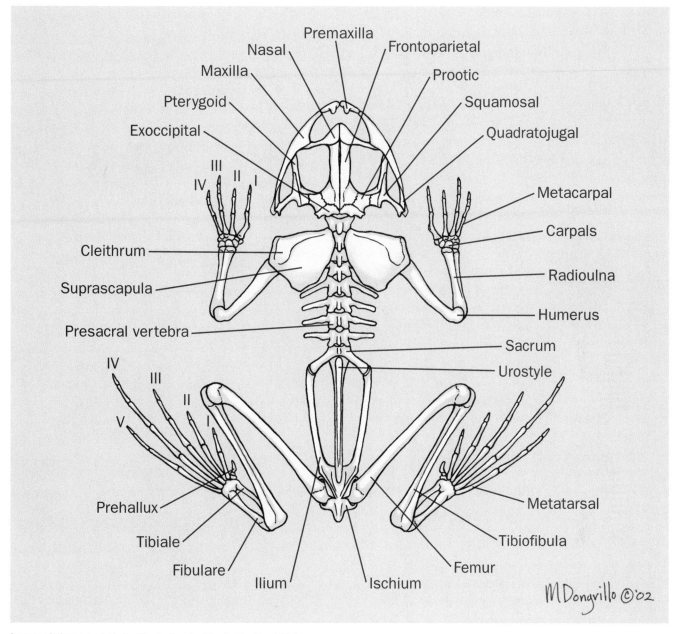

Anuran skeleton and skull. (Illustration by Marguette Dongvillo)

The pituitary is closely associated with the ventral surface of the brain and consists of several discrete parts that secrete different hormones and directly or indirectly are controlled by the brain. Among the functions controlled by pituitary hormones are activities of the ovaries and testes, larval growth, production and control of pigment cells, and regulation of water loss and salt balance. The pineal body is located on the dorsal surface of the brain; it is light sensitive. During prolonged darkness, the pineal body releases the hormone melatonin, which triggers aggregation of melanosomes in the skin chromatophores and thereby lightens the color of the skin.

The pair of thyroid glands lies in the throat and produces two hormones—thyroxine (T_4) and triodothyronine (T_3). These hormones control morphological and functional changes during the metamorphosis of lissamphibian larvae to adults. In adults, they are thought to be involved in the control of metabolic activities, skin structure, and nerve function.

Several other glands are located in the neck region. There are two small, parathyroid glands that secrete calcitonin and parathyroid hormone, both of which control calcium metabolism. Similarly, the ultimobranchial bodies secrete a calcitonin-like substance that affects mineral me-

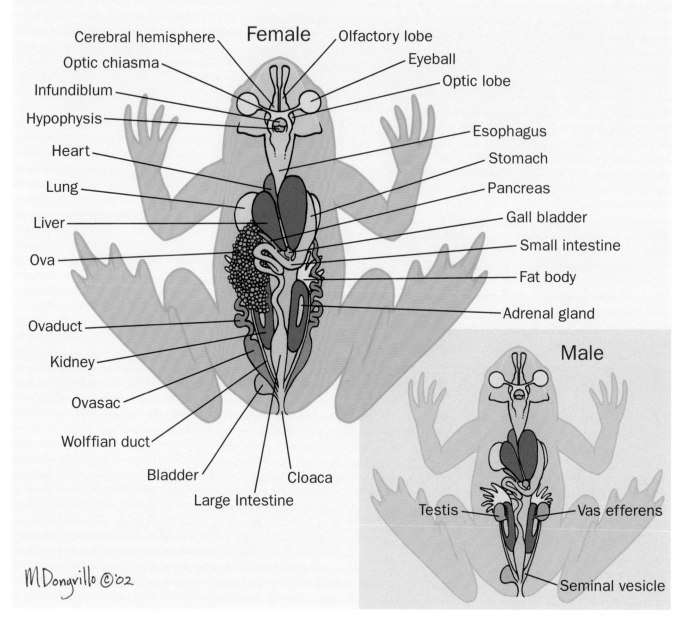

Frog internal organs, showing the differences between male and female. (Illustration by Marguette Dongvillo)

tabolism, especially during metamorphosis. The thymus secretes several substances, collectively known as thymosin, that stimulate production of lymph cells.

The pancreatic islets (islets of Langerhans) are pancreatic glands that develop in lissamphibian larvae and become active only at metamorphosis, when they begin to secrete insulin, glucagon, somatostatin, and pancreatic polypeptide. Insulin and glucagon are critical in carbohydrate metabolism; insulin facilitates the assimilation of sugar (glucose) into cells, whereas glucagon stimulates glucose levels to rise in the blood. Somatostatin promotes the growth of skeletal and soft tissues. Pancreatic polypeptide, which is released into the

blood after meals, promotes the flow of gastric juice (e.g., hydrochloric acid) in the stomach.

Each member of the pair of elongate adrenal glands is located on the underside of the kidney. The outer part of the gland produces corticosteroid hormones, which are involved in the control of water reabsorption and sodium transport in the kidney, metabolism of carbohydrates, and reproduction. The inner part of the gland produces catecholamines—epinephrine (adrenaline) and norepinephrine. Both of these affect the cardiovascular system and blood flow through the brain, liver, kidneys, and skeletal muscle, in addition to the rate of metabolism

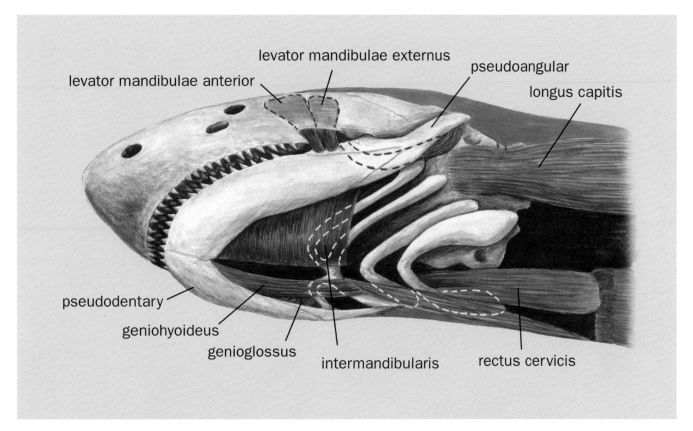

Ventrolateral view of cranial, hyoid, and anterior trunk musculature of a caecilian, *Dermophis mexicanus*. (Illustration by Dan Erickson)

of blood sugars. Their actions are antagonistic; thus, adrenaline dilates blood vessels to increase blood flow and increases the rate of sugar metabolism, whereas norepinephrine constricts blood vessels and decreases the rate of sugar metabolism.

In addition to producing gametes, the gonads produce and secrete hormones that regulate the reproductive cycle and development of secondary sex characters. The testes produce testosterone, which promotes sperm production and the appearance of male secondary sex characters such as nuptial excrescences. The ovaries produce estrogen, which promotes early development of eggs.

Nervous system and sensory organs

The vertebrate nervous system is composed of the central nervous system (brain and spinal cord) and the peripheral nervous system, which is composed of somatic and visceral nerves. Somatic nerves transmit information to and from skeletal muscle, skin, and derivatives of these structures, whereas visceral nerves serve involuntary muscles and glands and make up the autonomic nervous system. Specialized sensory organs, such as the olfactory organ, eye, ear, and lateral-line system, respond to chemical, electromagnetic, and mechanical stimuli to provide the organism with information about changes in its environment.

The lissamphibian brain is slightly more complex than that of fishes, but lacks the cerebral cortex of reptiles, birds, and mammals. The forebrain is composed of the dien-

cephalon (epithalamus, thalamus, and hypothalamus) and telencephalon (olfactory lobes and cerebral hemispheres); Cranial Nerve I (olfactory) arises from the forebrain. The midbrain, or mesencephalon, is composed of the optic lobes and a basal peduncular portion. Information from the eyes, ears, cerebellum, nose, and lateral-line system is processed in the midbrain. Three cranial nerves originate from this area of the brain—II (optic), III (oculomotor), and IV (trochlear); all are involved with receiving visual stimuli and controlling the eye. The hindbrain is composed of the cerebellum and medulla oblongata (continuous with the spinal cord); it is the center for motor coordination and is small in lissamphibians. The cranial nerves emerging from the hindbrain are V (trigeminal, serving the jaws and mouth), VI (abducens, serving eye muscles), VII (facial, serving lateral-line organ of head and taste buds), VIII (auditory, serving inner ear), IX (glossophryngeal, serving taste buds, pharynx, and lateral line), X (vagus, serving areas of mouth, pharynx, and viscera), and XI (spinal accessory, serving the muscle suspending pectoral girdle). Cranial Nerve XII (hypoglossus) innervates muscles associated with the tongue. In lissamphibians, this nerve is associated with the first and second spinal nerves; this contrasts to the condition in amniotes (reptiles, birds, mammals), in which the nerve emerges from the cranium.

The spinal cord is protected by the neural arches of the vertebral column. From it arise spinal nerves that innervate each body segment. The number of pairs of spinal nerves

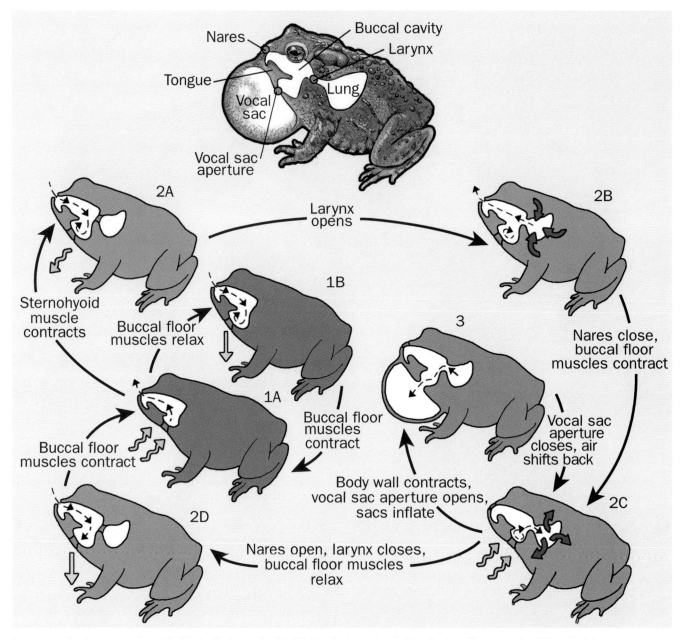

Anuran vocalization and airflow: 1A–1B. Oscillation cycle; 2A–2B. Ventilation cycle; 3. Vocalization. (Illustration by Gillian Harris)

varies according to the number of body segments. Spinal nerves coalesce to form complex networks in the thoracic region (brachial plexus) and sacral region (sciatic or crural plexus) to control movement of the fore and hind limbs, respectively.

The eyes of lissamphibians differ significantly from those of fishes and amniotes. The lissamphibian eye is focused by moving the lens; in amniotes, the lens is deformed. In lissamphibians the lens is moved distally (outward) to accommodate distal vision, whereas in fishes it is moved proximally (inward). Lissamphibians are unique in having specialized receptor cells (green rods) in the retina, in addition to the three other receptor cells possessed by other vertebrates. In addi-

tion to the eye, lissamphibians sense electromagnetic stimuli through the pineal end organ located on the top of diencephalons. The receptor cells in this organ help them to synchronize their daily and seasonal activity cycles and orient themselves spatially.

Chemosensory cues are important to lissamphibians and are processed by two different systems: the nose (olfactory system) and the vomeronasal organ. The olfactory organ is located in the snout and consists of a series of sacs lined with sensory epithelium that receives information about the chemical makeup of air that is inhaled through the external nares and the various olfactory chambers into the buccal cavity. The vomeronasal organ is an accessory olfactory system that is lo-

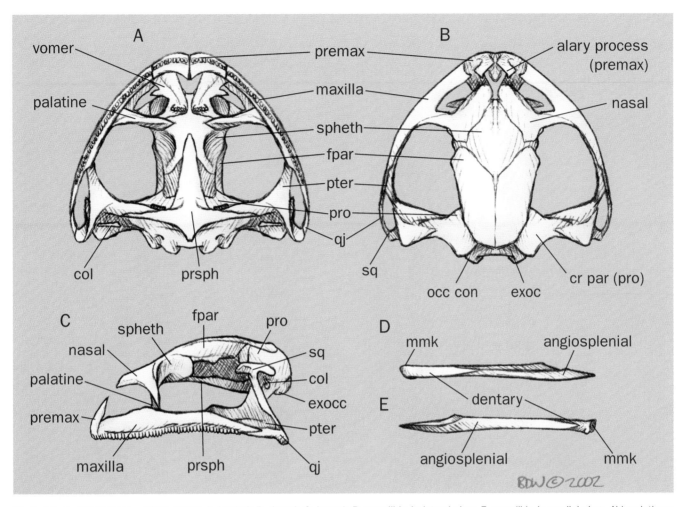

Skull of frog, *Gastrotheca walkeri*. Views: A=ventral; B=dorsal; C=lateral; D=mandible in lateral view; E=mandible in medial view. Abbreviations: pro=prootic; col=columella; prsph=parasphenoid; premax=premaxilla; spheth=sphenethmoid; fpar=frontoparietal; pter=pterygoid; qj=quadratojugal; sq=squamosal; occ con=occipital condyle; exoc=exoccipital; cr par=crista parotica; mmk=mentomeckelian bone. (Illustration by Bruce Worden)

cated within the larger olfactory system and separately innervated. The precise function of this organ is unknown, but in lissamphibians it seems to be important in social and reproductive behavior.

The lissamphibian auditory apparatus is unique among vertebrates because its structure functions in transmission of substrate vibrations, as well as sound waves in some. The vertebrate ear consists of three parts: the inner, middle, and outer ears. Most lissamphibians have inner and middle ears, but some only have inner ears. The inner ear consists of a series of canals that contain fluid and are suspended in the otic capsule. Specialized receptor cells are stimulated by the movement of the fluid contained in the inner ear; this provides information about sounds, vibrations, and balance or equilibrium. One of these patches of receptor cells, the papilla amphibiorum, is unique to lissamphibians and receives acoustic signals less than 1,000 Hertz in frequency. Although not all lissamphibians have a middle ear, most have a middle ear bone or stapes that is associated with a small opercular bone in the oval window of the otic capsule and extends outward to articulate with the lateral part of

the skull or the external ear. Vibrations from the air or the substrate are transmitted from the external ear or the side of the skull, along the stapes, to the oval window; vibrations against the oval window disturb the fluid of the inner ear, thereby stimulating the various receptor cells that are found there.

The diversity of lissamphibian body plans

Salamanders

Of the three living orders of lissamphibians, salamanders are the generalists. Typically they have elongate bodies, small heads, four limbs, a tail, and a sprawling gait. The rather arched, narrow skulls of most terrestrial salamanders are not well roofed; nonetheless, they have more bones than those of frogs and caecilians. A salamander skull also bears an additional articulation (total of three, instead of two) with the vertebral column. It is thought that this provides extra support for the head, for these animals lack the specialized trunk musculature that supports the heads of frogs and caecilians.

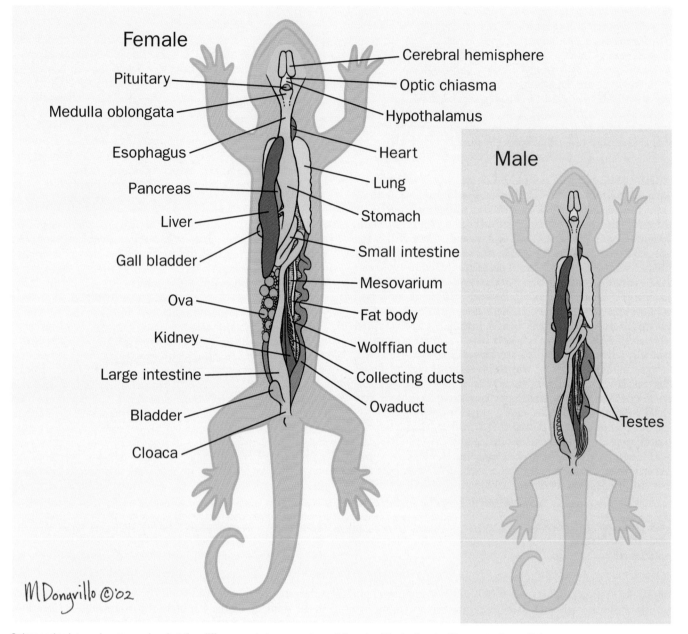

Salamander internal organs, showing the differences between male and female. (Illustration by Marguette Dongvillo)

Most salamanders have a rather simple hyoid apparatus, and nonspecialized jaw musculature and teeth. This simple architecture allows them to roll their fleshy tongue forward over the margin of the lower jaw to procure prey, which they transport to the mouth and manipulate with their teeth and tongue. In salamanders that lack lungs (plethodontids, the salamandrids *Chioglossa* and *Salamandrina*, and the hynobiid *Onychodactylus*), the hyobranchial apparatus no longer functions as a buccal pump. In these animals, the hyobranchium is used to project the tongue from the mouth; some of these salamanders can capture prey at distances equal to 4–80% of their body lengths.

Given their complex courtship and mating behavior and the ways that salamanders feed, vision and smell are particularly important. Thus all salamanders (except cave dwellers) have large, well-developed eyes, which are protected by eyelids in all but obligate neotenic salamanders (those that carry some larval traits into adulthood), such as axolotls. Salamanders use chemosensory cues in courtship and, in some cases, to return to the same breeding ponds each year. All have large, but relatively simple, olfactory organs and vomeronasal organs; these structures are best developed in terrestrial species and least developed in aquatic species. Vocalization is not important to salamanders; therefore, they have poorly developed ears and lack an external eardrum. They respond to low-frequency sounds that are conducted from the substrate through the forelimbs and girdle to the inner ear.

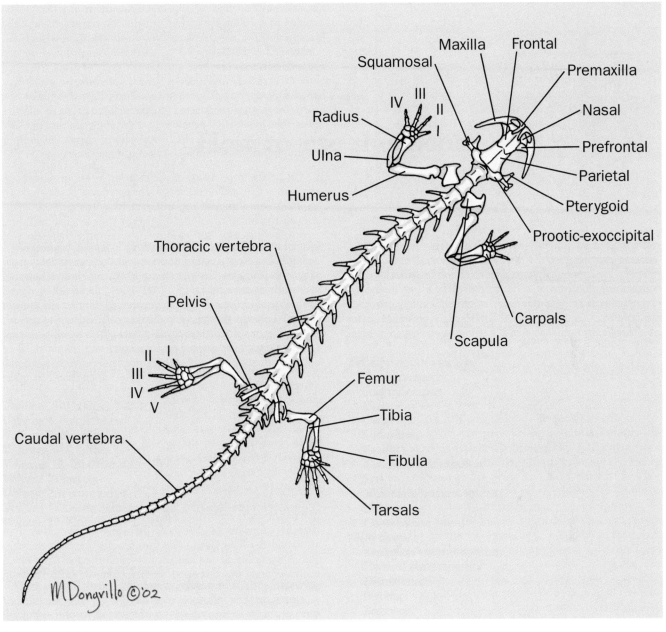

Salamander skeleton. (Illustration by Marguette Dongvillo)

The musculoskeletal system of most salamanders is rather generalized, having a relatively undifferentiated axial skeleton and poorly developed girdles that are not firmly attached to the vertebral column; the trunk musculature, however, is well developed. When startled, most salamanders undulate their body to move across the substrate. Otherwise, they throw their body into alternate curves to advance the stride of each forelimb as they move deliberately across the substrate. Aquatic salamanders deviate the most from this generalized body plan. Because these animals are supported by the water in which they live, they tend to be much larger than their terrestrial counterparts. Some retain their limbs, which they use to crawl across the bottom of ponds and streams, whereas oth-ers have lost or reduced their limbs and propel themselves through the water with undulatory motions.

Anurans

Frogs and toads have used their locomotory, feeding, and reproductive specializations to exploit habitats unavailable to most salamanders and caecilians; this doubtless accounts for their greater numbers of taxa and broader distributions. With its broad, flat, fenestrate head and short, inflexible trunk, an anuran body can be thought of as a projectile that is thrust forward in leaps from one place to another powered by strong hind limbs. The effectiveness of the hind limbs in propulsion is improved because anurans have two elongate ankle bones

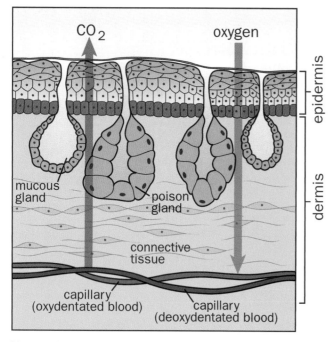

Diagram showing gas exchange through the skin—amphibian skin is very thin, enabling easy transfer of gases while mucous glands keep the surface damp. (Illustration by Patricia Ferrer)

CO₂ [CO$_2$]

oxygen

epidermis

dermis

mucous gland

poison gland

connective tissue

capillary (oxydentated blood)

capillary (deoxydentated blood)

(astragalus and calcaneum) which lengthen the hind limb and add another folding segment to it. As the limb unfolds the body is thrust up and forward, the anuran retracts its eyes into its large orbits, and it pulls its front limbs back alongside its body so that the head and trunk are as fusiform (tapered at each end) as possible. As the animal reaches the apogee (highest point) of its jump, the forelimbs are rotated forward and the eyes are opened. Because the animal lands on its forelimbs, an especially complex and elastic or strong pectoral-girdle mechanism is necessary to absorb the shock of landing. Likewise the pelvic girdle is modified with exceedingly long ilial shafts and a large acetabulum, and the end of the vertebral column is a long, bony rod to accommodate the complex, robust muscles involved in jumping.

A few basal, primitive frogs (e.g., ascaphids, bombinatorids) feed in much the same way as unspecialized salamanders. However, most anurans have developed a complex hyobranchial apparatus and associated musculature that permits them to catapult their tongues from their mouths to pick up prey. Once the prey is in the mouth, anurans retract their eyes into the orbital opening to help push the food into the pharynx and esophagus. Visual acumen is critical to animals that feed in this way and are saltatorial (jumping); thus, all frogs have eyes and the eyes usually are large.

Because anurans move by jumping, they cannot leave a continuous scent trail as most salamanders can. They must rely on another mechanism to advertise their presence to others of their kind for courtship and territorial behavior—they vocalize. The larynx is a cartilaginous capsule that contains the vocal cords and is located between the lungs and the buccal cavity. Air moving from the lungs to the buccal cavity

passes over the vocal cords and causes them to vibrate. The quality of the sound produced depends on the structure of the larynx and the nature of the vocal sac, which acts as a resonating chamber in male frogs; females can produce limited sounds because they have vocal cords, but they lack the vocal sacs of males. Some anurans do not call and the males lack vocal sacs. Among those that have them, the sacs of most are single or double and located under the floor of the mouth between the lower jaws, or laterally at the angles of the jaw. The sacs are connected with the buccal cavity via slits in the floor of the mouth on either side of the tongue.

Given the importance of vocalization in anurans, it is not surprising that they have well-developed ears that consist of an inner ear, a middle ear, and usually an external ear (tympanum or eardrum). Sound waves impinge on the eardrum, which is a piece of skin connected to an underlying ring of cartilage (tympanic annulus), and are conducted through the middle ear by a slim column of bone (stapes) to the inner ear. It is not unusual for anurans to lack the eardrum, but to retain the stapes and middle ear. In some anurans (e.g., some burrowers and montane stream breeders), the external and middle ear may be absent. Presumably in these anurans vibrations are conducted through the forelimbs to the inner ear in the same way they are in salamanders.

Despite their anatomical and behavioral specializations, anurans have exploited a wide range of habitats and have suites of morphological traits that are associated with specialized lifestyles. Toads typically have short hind limbs that allow them to hop only short distances and burrow; their skulls usually are heavily ossified and their skin verrucose (heavy and warty). These traits are suited to the arid or seasonally arid areas in which they frequently are found. Other anurans are adapted to life in bushes and trees. Generally, these frogs have lightweight skulls, long limbs, and suction-like pads on the ends of the digits. Frogs that spend most of their lives in water tend to have their eyes on top, rather than at the sides, of the head. The toes usually are fully webbed, and in some the bodies are flattened and the limbs sprawled at the sides of the bodies; these frogs cannot jump and hop on land, and move only with difficulty by "swimming" over the substrate.

Caecilians

The snakelike, limbless caecilians probably are the most bizarre of the living lissamphibians. Most of these peculiar animals are fossorial, living in subterranean burrows, but a few are aquatic. Their heads are blunt and their tails are short, if present at all. Caecilians lack any vestige of pectoral or pelvic girdles, but have a highly flexible vertebral column and exceedingly strong trunk musculature. The overlying skin is immovable on the underlying musculature. There are some modifications of the viscera, which are correlated with the snakelike morphology of caecilians. Both lungs are elongate, but the left one usually is reduced, and the testes are elongate. Typically caecilians have grooves (annuli) in the skin that correspond to the number of vertebrae (body segments) and extend over the length of the body. Buried within some or all of these annuli are minute dermal scales, which are thought to provide a frictional surface that aids the animal in burrowing through the soil.

All caecilians have well-ossified, compact, long, narrow skulls, which are well adapted to their habit of burrowing in soil. The skull structure constrains the development of jaw musculature, and caecilians have developed a unique dual jaw-closing mechanism that involves special adductor musculature that is attached to a process on the back end of the lower jaw. They have only a rudimentary tongue and robust, recurved, fanglike teeth, with which they seize their prey and then twist the body to shear off bites of food. The eyes of all caecilians are reduced and in some covered by bone. In contrast, their olfactory organs are elaborate. In addition to a well-developed nasal organ, caecilians have a tentacle on each side of the head in front of the eye. The tentacle can be extended and retracted; thus, when the animals are burrowing (and presumably their nostrils are closed), they can extrude the tentacle and pick up chemosensory cues. Hearing in caecilians probably is less acute than in anurans and salamanders. They lack external and middle ears, but retain a stapes (middle ear bone). This compact, heavy bone extends from the inner ear to the quadrate (a jaw bone) and probably functions only to receive low-frequency vibrations in the substrate.

Resources

Books

Duellman, William E., and Linda Trueb. *Biology of Amphibians.* New York: McGraw Hill, 1986.

Linda Trueb, PhD

·····

Reproduction

An essential attribute of any species or population is the ability to produce a succeeding generation. As the first vertebrates to set foot on land, amphibians were faced with new reproductive challenges. The primitive reproductive behavior involves terrestrial adults moving to water. There the eggs are deposited, fertilized externally, and develop into larvae that obtain necessary nutrients from the aquatic environment; the larvae grow and change into adults with a body form adapted for life on land—a process known as *metamorphosis*. Early European naturalists observed this kind of reproductive behavior in local frogs, toads, and newts, and for more than a century, amphibians were characterized as having a biphasic (two-stage) life cycle (as implied by the name *Amphibia*).

One of the most fascinating aspects of amphibians is that their successful exploitation of a great variety of habitats necessitated the evolution of diverse reproductive modes; these modes made use of existing environmental resources in mixed climatic conditions and enhanced the survival of their young. It generally is conceded that the ancestral reproductive mode is the deposition of eggs that are fertilized externally and that development takes place in water or in a moist terrestrial or arboreal (tree) environment; this mode is known as *oviparous*. During their nearly 300 million years of reproductive experimentation, different groups of amphibians independently evolved terrestrial eggs, many of which undergo direct development into miniatures of the adults and bypass the free-living aquatic larval stage. Various amphibians exhibited different degrees of parental care, not only attending eggs or larvae or both but in some cases also transporting them; others evolved ways to fertilize eggs internally. In the latter case, the result is that embryos derive nutrients from the yolk for development (ovoviparous) or obtain nutrients from maternal tissues (viviparous) in a manner reminiscent of placental mammals. Both ovoviviparity and viviparity result in the birth of living young that are miniatures of the adults; again, there are no intermediate aquatic larval stages.

Within these general evolutionary trends, there are many specializations restricted to a few species (e.g., stomach-brooding and carrying larvae in pouches) and some deviations that are counter to general trends (e.g., nonfeeding larvae in terrestrial nests in humid regions). However, the overall pattern clearly is toward increased terrestriality. Thus, the existence among amphibians of manifold ways to reproduce is an example of multiple evolutionary success stories—amphibians have adopted disparate life-history strategies to cope with a variety of environmental regimes. The diversity of these strategies within the group as a whole and their flexibility within species and even within populations reflect the evolutionary and ecological diversity of amphibians, the vertebrate pioneers of the terrestrial environment.

There are costs and benefits associated with different reproductive strategies in amphibians. The presumed primitive strategy is to produce many small eggs with a small amount of yolk and deposit them in water; these eggs hatch into small larvae that obtain nutrients from the environment (exogenous larvae). Parental investment (energetic expenditure) per offspring is minimal, but survivorship is low. The large numbers of potential offspring maintain populations; this is the strategy common to ambystomatid salamanders and many families of frogs, (e.g., bufonids, hylids, and ranids). Chances of survivorship improve when larger eggs with more nutrients are produced; the embryos hatch as more advanced larvae. These larvae can survive in more strenuous environments, such as mountain streams (e.g., salamanders, such as *Dicamptodon* and *Rhyacotriton*, and anurans, such as *Atelopus*, *Ptychohyla*, and *Scutiger*), or require less time to complete metamorphosis (e.g., some species of marsupial frogs). This strategy requires greater investment per offspring by the female. The next step is production of fewer eggs with sufficient nutrients for the completion of development as nonfeeding larvae or as miniatures of the adults. Maternal investment per offspring and survivorship are high. Survivorship is enhanced by many kinds of parental care, but in many of these species fecundity is low.

Direct development not only has evolved independently in different lineages of anurans, but it also has taken place in distinct ways. In most direct-developing anurans, the early larval stages are absent; thyroid hormones that are essential to triggering metamorphosis in tadpoles also influence later developmental stages in frogs, such as *Eleutherodactylus coqui*, and bring about the metamorphic climax shortly before hatching. In contrast, in hemiphractine hylids, the usual larval stages are present within the egg capsules of those species with direct development. Possibly, the production of tadpoles in some species of marsupial frogs (*Gastrotheca*) is an example of

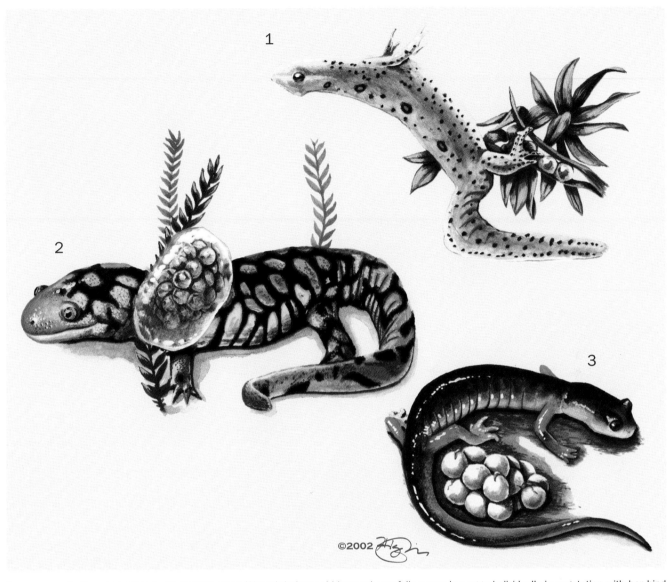

Anuran reproductive strategies: 1. An eastern newt (*Notophthalmus viridescens*) carefully wraps her eggs individually in vegetation with her hind feet; 2. A tiger salamander (*Ambystoma tigrinum*) guards her aquatic clump of eggs; 3. An Appalachian woodland salamander (*Plethodon jordani*) is coiled around its terrestrial clutch. (Illustration by Jonathan Higgins)

arrested development or simply suggests that there are insufficient amounts of nutrients in the eggs to complete development.

Courtship and mating

With the possible exception of some poison frogs of the genus *Dendrobates*, in which pairs apparently are bonded throughout a breeding season, amphibians are polygamous. Most salamanders reach sexual maturity during their second year, with females usually maturing later than males. Females of the aquatic cryptobranchids and proteiids do not breed before six years; in contrast, both sexes of the aquatic plethodontid *Eurycea multiplicata* reach sexual maturity shortly after metamorphosis at an age of five to eight months. Most anurans reach sexual maturity in six months to one year, but species inhabiting cool climates require much longer, up to four years in *Ascaphus* and many *Rana*. The limited data on caecilians suggest that sexual maturity is reached in two to three years.

Reproductive cycles are controlled by hormones, the actions of which are correlated with environmental variables as well as constraints of habitat, size, reproductive mode, and parental care. Caecilians reproduce biennially, and salamanders reproduce annually or biennially. In the wet tropics, anurans commonly reproduce continually and may deposit several clutches of eggs per year, but in seasonally dry or cold regions, the number of clutches may be limited to one per year or one every other year. In temperate regions, breeding coincides with higher temperatures and spring rains, whereas in semiarid regions and deserts, breeding activity is initiated by rains that result in the formation of temporary ponds. Thus,

Anuran reproductive strategies: 1. Rhacophorids create an arboreal foam nest; 2. *Leptodactylus* creates an aquatic foam nest; 3. A centrolenid with his clutch on the underside of a leaf, usually over a stream; 4. *Eleutherodactylus* creates a terrestrial clutch; 5. *Bufo* lays eggs in strings, one from each oviduct; 6. *Rana* lays a single, massive, aquatic clutch. (Illustration by Jonathan Higgins)

breeding activities may be limited to only a few days or weeks in any given year.

In many amphibians, especially those laying terrestrial eggs, courtship and mating take place within their normal home ranges, but most of those species that deposit their eggs in water migrate to breeding sites, and large numbers of individuals often congregate at these sites. Several species of anurans and salamanders are known to return to the same breeding sites in successive years; in some cases, this is the site where they developed as larvae. Olfaction seems to be the primary method used by salamanders and some frogs to locate breeding sites, but vocalization plays the dominant role

among most frogs. Calling by aggregations of anurans attracts not only females but also other males to the breeding site. Little is known about courtship in the secretive, subterranean caecilians; it is thought that olfaction is important for the location of mates in burrows.

Release of hormones (principally gonadotropin) during early phases of the reproductive cycle results in the development of many secondary sexual characters that can persist throughout adult life or might be transitory during the reproductive season. Among the former features are body size and skin texture. Females usually are larger than males. Males of some salamanders, especially newts (Salamandridae), de-

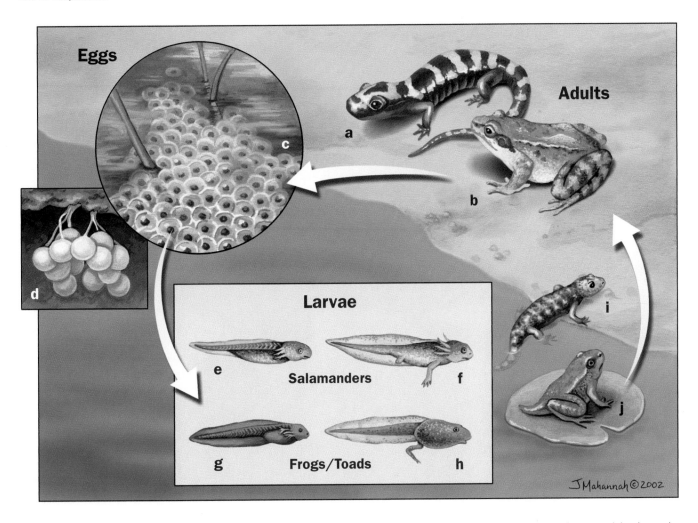

Life cycle of a salamander (*Ambystoma opacum*) and frog (*Rana temporaria*); a and b—adults; c—eggs laid in water; d—terrestrial salamander eggs laid in a moist area on land; e, f, g, h—larval stage; i and j—juvenile stage. (Illustration by Jacqueline Mahannah)

velop more intense coloration, and in many kinds of male frogs, vocal sacs become more brightly colored during the breeding season. Courtship glands develop on the chins of males of many kinds of salamanders, especially plethodontids, and on the chins and bellies of various kinds of frogs. The most conspicuous secondary sexual characteristics are horny growths known as *nuptial pads* or excrescences in males. Males use these growths to better grip the female during breeding. Most male anurans clasp females from above (amplexus); the clasp is around the waist (inguinal amplexus) in some frogs, but more often it is just behind the arms (axillary amplexus). Nuptial excrescences are present at the base of the thumb and sometimes on the fingers and the chest; typically, the excrescence is a roughened pad, but in some species it takes the form of one to many spines. Nuptial excrescences also develop on the insides of the arms or legs of some salamanders that breed in water.

Fecundity and egg deposition

Generally, larger species deposit more eggs than smaller species, and eggs placed in water are smaller and more nu-

merous than those laid on land or carried by a parent. For example, the oviparous caecilian *Ichthyophis glutinosus* may lay as many as 54 eggs in a clutch, but the viviparous *Geotrypetes seraphini* gives birth to only one to four young. Fecundity is higher in salamanders; again, clutches deposited in water are the largest. The tiger salamander (*Ambystoma tigrinum*) can lay as many as 500 eggs, and the hellbender (*Cryptobranchus alleganiensis*) and the greater siren (*Siren lacertina*) deposit clutches of about 450 and 500 eggs, respectively. In contrast, clutches of the Olympic torrent salamander (*Rhyacotriton olympicus*) contain as few as eight relatively large eggs laid in mountain streams. Many plethodontid salamanders lay their eggs on land; clutches may contain as few as nine and as many as 40 eggs. The live-bearing salamander (*Salamandra atra*) gives birth to only two young at a time.

Fecundity in anurans varies much more widely than in other groups of living amphibians. Many species, especially large species of *Bufo* and *Rana*, that deposit eggs in water have extremely large clutches containing thousands of eggs. Many small and medium-size species in the humid tropics lay clutches of only a few hundred eggs, but females return to

Egg-carrying anurans: 1. *Alytes obstetricans*; 2. *Hemiphractus johnsoni*; 3. *Pipa carvalhoi*; 4. *Colostethus subpunctatus*; 5. *Gastrotheca cornuta*. (Illustration by Jonathan Higgins)

breeding ponds in a matter of a week or two to deposit another clutch. Species (e.g., phyllomedusine hylids and hyperoliids of the genera *Afrixalus* and *Hyperolius*) that deposit their eggs on vegetation above water have smaller clutches, commonly fewer than 400 eggs, and the small glass frogs (Centrolenidae) usually deposit fewer than 40 eggs. Many kinds of frogs deposit clutches of six to 67 eggs on land; in some cases the eggs are deposited in chambers excavated by the parents. A few kinds of frogs give birth to no more than eight living young at a time.

Anurans deposit eggs in places other than in water or on land. Several kinds of hylids and microhylids lay their eggs in water-holding leaf axils, tree holes, or bromeliads; such clutches usually contain fewer than 60 eggs. Amplectant pairs of some leptodactylids and limnodynastids kick the eggs with their feet into a mixture of water, air, and secretions that form a foamlike mass floating on the water; the outer part of the foam nest hardens and protects the moist interior in which the eggs develop. Rhacophorid frogs also build foam nests on

leaves or branches over water, and some leptodactylids (*Adenomera*) deposit eggs in terrestrial foam nests, where the eggs and embryos obtain all their nutrients for development from the yolk. Fecundity decreases from as many as 1,000 eggs in aquatic foam nests to as few as 25 in terrestrial foam nests.

Fertilization

The eggs, which consist of an ovum within one or more gelatinous capsules, are fertilized externally in nearly all anurans and in salamanders of the families Cryptobranchidae, Hynobiidae, and, presumably, Sirenidae. Eggs are deposited in water (on land in the case of many anurans), and males exude sperm over them. However, in all caecilians and in most salamanders, success is enhanced by internal fertilization. In caecilians, males have a penis-like intromittent organ, the phallodeum, which is inserted into the female's cloaca. This unique structure in amphibians is a portion of the cloaca that is eversible (able to turn inside out). An analogous, but not

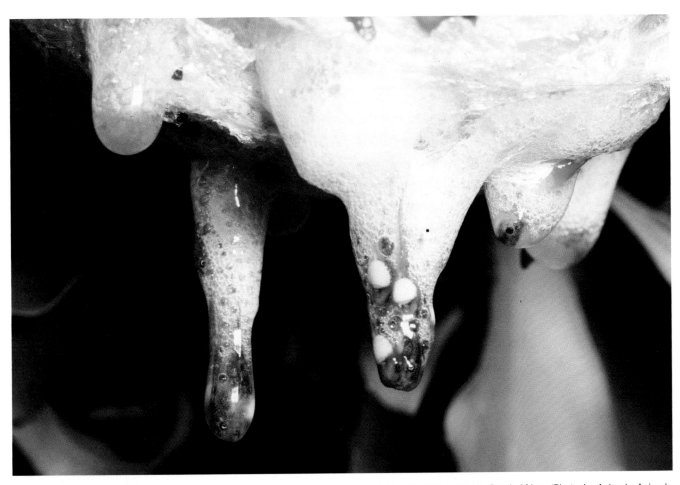

African gray treefrog (*Chiromantis xerampelina*) tadpoles drop from foam nests into seasonal ponds in South Africa. (Photo by Animals Animals ©Michael Fogden. Reproduced by permission.)

homologous, "tail" (a posterior extension of the cloaca) in frogs of the genus *Ascaphus* (Ascaphidae), which breed in fast-flowing streams, conducts sperm from the male into the female's cloaca. Internal fertilization in salamanders is accomplished by another unique feature, the spermatophore, which is a conical, gelatinous structure with a cap of sperm. During courtship, males deposit spermatophores on the substrate; females pick up spermatophores in their cloacas, and the sperm are stored in a small pouch, the spermatheca, off the cloaca. Subsequently, often many months later, the eggs are fertilized as they pass through the cloaca. Aside from *Ascaphus*, internal fertilization is known in a few other anurans, namely, African toads (*Nectophrynoides*) and some West Indian species of the leptodactylid genus *Eleutherodactylus*. Internal fertilization also is suspected in the African bufonid *Mertensophryne micranotis* and an East Indian ranid of the genus *Limnonectes*, because of modifications of the cloacal regions of males.

Development and hatching

Most aquatic eggs hatch as small larvae, whereas many terrestrial eggs undergo direct development and hatch as miniatures of the adults. In direct-developing eggs, the larval stages

are completed within the egg capsules, or the larval stage is suppressed. There is a positive correlation between ovum size and stage of hatching. The ova of salamanders that have aquatic larvae usually are 0.06–0.12 in (1.5–3.0 mm) in diameter, and anuran ova deposited in water are even smaller, 0.04–0.08 in (1.0–2.0 mm) in diameter. Such eggs contain small amounts of yolk that provide sufficient nutrients for only partial development. The larvae obtain nutrients from the environment for the rest of their development.

In those salamanders and frogs that undergo direct development, the ova contain all the nutrients necessary for growth into a small salamander or frog. Consequently, the ova are much larger—0.12–0.5 in (3.0–5.0 mm) in salamanders and 0.08–0.40 in (2.0–10.0 mm) in anurans—than eggs that hatch as tadpoles. A negative correlation exists between temperature and developmental rate. Aquatic eggs of salamanders develop in relatively cold water, and the duration of development ranges from 20 days in some newts (*Triturus*) to about nine months for the eggs of *Dicamptodon* in cold mountain streams. Likewise, most salamanders laying terrestrial eggs live in cool or temperate conditions; their eggs require 56–165 days to complete their growth to miniatures of the adults. The small aquatic eggs of many anurans hatch within one day of deposition, but those laid in cold water may require more than

African gray treefrogs (*Chiromantis xerampelina*) in cooperative foam nesting in a savanna habitat, South Africa. The frogs use their feet to beat the eggs and seminal fluid to form a foam nest. (Photo by Animals Animals ©Michael Fogden. Reproduced by permission.)

40 days to hatch. Direct-developing eggs of anurans need longer, usually about a month, but those of the small leptodactylid *Eleutherodactylus planirostris* complete their development in as few as 15 days. In a few salamanders (e.g., the plethodontid *Desmognathus aeneus*) and several frogs (e.g., the myobatrachid *Crinia nimbus*, limnodynastids of the genera *Kyarranus* and *Philoria*, leptodactylids of the genus *Adenomera*, bufonids of the genus *Pelophryne*, and some dendrobatids of the genus *Colostethus*), terrestrial eggs hatch as nonfeeding larvae that obtain the nutrients necessary to complete growth from yolk encased in their body cavities.

Parental care

Parental care in the form of protection and feeding typifies birds and mammals; although they are less universal, diverse kinds of parental care exist among amphibians as well. Parental care can be defined as any behavior exhibited by a parent toward its offspring that increases the offspring's chances of survival; this behavior, however, can reduce the parent's ability to invest in additional offspring. Among amphibians, parental care includes attendance of eggs, transportation of eggs or larvae, and feeding of larvae. Parental care occurs only in those species that deposit their eggs in

single clusters, never among species that scatter their eggs in aquatic situations. Nest construction and retention of eggs in the oviducts are not considered to be parental care.

Egg attendance

Egg attendance is the most common and taxonomically widespread type of parental care. In most cases, the eggs simply are guarded against potential predators, but some species of salamanders have been observed to rotate and possibly aid in aeration of aquatic eggs by creating water currents with their gills or tails. Guarding seems to be the principal function of terrestrial salamanders that coil about their clutches, but by osmotic transfer of moisture they also may help prevent desiccation of the eggs; some species also have been seen to rotate the eggs, which aids in the elimination of pathogenic fungi. A few anurans attend clutches of aquatic eggs, but many attend terrestrial or arboreal clutches.

Stream-inhabiting salamanders of five families are known to attend egg clutches attached to objects in streams; attendance is by males in *Andrias*, *Cryptobranchus* (Cryptobranchidae), and several species of *Hynobius* (Hynobiidae) but by females in two species of *Dicamptodon* (Dicamptodontidae), several species of plethodontids (*Desmognathus*, *Eurycea*, *Gyrinophilus*, and *Pseudotriton*), and *Necturus maculosus* (Proteiidae). Attendance is by either parent in the subterranean proteiid *Proteus anguinus*. Generally, the adults remain with the eggs from the time of deposition until hatching; the duration of this attendance varies from about six weeks in *Necturus maculosus* to about 13 weeks in *Desmognathus marmoratus*.

Attendance by females at terrestrial nests is known among many salamanders—the ambystomatid *Ambystoma opacum* and several genera of plethodontids (e.g., *Batrachoseps, Bolitoglossa, Ensatina, Desmognathus, Hemidactylium,* and *Plethodon*). Because egg deposition may occur many months after insemination by spermatophores, males of these salamanders may no longer be in the vicinity and do not attend nests. The duration of female attendance varies from about six weeks in *Ambystoma opacum* to nearly six months in *Bolitoglossa rostrata*. Females of some caecilians (*Ichthyophis* and *Idiocranium*) and one salamander (*Amphiuma*) are known to coil around subterranean clutches of eggs, presumably to minimize desiccation, or loss of moisture.

With the exception of males of the hairy frogs (Arthroleptidae: *Trichobatrachus robustus*), which sit on eggs in streams, and males of the moustache toads (Megophryidae: *Vibrissaphora*), which guard eggs under boulders at the edges of streams, attendance of aquatic eggs among anurans is known only in species that lay eggs in foam nests (Limnodynastidae: *Adelotus*) or in basins constructed by males (Bufonidae: *Nectophryne*; Hylidae: *Hyla rosenbergi*). In these cases, the eggs are in territories defended by males, who secondarily guard eggs. Eggs are not attended very long in these species—two to three days in *Hyla rosenbergi*, six days in *Adelotus brevis*, and 35 days in *Nectophryne afra*. Females of some species of *Leptodactylus* guard aquatic foam nests and subsequently remain with the schools of tadpoles, defending them from potential predators.

Male attendance of egg clutches on vegetation over water is common among territorial centrolenids, but it is unknown in the arboreal-nesting phyllomedusine hylids. Centrolenids not only guard the eggs from potential parasitic insects but also keep the eggs moist by perching on top of them by day. Species in three genera of microhylids (*Anodonthyla, Platypelis,* and *Plethodontohyla*) in Madagascar deposit their eggs in water-filled leaf axils; males attend the eggs for 26–35 days, until hatching. Females of the African ranid *Phrynodon* attend arboreal eggs, as do females of at least two hyperoliids (*Alexteroon obstetricans* and *Hyperolius spinigularis*), who not only guard their arboreal egg clutches but also moisten them by eliminating water from their bladders on the eggs.

Attendance of eggs is common among species that deposit their eggs on the ground and in burrows. The eggs of African *Hemisus* (Hemisotidae) and *Breviceps* (Microhylidae) are deposited in subterranean burrows and attended by females, who presumably moisten the eggs. Females of the former genus burrow headfirst from the chamber to a nearby pond, thereby releasing tadpoles into the water.

Several species of frogs in different families and one salamander, the plethodontid *Desmognathus aeneus*, have terrestrial eggs that hatch as nonfeeding larvae and derive all their nutrients from the yolk encased in their bodies. Attendance is by females in the salamander and in the leptodactylid frog

Wood frogs (*Rana sylvatica*) copulate among masses of previously laid frog eggs. (Photo by Gregory K. Scott/Photo Researchers, Inc. Reproduced by permission.)

Zachaenus parvulus, but the eggs are attended by males in the leptodactylid *Thoropa petropolitana*, the bufonid *Nectophrynoides malcolmi*, three species of the ranid genus *Petropedetes*, and at least two microhylids (*Breviceps adspersus* and *Synapturanus salseri*).

Clutches of terrestrial eggs undergo direct development into froglets in many different lineages of anurans. This mode of development is characteristic of all arthroleptines and brachycephalids, three species of *Leiopelma*, two genera of myobatrachids (*Arenophryne* and *Myobatrachus*), a few bufonids and ranids, and leptodactylids of the genus *Eleutherodactylus* (and relatives). The mode also characterizes about 50% of the Microhylidae (all asterophryines, brevicipines, and genyophrynines and at least one microhyline, *Myersiella microps*). In some cases females attend the clutches, but in others attendance is by males, especially territorial species, such as some *Eleutherodactylus*. The known duration of attendance is 17–100 days.

Transportation of eggs and larvae

Adults of diverse species of anurans that attend developing clutches of eggs subsequently transport eggs, larvae, or

The eggs of the Natal forest frog (*Natalobatrachus bonebergi*) are attached to twigs above small streams in the forest. (Photo by Alan Channing. Reproduced by permission.)

Larvae of glass frogs in Virolin, Colombia. Like many frogs who live near running water, their eggs are laid on leaves over the stream. When they hatch, the tadpoles simply drop into the water. (Photo by Animals Animals ©Juan Manuel Renjifo. Reproduced by permission.)

both. The European midwife toads of the genus *Alytes* exhibit the simplest form of this type of parental care. As the strings of eggs are deposited and fertilized in shallow water, they adhere to the hind limbs of the male; he carries them with him and enters water when the eggs are ready to hatch, at which time the egg membranes disintegrate, and the tadpoles swim away.

Most instances of larval transport are associated with terrestrial eggs. In all dendrobatids except *Aromobates*, an adult sits in the disintegrating gelatinous material, and the hatchling tadpoles wriggle up the legs and onto the back of the adult. The larvae do not hold on to the adult with their mouths; instead, their bellies adhere to the skin on the dorsum of the parent by means of a gluelike substance (mucopolysaccharide) that dissolves in water once the parent transports them to a stream, small pond, or water-holding plant. In some *Dendrobates*, adults transport tadpoles from terrestrial nests to arboreal bromeliads that may be as high as 100 ft (30 m) above the ground. Similar transportation of larvae from terrestrial nests to aquatic sites for tadpole development is known in two genera (*Aphantophryne* and *Liophryne*) of genyophrynine microhylids and in some species of the ranid genus *Limnonectes*. Terrestrial eggs of the sooglossid *Sooglossus sechellensis* hatch as nonfeeding larvae that wriggle onto the back of the attending female, where they complete their growth. Transportation of hatchling froglets occurs in three of the four species of *Leiopelma* in New Zealand; the hatchlings climb on the back of attendant males.

Males of the small myobatrachid frog in Australia (*Assa darlingtoni*) have an inguinal pocket on each side of the body. The male sits in a clutch of 10 or 11 terrestrial eggs; upon hatching, the nonfeeding tadpoles wriggle onto the male and into the inguinal pockets, where they complete their development and emerge as froglets about two months later. Males of the southern South American mouth-brooding frogs of the family Rhinodermatidae attend terrestrial clutches of eggs. Male *Rhinoderma rufum* transport the tadpoles in the mouth to water, where they complete their development. Male *Rhinoderma darwinii* pick up hatchling tadpoles in the mouth; the tadpoles enter the vocal sac via the vocal slits in the floor of the mouth. The male carries the tadpoles in his vocal sac for 50–70 days, at which time fully developed young crawl through the vocal slits and emerge from the mouth. Some evidence suggests that nutrients are provided by the epithelial lining of the vocal sac.

Two groups of anurans exhibit highly specialized modes of transport of eggs and developing embryos under entirely different environmental conditions. During inguinal amplexus in aquatic frogs of the genus *Pipa* (Pipidae), females exude eggs into the water, and males sweep them with the feet onto the backs of the females, where the eggs become imbedded in the females' skin. In all species the eggs hatch as tadpoles. In some species (e.g., *Pipa carvalhoi* and *P. myersi*), the tadpoles leave the chambers and complete development as free-swimming tadpoles. In other species (e.g., *Pipa aspera* and *P. pipa*), the tadpoles complete their growth within the chambers and emerge as froglets. Females of the terrestrial and arboreal hemiphractine hylid frogs transport eggs or

Tokyo salamander (*Hynobius tokyoensis*) egg sac. (Photo by henk.wallays@skynet.be. Reproduced by permission.)

Female common European toad (*Bufo bufo*) carrying male to the place where she'll lay her eggs. (Photo by Animals Animals ©Robert Maier. Reproduced by permission.)

tadpoles on the dorsum (back) or in a dorsal pouch; the eggs are enclosed at least partially in bell-shaped external gills. The eggs adhere to the dorsum in *Cryptobatrachus*, *Stefania*, and *Hemiphractus* and hatch as froglets.

In *Flectonotus*, the eggs reside in a basinlike structure (which may be open or closed by lateral folds of skin) on the female's back; the eggs develop into nonfeeding tadpoles that are deposited in water in bromeliads or tree holes, where they complete their development in a few days. During amplexus, male marsupial frogs of the genus *Gastrotheca* push eggs into the opening of a pouch on the back of the female. In most species (e.g., *Gastrotheca ceratophrys*, *G. guentheri*, and *G. plumbea*), the large eggs develop directly into froglets that emerge from the pouch. In several species inhabiting high elevations of the Andes, the eggs hatch as tadpoles, at which time the females sit in shallow ponds and spread out the pouch opening with their toes to allow the tadpoles to escape into the water, where they feed and complete their growth.

Perhaps the most unusual mode of transport was in two species of Australian gastric-brooding frogs (Myobatrachidae: *Rheobatrachus*) that lived in mountain streams in northeastern Australia and are now presumed to be extinct. The female swallowed the fertilized eggs; the eggs or embryos secreted a hormone, prostaglandin E_2, that inhibited the usual production of digestive enzymes and acids by the epithelial tissue of the stomach. Thus, for a period of six to seven weeks, the female did not feed, because her digestive system had been shut down and her stomach converted to a gestation chamber. The tadpoles obtained all nutrients for development from the large amount of yolk contained in the eggs. Young were expelled from the mouth by the mother's propulsive vomiting. Within a few days after giving birth, the female's stomach resumed its digestive function, and the female began to feed.

Feeding of tadpoles

Females of frogs in four families are known to provide eggs as nutrients for developing tadpoles; all examples of this kind

of maternal behavior occur in cases in which the tadpoles are in confined or constrained situations, such as water in bromeliads or tree holes or in foam nests. After deposition of eggs or transportation of tadpoles, the female returns to the site and deposits eggs on which the tadpoles feed. Insofar as is known, tadpoles of some species are obligatorily oophagous (egg eating), whereas those of other species also feed on detritus or insect larvae, which may be present in the water in bromeliads or tree holes.

The simplest expression of this type of parental behavior is seen in a species of *Leptodactylus* (*L. fallax*) in the Lesser Antilles and several species of hylid frogs in Jamaica and Central and South America. The eggs of *L. fallax* are deposited as a foam nest in a shallow basin; the hatchlings remain in a disintegrating foam nest and produce secretions that mix with moisture to create additional foam in which they develop. The female periodically inserts her cloaca into the foam and exudes eggs, on which the tadpoles feed.

Likewise, females of several hylids that deposit eggs in bromeliads or tree holes provide eggs for their larvae; these species include *Anotheca spinosa*, *Hyla picadoi*, and *H. zeteki* in Central America; *Osteocephalus oophagus* and some species of *Phyllodytes* in South America; and *Osteopilus brunneus* in Jamaica. Because presumably conspecific (i.e. of the same species) frog eggs have been found in the stomachs of some other tadpoles, the females of those species are thought to provide eggs as nutrients for their tadpoles. These species include the microhylid *Hoplophryne rogersi* in Africa, a species of *Philautus* (Rhacophoridae) in Asia, *Phrynohyas resinifictrix* in South America, and *Calyptahyla crucialis*, *Hyla marianae*, and *H. wilderi* in Jamaica. With additional studies in the field, we should expect to find many more examples of this kind of parental care.

In some species of *Dendrobates*, females transport tadpoles individually from the terrestrial nest to an aquatic microhabitat (bromeliad, tree hole, or the husk of a Brazil nut). Subsequently, the female returns to each of the sites of tadpole development and deposits unfertilized eggs for the tadpoles to eat. In some other species of *Dendrobates* (e.g., *D. vanzolinii*), however, the male transports the tadpoles and subsequently leads the female to each deposition site so that she can feed the tadpoles.

Live birth

Ovoviviparity (in which all nutrients during development are provided by yolk) occurs facultatively (i.e. in some conditions but not others) in two salamanders (*Mertensiella caucasica* and *Salamandra salamandra*) and two frogs (*Nectophrynoides*

tornieri and *N. viviparus*). True viviparity (maternal provision of nutrients during development in the oviducts) is known in several caecilians of three different families and in two salamanders (some populations of *Mertensiella luschani* and *Salamandra atra*) and one anuran (*Nectophrynoides occidentalis*). During their development, fetuses of the caecilians quickly exhaust their yolk supply, escape from the embryonic membranes, and obtain nourishment from the female by ingesting secretions and epithelial tissue from the lining of the oviducts; fetal caecilians have deciduous teeth that are specialized for scraping the lining of the oviduct. Maternal nutrients also are supplied from the walls of the oviduct in *Salamandra atra* and by epithelial secretions in the oviducts of *Nectophrynoides occidentalis*.

Resources

Books

Crump, Martha L. "Parental Care." In *Amphibian Biology*, Vol. 2, *Social Behaviour*, edited by Harold Heatwole. Chipping Norton, Australia: Surrey Beatty and Sons, 1995.

Duellman, William E., and Linda Trueb. *Biology of Amphibians*. Baltimore: Johns Hopkins University Press, 1994.

Taylor, Douglas H., and Sheldon I. Guttman, eds. *The Reproductive Biology of Amphibians*. New York: Plenum Press, 1977.

Tyler, Michael J., ed. *The Gastric Brooding Frog*. London: Biddles, 1983.

Periodicals

Callery, Elizabeth M., Hung Fang, and Richard P. Elinson. "Frogs Without Polliwogs: Evolution of Anuran Direct Development." *BioEssays* 23 (2001) 233–241.

Crump, Martha L. "Reproductive Strategies in a Tropical Anuran Community." *Miscellaneous Publications, Museum of Natural History, University of Kansas* 61 (1974) 1–68.

Duellman, William E. "Reproductive Strategies of Frogs." *Scientific American* 267 (1992) 80–87.

Elinson, Richard P. "Direct Development: An Alternative Way to Make a Frog." *Genesis* 29 (2001) 91–95.

Wassersug, Richard J., and William E. Duellman. "Oral Structures and Their Development in Egg-Brooding Hylid Frog Embryos and Larvae: Evolutionary and Ecological Implications." *Journal of Morphology* 182 (1984) 1–37.

William E. Duellman, PhD

• • • • •

Larvae

The majority of amphibian species have free-living larvae that are temporary residents in aquatic habitats. Although direct development of terrestrial eggs has evolved many times among modern amphibians, most species still retain a larval stage and for good reason. In parts of the world where the seasons change and ponds vary in their longevity and productivity, there are clear benefits in having an aquatic stage of development. Most amphibians lay their eggs at a time of year (spring in the temperate world and the rainy season in the tropics) when food is abundant for the larvae. The larvae then metamorphose later in the season, when competition, predation, or physical degradation of the environment (e.g., drying up of ponds) makes those habitats unsafe. Amphibian species that have lost the larval stage tend to live in tropical environments with little seasonality. Indeed, the majority of caecilians (approximately 75%) are viviparous (born alive) without a free-living larval stage. Most of the remaining caecilian species, which are oviparous (egg laying), have direct development of terrestrial eggs.

Morphologic characteristics

Among the caecilians, free-living larvae are found in some species of Caeciliidae, Rhinatrematridae, and Ichthyophiidae. These larvae are morphologically similar to the adults but have open gill slits and relatively long, filamentous, external gills. They also have a vertical tail fin and thin skin and lack scales. They are carnivorous, feeding largely on small aquatic invertebrates. Caecilian larvae are secretive, nocturnal, and little is known about their behavior and ecology.

There are free-living larvae in all salamander families, though they are absent in most plethodontid genera. Common external features that distinguish larval salamanders from adult salamanders are open gill slits and external gills, a tail fin, and specialized dentition. In certain taxa (e.g., *Ambystoma*), flaps of skin at the corners of the mouth help make the mouth rounder when it is open. This facilitates suction feeding, which is important in catching active prey.

At metamorphosis both caecilian and salamander larvae lose their gills, gill slits, and tail fins. There are also changes in the bones of the skull. The alimentary tract, however, alters little relative to the changes seen at metamorphosis in frogs and toads. The caecilian and salamander larvae, like the

adults, are predators, feeding on aquatic invertebrates and other amphibian larvae, including in certain circumstances members of their own species.

The larvae of frogs and toads are known as tadpoles. Tadpoles differ far more from adults than do the larvae of caecilians or salamanders. The most conspicuous features that tadpoles share with adult anurans are a wide head, a short vertebral column, and no neck. As a result, tadpoles have a round combined head and body, with a laterally compressed tail appended to it. This "lollipop" shape, as seen from above, distinguishes tadpoles not only from other amphibian larvae but also from virtually all fishes. Tadpoles also differ from other amphibian larvae in that their gills and forelimbs develop under a fold of skin, the opercular fold. This fold may not cover the gills fully at hatching, but the external gills shrink, and the opercular fold grows quickly backward from the throat region to cover the gills by the time the tadpoles are freely swimming and feeding.

Compared with frogs, tadpoles have small mouths externally. This mouth is directed ventrally (downward) or anteroventrally (forward and downward) in the majority of tadpoles, which are bottom feeders. A few tadpoles graze on particles floating at the water's surface, and their mouths are directed dorsally (upward). Tadpoles in the families Microhylidae (with the exception of the Peret' toad, *Otophryne*), Pipidae, and Rhinophrynidae lack keratinized structures (hard tissue such as human nails) surrounding their mouths. Keratinized mouthparts are also absent in a few species with obligatorily carnivorous tadpoles that feed only on large prey (including other tadpoles of their own species, e.g., *Lepidobatrachus*) as well as some genera with nonfeeding larvae that survive until metamorphosis solely on yolk reserves.

Free-living tadpoles in all the other families have complex external oral features surrounding their mouths. The most prominent are the jaw sheaths (or beaks), formed of hard, darkly pigmented, keratinized tissue. The margins of these sheaths commonly are embellished with fine serrations, which make the jaws efficient at scraping and biting into soft material during feeding. In some carnivorous tadpoles (e.g., *Ceratophrys*), the jaw sheaths are sharply pointed. In certain stream-associated tadpoles from Southeast Asia, the sheaths are divided on the midline and are peglike in structure. This

Grzimek's Animal Life Encyclopedia   **39**

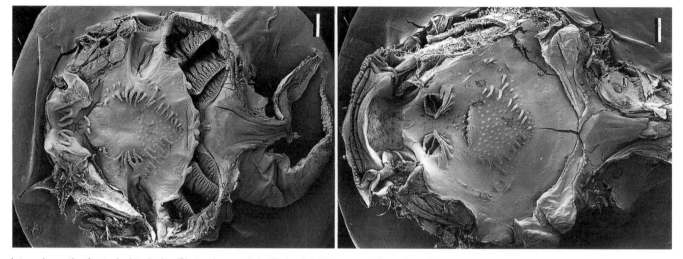

Internal mouth of a typical tadpole. (Photomicrograph by Richard J. Wassersug. Reproduced by permission.)

shape presumably is an aid to holding on to irregular rocky surfaces. In some other stream-associated species, many keratinized spikes replace the sheaths. Most pond- and stream-dwelling tadpoles have a fleshy oral disc surrounding the jaw sheaths. The free border of this disc may be partly or fully covered with tiny finger-like projections (papillae). Between the marginal papillae and the jaw sheaths run transverse rows of labial teeth (also called denticles or keradonts). These teeth are formed of the same keratin that stiffens the jaw sheaths. Seen under the microscope, the individual teeth may end in a blunt point, or they can be multi-cusped, depending on the species. In the majority of tadpoles, which feed by scraping food off surfaces, the labial teeth are multi-cusped.

Tadpoles of most species from around the world have two rows of labial teeth in front of (rostral to) the upper jaw sheath and three rows of denticles behind (caudal to) the lower jaw sheath. The number of labial tooth rows ranges greatly from species to species, even among closely related taxa. For example, tadpoles of the Central American treefrog *Hyla microcephala*, which feed on large food particles in ponds, have small jaws set back in an oral tube and no labial teeth. Tadpoles of a certain tropical stream-dwelling *Hyla*, in contrast, have the most number of rows known: 17 upper and 21 lower rows.

Variations in the size and shape of the oral disc, the papillae at the margins of the oral disc, the shape of the jaws, the numbers of denticle rows, and any gaps in those rows are all important features in identifying tadpoles of different species. The ways in which these structures actually function has received little study, however. It is clear that large oral discs with many denticle rows are common among stream-dwelling tadpoles exposed to currents. The larvae use these structures to hold on to surfaces and resist being swept downstream. Jaw sheaths that have sharp edges characterize many tadpoles that feed on active prey.

High-speed video of feeding North American bullfrog (*Rana catesbeiana*) larvae, which have the common pattern of two upper and three lower tooth rows, show that tadpoles use their labial teeth to anchor the oral disc to surfaces while their

jaws bite at the substrate. When grazing on algae, the mouth can open and close rapidly, more than six times per second at room temperature. As the jaws close, the labial teeth release their grip and then rake the surface toward the mouth. This action produces a suspension of fine material that can be sucked into the mouth. The sucking itself is achieved by the pulsatile raising and lowering of cartilaginous plates that lie under the front, or the buccal region, of the oral cavity. Pulsations of the buccal floor draw water into the mouth, aiding both feeding and aquatic breathing (i.e., gill irrigation).

Although the mouths of tadpoles are externally small, inside they are relatively large and structurally complex. Typical pond and stream tadpoles have sensory papillae near the front of the buccal cavity and additional rows of papillae on the buccal floor and roof that are used to trap larger food particles and funnel them toward the esophagus. Most tadpoles have a flap of skin that acts as a valve and separates the buccal cavity anteriorly from the pharynx region posteriorly. This valve ensures that flow through the mouth is one way, that is, backward into the pharynx. In the pharynx there are mucus-secreting organs that trap finer particles that get past the papillae of the buccal cavity. They work in conjunction with ruffled gill filters, which extend inward and upward from gill bars to catch the smaller particles that the tadpoles draw into their mouths.

These internal oral features of tadpoles vary among taxa and can be correlated with their diets. Most stream-dwelling tadpoles, for example, feed on a rather coarse suspension of material that they generate by scraping their food off of algal-covered surfaces. Such tadpoles tend to have many large papillae on the buccal floor and roof, for coarse sieving of food particles, but smaller and less dense gill filters in the pharynx. Pond tadpoles that live midwater and feed largely on single-cell organisms already in suspension (i.e., microphagous tadpoles) have few or no papillae in their mouths but comparatively large, dense gill filters. Obligatorily macrophagous tadpoles—ones that feed solely on big items, like frogs' eggs or other tadpoles—have neither elaborate buccal papillation nor large and dense gill filters. They also lack the mucus-secreting food traps. The muscles that depress the buccal floor are massive, however, which is consistent with the powerful sucking forces they must generate during feeding to capture active prey.

Water that passes the gill filters of tadpoles washes through the gill slits and around the gill filaments that lie on the superficial side of the gill bars. This water must exit the gill chamber, which is covered by an opercular fold. There may be one or two openings, called spiracles, for expelling water from the opercular cavity. The pattern of these openings has been important in the superfamilial taxonomy of anurans. Tadpoles of Ascaphidae, Leiopelmatidae, Discoglossidae, Bombinatoridae, and Microhylidae (with the exception of *Otophryne*) have a single, midline spiracle. Those of Pipidae and Rhinophrynidae (plus the carnivorous leptodactylid tadpole *Lepidobatrachus*) have two spiracles, one on each side of the tadpole. By far the most common pattern, found in the free-living ectotrophic tadpoles of all other anuran families, is a single sinistral spiracle.

In most tadpoles this single spiracle lies halfway between the ventral (front) and dorsal (back) surfaces of the tadpole, about halfway between the snout and the end of the head/body. For tadpoles of the leaf frogs (phyllomedusine hylids), which are largely midwater feeders, the otherwise sinistral spiracle lies close to the midline of the belly. In some microhylid tadpoles the branchial chamber extends all the way to the end of the body, and thus the spiracle opens near the vent. In *Otophryne* the sinistral spiracle is at the end of a long, flexible, free tube that extends caudally halfway to the tip of the tadpole's tail. This strange appendage is believed to help these tadpoles expel water when they are buried below the surface in the sandy bottoms of streams in northern South America.

The body cavity of tadpoles is filled mostly with an elongated and coiled intestine. Except in a few carnivorous tadpoles, the foregut of tadpoles is undeveloped, and the region of the gut tube that later becomes the stomach does not expand into a sac, as in most vertebrates, and does not secrete acid. The intestines of many tadpoles can be more than 10 times the length of the tadpole's head and body, though it is shorter in strictly carnivorous species. Most tadpoles are omnivorous grazers and at the same time suspension feeders. One typically finds silt and fragments of plant matter packed in the intestines of bottom-dwelling pond tadpoles. The microscopic animals living within that material may be disproportionately important as a source of protein for these larvae. Tadpoles of the clawed frogs (genus *Xenopus*) and the microhylid tadpoles are obligatory midwater suspension feeders, which is consistent with their lack of hard mouthparts for grazing on surfaces. Their guts are filled with a mixture of the various planktonic organisms that live with them in the water column.

Tadpoles vary in the timing of lung development. Most tadpoles that live in lentic (still) water fill their lungs for the first time shortly after hatching. From then on, they supplement aquatic respiration with aerial respiration. For some tadpoles that live in turbid (muddy) water, such as those of *Xenopus*, occasional air breaths are essential for normal growth and development. At the other extreme, tadpoles that live in lotic (flowing) water tend to be negatively buoyant and do not in-

The barking treefrog (*Hyla gratiosa*) tadpole. The black spot on the tail, which disappears as the tadpole grows, may aid in disrupting the tadpole's outline, making it difficult for predators to detect it. (Photo by Jan Caldwell. Reproduced by permission.)

flate their lungs until shortly before metamorphosis. Most, if not all, bufonid tadpoles fill their lungs just before metamorphosis.

As tadpoles grow in size, their head/body and tail change little in shape, but conspicuous hind limbs develop. The limbs start as simple rounded protuberances or limb buds at the junction of the body with the tail. By the time the tadpole is ready to metamorphose, those limbs are large and functional, assisting the tadpole in locomotion. The forelimbs develop at the same time as the hind limbs, but they do so under the opercular cover and thus are not seen externally until the moment of metamorphosis, when within a day or so they erupt through the operculum.

Behavior and ecology

Where a tadpole lives is determined largely by where its mother lays her eggs. Indeed, there is evidence that adult frogs can sense the presence of potential aquatic predators and even the intermediate hosts of some parasites that might harm their tadpoles, and, given a choice, they avoid depositing their eggs in those dangerous places. Once the eggs hatch, most tadpoles in ponds and streams are on their own. The most common defense that tadpoles have against predators is their cryptic coloration and secretiveness. Most tadpoles that live in ponds hide among vegetation. Those that live in streams may hang on to rocks in torrents, where they are similarly difficult to see. Other stream-dwelling tadpoles may sequester themselves between rocks at the bottom of streams or in vegetation at the stream margins. The daily activity cycles of tadpoles have not been well studied, but pond-dwelling tadpoles of species such as the green frog, *Rana clamitans*, change their location throughout the day. Temperature, oxygen concentration, and predation risk all may be factors affecting the microhabitat selection of tadpoles at any hour on any day.

When tadpoles swim rapidly, they produce high-amplitude waves in their tails, and their snouts oscillate accordingly from side to side. This wobbly swimming may appear grossly inefficient. Indeed, a computer simulation of tadpole swimming has shown that tadpoles are less efficient than more streamlined fishes of similar sizes when swimming in a straight line. Those same simulations also show that the tadpole's kinematics and shape work in concert to produce a region behind the body where the hind limbs can develop without handicapping their swimming. Thus, although the shape and swimming style of tadpoles are not graceful compared with those of fish, they allow tadpoles to grow hind limbs in preparation for metamorphosis with far less loss of efficiency than a fish-shaped animal would experience. In that regard, the tadpole shape and swimming style may not be ideal for the aquatic environment, but it fits well with tadpoles' ability to transform into something quite different—a frog or a toad.

Because of their highly flexible tails, tadpoles can turn rapidly; that is, they have high angular acceleration, with a short turning radius. Those features, rather than simple speed or endurance, may be most important in terms of escaping predatory insects, fish, and wading birds. In general, though, tadpoles do not do well in large, open bodies of water, particularly if large predatory fishes are present. Most tadpoles live in temporary ponds or isolated lakes that, in the absence of active stocking programs, would not have resident populations of large fishes. Those tadpoles that live in larger and more permanent waters are found most often in the grassy margins or shallow reaches. A few tadpoles that live in the open in permanent ponds with fishes (e.g., *Rana catesbeiana*) that are unpalatable to some predators.

Toad tadpoles (genus *Bufo*) from around the world are black and particularly toxic. They form large schools with hundreds to thousands of individuals. *Bufo* tadpoles can distinguish siblings from nonsiblings, suggesting that school structure may be influenced by the genetic relationships of the individual tadpoles. Schooling is seen in other anuran larvae from diverse genera around the world. In a few species (e.g., the genus *Leptodactylus*), schooling tadpoles may even follow an adult frog, which is presumed to be guarding them, around the pools where they live.

There is increasing evidence that amphibian larvae are cognizant of other animals in their environment besides conspecifics. Salamander larvae (*Ambystoma*) in ponds with fish, for example, avoid the open water much more than those in similar ponds without fish. Tadpoles of many species minimize their activity and stay near the bottom when housed in aquariums with fish or predatory insects, even when the predators are screened off and thus pose no real risk to the tadpoles. Tadpoles also can exhibit phenotypic plasticity and change their form in subtle ways in response to environmental stresses. These changes are best documented in the shape of their tails, which become more efficient for swimming when the tadpoles are raised in the presence of potential predators. There are, however, trade-offs in these situations. If tadpoles change their behavior and morphologic features in response to predators, they pay for it in the time that they can spend feeding and in the morphologic characteristics they have dedicated to food capture. As a result, the threat of predation can reduce the growth rates of tadpoles. The way tadpoles sense other species is not well studied. Schooling species

respond to the visual presence of other tadpoles, but for sensing nearby predators olfaction appears to be most important.

Ecomorphological types

The feature of salamander larvae that varies most with habitat is gill size. Species that live in lotic environments have proportionately smaller gills than those species that live in lentic environments. Various researchers have divided tadpoles into a wealth of categories based on ecomorphologic factors, but there is clearly a continuous spectrum of tadpole types. Anuran larval diversity is greatest in the tropics, whereas salamander larval diversity is highest among temperate taxa (most tropical salamanders, in fact, are direct developers). In the wet tropics, one can find tadpoles in aquatic habitats that are as meager as the axil bases of bromeliads or cattle footprints or as vast as a torrential stream.

In general, tadpoles that live in still water and off the bottom have tall tail fins, compared with similar species of bottom-dwelling tadpoles. Tadpoles in several families that are midwater specialists have tails that terminate in an elongated filament that can oscillate rapidly. This allows the tadpoles to hold their positions or move slowly through the water without the whole-body movements that occur when tadpoles use the entire tail for locomotion. Tadpoles that live in flowing water have proportionately longer tails with more axial musculature. The few semiterrestrial tadpoles that live on wet, rocky surfaces have long, thin tails with reduced fins. Fossorial tadpoles—whether they live in wet leaves along the edge of tropical streams or among the axils of bromeliads—also tend to have long, thin tails.

Metamorphosis

Tadpoles vary greatly in their size at metamorphosis. The tadpoles of some small treefrogs (genus *Hyla*) leave their aquatic environment when they are less than 0.79 in (20 mm) in length, whereas tadpoles of the paradox frog, *Pseudis paradoxa*, can grow to 9.8 in (25 cm) before they transform. How close a tadpole's size at metamorphosis is to the size of the mature adult varies from family to family. Thus, for example, in the Ranidae and Leptodactylidae, tadpoles that transform at a large size typically become large frogs. In the family Bufonidae, however, the tadpoles always transform at a small size regardless of whether the adult is the 1.2-in-long (30-mm-long) oak toad (*Bufo quercicus*) or the 9-in-long (23-cm-long) marine toad (*Bufo marinus*).

Metamorphosis for anurans is very rapid compared with the length of their larval life. Whereas some temperate tadpoles may take more than two years to reach metamorphosis (e.g., *Ascaphus truei* and *Rana catesbeiana*), most tadpoles can go from emergence of the forelimbs to complete loss of the tail in just a few days. At metamorphosis the forelimbs emerge, the tail is resorbed, and the head changes shape. The rapid loss of the tail is facilitated by the absence of vertebrae, except at the base. Those few caudal vertebrae fuse at the end of metamorphosis to form the urostyle, which is a long thin bone that extends backward between the hip bones of the frog and provides surfaces for the attachment of muscles used in jumping. The major change in the head is associated with the shift from a small-mouthed tadpole to a big-mouthed adult. The oral disc, labial teeth, and jaw sheaths are lost. The corners of the mouth move backward as the jaws themselves elongate. The tongue develops, except in the tongueless frogs (family Pipidae). All the internal oral features involved in the capture of food particles are lost, as are the gill filaments and gill slits. The foregut expands into a stomach, and the intestines shorten greatly as the gut prepares for a strictly carnivorous diet. All these changes testify to the great difference in the way of life of a tadpole versus an adult anuran.

Resources

Books

Anstis, Marion, ed. *Tadpoles of South-eastern Australia: A Guide with Keys.* Sydney, Australia: New Holland Publishers, 2002.

McDiarmid, Roy W., and Ronald Altig, eds. *Tadpoles: The Biology of Anuran Larvae.* Chicago: University of Chicago Press, 1999.

Sanderson, S. Laurie, and Sarah J. Kupferberg. "Development and Evolution of Aquatic Larval Feeding Mechanisms." In *The Origin and Evolution of Larval Forms,* edited by Brian K. Hall and Marvalee H. Wake. San Diego: Academic Press, 1999.

Wassersug, Richard J. "Assessing and Controlling Amphibian Populations from the Larval Perspective." In *Amphibians in Decline: Canadian Studies of a Global Problem,* edited by David Green. *Herpetological Conservation,* Vol. 1. St. Louis: Society for the Study of Amphibians and Reptiles Publications, 1997.

Zug, George R., Laurie J. Vitt, and Janalee P. Caldwell, eds. *Herpetology: An Introductory Biology of Amphibians and Reptiles.* 2nd edition. San Diego: Academic Press, 2001.

Periodicals

Liu, Hao, Richard J. Wassersug, and Keiji Kawachi. "The Three Dimensional Hydrodynamics of Tadpole Locomotion." *Journal of Experimental Biology* 200, no. 20 (1997): 2807–2819.

Liu, Hao, Richard J. Wassersug, Keiji Kawachi, and Masamichi Yamashita. "Plasticity and Constraints on Feeding Kinematics in Anuran Larvae." *Comparative Biochemistry and Physiology A: Molecular Integrative Physiology* 131, no. 1 (2001): 183–195.

Relyea, Rick A. "Morphological and Behavioral Plasticity of Larval Anurans in Response to Different Predators." *Ecology* 82, no. 2 (2001): 523–540.

Van Buskirk, J., and S. A. McCollum. "Functional Mechanisms of an Inducible Defence in Tadpoles: Morphology and Behavior Influence Mortality Risk from Predation." *Journal of Evolutionary Biology* 13 (2000): 336–347.

Richard J. Wassersug, PhD

· · · · ·

Behavior

Amphibians are not by nature especially social creatures. Most live solitary lives, and even when they form temporary aggregations, they tend to ignore one another. Some tadpoles form large schools that protect them from predators or enhance feeding, and some exhibit a preference for aggregating with closely related individuals. There is little evidence of such cooperative behavior in adult amphibians. Most social interactions are competitive, and most competition is related to acquisition of mates. Sometimes such competition is relatively benign, with males scrambling for access to females, but in some species, males fight violently for individual females or for territories that contain resources that are attractive to females.

Modes of communication

Any type of social interaction between individuals involves an exchange of communication signals based on chemical, visual, acoustic, or tactile cues. The three major lineages of amphibians have undergone millions of years of independent evolution, and, not surprisingly, their modes of communication are different. Little is known about the communication and social behavior of caecilians. We do not know, for example, how males and females of any species of caecilian locate one another. Because most caecilians spend their lives underground, are entirely or nearly blind, and are not known to produce sounds, it is likely that chemical signals are used for mate location and courtship.

Chemical communication in salamanders

The ancestral mode of communication in salamanders appears to be chemical. Salamanders have a variety of specialized glands that produce chemical signals (pheromones) that convey messages of aggression or attraction to other individuals. The use of chemical signals in aggressive interactions has been studied best in the North American red-backed salamander (*Plethodon cinereus*). Both males and females defend feeding territories under logs and other cover objects (objects used for cover) outside the breeding season. Territory owners mark their territories with fecal pellets containing pheromones produced by glands near the cloaca. Other individuals avoid areas marked by territorial salamanders. During the breeding season, females apparently use the same chemical cues to assess the quality of potential mates. In lab-

oratory experiments, females were more likely to enter territories of males marked with fecal pellets containing termites, a high-quality food, than those marked with pellets containing ants, a low-quality food.

For many other salamanders, chemical cues are used in the initial identification of potential mates as members of the same species. Studies of several closely related species in the terrestrial genus *Plethodon* have shown that males court only females of their own species and prefer both airborne and substrate-borne chemical cues from conspecific females to those of other species. Similar results have been obtained in studies of dusky salamanders in the genus *Desmognathus*. In both of these groups, hybridization (mating between species) is relatively common in areas where populations have diverged only recently, and behavioral experiments have shown that discrimination of chemical cues is most accurate in populations where hybridization does not occur.

Male salamanders also use chemical cues during courtship to increase the receptivity of females. Many salamanders have elaborate courtship behavior that involves the transfer of pheromones from the male to the female. The ancestral condition appears to be the production of pheromones by glands in the cloacal region. In mole salamanders (Ambystomatidae), such as the tiger salamander (*Ambystoma tigrinum*), the female follows the male in a tail-nudging walk with her snout pressed against the male's cloacal gland, presumably receiving some chemical stimulation from the male. More derived salamanders in the families Salamandridae and Plethodontidae have courtship glands on the chin or head. In the North American eastern newt (*Notophthalmus viridescens*), the male clasps the female around her neck with his hind legs and rubs the side of his head against her snout, transferring pheromones from glands on his cheeks. In large species of the genus *Plethodon*, such as the red-legged salamander (*P. shermani*), the male leads the female in a tail-straddling walk, with the female walking over the male's tail and resting her chin at the base of the tail. Periodically, the male turns and slaps the female's snout with a large padlike gland on his chin. This gland produces a protein-based pheromone that has been shown experimentally to increase the sexual receptivity of the female.

In the dusky salamanders of the genus *Desmognathus*, males often have enlarged front teeth that are used to transfer

Amphibian morphological defense mechanisms; a. Darwin's frog (*Rhinoderma darwinii*) uses camouflage and cryptic structure; b. *Pseudotriton ruber* and *Notophthalmus viridescens* display mimicry; c. *Bufo americanus* has poison parotid glands; d. Strawberry poison frog (*Dendrobates pumilio*) has warning coloration; e. *Physalaemus nattereri* has "eyespots" on its hindquarters. (Illustration by Jacqueline Mahannah)

pheromones from small glands at the tip of the chin. In most species, the male rakes his teeth across the skin of the female, dragging the chin gland across the wound to introduce the chemical secretions directly into the bloodstream. In two very small species, *Desmognathus wrighti* and *D. aeneus*, this somewhat violent form of courtship is carried a step further, and the male actually bites the female to deliver the pheromone into the bloodstream. Remarkably, this unusual form of courtship appears to have evolved independently in these two species (a phenomenon known as convergence), which are not closely related.

Visual communication in salamanders

Some salamanders also make use of visual displays during courtship, often in conjunction with pheromone delivery. This form of communication is best developed in the aquatic Old World newts (*Triturus*). Males do not clasp females but display near them. *Triturus* males have wide tail fins and crests extending over most of the back, and the fins, crests, and sides of the body are marked with bright colors and dark spots and blotches. A male uses his tail fin to waft pheromones produced in cloacal glands toward the female, but this display probably provides visual stimulation as well. In the largest species of newts, including the great crested newt (*Triturus cristatus*) and the marbled newt (*T. marmoratus*), components of courtship involved in chemical signaling are reduced, while visual displays have become elaborated, with the male exhibiting his bright coloration in broadside displays to the female.

Chemical communication in frogs and toads

Chemical communication is poorly developed in most frogs and toads, although there is evidence that males of some species emit chemical signals that are attractive to females. In dwarf African clawed frogs (*Hymenochirus*), males have glands behind their front legs that become greatly enlarged during the breeding season. Experimental studies have shown that females are attracted to water containing breeding males or

Amphibian behavioral and physiological defense mechanisms; a. Marine toad (*Bufo marinus*) inflates its lungs and enlarges; b. Two-lined sala-mander (*Eurycea bislineata*) displays tail autotomy (tail is able to detach); c. *Eleutherodactylus curtipes* feigns death; d. *Echinotriton andersoni* protrudes its ribs; e. *Bombina* frog displays unken reflex. (Illustration by Jacqueline Mahannah)

to extracts from the glands but not to males from which the glands have been removed surgically. The use of chemical signals by an aquatic frog such as *Hymenochirus* is not surprising, because pheromones are dispersed readily through water. More surprising is the finding that males of a terrestrial frog, the magnificent treefrog (*Litoria splendida*) from Australia, also produce a courtship pheromone, called splendipherin, that is attractive to females.

Visual communication in frogs and toads

Males of some species of frogs and toads have bright coloration that develops during the breeding season and probably serves as a visual signal to other males. In the North American green frog (*Rana clamitans*), breeding males have bright yellow throats that probably advertise ownership of territories to other males. Males of several species of frogs that breed in fast-running streams near noisy waterfalls have independently evolved foot-flagging displays, in which a hind foot is raised above the head or extended sideways, often dis-

playing bright white or blue webbing between the toes. These displays are used both for territorial display to other males and to attract females. The displays provide a conspicuous visual signal in a noisy environment, where calls are difficult to hear. Very similar foot-flagging displays have evolved in frogs from Malaysia (*Staurois latopalmatus*, Ranidae), Brazil (*Hylodes asper*, Leptodactylidae), Venezuela (*Hyla parviceps*, Hylidae), and Australia (*Taudactylus eungellensis*, Myobatrachidae, and *Litoria genimaculata*, Hylidae). Some frogs also use postural displays to appear larger, often elevating the body during aggressive encounters with other males.

Acoustic communication in frogs and toads

Frogs and toads are unique among amphibians in having evolved elaborate acoustic signals that are used both in aggressive interactions with other males and to attract females. Indeed, frogs probably were the first vocal vertebrates, and their calls are a familiar sound to anyone who lives near a swamp or pond. Frog calls are produced by contractions of muscles in the trunk region that force air out of the lungs, through the vocal cords, and, in most species, into a thin vocal sac that expands to radiate sound to the surrounding air. Vocal sacs of some species are balloon-like structures in the throat region, whereas in other species they expand from slits in the sides of the head.

The muscles involved in call production differ from other muscles in the body, having specialized anatomical and physiological features that allow them to contract hundreds of times per hour for hours at a time without becoming exhausted. This type of sound production is energetically expensive, and some tree frogs' metabolic rates while calling are more than 25 times their resting metabolism. In such species as the North American spring peeper (*Pseudacris crucifer*), which call at high rates in cold weather, calling is supported by huge stores of fat that accumulate in the trunk muscles in the fall, before the frogs go into hibernation. Consequently, the length of time a male can remain in a chorus may depend on energy reserves that were accumulated months earlier. This, in turn, can affect a male's ability to acquire mates.

Most frogs have a repertoire of several kinds of calls. The most commonly heard are advertisement calls, which serve not only to attract females but also to communicate a male's ownership of a territory to other males. Experiments with many different species have shown that females are attracted only to the calls of their own species, and this ensures that females do not waste their reproductive effort on matings that cannot produce viable offspring. Often, a relatively simple feature of the call is sufficient for females to discriminate between members of their own species and those of closely related species.

For example, two species of North American gray treefrogs are closely related and sometimes breed in the same ponds. One species, *Hyla chrysoscelis*, has a normal diploid complement of chromosomes, whereas the other, *Hyla versicolor*, has a double set of chromosomes (that is, it is a tetraploid animal) and evolved from *Hyla chrysoscelis*. These species look almost identical, and their calls have the same frequency structure (pitch). The calls consist of a series of repeated pulses of sound, but they differ in the rate at which pulses are produced. The pulse rate of *Hyla chrysoscelis* is about twice that of *Hyla versicolor*, and females readily approach males of their own species and reject males of the other species, even when the calls of the wrong species are much louder. The ability of females to find males of their own species in a noisy chorus of several kinds of frogs prevents wasted matings that would result in inviable hybrid offspring.

Many frogs also have courtship calls that are used in close-range interactions with females. In some species, the courtship call is simply a more rapidly repeated version of the advertisement call, which provides a better directional signal to females trying to locate males. In other species, a male gives a distinctly different call that is softer than the advertisement call, probably to avoid attracting nearby males that might attempt to intercept the approaching female. In some species, females even answer males with calls of their own. These calls invariably are very soft, because female frogs lack vocal sacs

Two North American green frog (*Rana clamitans*) males wrestling for possession of a territory in Ithaca, New York. (Photo by Kentwood D. Wells. Reproduced by permission.)

to amplify their calls, but they probably allow the male and female to approach each other more efficiently. In some frogs, such as European midwife toads (*Alytes*), males and females call on the ground away from water and engage in duets as they approach one another. Similar duets have been recorded in African clawed frogs (*Xenopus*), which call entirely underwater and have a completely different mechanism of call production from other frogs. Their calls consist of a series of simple clicks, and females respond to males with clicks of their own. This calling probably enables males and females to find one another in the muddy pools where these frogs normally breed.

Most frogs also have aggressive calls that are used to challenge intruders into male territories and in actual fights with

Calling male of *Physalaemus pustulosus* (Leptodactylidae) from Gamboa, Panama. This species has an unusually large external vocal sac. (Photo by Kentwood D. Wells. Reproduced by permission.)

Mixed breeding aggregation of common European toads (*Bufo bufo*) and brown frogs (*Rana temporaria*). Both species have explosive mating aggregations with scramble competition among males for the females, and males may clasp females of the wrong species in their attempts to find mates. (Photo by Walter Hödl. Reproduced by permission.)

other males. Usually, these calls are quite distinct from the advertisement call, but in some species, such as North American cricket frogs (*Acris*), aggressive calls grade into advertisement calls and differ mainly in the number and timing of repeated pulses. Some frogs have graded aggressive calls that vary in structure as a function of the intensity of the aggressive interaction. For example, in a tiny treefrog from Panama, *Hyla ebraccata*, males produce aggressive calls that are similar to advertisement calls, but call notes are much longer and have higher pulse rates. As males approach each other in fights, these calls become progressively longer, signaling an increase in aggressiveness.

Mating systems and sexual selection

Much of the exchange of communication signals in amphibians occurs during mate attraction and competition. As

is the case for most animals, males tend to compete for access to females rather than the other way around. This is because males can fertilize the eggs of many females, so the availability of females limits male reproductive success. This situation results in intense competition among males for the available females. The exact nature of this competition depends on the length of time females are available and the degree to which they are aggregated in a limited area.

Scramble competition

Many amphibians have explosive breeding periods that last only a few days. This is characteristic of many desert-dwelling amphibians, which rely on temporary rain pools for reproduction, and of many species that breed in temporary ponds in early spring. In both cases the breeding season is short, because it is critical for eggs to be laid quickly and larvae to develop and get out of the ponds before they dry up. These

Foot-flagging display of a male *Hylodes asper* (Leptodactylidae) from the Atlantic forest of southeastern Brazil. The male is calling with paired lateral vocal sacs inflated while giving the visual foot-flagging display. These frogs call on rocks in streams or near noisy waterfalls. (Photo by Walter Hödl. Reproduced by permission.)

conditions generally result in dense aggregations of males and females and lead to a mating system known as scramble competition. In North American spotted salamanders (*Ambystoma maculatum*), males gather in large numbers in early spring and engage in group courtship of females. Fertilization is internal and is accomplished by means of spermatophores, or sperm packets, deposited by males on the bottom of a pond. Males often interfere with the mating of other males by placing their spermatophores on top of those already deposited by other males. When a female picks up spermatophores with the lips of her cloaca, she is likely to get only that which is placed on top of the pile.

The European brown frog (*Rana temporaria*) and the similar North American wood frog (*Rana sylvatica*) both form "explosive" mating aggregations. Males search the pond for mates, grabbing anything that resembles a female frog. Often, several males pile onto a single female and struggle to be the one to fertilize her eggs. These mating balls can be dangerous to females, and many are crushed or drowned by the competing males. Similar scramble competition occurs in some African treefrogs (genus *Chiromantis*) that lay eggs in foamy masses on tree branches over temporary ponds. It also is characteristic of some Central and South American treefrogs (*Agalychnis* and *Phyllomedusa*) that lay eggs in jelly masses over water. In these species more than one male sometimes remains on the back of the female when she lays her eggs, so more than one male can fertilize her eggs.

Mate searching and mate guarding

When breeding seasons are relatively long, the arrival of females is less predictable. In the case of many species of salamanders, males search for mates and court females individually. This is the mating system of many newts, including *Triturus* in Europe and *Notophthalmus* in North America. Males do not produce chemical signals that attract females from long distances but instead move about the pond bottom

in search of suitable mates. Males of the genus *Triturus* court females but do not physically restrain them. In *Notophthalmus* and many other genera in the family Salamandridae, males clasp females during courtship. This is a form of mate guarding behavior that prevents other males from courting the same female. Because most frogs and toads use vocalizations to attract females, males usually do not search for mates, but some species engage in prolonged mate guarding. South American toads of the genus *Atelopus* sometimes remain in amplexus for weeks or months, presumably because females are encountered infrequently.

Leks and choruses

Among some of the larger European newts, such as the great crested newt (*Triturus cristatus*), males gather in groups and defend small territories, where they display to females. This mating system resembles the leks of many birds and mammals. A lek is a traditional display ground on which males gather to attract females. They defend territories used as display sites, but these territories do not contain resources that are attractive to females. Female choice in this type of mating system is based on behavioral or morphological characteristics of the males, and for this reason sexual dimorphism in size and coloration often is pronounced. Newts with lek mating systems are among the most sexually dimorphic of all salamanders.

Many frogs that gather in large choruses also have lek-like mating systems, with males defending a small space around a calling site that is used to attract mates; once a female arrives, however, she carries the male in amplexus to another site to lay eggs. Females use the rate at which males call or other aspects of their vocal displays to assess the quality of potential mates, usually choosing the ones with the most vigorous displays. In many species, however, it is simply persistence that pays off; males that spend the most time in a chorus tend to be the ones that mate most frequently. For many species time in the chorus probably is limited by energy reserves to support their vigorous calling.

Resource defense

Some amphibians attract females by defending resources, such as egg-laying sites, as territories; males with the most attractive territories obtain the most mates. This type of mating system is rare among salamanders, but does occur in North American hellbenders (*Cryptobranchus allegeniensis*) and closely related members of the same family, the Japanese and Chinese giant salamanders (*Andrias*). In these species, males defend cavities under rocks on the bottom of rivers as territories. Other males are excluded with biting and other aggressive behavior, but females are allowed to enter the territory to mate. Males with large cavities often mate with several females, which place their eggs in large groups under rocks. Some male frogs also defend egg-laying sites. This type of mating system is characteristic of North American bullfrogs (*Rana catesbeiana*) and green frogs (*Rana clamitans*), and males often fight violently for possession of choice territories. Males with the best territories may mate five or six times in a single breeding season and fertilize as many as 100,000 eggs, whereas males with poor-quality territories often do not mate at all.

In South America, males of the large treefrogs known as gladiator frogs, such as *Hyla boans* and *Hyla faber*, build mud nests at the edges of streams and defend them against other males. These frogs are equipped with sharp spines in the thumb region that are used to slash and stab other males in fights. Some males are seriously injured, but those with especially good nests are most likely to mate and produce offspring. Territorial males sometimes continue to guard their nests after eggs are laid, to prevent other males from destroying the egg masses.

Resources

Books

Dawley, Ellen M. "Olfaction." In *Amphibian Biology*. Vol. 3, *Sensory Biology*, edited by Harold Heatwole and E. M. Dawley. Chipping Norton, Australia: Surrey Beatty and Sons, 1998.

Griffiths, Richard A. *Newts and Salamanders of Europe*. San Diego: Academic Press, 1996.

Halliday, Tim R. "Sperm Competition in Amphibians." In *Sperm Competition and Sexual Selection*, edited by T. R. Birkhead and A. P. Møller. San Diego: Academic Press, 1998.

Halliday, Tim R., and Miguel Tejedo. "Intrasexual Selection and Alternative Mating Behaviour." In *Amphibian Biology*. Vol. 2, *Social Behaviour*, edited by Harold Heatwole and B. K. Sullivan. Chipping Norton, Australia: Surrey Beatty and Sons, 1995.

Hödl, W., and A. Amezquita. "Visual Signaling in Anuran Amphibians." In *Anuran Communication*, edited by M. J. Ryan. Washington, DC: Smithsonian Institution Press, 2001.

Jaeger, R. G., M. E. Peterson, and J. R. Gillette. "A Model of Alternative Mating Strategies in the Redback Salamander, *Plethodon cinereus*." In *The Biology of Plethodontid Salamanders*, edited by Richard C. Bruce, Robert G. Jaeger, and Lynne D. Houck. New York: Kluwer Academic/Plenum Press, 2000.

Mathis, A., R. G. Jaeger, W. H. Keen, P. K. Ducey, S. C. Walls, and B. W. Buchanan. "Aggression and Territoriality by Salamanders and a Comparison with the Territorial Behaviour of Frogs." In *Amphibian Biology*. Vol. 2, *Social Behaviour*, edited by Harold Heatwole and B. K. Sullivan. Chipping Norton, Australia: Surrey Beatty and Sons, 1995.

Sullivan, B. K., M. J. Ryan, and P. A. Verrell. "Female Choice and Mating System Structure." In *Amphibian Biology*, Vol. 2, *Social Behaviour*, edited by Harold Heatwole and B. K. Sullivan. Chipping Norton, Australia: Surrey Beatty and Sons, 1995.

Verrell, P., and M. Mabry. "The Courtship of Plethodontid Salamanders: Form, Function, and Phylogeny." In *The Biology of Plethodontid Salamanders*, edited by Richard C. Bruce, Robert G. Jaeger, and Lynne D. Houck. New York: Kluwer Academic/Plenum Press, 2000.

Wells, K. D. "The Energetics of Calling in Frogs." In *Anuran Communication*, edited by M. J. Ryan. Washington, DC: Smithsonian Institution Press, 2001.

Periodicals

Gerhardt, H. C. "Acoustic Communication in Two Groups of Closely Related Treefrogs." *Advances in the Study of Behavior* 30 (2001): 99–167.

Martins, M., C. F. B. Haddad, and J. P. Pombal. "Escalated Aggressive Behaviour and Facultative Parental Care in the Nest Building Gladiator Frog, *Hyla faber*." *Amphibia-Reptila* 19, no. 1 (1998): 65–73.

Pearl, C. A., M. Cervantes, M. Chan, U. Ho, R. Shoji, and E. O. Thomas. "Evidence for a Male-Attracting Chemosignal in the Dwarf African Clawed Frog *Hymenochirus*." *Hormones and Behavior* 38, no. 1 (2000): 67–74.

Rollmann, S. M., L. D. Houck, and R. C. Feldhoff. "Proteinaceous Pheromone Affecting Female Receptivity in a Terrestrial Salamander." *Science* 285, no. 5435 (1999): 1907–1909.

Wabnitz, P. A., J. H. Bowie, M. J. Tyler, J. C. Wallace, and B. P. Smith. "Differences in the Skin Peptides of the Male and Female Australian Tree Frog *Litoria splendida*. The Discovery of the Aquatic Male Sex Pheromone Splendipherin, Together with Phe8 Caerulein and a New Antibiotic Peptide Caerin 1.10." *European Journal of Biochemistry* 267, no. 1 (2000): 269–275.

Kentwood D. Wells, PhD

Amphibians and humans

Amphibians have figured in the lives of humans since antiquity. Frogs and salamanders are richly represented in mythology, culture, art, and literature, and even today they are seen as attractive characters in commercial advertising and as whimsical stars on television. In contrast to these roles, which in some cases are rather superficial, amphibians are of real importance as food and the source of compounds of medicinal value. They are key organisms in research and teaching and for purposes of natural control of insects. Despite the importance of amphibians, humans' actions have negative effects on them in numerous direct and indirect ways—through introductions of exotic species, in the loss or alteration of habitats, and even by overcollecting. Several species are believed to have become extinct within the past two decades, probably owing to human activities.

Mythology and culture

Frogs and salamanders have appeared in the legends and folklore of many cultures throughout history. Certain beliefs, such as the connections of amphibians to water, rainfall, earth, and the underground, recur in diverse cultures, but other beliefs are more localized. Such linkages often are depicted in indigenous art. For example, the Zuñis of New Mexico even today decorate their water-holding pots with frog tadpoles. In ancient Egypt frogs were associated with water and mud because of their sudden reappearance and reproductive activity following the annual flooding of the Nile River Valley. Thus, frogs came to symbolize birth and resurrection. Among the hieroglyphics found on the walls of the Egyptian funerary temple of Hatshepsut (queen of Egypt during the fifteenth century B.C.) are images of the god of creation, Khnum, and his wife, the frog-headed Heqet, forming children on a potter's wheel. Indeed, several Egyptian gods were depicted with the heads of frogs.

Half a world away, in the Mayan culture of the Yucatán Peninsula of Central America, frogs and toads were believed to announce the rains with their choruses. Today's Maya still perform rain dances, rituals that are thought to be of great antiquity. At one point in the ceremony, four boys are tied to the altar and mimic the calls of two different species (*Bufo marinus* and *Rhinophrynus dorsalis*). The Maya also associated frogs with agriculture. The Madrid Codex, a fifteenth-century Mayan almanac painted on plaster-coated bark paper, shows frogs making furrows with sticks and sowing seeds. One frog, which the Maya called the *uo*, was thought to come from the sky with green corn grains in its intestines. The *uo* was probably *Rhinophrynus*, which breeds only during heavy rains. The name *uo* is onomatopoetic—the name represents the sound of the frog's call. *Uo* is also the name of the Mayan month of greatest rainfall.

In other cultures amphibians were believed to have mystical powers, and shamans used their images in various rituals. Several ancient cultures in Egypt, Greece, Turkey, and Italy had images of frogs as amulets for good luck or to ward off evil; this is true even in present-day Myanmar (Burma). The Itelmens, aboriginal people of the Kamchatka Peninsula of eastern Siberia, considered hynobiid salamanders (*Salamandrella keyserlingii*) to be spies sent by Gaech, lord of the underground, to find and capture them for their master. Another Siberian nation, the Selkups, thought that frogs protected them from evil spirits, and frog images thus were used widely by shamans in ceremonies. In medieval Europe, much interest was attached to the toad because of its poison glands, and extracts from the glands were employed in witchcraft. It was believed that the toad could withstand its own poison by carrying around an antidote in the form of a stone located in its head. Shamans used so-called toad stones—in practice, any stone the size and shape of a toad—to neutralize poisons from snakebites or bee stings.

Another common belief in European culture, since at least the time of Pliny the Elder in the first century A.D., is the myth of the invulnerability of salamanders to fire. The common name of the European *Salamandra salamandra*—the fire salamander—is directly traceable to this legend. Images of salamanders emerging from fire have led to an otherwise inexplicable association among the common dictionary definitions of the term *salamander*. Among these definitions are "a mythical being thought to live in fire," "a portable stove or burner," and "the mass of iron that accumulates at the bottom of a blast furnace." When asbestos, an incombustible mineral, was discovered, it was believed to be the hair of the salamander and sometimes was referred to as "salamander's wool." The basis for the association between salamanders and fire is thought to be that salamanders seemed to emerge from the flames when the logs in which they were hiding were

Humans have long had an affinity for frogs, as represented by Kermit the Frog, a favorite character. (Photo by Reuters/Fred Prouser. Reproduced by permission.)

Frog carving on Chief Kadashan totem pole on Chief Shakes Island, Wrangell, Alaska, USA. (Photo by Pat O'Hara/Corbis. Reproduced by permission.)

thrown onto a fire. Recent observations of California newts (*Taricha torosa*), which cover their bodies with slime secreted by their glands and then walk unaffected through the flame fronts of brush fires, demonstrate that these amphibians have a greater tolerance for fire than was previously understood.

Art and literature

Traditional art often has incorporated the likeness of frogs. Among the weaving patterns of the Native South Americans of northeastern South America, frogs appear regularly, often in a highly stylized form that resembles a dumbbell. The pre-Columbian Hopewell Native Americans of the Upper Ohio River Valley smoked ceremonial pipes in the shape of frog effigies. The fleur-de-lis, the traditional iris symbol of French kings and of France, originally was depicted as a group of three frogs.

Amphibians also have figured regularly in literature. *The Frogs*, a Greek satirical play, was first performed in Athens in 405 B.C. In this play Aristophanes used frogs to make fun of humans when the chorus repeatedly sings out to the god Dionysus, the patron of drama, as he crosses the River Styx to enter Hades and bring back the playwright Euripedes. The call, "Brekekekex, co-äx, co-äx," is thought to be the first use of phonetic imitations of animal sounds in literature. Many of Aesop's famous animal fables dealt with frogs, and the traditional fairy tale of the prince turned into a frog by a wicked witch, only to be restored by the kiss of a beautiful princess, is widely known.

Shakespeare regularly used frog and salamander references in his plays. In *Richard III*, he derisively referred to the king as "that bottled spider, that foul bunch-back'd toad." The three witches in *Macbeth* chant, "Eye of newt, and toe of frog," as they stir those ingredients into their evil brew. In *As You Like It*, Shakespeare made yet another of his many toad metaphors: "Sweet are the uses of adversity, / Which, like the toad, ugly and venomous, / Wears yet a precious jewel in his head." The jewel, often thought to signify the toad's beautiful eye, may well refer to a toad stone.

Later literary references to amphibians include those in Mark Twain's first story, "The Celebrated Jumping Frog of Calaveras County," which featured a frog by the name of Dan'l Webster, and Karel Čapek's science fiction thriller *War with the Newts*. Through the ages amphibians have held a secure place in mythology, art, and literature, as they still do in today's culture. Frogs are used regularly in commercial advertising—whether for beer or shoe polish—and arguably the most famous anuran of them all, Kermit the Frog, the muppet star of *Sesame Street* on public television, is loved by children around the world.

A contestant (bullfrog) in the Jumping Frog Jubilee in Calavaras County, California, USA. (Photo by Tim Davis/Photo Researchers, Inc. Reproduced by permission.)

Frogs have appeared in literature, sometimes acting somewhat like humans and dressed in clothing. This is a scene from *A Frog He Would a Wooing Go* by Randolph Caldecott from the early twentieth century. (Photo by Stapleton Collection/Corbis. Reproduced by permission.)

Medical and research uses

Amphibians have been employed for medicinal purposes for millennia. The Chinese brown frog (*Rana chensinensis*) has long been used in traditional medicine in the three northeastern provinces of the country. An oil called "Ha Shi Ma Yu," derived from the dried oviducts, is believed to cure nervous exhaustion. Until the 1970s, as many as 72 million frogs were collected annually for the purpose of obtaining this oil, but the yields have now dropped below five million as the result of habitat loss and overcollecting. Until a century ago frog egg clutches were used as plaster in Russia, frog meat was put on snakebite wounds in western Siberia, and teas made from dried and powdered hynobiid salamanders (*Ranodon sibiricus*) were used to treat bone fractures and malaria in northwestern China. Amphibians continue to be an important part of traditional medicine in many parts of the world.

More than 200 psychoactive alkaloids have been extracted from the skin of frogs and toads. For these amphibians, they act as natural chemical defenses by affecting the muscles and nerves of would-be predators. Scientists have been able to synthesize many of these alkaloids for research. One of them, batrachotoxin (found only in the skin of the dendrobatid frog *Phyllobates*), causes ion channels in nerve and muscle cells to fail, resulting in heart failure; when it is labeled radioactively the toxin becomes a very useful tool for medical research. Another alkaloid, epibatidine (from the skin of another dendrobatid, *Epipedobates tricolor*), is a highly effective painkiller; it is 200 times stronger than morphine, but it is not addictive and has no sedating effects. Epibatidine is produced synthetically and is being tested as a drug for humans. Skin secretions from the green treefrog (*Litoria caerulea*), called caeruletide, stimulate activity in the pancreas and intestine, and synthetic versions of it are commercially available for human use for these purposes.

The large parotid glands of toads of the genus *Bufo*, located just behind the eyes, produce two substances—bufogenin and bufotoxin—that affect the adrenal and cardiovascular systems in humans. A third parotoid secretion—bufotenin, an alkaloid—is a powerful hallucinogen. The Colorado River toad (*Bufo alvarius*) possesses the specific enzyme for production of this substance, and the parotoids, which can contain large amounts of the hallucinogen, can produce hallucinations when the skin is dried and smoked. The hallucinogenic properties of toad parotoid glands were well known to the native peoples of Central America, and images of toads with prominent parotoid glands are a common feature on bowls and other objects found at archeological sites.

Frog skin secretions also can have powerful antimicrobial properties. The skin of African clawed frogs (*Xenopus laevis*) produces peptides called magainins that assist in the natural healing of cuts and bruises. These peptides have potential as a new class of antibiotics. Glues extracted from frog skin can be used to fix crockery, and research suggests that skin secretions may help repair human internal organs.

Among the many medical and research applications of amphibians, frogs and salamanders have been standard laboratory preparations for studies in embryology and physiology. Amphibians are also highly useful model organisms for many field studies of behavior and ecology. *Xenopus* frogs were the first test organisms to be used for determining pregnancy in humans. Frogs and salamanders are commonly found in biology teaching laboratories throughout the world.

Rattle in the form of a raven with the head of a hawk used in shamanistic dances and rituals. On the back is a human face and human figure with a tongue protruding into the mouth of a frog. Shamans may have used frog poison in their rituals. (Photo by Werner Forman/Corbis. Reproduced by permission.)

Other human uses of amphibians

Frogs—mainly their large, muscular legs—are eaten regularly by many indigenous peoples, especially in impoverished societies, where they constitute an important source of protein. In the Rift Valley of eastern Africa, clawed frogs (*Xenopus*) are netted in huge numbers as seasonal supplements to human diets. In affluent countries, frogs' legs are consumed as a delicacy but also as a meat alternative during Lent. The international trade in frogs' legs is enormous and mostly originates in southern Asia and the East Indies. The wild capture of so many frogs, which are insectivores, has resulted in growing populations of mosquitoes and other insects in these countries. Salamanders are eaten rarely, but a major exception is the Chinese giant salamander (*Andrias davidianus*). This animal, which reaches a length of 5.25 ft (1.6 m) and a weight of 143 lb (65 kg), is raised in farm ponds in China for food. Because of its importance as a food item, an institute devoted to the biology of this species has been established in Hunan Province.

Amphibians are key elements in many ecosystems. They feed primarily on invertebrate prey, especially insects, and thus represent an all-important trophic link between their prey and the larger animals that, in turn, feed on them. Because of their insectivorous nature, frogs and toads, among them, North American bullfrogs (*Rana catesbeiana*), have been introduced to many parts of the world to control insect populations. A poison frog species, *Dendrobates auratus*, was introduced to Hawaii from Panama in 1932 to help control mosquitoes. The most famous, and ill-advised, introduction was that of the marine or cane toad (*Bufo marinus*) to Australia, ostensibly to get rid of a beetle that infested sugarcane. In 1935, 102 toads were released in Queensland. The experiment failed to control the beetle population. Moreover, the toads ate a wide variety of prey, including native frogs; reproduced rapidly in the absence of natural predators; and expanded their range enormously. Today the toads represent a major challenge in themselves and have created a new and destabilizing relationship between amphibians and humans.

Over the years, beginning in Europe in the late eighteenth century, amphibians have become popular as terrarium pets. Many species are regularly kept, including newts, colorful hylids, neotropical poison frogs, Madagascan mantellas and tomato frogs, and aquatic caecilians. Terrarium keepers often have made observations that are of importance to science. Some species have been bred successfully, and an entire industry has developed around amphibians as pets, including public expositions, wholesale and retail dealers in live specimens, veterinarians that specialize in their care, amphibian keepers' magazines, and texts on medicine and husbandry. It seems that we have reached the ultimate in relationships: amphibians as human companions.

Resources

Books

Hofrichter, Robert, ed. "Amphibians in the Cultural Heritage of Peoples Around the World." In *Amphibians: The World of Frogs, Toads, Salamanders and Newts*. Buffalo, NY: Firefly Books, 2000.

Kuzmin, S. "History." In *Amphibians of the Former Soviet Union*, by S. Kuzmin. Sofia, Bulgaria, and Moscow, Russia: Pensoft, 1999.

Lee, J. C. "Ethnoherpetology in the Yucatán Peninsula." In *Amphibians and Reptiles of the Yucatán Peninsula*, by J. C. Lee. Ithaca, NY: Cornell University Press, 1996.

Stebbins, R. C., and N. W. Cohen. "Contributions of Amphibians to Human Welfare." In *Natural History of Amphibians*, by R. C. Stebbins and N. W. Cohen. Princeton: Princeton University Press, 1995.

Kraig Adler, PhD

Conservation

The world's amphibians face a variety of threats to their continued existence. Since the late 1980s herpetologists have become increasingly concerned about dramatic population declines among amphibians throughout the world. In many places, these declines reflect the global deterioration of the environment and have led to the extinction of species. Amphibians are by no means unique; there is just as much concern about birds, reptiles, and all other forms of life. In relation to amphibians, it has been of particular concern that declines and extinctions have occurred in nature reserves, national parks, and other supposedly protected areas set aside to preserve biodiversity. A notable example is the loss of several amphibian species endemic to the Monteverde Cloud Forest Reserve in Costa Rica, including the golden toad (*Bufo periglenes*), which has become the symbol of amphibian decline.

Reasons for amphibian decline and extinction

The decline and loss of amphibians in protected areas rule out habitat loss as the immediate cause, but there is no doubt that this is the reason for such declines over much of the world, where amphibians are threatened by such consequences of human population growth and development as deforestation, industrialized agriculture, and pollution. Common features of amphibian population declines in protected areas in such widely separated parts of the world as eastern Australia, the Pacific Northwest of the United States, and Central and South America are twofold. First, they have been very sudden, with species vanishing over two or three years, and, second, they have affected some amphibian species but not others. This has stimulated research to find one or more environmental factors that affect amphibians on a global scale but to which some species are more susceptible than others.

One such factor is the increase in the amount of ultraviolet B (UV-B) radiation that now reaches the earth's surface as a result of the thinning of the ozone layer by atmospheric pollutants. Research carried out both in the field and in the laboratory has shown that the eggs, embryos, and larvae of most amphibians are generally highly sensitive to elevated UV-B, which breaks up their DNA and thus causes them to develop abnormally and die. Nonetheless, some species were found to be unaffected by increased UV-B, raising hopes that the global factor that affects only some amphibians had been isolated. This optimism was short-lived.

Many amphibians have declined, especially in the tropics, in localities where levels of UV-B radiation have not increased and in species whose eggs and embryos are not exposed to sunlight. While this rules out UV-B as the cause of all amphibian declines, it is a significant threat to some species, particularly those that breed at high altitude and in shallow water, where levels of UV light are higher. Recent research also indicates that, whereas elevated UV-B does not always cause death, it does have a harmful effect on developing amphibians, reducing their growth and causing physical deformities, thus limiting the reproductive output of populations.

In many parts of the world, amphibians are threatened by one or more human-made chemical compounds, released into the environment as herbicides, pesticides, and fertilizers or as the by-products of industrial processes. The list of compounds known to be harmful to amphibians is very long. Of particular concern are nitrates, which are used as agricultural fertilizers and accumulate in ponds and streams, and a variety of compounds called endocrine disrupters, which interfere with amphibians' natural hormones. These have two major harmful effects. First, they can cause amphibians to develop abnormally, with deformed mouthparts or, in extreme cases, missing or additional limbs. Second, they can have a feminizing effect on males, reducing their reproductive success. The herbicide atrazine, widely used throughout the world in agricultural areas, has been shown to have a feminizing effect on male frogs, even at very low concentrations.

Deformities among amphibians have excited a great deal of public and media interest in the United States, but their relevance to amphibian populations is unclear. They tend to be concentrated in particular areas; Minnesota, in particular, is a deformed frog hot spot. Deformities are caused by several factors, some of which are entirely natural. They can be the result of predatory attacks, and there are parasites that burrow into the limb buds of frog tadpoles, causing two or more legs to develop where there should be only one. Nonnatural causes of deformities, usually missing limbs or parts of limbs, are several human-made chemicals, increased UV radiation, and inbreeding in very small, isolated populations.

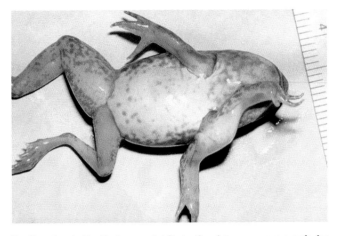

Fertilizer levels the Environmental Protection Agency says are safe for human drinking water can kill some species of frogs and toads, according to a study by Oregon State University researchers. They found some tadpoles and young frogs reduced their feeding activity, swam less vigorously, experienced disequilibrium, developed physical abnormalities (shown in photo), suffered paralysis, and eventually died. In control tanks with normal water, none died. (Photograph. AP/Wide World Photos. Used by permission.)

Male glass frog hydrating its eggs in Monteverde Cloud Forest Preserve, Costa Rica. The dramatic loss of several frog species in this preserve has been linked to a succession of El Niño events that have resulted in a marked reduction in the amount of land that becomes enveloped by low clouds each year. (Photo by Rita Nannin/Photo Researchers, Inc. Reproduced by permission.)

Deformities are sometimes common in individual populations and so may have a negative impact on amphibian numbers at a local level. They may represent a response to sublethal levels of environmental factors than can kill amphibians. One study found that exposure to low levels of a pesticide increases the susceptibility of frog tadpoles to a deformity-causing parasite.

In many parts of the world, industrial activity creates acid rain, which can fall hundreds of miles from the immediate source of the pollution. For example, burning of fossil fuels in the United Kingdom (UK) is a major cause of acid rain in Scandinavia. Acidification of water has a negative effect on the egg and embryo stages of amphibians and can cause amphibian population declines over wide areas. Many amphibians are highly dependent on ephemeral ponds or streams for breeding, and their mating activity is linked closely to climatic changes that herald the advent of suitable conditions. Amphibians in Britain are now breeding several weeks earlier in the year than they were 20 years ago, a trend commonly seen as a symptom of "global warming." Climate change can affect amphibians in many different ways and has been implicated in several instances of population decline. Notably, the dramatic loss of several frog species at Monteverde, Costa Rica, has been linked to a succession of El Niño events that have resulted in a marked reduction in the amount of land

that becomes enveloped by low cloud cover each year. It has been suggested that the drier conditions that have resulted from the limited cloud cover have forced amphibians to concentrate in fewer underground hiding places, increasing the spread of parasites and diseases.

Disease had the most dramatic impact on amphibians in the last 10 years of the twentieth century. In the 1990s, there were mass deaths among brown frogs (*Rana temporaria*) over a wide area of the southern UK caused by viral infections. Of much greater concern has been an apparently global outbreak of the disease chytridiomycosis, caused by a single-cell fungus called a chytrid. The fungus invades the skin of amphibians and appears to have been responsible for the dramatic collapse of amphibian faunas in Central America, eastern Australia, and parts of the western United States. First described among captive animals, chytridiomycosis has been found on nearly every continent of the world. It is not yet clear whether a new strain of what is presumably a well-established disease has appeared or whether, for a variety of reasons, amphibians have become susceptible to a disease with which they were previously able to coexist.

Much of the research carried out to investigate possible causes of amphibian declines inevitably involves considering one factor in isolation, although, in reality, amphibians are threatened by many different factors. Some research has looked at interactions between two or more factors and has shown that there can be significant synergistic effects between them. For example, in the western United States, climate change, increased UV-B radiation, and disease have acted together to cause amphibian declines. Climate change has reduced water levels in breeding ponds, with the result that amphibian eggs are less protected by deep water from UV light. This, in turn, makes the eggs more susceptible to the pathogenic fungus *Saprolegnia*, which invades and kills amphibian eggs.

A green and golden bell frog (*Litoria aurea*) wears a radio tracking device, consisting of a harness, batteries, and a 6-in (15-cm) aerial made of piano wire. The frog is Near Threatened (IUCN Red List) and was found in a brick pit at the Homebush Sydney 2000 Olympic site in Australia. (Photograph. AP/Wide World Photos. Used by permission.)

The eggs and larvae of most amphibians have poor defenses against such predators as fish, and many amphibian populations have been devastated by the artificial introduction of fishes to ponds, lakes, and streams. For example, mosquitofish (*Gambusia affinis*) have been released into many parts of the world in an attempt to control malaria-carrying mosquitoes, and trout are commonly introduced to provide sport. Both kinds of fish find amphibian larvae easy and attractive prey. The loss of several amphibian species from mountain lakes in California is largely due to predation by introduced trout. Fish are not the only introduced enemies of amphibians; even amphibians, when they are moved to places where they do not belong, can threaten native species. The North American bullfrog (*Rana catesbeiana*) has been introduced to many parts of the world to sustain a trade in frog legs. Its larvae grow to enormous size and often compete with and win over the larvae of native species. Most famously, the introduction of the marine toad (*Bufo marinus*) into Australia—where it is called the cane toad because it was hoped that it would control sugarcane pests—has devastated many local frog species, through larval competition and by predation on adult frogs.

The pressures of the ever expanding human population generate an insatiable demand for land that results in the destruction of the natural habitat of plants and animals. This process is offset, to a very small degree, by the creation of nature reserves, but these reserves can become prisons rather than havens for such animals as amphibians. Many amphibians live in small, local populations, the long-term survival of which depends on the occasional immigration of animals from other such populations. Increasingly, amphibians are being forced to live in fragmented landscapes in which roads, land development, and agriculture separate one population from another. There is growing evidence that this isolation leads to inbreeding and a consequent loss of genetic diversity, manifested by decreased survival and an increased incidence in anatomical deformities. As animals become rare, their value in the international pet trade grows, and collecting can become another serious threat to their survival. Collecting poses a risk to several of the world's most colorful frogs, such as the poison frogs and harlequin frogs of Central and South America and the mantellas of Madagascar.

Although amphibian population declines have attracted a great deal of scientific and media interest, there is no reason to think that they are unusual or unique. All the factors that adversely affect amphibians pose a threat to other forms of wildlife as well. In particular, the kinds of freshwater habitats upon which many amphibians depend—ponds, marshes, and wetlands—are under severe threat all over the world, with serious consequences for countless fish, insects, and other animals that frequent them. What may be special about amphibians is that they are providing an early warning of an ecological disaster that is just beginning. Amphibians possess a number of features that make them especially sensitive to a wide variety of environmental insults. As eggs, larvae, and adults, they lack any kind of protective body surface that could shield them from radiation or chemical pollution. In the early stages of growth, they often lack protection against predators and can develop safely only in ephemeral water bodies threatened by climate change and habitat destruction. Compared with many animals, amphibians have very poor powers of dispersal, with the result that habitat fragmentation prevents the exchange of genetic diversity on which the long-term survival of individual populations depends.

Efforts to protect amphibians

The geographic scale at which the many threats to amphibians are relevant ranges from global phenomena, such as climate change, to local factors, such as toads being killed by traffic as they cross a road on their way to a breeding pond. When it is asked what can be done to protect amphibians and by whom, the answers depend on the scale at which a conservation initiative is being directed. If it is the case that amphibians are declining because of climate change, elevated UV radiation, or acid rain, the solution lies in the hands of politicians and global organizations who must seek the appropriate remedies through international treaties and agreements. There is little that individuals or local conservation groups can do to counter such threats, other than adding their voices to the pressure on political leaders to move environmental issues closer to the top of the political agenda. At the local level there is a great deal that small groups of people can do to protect and encourage amphibians. In many parts of the UK, mainland Europe, and North America, groups go out at night in spring to protect migrating amphibians as they cross busy roads. In some places, such groups have succeeded in persuading local authorities to close stretches of road at the appropriate time. Another strategy that addresses the same threat is the construction of tunnels under roads, which, if they are appropriately designed and positioned, enable amphibians to reach their breeding sites in safety.

Habitat loss can be offset to a small extent by habitat creation or restoration. Research carried out in the UK and the United States has shown that new ponds created on agricultural land are quickly colonized by newts, frogs, and toads. Even tiny ponds in gardens will support good populations of amphibians, provided that they are not also stocked with fish, and it is estimated that a larger proportion of the UK's common frog population now lives in garden ponds than in natural habitats. Amphibians can be a bonus in gardens; the common toad has been called the gardener's friend because of its appetite for slugs and insect pests. Conservationists must remember, however, that most amphibians spend only a small proportion of their lives in water and that the creation of suitable terrestrial habitat is just as important as making new ponds. Unfortunately, the ecology of terrestrial amphibians is poorly known; thus, creating suitable habitat for amphibians is often a matter of guesswork.

Leopard frogs with missing, deformed, or extra legs started appearing near St. Albans Bay of Lake Champlain in St. Albans, Vermont. Biologists are not sure if pollution, a parasite, disease, or something else is causing the frogs to develop abnormally. (Photograph. AP/Wide World Photos. Reproduced by permission.)

In many developed countries, endangered amphibian species are afforded varying levels of legal protection. In the UK, for example, it is illegal to collect or kill a great crested newt (*Triturus cristatus*) or a natterjack toad (*Bufo calamita*). More important, their breeding sites often are protected, and developers who wish to destroy a pond have to pay for mitigation measures, such as the creation of a pond elsewhere, to which the threatened population can be moved. Some amphibians have been conserved successfully by programs involving captive breeding and the release of animals back into the wild. Because of their high fecundity, this has considerable potential for many amphibians, provided that it is combined with measures to protect their natural habitat. In captivity, it is possible to prevent the heavy mortality rates from predation that are typical in nature, with the result that very large numbers of captive-bred animals can be produced. The Majorca midwife toad (*Alytes muletensis*) has been conserved in this way, and, in Australia, a similar project seeks to protect the highly endangered Corroboree toadlet (*Pseudophryne corroborree*).

Scientists are studying such species as this endangered golden frog in an effort to prevent species extinction. (Photo by JLM Visuals. Reproduced by permission.)

Disease as a cause of amphibian declines requires its own set of conservation measures. For example, individual amphibians infected with the fungal disease chytridiomycosis can be cured with a preparation that is used to treat athlete's foot in humans. This is unlikely to be of any help, however, in protecting natural populations. There is a real possibility that herpetologists, the very people who seek to conserve amphibians, have helped to spread diseases by carrying spores on their rubber boots or collecting gear. Many organizations, including the Declining Amphibian Populations Task Force, have issued guidelines to try to prevent the local spread of amphibian diseases. At the international level, there are moves to control and limit the movement of amphibians around the world, in an effort to reduce the chance that diseases will be spread from one country or one continent to another.

Nature reserves are, of course, an obvious way of conserving amphibians, but this does not protect them from many of the threats that they face. An important issue here is how protected areas should be designed to provide optimal conditions for amphibians. It is clear that populations based on a single breeding site are likely to face eventual extinction despite protection, because they become inbred. Many amphibians seem to require a network of breeding sites, connected by habitat that they can cross reasonably easily, so that the population can continue to maintain a high level of genetic variation.

While much can be done and is being done to conserve amphibians on local, national, and international scales, much of it is carried out more in hope than in the expectation of success. Successful conservation requires a deep understanding of ecology, and, sadly, there are many aspects of the ecology of amphibians about which we remain profoundly ignorant. For most amphibians we do not know the answer to one simple question: Where do they go when they are not breeding?

Resources

Books

Crump, M. *Amphibians, Reptiles, and their Conservation.* North Haven, CT: Linnet Books, 2002.

Lannoo, M. J., ed. *Status and Conservation of U.S. Amphibians.* Berkeley: University of California Press, 2003.

Periodicals

Alford, R. A., and S. J. Richards. "Global Amphibian Declines: A Problem in Applied Ecology." *Annual Review of Ecology and Systematics* 30 (1999): 135–165.

Daszak, P., L. Berger, and A. A. Cunningham, et al. "Emerging Infectious Diseases and Amphibian Population Declines." *Emerging Infectious Diseases* 5 (1999): 735–748.

Houlahan, J. E., C. S. Findlay, and B. R. Schmidt, et al. "Quantitative Evidence for Global Amphibian Population Declines." *Nature* 404 (2000): 752–755.

Other

DAPTF (Declining Amphibian Populations Task Force). (September 30, 2002) <http://www.open.ac.uk/daptf/>

Tim Halliday, PhD

Anura

(Frogs and toads)

Class Amphibia

Order Anura

Number of families 28

Number of genera, species 352 genera; 4,837 species

Photo: The common European toad (*Bufo bufo*) secretes poison from its parotoid gland. (Photo by George McCarthy/Corbis. Reproduced by permission.)

Evolution and systematics

Anurans (frogs and toads) usually are divided into two informal groups. Of these groups, the archaeobatrachians include the basal living families Ascaphidae, Leiopelmatidae, Bombinatoridae, Discoglossidae, Pipidae, Rhinophrynidae, Megophryidae, Pelobatidae, and Pelodytidae as well as the fossil family Paleobatrachidae. Some authors consider Megophryidae, Pelobatidae, Pelodytidae, Pipidae, and Rhinophrynidae to constitute an "intermediate" group, the mesobatrachians; all other families are placed among the neobatrachians.

Even the most experienced herpetologists can have difficulty ascertaining the family to which a frog belongs by examining only the external features, because many species closely resemble other species in unrelated families. The most important morphological characters are features of the internal anatomy, especially the skeleton. In the nineteenth century, biologists discovered that some frogs lacked tongues and so divided the Anura into two suborders—tongued frogs (Phaneroglossa) and tongueless frogs (Aglossa). Subsequently, two basic types of pectoral girdles were recognized—two halves overlapped ventrally (arciferal condition) and two halves fused midventrally (firmisternal condition).

Early in the twentieth century many additional suites of characters were discovered, including different kinds of vertebral articulations, presence of free ribs, dentition, and thigh musculature. By the middle of the twentieth century, classification of anurans commonly consisted of five suborders (Amphicoela, Anomocoela, Diplasiocoela, Opisthocoela, and Procoela) based on the nature of the articulating surfaces of the vertebrae and the intervertebral elements, but contemporary herpetologists no longer accept this arrangement. Later in the century, larval characters, developmental patterns, nature of the mating embrace (amplexus), and pupil shape were added to the growing number of characters used in classifi-

cation. By the end of the twentieth century, molecular data sets provided support for some but not all arrangements based on morphological features; also, by this time, rigorous analyses were used to propose testable hypotheses of phylogenetic relationships of both living and extinct anurans.

The resulting phylogenies and classifications place the Triassic *Triadobatrachus* as the sister taxon to anurans and establishes the monophyly (descendents of a single ancestor) of Anura, within which the basal familes Ascaphidae, Leiopelmatidae, Bombinatoridae, and Discoglossidae form a grade. The assumed sister relationships of Pipoidea (Pipidae and Rhinophrynidae) and Pelobatoidea (Megophryidae, Pelobatidae, and Pelodytidae) are the subject of controversy. Many unresolved problems exist within the neobatrachians, but most evidence supports a clade (all descended from one ancestor) usually referred to as the ranoids (Arthroleptidae, Hemisotidae, Hyperoliidae, Microhylidae, Ranidae, Rhacophoridae, and Scaphiophrynidae). A group of Madagascar frogs recognized by some workers as Mantellidae has been placed in the Rhacophoridae or Ranidae by various researchers (covered here in Rhacophoridae).

The remaining neobatrachians, sometimes referred to as bufonoids, may be viewed as a grade between archaeobatrachians and ranoids. Among the bufonoids, no evidence supports the monophyly of Leptodactylidae; morphological and molecular data support the monophyly of one group of families—Allophrynidae, Centrolenidae, and Hylidae (including Pseudidae). Morphological data associate Sooglossidae, endemic to the Seychelles Islands group, with the Australo-Papuan Limnodynastidae and Myobatrachidae, but molecular evidence places Sooglossidae as the sister taxon to ranoids. Also, Dendrobatidae has been placed in the ranoids by some authors, but molecular evidence does not support that arrangement. Relationships of Bufonidae, Brachycephalidae, Heleophrynidae, Limnodynastidae, Myobatrachidae, and Rhinodermatidae have

Unken reflex in *Bombina variegata*. (Illustration by Wendy Baker)

yet to be determined with any degree of certainty, but some molecular evidence supports the relationships of Bufonidae and Rhinodermatidae and of Heleophrynidae and Myobatrachidae.

Physical characteristics

Anurans are unique among amphibians and all other vertebrates in having a broad head, large mouth, large protuberant eyes, short body, and no tail. The hind limbs are long and modified for jumping by having an extra segment composed of elongated "ankle" bones—fibulare and tibiale (astragalus and calcaneum, respectively). The vertebral column is short and consists of no more than nine (usually eight, but 10 in the Jurassic *Notobatrachus*) presacral vertebrae; the pre-

sacral vertebrae are articulated firmly so as to allow only slight lateral and dorsoventral flexure, and the postsacral vertebrae are fused into a bony rod, the urostyle (coccyx).

Although most anurans have snout-vent lengths of about 1.5–3.0 in (35–75 mm), many are much smaller, and a few are much larger. The smallest frogs are the Brazilian two-toed toadlet (*Psyllophryne didactyla*) and the Cuban Iberian rain frog (*Eleutherodactylus iberia*), which have lengths of 0.42 in (10.2 mm) and 0.43 in (10.5 mm), respectively. By far the largest anuran is the West African ranid, the Goliath frog (*Conraua goliath*), which reaches a length of 13 in (32 cm).

Larval anurans (tadpoles) are unlike the aquatic larvae of other amphibians. Tadpoles have short, globular bodies and long tails. The mouth is a unique structure usually containing keratinized rows of labial teeth and jaw sheaths supported by cartilaginous elements. During metamorphosis the tail is absorbed, and the mouthparts and their support structures change dramatically to produce the adult condition.

Distribution and habitat

Frogs and toads are nearly worldwide in distribution, except for Antarctica, Greenland, Arctic regions of North America and Eurasia, and some oceanic islands. In desert regions, such as the Sahara, they are restricted to oases. Few species live at high latitudes. The ranges of only three species extend north of the Arctic Circle; these are the brown frog (*Rana temporaria*) and the moor frog (*Rana arvalis*) in Eurasia and the wood frog (*Rana sylvatica*) in North America. The southernmost frog is the gray four-eyed frog (*Pleurodema bufonina*), which reaches the Straits of Magellan. Most frogs and toads live at low to moderate elevations, but a few are found at high elevations. The highest known record is for the Pakistani toad (*Bufo siachinensis*) at an elevation of 16,971 ft (5,238 m) in the Himalaya Mountains of Pakistan. In South America, the range

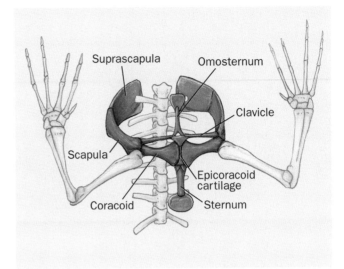

Pectoral girdle. (Illustration by Joseph E. Trumpey)

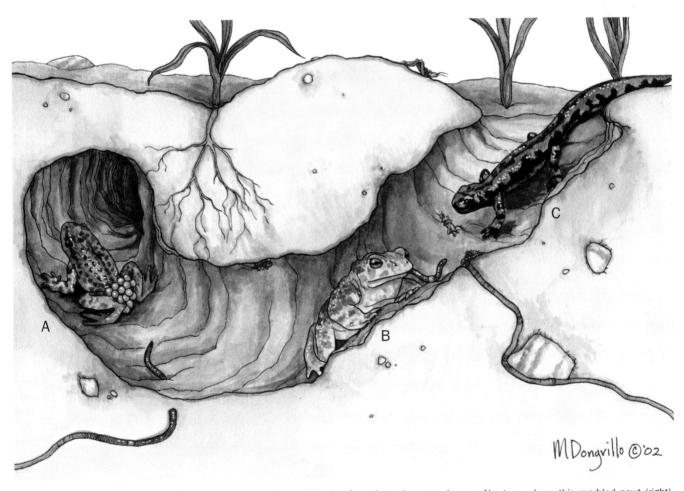

In Spain or southern France, midwife (A) and natterjack (center) toads often share the same burrow. Newts, such as this marbled newt (right), can also take advantage of the safe retreat and ready-made burrow with a food supply of earthworms, spiders, and beetles. The burrows may be up to 26 ft (8 m) long. (Illustration by Marguette Dongvillo)

Defensive posture of *Physalaemus nattereri,* displaying "eyespots" on its hindquarters. (Illustration by Gillian Harris)

of the puna frog (*Pleurodema marmorata*) extends to 16,200 ft (5,000 m) in the Andes of Peru.

Far more anurans live in the tropical parts of the world than in the northern temperate climates. Only 90 species live in North America and 116 species in temperate Eurasia. The greatest diversity of anurans is in the neotropical region (Central America, South America, and the West Indies), which is home to about 2,200 species. This number is about three times those in tropical Asia or tropical Africa and about five times that in the Australo-Papuan region.

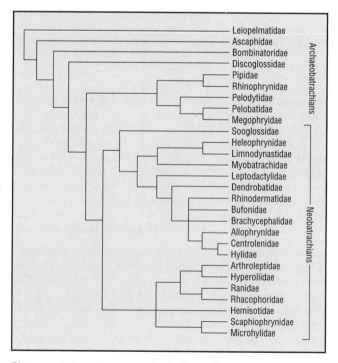

Phylogenetic tree of anurans. (Illustration by Argosy. Courtesy of Gale.)

Different historical patterns are evident among anurans. The basal living families Ascaphidae (northwestern North America) and Leiopelmatidae (New Zealand) apparently are relics of former widespread Pangaean distributions and presumably are related to *Notobatrachus* from the Middle to Upper Jurassic boundary in Argentina. With the breakup of Pangaea in the Triassic, the ancestors of the archaeobatrachian families Bombinatoridae, Discoglossidae, Megophryidae, Pelobatidae, and Pelodytidae were in Laurasia, whereas the ancestors of the other families of archaeobatrachians and the neobatrachians were in Gondwana.

Behavior

Because of their thin skin, through which water is lost, most frogs live in humid regions or are active only during rainy seasons of the year. During dry times of the year, anurans estivate, usually below ground. Likewise, in temperate regions, anurans hibernate below the frost line. Despite these physiological limitations, anurans display a wide variety of activity, mostly at night, when they feed and breed. Most respiration is cutaneous (through the skin) and is facilitated by dermal mucous glands that secrete a moist coating.

The long hind limbs facilitate a saltatorial (jumping) locomotion; most frogs can leap two to 10 times their body length, and a few can approach 30 times their body length. A few anurans (e.g., species in the Andean bufonid genus *Osornophryne*) have relatively short hind limbs and slowly walk instead of jumping. Because of their saltatorial locomotion, anurans do not leave a scent trail, and females do not locate males by chemosensory means. Instead, male frogs vocalize; air is forced from the lungs over the vocal cords and is resonated by a single or paired vocal sacs. Acute hearing allows females (and other males) not only to recognize the unique vocalization of their species but also to locate the calling males.

Amplectic positions in anuran males (orange) and females (blue): 1. axillary; 2. cephalic; 3. glued; 4. independent; 5. inguinal; 6. straddle. (Illustration by Dan Erickson)

Northern leopard frog (*Rana pipiens*) in mid-jump. (Illustration by Barbara Duperron)

Predation of anurans

Throughout all stages of their life cycle anurans are preyed upon by a great variety of animals, and small frogs even fall prey to a carnivorous plant—the Venus flytrap. Aquatic eggs are eaten by fish and various aquatic invertebrates; the arboreal eggs of centrolenids are consumed by various orthopterns and parasitized by wasps and flies, and the arboreal eggs of phyllomedusines are eaten by noctural colubrid snakes of the genus *Leptodeira*. Tadpoles are eaten by fishes, snakes, wading birds, and aquatic insects, such as diving beetles, water bugs, water scorpions, and dragonfly larvae. Some salamander larvae also feed on tadpoles, and the large tadpoles of species of *Ceratophrys* and *Leptodactylus pentadactylus* consume the smaller tadpoles of other species. Adult African clawed frogs (*Xenopus laevis*) also feed on tadpoles. Practically anything will eat anurans, especially newly metamorphosed individuals. Frogs are consumed by a variety of birds and mammals. Many snakes feed almost exclusively on frogs, and several species of carnivorous frogs include anurans in their diets. When in water, anurans fall prey to fishes, turtles, and crocodilians. Spiders are the major invertebrate predators on small anurans.

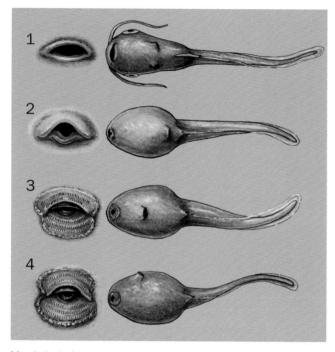

Morphological types of tadpoles showing differences in mouth structure and position of spiracle(s). (Illustration by Joseph E. Trumpey)

A frog jumping contest candidate. (Photo by Robert Holmes/Corbis. Reproduced by permission.)

Anurans have evolved a variety of defense mechanisms to escape predation. The most obvious of these methods is the jumping ability of most anurans; by leaping away, anurans leave no scent trail that can be followed by a potential predator using chemosensation in tracking its prey. This kind of escape behavior may involve a long leap to shelter (e.g., from land to water, as employed by many species of *Rana*); a single leap and subsequent immobility, with the anuran depending on cryptic coloration to avoid subsequent discovery (e.g., many cryptically colored terrestrial frogs, such as some species of *Eleutherodactylus*); a leap from one branch to another, as is characteristic of most treefrogs; a series of long leaps that carry the frog a sufficient distance from the predator (e.g., the rocket frog, *Litoria nasuta*, in Australia); or a series of multidirectional hops, such as are employed by cricket frogs of the genus *Acris* and dendrobatid frogs of the genus *Colostethus*.

Many anurans have cryptic or disruptive coloration, so that they are difficult to detect visually by potential predators. Other anurans are structured cryptically so that they blend into the substrate. This is a common feature of "dead-leaf mimics" that live on the forest floor. Examples are species of *Bufo, Ceratobatrachus, Edalorhina, Hemiphractus,* and *Megophrys*, all of which have disruptive structures, such as projecting snouts or posterolateral corners of their skulls, dermal flaps, or dermal ridges.

Many anurans exhibit defensive behavior when faced by a potential predator. Some treefrogs (Hylidae) feign death by tucking the limbs close to the body and remaining motionless on their backs. A common defensive behavior among heavy-bodied anurans is the inflation of the lungs, thereby puffing up the body and presenting a larger image to a potential predator. Other species modify their posture to display aspects of their coloration. Some leptodactylid frogs of the genera *Physalaemus* and *Pleurodema* have large, elevated inguinal glands, which are displayed prominently in a defensive posture when the head is lowered and the pelvic region is elevated, thereby emphasizing the glands to a potential predator. The markings on the glands have been interpreted as "eyespots," with the suggestion that the broad pelvic region with elevated "eyes" gives an image of a much larger frog.

Some anurans avoid predation by being unpalatable to potential predators. Granular (or poison) glands may be distributed throughout the integument or concentrated in certain areas, such as the parotoid glands behind the eyes in toads of the genus *Bufo*, and secrete substances that are noxious or even toxic. Consequently, potential predators soon learn to avoid grabbing such anurans. Poison frogs of the genera *Dendrobates, Epipedobates,* and especially *Phyllobates* have extremely toxic steroidal alkaloids in the skin; these frogs also have bright aposematic (warning) coloration and usually are avoided by predators. Defensive postures may be assumed to direct poison glands toward a potential predator. This is obvious in toads of the genus *Bufo* when they elevate the posterior part of the body and lower the head directly at the

This green frog is about to become a meal for an eastern screech owl (*Otus asio*). (Photo by Joe McDonald/Corbis. Reproduced by permission.)

Malaysian painted toad (*Kaloula pulchra*) has a deep booming voice that is heard after heavy rains in Malaysia. (Photo by Joe McDonald. Bruce Coleman Inc. Reproduced by permission.)

predator. Some frogs (e.g., *Bombina* and *Melanophryniscus*) display their brightly colored venters in an unken reflex consisting of arching the back and elevating the head and posterior part of the body while remaining motionless. Even the eggs of some species of bufonids of the genera *Atelopus* and *Bufo* have noxious properties, as do the larger tadpoles of other anurans, such as *Rana chalconota* and *Gastrophryne carolinensis*.

Feeding ecology and diet

Most anurans adopt the sit-and-wait foraging strategy; that is, they perch in one place and wait for suitable prey to appear. In most anurans, vision is important in detecting potential prey, and anurans respond positively to movement of prey. Anurans can distinguish different colors, and visual cues are used to identify different kinds of prey, such as those that may be optimal in energy content or those that are distasteful. Some frogs, such as *Bufo boreas*, *B. marinus*, and *Rana pipiens*, are capable of locating prey solely by olfaction, and some species of *Bufo* are known to be able to locate prey by audi-

tory detection. The aquatic Pipidae, however, have poor vision and detect prey by olfaction; they also can detect movements of potential prey by the sensitive lateral-line organs.

Prey are captured with the tongue, which is equipped with glands that produce a sticky substance. Prey capture involves a lingual flip, during which the posterodorsal surface of the retracted tongue becomes the anteroventral surface of the extended tongue; adhesion to the prey permits retraction of the prey into the mouth. Food is not chewed but swallowed whole. In this manner, anurans feed on a great variety of insects, spiders, and other small invertebrates. It seems that most anurans feed on a variety of prey, determined by the animal's gape and corresponding size of the prey. Several small frogs specialize on small prey, especially ants (e.g., dendrobatids and many microhylids) and termites (e.g., members of the leptodactylid genus *Physalaemus* and fossorial frogs of the genera *Hemisus* and *Rhinophrynus*). Some large frogs, such as the African ranid (*Pyxicephalus*) and the South American leptodactylid (*Ceratophrys*) feed on small vertebrates, including other frogs, snakes, lizards, rodents, and birds. But some anurans feed in other ways. The diurnal dendrobatids use the same mechanism for feeding, but they are active foragers on the ground, where they feed on small prey, such as ants and small beetles. A few frogs feed on ants and termites underground; at least one of these frogs, the Mesoamerican burrowing toad, *Rhinophrynus dorsalis*, does not flip its tongue but protrudes it forward from the small mouth. Pipid frogs are completely aquatic and lack tongues; feeding is accomplished by transportation of food into the mouth with water currents produced by pumping movements of the throat, but larger prey are pushed into the mouth by the fingers.

A southern ground-hornbill (*Bucorvus leadbeateri*) finds frogs make a tasty meal. (Photo by Karl Ammann/Corbis. Reproduced by permission.)

Reproductive biology

Except for Dendrobatidae and a few Ranidae, mating typically takes place by males grasping females from above (am-plexus). In archaeobatrachians, Myobatrachidae, Sooglossidae, and a few Leptodactylidae, the male grasps the female around the waist (inguinal amplexus), whereas males of most neobatrachians grasp the female just behind the forelimbs (axillary amplexus). In the globular-bodied microhylids (e.g., *Breviceps,*), the small males are "glued" by dermal secretions of the male to the posterior part of the body of the much larger females. In a few Ranidae (e.g., *Nyctibatrachus* and some species of *Mantidactylus*), males simply straddle females. In some Dendrobatidae, amplexus is cephalic, other dendrobatids and some *Eleutherodactylus* do not amplex but solely juxtapose their cloacas. In these various positions, the female deposits eggs that are fertilized externally.

Most species deposit their eggs in water, but many Leptodactylidae (*Eleutherodactylus* and relatives), some Arthroleptidae, Microhylidae, and Ranidae, among others, deposit eggs in moist places on the ground, and these eggs undergo development directly into froglets; the aquatic tadpole stage is bypassed. Internal fertilization is known for a few anurans. Males of the stream-dwelling *Ascaphus* have a "tail," an extension of the cloaca, that during inguinal amplexus is inserted into the cloaca of the female. Fertilization is internal and accomplished by cloacal apposition in *Eleutherodactylus jasperi* and some species of *Nectophrynoides.*

Resources

Books

Duellman, William E., ed. *Patterns of Distribution of Amphibians: A Global Perspective.* Baltimore: Johns Hopkins University Press, 1999.

Duellman, William E., and Linda Trueb. *Biology of Amphibians.* Baltimore: Johns Hopkins University Press, 1994.

Lynch, John D. "The Transition from Archaic to Advanced Frogs." In *Evolutionary Biology of Anurans,* edited by James L. Vial. Columbia: University of Missouri Press, 1973.

Savage, Jay M. "The Geographic Distribution of Frogs: Patterns and Predictions." In *Evolutionary Biology of Anurans,* edited by James L. Vial. Columbia: University of Missouri Press, 1973.

Tyler, Michael J. *Frogs.* Sydney, Australia: Collins, 1976.

Periodicals

Bossuyt, Franky, and Michel C. Milinkovitch. "Convergent Adaptive Radiations in Madagascan and Asian Ranid Frogs Reveal Co-variation Between Larval and Adult Traits."

Proceedings of the National Academy of Sciences of the United States of America 97 (2000): 6585–6590.

———. "Amphibians as Indicators of Early Tertiary 'Out-of-India' Dispersal of Vertebrates." *Science* 292 (2001): 93–95.

Ford, Linda S., and David C. Cannatella. "The Major Clades of Frogs." *Herpetological Monographs* 7 (1993): 94–117.

Hay, Jennifer M., et al. "Phylogenetic Relationships of Amphibian Families Inferred from DNA Sequences of Mitochondial 12S and 16S ribosomal RNA genes." *Molecular Biology and Evolution* 12 (1995): 928–937.

Maglia, Anne S., L. Analía Pugener, and Linda Trueb. "Comparative Development of Anurans: Using Phylogeny to Understand Ontogeny." *American Zoologist* 41 (2001): 538–551.

Orton, Grace L. "The Systematics of Vertebrate Larvae." *Systematic Zoology* 2 (1953): 63–75.

William E. Duellman, PhD

New Zealand frogs

(Leiopelmatidae)

Class Amphibia

Order Anura

Family Leiopelmatidae

Thumbnail description
Medium-small and brown, reddish, or green frogs with broad heads, rounded pupils, and smooth soles of the feet

Size
7.9–2.0 in (20–50 mm)

Number of genera, species
1 genus; 4 species

Habitat
Humid forest, banks of rocks, and streams

Conservation status
Vulnerable: 2 species; Lower Risk/Near Threatened: 1 species

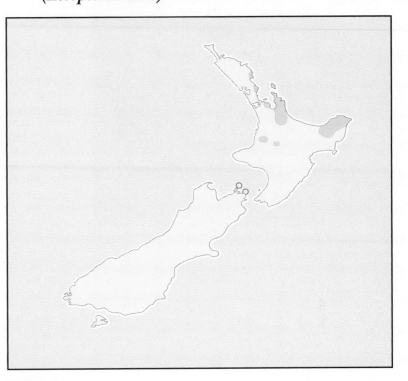

Distribution
New Zealand

Evolution and systematics

Leiopelmatidae is among the most primitive of living frogs. These frogs have skeleton characters otherwise known only in the tailed frogs, family Ascaphidae, and in the most primitive known fossil frogs, which date from the Jurassic. Because of their shared primitive morphologic features, the tailed frogs, genus *Ascaphus*, on some occasions have been grouped into the same family. All that the frogs in these families share are primitive features, however, and so there is no sound basis for considering them to be closely related. Leiopelmatidae is certainly an ancient group and dates from the time when New Zealand broke off from the rest of the continental landmasses sometime during the Mesozoic era. Within the genus, the species *Leiopelma archeyi*, *L. hamiltoni*, and *L. pakeka* are closely related, to the exclusion of *L. hochstetteri*. Of three additional subfossils, extinct species from the North Island of New Zealand, *L. auroraensis* and *L. markhami*, resemble *L. hochstetteri*, whereas *L. waitomoensis* seems to be related to the other extant species. No subfamilies are recognized.

Physical characteristics

These are small frogs, less than 2.0 in (50 mm) in length, with characteristically broad heads and smooth skin on the soles of the feet. They have rounded pupils, no visible eardrum, and little or no webbing between the toes. Parotoid glands (glandular swellings in the skin on the back of the head behind each eye) are present. Mostly, these frogs are various shades of brown, but some individuals have a reddish tint or are green. Skeletally, they have nine presacral vertebrae, of which the third, fourth, and fifth have free ribs; they also have inscriptional (abdominal) ribs between the blocks of muscle in the belly, terminating posteriorly with a broad, flat prepubic bone.

Distribution

The species are known from North Island, Maud Island, Stephens Island, and Great Barrier Island, New Zealand.

Habitat

The frogs inhabit humid native forest except on Stephens Island, where there is no forest remaining and the frogs are restricted to a fog-enshrouded and shrub-covered rock pile at the summit. *Leiopelma hochstetteri* is usually found alongside forest streams, particularly in rocky stretches of stream near cascades or even waterfalls. The other species are not restricted to standing or flowing bodies of water.

There are only four living native species of New Zealand frogs. *Leiopelma hochstetteri* is found only in a few isolated areas on the North Island of New Zealand. It is thought that spreading human land use has reduced the frog's native habitiat. (Photo by R. Wayne Van Devender. Reproduced by permission.)

Behavior

Largely nocturnal and cryptic, these frogs forage on the forest floor or along the stream banks. During the day they generally hide underneath rocks or fallen wood or under other debris on the forest floor or by the streamside. When disturbed, all species sit motionless and rely upon their cryptic coloration to avoid detection. If physically molested, however, the frogs squeak, even though they otherwise make no sounds. If they become particularly agitated by a potential predator, the frogs stand high on their four legs, head down and turned in the direction of their persecutor. The frogs swim using alternating kicks of their legs rather than with a synchronous motion of the two legs, as in most other frogs.

Feeding ecology and diet

The diet consists of small invertebrates that are captured by lunges and bites, inasmuch as the frogs do not have protrusible tongues.

Reproductive biology

The New Zealand native frogs do not call and presumably find mates by olfaction. During mating, the male clasps the female around the waist with his arms in inguinal amplexus. The eggs are fertilized externally. All species of *Leiopelma* lay small clutches of up to two to three dozen unpigmented, yolky eggs. Those of *Leiopelma hochstetteri* hatch into whitish, free-living tadpoles that tend to remain hidden under stones or other cover in seepages and stream edges. The other species lay terrestrial eggs and have no free-living larvae, passing through complete development within the egg to emerge as tiny frogs. The male parent guards the eggs and, for a time, the newly emerged froglets.

Conservation status

Two of the species of *Leiopelma* are among the rarest frogs in the world. The IUCN considers Archey's frog to be Lower Risk/Near Threatened and Hamilton's frog (encompassing both *L. hamiltoni* and *L. pakeka*) as Vulnerable. All species are protected in New Zealand under the Wildlife Act and may not be disturbed except by permit.

Significance to humans

These frogs are of scientific interest because they are among the most primitive of living frogs, as well as for a peculiar system of chromosomal sex determination in Hochstetter's frog. The North Island populations of this species, uniquely, have a single, female-determining sex chromosome that exists without a homologue. Otherwise, these frogs are unseen by most people, since they are small, cryptic, and nocturnal. To see them, you have to go looking for them, and they may have been almost completely unknown even to the Maori people of New Zealand.

1. Hamilton's frog (*Leiopelma hamiltoni*); 2. Archey's frog (*Leiopelma archeyi*); 3. Maud Island frog (*Leiopelma pakeka*); 4. Hochstetter's frog (*Leiopelma hochstetteri*). (Illustration by Brian Cressman)

Species accounts

Archey's frog
Leiopelma archeyi

TAXONOMY
Leiopelma archeyi Turbott, 1942, Tokatea, near Coromandel, New Zealand. Although it was encountered by naturalists as early as 1862 and its behavior was characterized in 1922, this species was not formally described until 1942.

OTHER COMMON NAMES
None known.

PHYSICAL CHARACTERISTICS
This is the smallest species of *Leiopelma*: it reaches less than 1.6 in (40 mm) in length in the Coromandel Peninsula, although larger specimens are common in the Waikato district. Females tend to be larger than males, but the sexes are otherwise indistinguishable. This is the most boldly patterned species of *Leiopelma*, with dark blotches and a dorsolateral glandular ridge elegantly underlined with dark spots. The skin glands are arranged in parallel longitudinal rows down the back. Parotoid glands are present behind each eye. Archey's frogs vary considerably in color. Some individuals are green, others may be largely red, and still others are an assortment of shades of tan or brown. There is a pale patch on the snout. The upper part of the iris is bright gold. The hind limbs are fairly short, and the feet are not webbed.

DISTRIBUTION
The range is restricted to North Island, New Zealand, where it occurs at elevations above 1,300 ft (400 m) in the Coromandel Peninsula and in the western Waikato district.

Leiopelma archeyi
Resident

HABITAT
This species inhabits cool, moist, native forest or mist-enshrouded mountain ridges where there are adequate rocks, logs, or other fallen debris as daytime cover.

BEHAVIOR
Largely terrestrial and nocturnal, the frogs hide during the day and forage at night, sometimes climbing to moderate heights in tree ferns and other moisture-holding plants. They squeak if disturbed, but they tend not to try to escape as readily as Hochstetter's frogs. Predators are not known.

FEEDING ECOLOGY AND DIET
The diet consists of small insects and other invertebrates.

REPRODUCTIVE BIOLOGY
Up to a dozen large, yolk-filled eggs are laid under cover in cool, damp terrestrial sites. The larvae undergo development within the egg capsule and hatch when fully metamorphosed. The male attends the eggs and froglets, which clamber onto his back and legs.

CONSERVATION STATUS
This species is designated Lower Risk/Near Threatened according to the IUCN and is protected under New Zealand's Wildlife Act.

SIGNIFICANCE TO HUMANS
None known. ◆

Hamilton's frog
Leiopelma hamiltoni

TAXONOMY
Leiopelma hamiltoni McCullough, 1919, Stephens Island, New Zealand.

OTHER COMMON NAMES
German: Hamilton-Frosch

PHYSICAL CHARACTERISTICS
This species is virtually indistinguishable from the Maud Island frog, *Leiopelma pakeka*, but differs by usually being paler. Like the Maud Island frog, it reaches 2.0 in (50 mm) in length; females tend to be larger than males, but the sexes are otherwise identical. Dorsolateral glandular ridges are underlined with dark spots, and parotoid glands are present behind each eye. These frogs generally are pale brown or tan with a pale patch on the snout. The upper part of the iris is bright gold. The hind limbs are fairly short, and the feet are not webbed.

DISTRIBUTION
The species is restricted to a deforested, 6,460 ft² (600 m²) bank of rocks at an elevation of 900 ft (275 m) near the summit of Stephens Island, which is located in the Cook Straight off the northern tip of the South Island of New Zealand.

HABITAT
The frogs inhabit a bank of rocks, which, though previously bare, lately has been allowed to become overgrown with grass

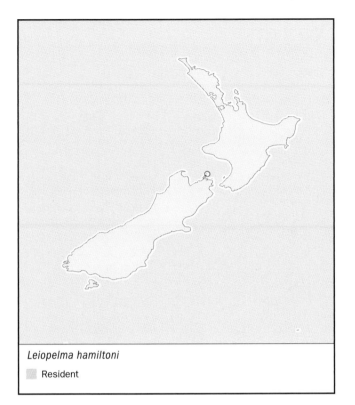

Leiopelma hamiltoni
▨ Resident

and shrubs. The interior of the rock pile maintains conditions cool and moist enough to sustain the frogs.

BEHAVIOR
This nocturnal frog squeaks repeatedly if molested. There is evidence that it may be preyed upon by tuatara (*Sphenodon punctatus*).

FEEDING ECOLOGY AND DIET
This frog most certainly eats small insects and other terrestrial arthropods.

REPRODUCTIVE BIOLOGY
Hamilton's frog lays five to nine terrestrial eggs. The white embryos undergo virtually all their development within the egg and emerge as small frogs, about 0.4 in (11 mm) long. Males attend the eggs and hatchlings, which climb onto its back and legs.

CONSERVATION STATUS
This is one the rarest and most localized species of frogs in the world. Its habitat and population status, as well as the whole of Stephens Island, are monitored and patrolled carefully. The frog is ranked Vulnerable by the IUCN (encompassing both *Leiopelma hamiltoni* and *L. pakeka*), and it is protected under New Zealand's Wildlife Act.

SIGNIFICANCE TO HUMANS
None known. ◆

Hochstetter's frog
Leiopelma hochstetteri

TAXONOMY
Leiopelma hochstetteri Fitzinger, 1861, Coromandel, New Zealand.

OTHER COMMON NAMES
None known.

PHYSICAL CHARACTERISTICS
This is a rather stout, wide-faced frog up to 2.0 in (50 mm) long. Females are larger than males, but males have decidedly more robust forearms. The feet are webbed. The skin of the back is rugose, with many scattered glandular tubercules but with neither distinct parotoid glands behind the eyes nor dorsolateral glandular ridges. A small tubercle adorns the top of each eyelid. The belly skin is smooth and pink. Hochstetter's frogs may be dark olive brown to reddish tan, with some individuals being distinctly green. There are oblique dark bands on the legs and indistinct dark patches on the back. There is a pale patch on the snout from between the eyes to the nostrils.

DISTRIBUTION
This species is known only from scattered hilly localities on the northern half of the North Island of New Zealand, including the Coromandel, Waitakere, Dome, and Hunua ranges and East Cape, Mount Ranginui, western Waikato, and the vicinity of Waipu as well as the northern section of Great Barrier Island.

HABITAT
This frog normally is near streams, commonly beside cascading and rapidly flowing water, where there are rocks at the stream edge or a splash zone for refuge. It also inhabits rock-strewn trickles and seeps in damp forest.

BEHAVIOR
These nocturnal frogs hide during the day under rocks but emerge to forage at night and wander some distance during rains. They are the most aquatic of the *Leiopelma*. If disturbed, a frog often jumps from under a rock into the stream and swims furiously underwater to hide under a submerged rock. In 20 minutes to half an hour, the frog reemerges and hides once more under a rock out of the stream, often the same one it was

Leiopelma hochstetteri
▨ Resident

under before. The frogs squeaks if disturbed and, under extreme duress, secretes a noxious white substance from its numerous skin glands. Its predators are not known but probably consist of native stream fishes, centipedes, and birds.

FEEDING ECOLOGY AND DIET
Food consists of small insects and other invertebrates.

REPRODUCTIVE BIOLOGY
Females lay 10–15 large, yolky white eggs under rocks or fallen vegetation in seepages or the damp sides of streams. The eggs hatch into white, free-swimming tadpoles, which generally remain hidden under cover. They retain a large amount of yolk and may not feed during larval development. The forelimbs are not covered completely by an operculum. There is neither egg attendance nor protection of the tadpoles by the adults.

CONSERVATION STATUS
This is the most widespread, most common, and least threatened of the native New Zealand frogs. Hochstetter's frog nevertheless is protected under New Zealand's Wildlife Act.

SIGNIFICANCE TO HUMANS
The frog is of scientific interest because of its uniqueness, primitive morphologic features, and high chromosomal variability. It appears to have been little known to the Maori. ◆

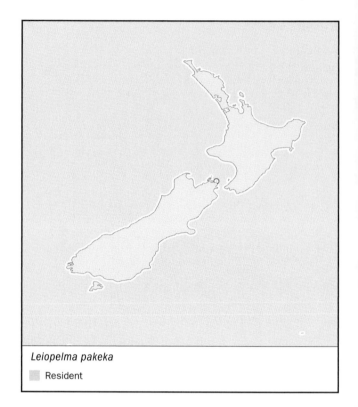

Leiopelma pakeka
■ Resident

Maud Island frog
Leiopelma pakeka

TAXONOMY
Leiopelma pakeka Bell Daugherty, and Hitchmough, 1998, Maud Island, New Zealand.

OTHER COMMON NAMES
None known.

PHYSICAL CHARACTERISTICS
This is a larger, duller version of Archey's frog, reaching 2.0 in (50 mm) in length. Females tend to be larger than males, but the sexes are otherwise indistinguishable. Dorsolateral glandular ridges are underlined with dark spots, and parotoid glands are present behind each eye. These frogs are generally brown with a pale patch on the snout. The upper part of the iris is bright gold. The hind limbs are fairly short, and the feet are not webbed.

DISTRIBUTION
Restricted to Maud Island in the Marlborough Sounds off the north coast of the South Island of New Zealand, this frog inhabits a remnant patch of forest about 0.06 mi² (0.15 km²) in extent.

HABITAT
The patch of forest on Maud Island is at an elevation of 295–980 ft (90–300 m) on an eastward-facing hillside. There are numerous boulders, logs, and rocks toward the lower, damper, less steeply sloping part of the forest, where the frogs are most abundant. There are no permanent streams or seepages.

BEHAVIOR
The frogs emerge from hiding places after dark and sit on rocks or logs or forage slowly on the forest floor.

FEEDING ECOLOGY AND DIET
The diet consists of terrestrial insects.

REPRODUCTIVE BIOLOGY
Females lay up to 20 large, yolk-filled eggs in cool, damp depressions under cover on the ground. The whitish larvae undergo development within the egg capsule and hatch when fully metamorphosed. The male attends the eggs and froglets, which clamber onto his back and legs.

CONSERVATION STATUS
This species is Vulnerable according to IUCN (encompassing both *Leiopelma hamiltoni* and *L. pakeka*) and is protected under New Zealand's Wildlife Act. This rare frog may be reasonably secure in its isolation: Maud Island is monitored carefully to keep out invasive mammals, and the frogs persist in fairly high density.

SIGNIFICANCE TO HUMANS
None known. ◆

Resources

Books

Grigg, G., R. Shine, and H. Ehmann, eds. *The Biology of Australasian Frogs and Reptiles.* Chipping Norton, Australia: Surrey Beatty and Sons, 1985.

Robb, Joan. *New Zealand Amphibians and Reptiles in Color.* Auckland: Collins Publishers, 1980.

Resources

Periodicals

Abourachid, A., and D. M. Green. "Origins of the Frog-Kick? Alternate-Leg Swimming in Primitive Frogs, Familes Leiopelmatidae and Ascaphidae." *Journal of Herpetology* 33, no. 4 (1999): 657–663.

Bell, Ben D. "A Review of the Status of New Zealand *Leiopelma* Species (Anura: Leiopelmatidae), Including a Summary of Demographic Studies in Coromandel and on Maud Island." *New Zealand Journal of Zoology* 21, no. 4 (1994): 341–349.

————. "The Amphibian Fauna of New Zealand." *New Zealand Wildlife Service Occasional Publications* 2 (1982): 27–89.

Green, D. M., and D. C. Cannatella. "Phylogenetic Significance of the Amphicoelous Frogs, Ascaphidae and Leiopelmatidae." *Ecology, Ethology, and Evolution* 5, no. 2 (1993): 233–245.

Worthy, T. H. "Osteology of *Leiopelma* (Amphibia: Leiopelmatidae) and descriptions of three new subfossil *Leiopelma* species." *Journal of the Royal Society of New Zealand* 17, no. 3 (1987): 201–251.

Organizations

Society for Research on Amphibians and Reptiles in New Zealand (SRARNZ). SBS, Victoria University of Wellington, PO Box 600, Wellington, New Zealand.

David M. Green, PhD

Tailed frogs

(Ascaphidae)

Class Amphibia
Order Anura
Family Ascaphidae

Thumbnail description
Small, stream-dwelling frogs with broad heads, vertical pupils, no visible eardrum, long hind legs, rugose skin, and webbed hind feet with a thickened outermost toe; males have a tail-like extension of cloaca; tadpoles are dark and streamlined and have enlarged, ventral, suctorial lips bearing many rows of small denticles

Size
1.2–2.0 in (30 to 50 mm)

Number of genera, species
1 genus; 2 species

Habitat
Streams on forested mountains

Conservation status
Not threatened

Distribution
Northwestern North America

Evolution and systematics

Tailed frogs are considered to be among the most primitive of living frogs, rivaled in that position only by the leiopelmatid frogs of New Zealand. Their skeletal anatomy is remarkably similar to that of the earliest known fossil frogs from the Jurassic. The two species were distinguished from each other in 2001 based on genetic data, though at one time they were described as different subspecies. There is no fossil record. No subfamilies are recognized.

Physical characteristics

These are small brown or gray frogs with a pale patch on the snout. The head is broad, the nostrils are placed far apart, and the eyes have vertical, diamond-shaped pupils. A slight fold of skin curves back from the eye to the corner of the mouth. No tympanum is visible, and the dorsal skin is roughened with small tubercles. The toes of the hind feet are short, but the feet are webbed; the outermost toe of the hind foot is thicker than the rest of the toes. Adult males have a fleshy "tail," which is actually an extension of the

cloaca and not a tail in the usual sense. Females are slightly larger than males. These frogs have small lungs and rely upon their vascularized skin for much of their respiratory gas exchange. Such reduced lungs do not present a problem to frogs living in fast-flowing, well-oxygenated streams. There are nine prescacral vertebrae, and small, free ribs are associated with the third, fourth, and fifth vertebrae. All other living frogs, except the New Zealand native frogs, genus *Leiopelma*, have eight or fewer presacral vertebrae and ribs that are fused to the vertebrae. The tadpoles are dark gray with wide heads dominated by a large, ventral sucker surrounding the mouth.

Distribution

Tailed frogs are found in the Pacific coastal mountainous region of North America from northern California north to the Nass River in British Columbia, but not on Vancouver Island (*Ascaphus truei*), and in the Rocky Mountains of Idaho, Montana, northeastern Oregon, southwestern Washington, and extreme southeastern British Columbia (*Ascaphus montanus*).

The coastal tailed frog (*Ascaphus truei*) inhabits streams of the northwestern United States. (Photo by Animals Animals ©David M. Dennis. Reproduced by permission.)

Habitat

Tailed frogs inhabit small, clear, unsilted, permanent mountain streams surrounded by forest; they tend to avoid steep gradients and flat, still waters.

Behavior

Tailed frogs hide under rocks during the day and emerge at night, especially during rains, to forage for food. The tadpoles tend to cling to rocks in fast-flowing currents or inhabit pools and riffles in the stream.

Feeding ecology and diet

Tailed frogs eat terrestrial and aquatic insects or other invertebrates. The tadpoles scrape algae and diatoms from rocks.

Reproductive biology

In the fall breeding season males develop black pads on the hands and black tubercles on their arms and sides. They do not vocalize. During mating amplexus is inguinal, and the "tail" is used for internal fertilization. The female lays 35–100 unpigmented eggs in small clusters under rocks in streams. The tadpoles take up to seven years to metamorphose and another three to eight years to reach maturity.

Conservation status

Not threatened.

Significance to humans

These frogs are of scientific interest, because they are among the most primitive of living frogs. They are also indicators of environmental health, in that they inhabit small, clear, unsilted streams devoid of fish.

1. Coastal tailed frog (*Ascaphus truei*); 2. Rocky Mountain tailed frog (*Ascaphus montanus*). (Illustration by Dan Erickson)

Species accounts

Rocky Mountain tailed frog
Ascaphus montanus

TAXONOMY
Ascaphus truei montanus Mittleman and Myers, 1949, tributary of Lincoln Creek, Glacier National Park, Flathead County, Montana, United States. Recognized as a species in 2001.

OTHER COMMON NAMES
None known.

PHYSICAL CHARACTERISTICS
The body is brown or gray with distinct, dense, fine black speckling on the dorsal and ventral surfaces. There is a light patch between the eyes that extends forward across the snout, but there are no spots or blotches on the back. The belly is pink. The toes of the hind feet are short, but the webbing of the hind foot is even more extensive than it is in the coastal tailed frog. Adults are 1.2–2.0 in (30–50 mm) long; females are slightly larger than males. The tadpoles are slate gray and up to 1.2 in (30 mm) long, with large, suctorial mouthparts. Within the mouth disc are broad, horny upper jaws and large numbers of labial teeth.

DISTRIBUTION
The species occurs in western North America in the Rocky Mountains and Columbia Mountains of Idaho, western Montana, northeastern Oregon, and southeastern Washington to extreme southeastern British Columbia.

HABITAT
This frog inhabits small, permanent, mid-elevation mountain streams.

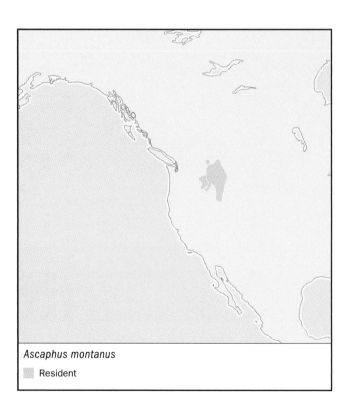

Ascaphus montanus
▪ Resident

BEHAVIOR
During the day Rocky Mountain tailed frogs hide under rocks beside the stream or in nearby rivulets. At night they forage in the surrounding forest. Tadpoles are found in the swiftest-flowing parts of streams, and metamorphosing tadpoles tend to occur in pools where there are large boulders. The frogs face predatory threats from snakes, fish, larger frogs, birds, predatory invertebrates, and small mammals.

FEEDING ECOLOGY AND DIET
Rocky Mountain tailed frogs eat terrestrial and aquatic insects and forage under water as well as on land. To feed, tadpoles scrape algae off rocks with their rows of small teeth.

REPRODUCTIVE BIOLOGY
Breeding occurs in the fall. In mid-summer females adhere 45–75 unpigmented eggs in small clusters to the undersides of rocks in streams. The tadpoles stay in pools until the suction-cup-like mouth develops fully. The tadpoles may take up to five years to metamorphose, and they do so usually in early spring to late summer. The newly transformed frogs typically do not reach sexual maturity for seven to eight years. These are long-lived frogs, estimated to live anywhere from 15 to 20 years.

CONSERVATION STATUS
Not threatened.

SIGNIFICANCE TO HUMANS
These frogs are interesting because of their primitive morphologic features and unusual life history; they are of some importance because of their reliance on undamaged forested streams that are generally too small to maintain fish. ◆

Coastal tailed frog
Ascaphus truei

TAXONOMY
Ascaphus truei Stejneger, 1899, Humptulips, Washington, United States.

OTHER COMMON NAMES
None known.

PHYSICAL CHARACTERISTICS
The dorsum usually is brown or gray with a pale patch on the snout between the eyes and nose. The belly is a translucent pink with scattered small white dots. The toes of the hind feet are short, but the feet are webbed, though slightly less extensively than in *Ascaphus montanus*. Adults are 1.2–2.0 in (30–50 mm) long; females are slightly larger than the males. The tadpoles are slate gray and up to 1.2 in (30 mm) long; they typically have a white spot on the tip of the tail.

DISTRIBUTION
This frog is distributed along the Pacific coast of North America from the Nass River in British Columbia south through the Coast Ranges and Cascade Range to northwestern California.

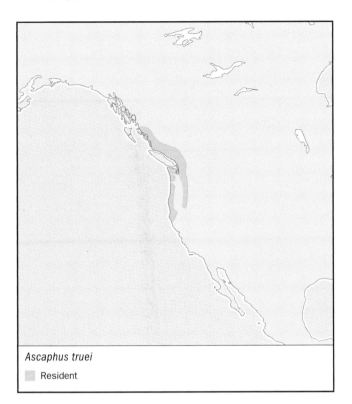

Ascaphus truei

■ Resident

HABITAT
This species inhabits cold, clear, unsilted streams from sea level up to subalpine meadows.

BEHAVIOR
Elusive and nocturnal and tending to hide under rocks or logs by the stream edge or in little rivulets, these frogs range into the forest on rainy days and in the evenings to forage for food. The tadpoles cling to rocks in fast-flowing currents by means of an enlarged, suction-cup-like mouth that keeps them from being washed away. Small tadpoles tend to be found in pools, and larger tadpoles inhabit riffles; sometimes they can be seen attached to exposed rocks in midstream. Predatory threats are from snakes, fish, larger frogs, birds, invertebrates, and small mammals.

FEEDING ECOLOGY AND DIET
The frogs' food includes terrestrial and aquatic insects. The tadpoles use their rasping mouthparts to scrape off algae and diatoms from rocks.

REPRODUCTIVE BIOLOGY
Breeding is in the fall. During mating, inguinal amplexus lasts 24–30 hours; the so-called tail is used for insemination. The sperm remains viable within the female's oviduct until egg laying takes place many months later. Females lay 45–75 unpigmented eggs in small clusters adhering to the underside of rocks in streams. The tadpoles take one to three years to metamorphose. The newly transformed frogs reach sexual maturity in three to five years. These frogs are estimated to live up to 15 years.

CONSERVATION STATUS
Not threatened.

SIGNIFICANCE TO HUMANS
The species is interesting because of its primitive morphologic features and unusual life history. The frogs are of some importance because of their reliance upon undamaged forested streams that are generally too small to maintain fish. ◆

Resources

Books

Nussbaum, R. A., E. D. Brodie, and R. M. Storm. *Amphibians and Reptiles of the Pacific Northwest.* Moscow: University of Idaho Press, 1983.

Wright, A. H., and A. A. Wright. *Handbook of Frogs and Toads of the United States and Canada.* Ithaca, NY: Comstock, 1949.

Periodicals

Bull, E. L., and B. E. Carter. "Tailed Frogs: Distribution, Ecology, and Association with Timber Harvest in Northeastern Oregon." *U.S. Department of Agriculture, Forest Service Research Paper* PNW-RP-497 (1996): 1–12.

Bury, R. B., and M. J. Adams. "Variation in Age at Metamorphosis Across a Latitudinal Gradient for the Tailed Frog *Ascaphus truei.*" *Herpetologica* 55, no. 2 (1999): 283–291.

Daugherty, C. H., and A. L. Sheldon. "Age-Determination, Growth and Life History of a Montana Population of the Tailed Frog (*Ascaphus truei*)." *Herpetologica* 38, no. 4 (1982a): 461–468.

Daugherty, C. H., and A. L. Sheldon. "Age-Specific Movement Patterns of the Tailed Frog *Ascaphus truei.*" *Herpetologica* 38, no. 4 (1982b): 468–474.

Diller, L. V., and L. R. Wallace. "Distribution and Habitat of *Ascaphus truei* in Streams on Managed, Young Growth Forests in North Coastal California." *Journal of Herpetology* 33, no. 1 (1999): 71–79.

Dupuis, L., and D. Steventon. "Riparian Management and the Tailed Frog in Northern Coastal Forests." *Forest Ecology and Management* 124, no. 1 (1999): 35–43.

Nielson, M., K. Lohman, and J. Sullivan. "Phylogeography of the Tailed Frog (*Ascaphus truei*): Implications for the Biogeography of the Pacific Northwest." *Evolution* 55, no. 1 (2001): 147–160.

Wallace, R. L., and L. V. Diller. "Length of the Larval Cycle of *Ascaphus truei* in Coastal Streams of the Redwood Region, Northern California." *Journal of Herpetology* 32, no. 3 (1998): 404–409.

David M. Green, PhD

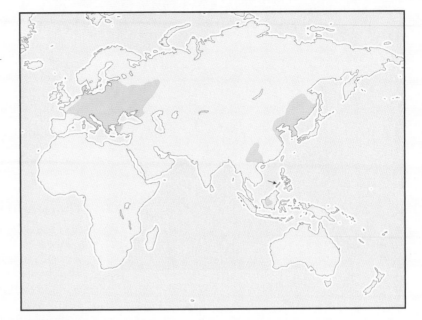

Fire-bellied toads and barbourulas

(*Bombinatoridae*)

Class Amphibia

Order Anura

Family Bombinatoridae

Thumbnail description
Often warty, aquatic toads with flattened bodies that may have a brightly colored venter

Size
1.6–3.9 in (40–100 mm)

Number of genera, species
2 genera; 10 species

Habitat
Usually found in marshes, ponds, stony mountain streams, shallow pools, or rock crevices

Conservation status
Vulnerable: 1 species; Lower Risk/Conservation Dependent: 1 species

Distribution
Much of Europe and eastern Asia

Evolution and systematics

The evolutionary relationships of the Bombinatoridae are debatable. Many authors believe that *Bombina* and *Barbourula* should be placed together with *Alytes* and *Discoglossus* in the family Discoglossidae. Others suggest that only *Alytes* and *Bombina* should be placed together and should be given the name Bombinatoridae (without consideration of *Barbourula*). At present, the most accepted hypothesis of relationships (and the one followed here) suggests that *Bombina* and *Barbourula* are each other's closest relative and should be grouped in the Bombinatoridae, whereas *Discoglossus* and *Alytes* are a separate, more distantly related group, the Discoglossidae.

Fossils of *Bombina* are known from the Pliocene to the Pleistocene. No subfamilies are recognized.

Physical characteristics

These medium-sized frogs have warty, almost "spiny" skin on the back. The color of the dorsum varies from brown-gray to greenish gray or bright green with dark spots. The belly, which is smooth, may be red, orange, or yellow with dark spots. There is no external eardrum (tympanic membrane), and the eyes have triangular pupils. Males have nuptial pads, enlarged bumps that help aquatic frogs hold on to females during breeding, on their first and second fingers.

Distribution

Bombinatorids occur in Europe east to Ukraine, western Russia, Turkey, eastern Russia and also in China, Korea, Vietnam, Borneo, and the Philippines.

Habitat

Frogs in the genus *Bombina* are aquatic and generally prefer slow-moving and open waters, such as swamps, ponds, and marshes. *Barbourula* typically are found in water in more mountainous regions, where they prefer streams and shallow pools, particularly those with stones and rocks. They often hide below rocks or in rocky crevices close to the edge of the water.

Behavior

Fire-bellied toads of the genus *Bombina* are diurnal and quite active in open areas during the day. These frogs have poisonous skin secretions that help protect them from predators. As is the case with many poisonous amphibians, their bright colors and distinct patterns help remind predators that they are toxic. If attacked or threatened by a would-be predator, fire-bellied toads will perform an arching "back bend" called the unken reflex; this maneuver exposes their brightly colored underbellies. Barbourulas, however, have more camouflaged color patterns and do not engage in the anti-predator behavior pat-

An oriental fire-bellied toad (*Bombina orientalis*) adopts a defensive display, showing the warning colors on its belly. (Photo by M.P.L. Fogden. Bruce Coleman Inc. Reproduced by permission.)

terns of the fire-bellied toads. They are highly secretive and spend most of their time hiding under rocks in streams. For this reason, little more is known about their behavior.

Feeding ecology and diet

Depending on the species, the diet may consist of different proportions of aquatic or terrestrial invertebrates, in-cluding worms, snails, beetles, and bugs. Tadpoles eat plants, fungus, and small invertebrates.

Reproductive biology

Fire-bellied toads from Europe breed from late spring to midsummer; males often call throughout the day and night. Most breeding occurs in the evening, and males grasp females around the waist. Females lay up to 200 eggs on immersed vegetation or directly on the bottom of the pond. Eggs hatch in about seven days, and tadpoles metamorphose within 45 days of hatching. Although little is known about the biology of barbourulas, it seems that females lay about 80 large eggs and place them under stones in streams.

Conservation status

The IUCN lists *Barbourula busuangensis* as Vulnerable, and *Bombina bombina* as Lower Risk/Conservation Dependent. Several species have disappeared from parts of their range, and one species is known only from a single locality. Others are critically threatened because of destruction of their habitats. A few species, however, seem to have been able to tolerate human modification of the environment and may even have increased in numbers in certain areas because of human influences.

Significance to humans

Fire-bellied toads are common laboratory animals, particularly for studies of embryology and physiology. They are also common in the pet trade, owing to their bright colors, interesting anti-predator behavior, and ease of care.

1. Oriental fire-bellied toad (*Bombina orientalis*); 2. Yellow-bellied toad (*Bombina variegata*); 3. Fire-bellied toad (*Bombina bombina*). (Illustration by Barbara Duperron)

Species accounts

Fire-bellied toad
Bombina bombina

TAXONOMY
Rana bombina Linnaeus, 1761, Europe and western Asia. No subspecies recognized.

OTHER COMMON NAMES
English: Firebelly toad; French: Sonneur á ventre feu; German: Rotbauchunke; Spanish: El sapillo de vientre de fuego.

PHYSICAL CHARACTERISTICS
The skin on the back of these frogs is covered with rounded warts and is dark gray to black, with large dark spots. Some individuals living in pools with a lot of vegetation are camouflaged by being bright green with sparse dark green spots. The belly is red or orange, with large bluish black spots and many white dots. There is no external eardrum (tympanic membrane), and the pupil of the eye is triangular.

DISTRIBUTION
These frogs are found in central and eastern Europe from Denmark and western Germany east to the Ural Mountains and south to the Caucasus Mountains. In the north they range to the Gulf of Finland. They also are found in Turkey. Some have been seen in Sweden, but these are most likely an introduced population.

HABITAT
Fire-bellied toads are aquatic in forests and wetlands. They live in dense vegetation as well as open areas, such as drainage ditches. They are also common to permanent freshwaters, such as river valleys, shallow stagnant lakes, ponds, swamps, bogs, ditches, flooded rice fields, and quarries. Sometimes they are found in slow-moving waters, such as springs, irrigation channels, rivers, and streams. In other areas, they seem to prefer stagnant water.

BEHAVIOR
These frogs are active mostly during the day when temperatures reach about 60°F (about 15°C). In the daytime they spend most of their time in the water or near the shore hunting for food. At night, when the humidity in the air is higher, they move onto land to continue foraging. During windy or cold weather, their activity levels decrease. From October to April they hibernate in mud at the bottom of ponds or on land. Although they are active primarily by day, males call mostly at dusk. As with other fire-bellied toads, this species displays the unken reflex when threatened. Despite this behavior and their toxic skin, they are still common prey for many animals.

FEEDING ECOLOGY AND DIET
Fire-bellied toads eat a variety of insects, but more than half of their diet is made up of aquatic prey. Of course, the more time they spend foraging on land, the more terrestrial insects they include in the diet. Terrestrial insects eaten most often include beetles, flies, and ants. The tadpoles may eat some aquatic insects as well, but they mainly eat algae and plants.

REPRODUCTIVE BIOLOGY
These toads breed from May to September, during which time males call either underwater or from a position floating on the water's surface. The male grabs the female around the waist, and she deposits up to 300 eggs. After about two months, eggs hatch, and tadpoles usually metamorphose before autumn. Toads become adults at about three years of age and live for about 12 years.

CONSERVATION STATUS
The IUCN lists this species as Lower Risk/Conservation Dependent. In western Europe this species is threatened or extinct in many areas. The destruction of wetland habitats seems to be the major cause of its decline. In other areas, it seems to be one of the most common toads.

SIGNIFICANCE TO HUMANS
As with other fire-bellied toads, this species is common in the pet trade and laboratory. ◆

◻ *Bombina variegata*
◼ *Bombina bombina*

Oriental fire-bellied toad
Bombina orientalis

TAXONOMY
Bombinator orientalis Boulenger, 1890, Chefoo (Yantai, Shandong, China). No subspecies recognized.

OTHER COMMON NAMES
English: Oriental firebelly toad, Oriental bell toad; German: Chinesische Rotbauchunke; Spanish: Sapo de vientre de fuego.

PHYSICAL CHARACTERISTICS
The skin on the back of these frogs is covered with pointed, even spiked warts. The dorsum is brownish gray, greenish gray,

☐ Bombina orientalis
▨ Barbourula busuangensis

or bright green with large dark spots. The belly is red or orange with large dark spots. The pupil of the eye is triangular.

DISTRIBUTION
These toads are found in the southern part of Primorsky Kraj (the Russian maritime territory), northeastern China (south to Jiangsu), Korea, and the Tsushima and Kyushu islands of Japan.

HABITAT
Oriental fire-bellied toads inhabit mixed coniferous or broad-leaved forests as well as spruce and pine forests, open meadows, river valleys, and swamps. They typically are found in slow-moving waters, such as lakes, ponds, swamps, streams, springs, ditches, and puddles. At the end of the summer, they are on land close to water.

BEHAVIOR
This toad is active in warmer temperatures. It hibernates from October to May, mostly on land in tree stumps, piles of stones, or leaves but also on stream bottoms. Up to six toads may hibernate together, presumably as a way to prevent water loss. As with other fire-bellied toads, this species displays the unken reflex when threatened.

FEEDING ECOLOGY AND DIET
Oriental fire-bellied toads eat a variety of insects, including beetles, flies, and ants. They also include worms and snails in their diet. The tadpoles mainly eat algae and plants but, as they age, increase the amount of aquatic and terrestrial insects.

REPRODUCTIVE BIOLOGY
These toads breed from May to August. Breeding and calling are similar to those of the fire-bellied toad, but a female may take many weeks to deposit all of her eggs. She deposits about 30 or so each week until finished and may deposit as many as 250 eggs. Eggs hatch in about two months, and tadpoles usually metamorphose before autumn. The maximum life span recorded for these toads is estimated to be about 20 years.

CONSERVATION STATUS
Not threatened.

SIGNIFICANCE TO HUMANS
As with other fire-bellied toads, this species is common in the pet trade and laboratory. ◆

Yellow-bellied toad
Bombina variegata

TAXONOMY
Rana variegata Linnaeus, 1758, Switzerland. No subspecies recognized.

OTHER COMMON NAMES
English: Yellowbelly toad; French: Sonneur á ventre jaune; German: Gelbbauchunke; Russian: Zheltobryukhaya zherlyanka; Spanish: El sapillo de vientre amarillo.

PHYSICAL CHARACTERISTICS
The skin on the back is covered with sharp warts. Unlike other fire-bellied toads, the skin on the belly also has warts, though fewer than on the back. These toads are also drabber than other fire-bellied toads, being dark olive with small dark spots. The belly usually is yellow with large dark spots, and the inner thighs and tips of the toes also are brightly colored. The pupil of the eye is triangular.

DISTRIBUTION
These toads occur in central and southern Europe (excluding the Iberian Peninsula, adjacent France, and Britain) southeast to the Carpathian Mountains in Ukraine.

HABITAT
Yellow-bellied toads inhabit all kinds of forests, meadows, grasslands, and glades, where they occur in lakes, ponds, swamps, rivers, streams (even those with fast currents), and springs. Apparently, the species has a fairly high tolerance for poor quality water, because it has been found in wetlands that are highly polluted with hydrogen sulfide and salts.

BEHAVIOR
As with fire-bellied toads, this toad is active in warmer temperatures. Hibernation begins in October and ends sometime between March and May, depending on the elevation. These toads hibernate on land in burrows or holes under stones and logs. In thermal springs with warm waters, they may stay active throughout the winter.

FEEDING ECOLOGY AND DIET
Yellow-bellied toads mainly forage for food on land and eat a variety of terrestrial arthropods, including beetles, spiders, flies, and ants.

REPRODUCTIVE BIOLOGY
In the spring these toads leave hibernation and migrate to waters. Mating is similar to that of other toads in the group; it begins within 10 days of entering the water and continues throughout the summer. Heavy rains often increase the intensity of spawning in populations. Sometimes heavy rains in summer are followed by intensive spawning in small wetlands. The mating call is similar to that of the fire-bellied toad, but quieter and higher. The clutch consists of 45–100 eggs deposited in portions, similarly to the oriental fire-bellied toad.

CONSERVATION STATUS
Although not listed by the IUCN, at least 13 local populations of this toad are now extinct, and others are in grave danger.

Despite their tolerance for poor water conditions, destruction of natural habitats and urbanization is the main threat to their survival.

SIGNIFICANCE TO HUMANS
Generally, the species is of no major significance to humans, but it may be found sporadically in the pet trade. ◆

Philippine barbourula
Barbourula busuangensis

TAXONOMY
Barbourula busuangensis Taylor and Noble, 1924, Philippines. No subspecies recognized.

OTHER COMMON NAMES
English: Busuanga jungle toad.

PHYSICAL CHARACTERISTICS
Barbourulas have cryptic color patterns that help them blend in with their surroundings. Usually these are drab colors, such as olive and brown, with some darker markings. Their hands and feet are fully webbed, which is an adaptation for a highly aquatic lifestyle.

DISTRIBUTION
The species occurs in the Busuanga and Palawan islands of the Philippines.

HABITAT
These frogs generally are found in water in mountains, where they prefer streams and shallow pools, particularly those with stones and rocks. They often are found below rocks or in rocky crevices close to the edge of the water.

BEHAVIOR
They are highly secretive and spend most of their time hiding under rocks in streams. For this reason, little more is known about their behavior.

FEEDING ECOLOGY AND DIET
Presumably, they actively forage for a variety of aquatic invertebrates, including insects. They also may include terrestrial invertebrates in their diet.

REPRODUCTIVE BIOLOGY
Although little is known about the biology of barbourulas, it seems that females lay about 80 large eggs and place them under stones in streams.

CONSERVATION STATUS
The IUCN lists this species as Vulnerable. Because these frogs are sensitive to water quality, pollution of streams on Busuanga severely limits the amount of available habitat. Therefore, they are threatened and likely subject to extinction.

SIGNIFICANCE TO HUMANS
None known. ◆

Resources

Books

Duellman, William E., and Linda Trueb. *Biology of Amphibians.* Baltimore: Johns Hopkins University Press, 1994.

Garcia Paris, Mario. *Los Anfibios de España.* Madrid: Ministerio de Agricultura, Pesca y Alimentación, 1985.

Gasc, Jean-Pierre, A. Cabela, J. Crnobrnja-Isailovic, et al., eds. *Atlas of Amphibians and Reptiles in Europe.* Paris: Societas Europaea Herpetologica and Muséum National d'Histoire Naturelle, 1997.

Herrmann, Hans-Joachim. *Terrarien Atlas.* Vol. 1, *Kulturgeschichte, Biologie, und Terrarienhaltung von Amphibien, Schleichenlurche, Schwanzlurche, Froschlurche.* Melle, Germany: Mergus Verlag, 2001.

Zug, George R., Laurie J. Vitt, and Janalee P. Caldwell. *Herpetology: An Introductory Biology of Amphibians and Reptiles.* 2nd edition. San Diego: Academic Press, 2001.

Other

Canatella, David. "Bombinatoridae." Tree of Life. (15 June 2002) <http://tolweb.org/tree>

Frost, Darrel R. Amphibian Species of the World: An Online Reference. V2.20. 1 Sept. 2000. (15 June 2002) <http://research.amnh.org/herpetology/amphibia>

AmphibiaWeb. (15 June 2002) <http://elib.cs.berkeley.edu/aw/index.html>

Anne M. Maglia, PhD

Midwife toads and painted frogs
(Discoglossidae)

Class Amphibia

Order Anura

Family Discoglossidae

Thumbnail description
Moderate-sized, terrestrial to semiaquatic frogs with thick, disc-shaped tongues

Size
1.6–3.0 in (40–75 mm)

Number of genera, species
2 genera; 10 species

Habitat
Banks of fast-flowing streams, small ponds and swamps, and densely wooded areas

Conservation status
Extinct: 1 species; Critically Endangered: 1 species; Vulnerable: 2 species

Distribution
Northwestern Africa, central and western Europe

Evolution and systematics

The evolutionary relationships of the Discoglossidae are debatable. Many authors consider the group to include four genera: *Discoglossus*, *Alytes*, *Bombina*, and *Barbourula*. However, others suggest that Discoglossidae only includes one genus, *Discoglossus*, and that *Alytes* and *Bombina* should be grouped together in the Bombinatoridae without consideration of *Barbourula*. The most accepted hypothesis of relationships (and the one followed here) suggests that *Discoglossus* and *Alytes* are each other's closest relatives and should be grouped in the Discoglossidae, whereas *Bombina* and *Barbourula* are a separate, more distantly related group, the Bombinatoridae.

Several fossil forms have been attributed to the Discoglossidae. Of these, several forms, including *Latonia* (Miocene of Europe) and *Eodiscoglossus* (Jurassic of Spain) may be removed from the group as more information about their phylogenetic relationships becomes available. Other fossil taxa include *Spondylophryne* (Pleistocene of Hungary), *Scotiophryne* and *Paradiscoglossus* (Cretaceous of the United States), *Prodiscoglossus* (Oligocene of France), and *Pelophilus* (Miocene of Germany). The name *Baleaphryne* was given to a fossil from the Pleistocene of Spain; it was later discovered to be the same species as the living *Alytes muletensis*. No subfamilies are recognized.

Physical characteristics

These frogs are generally small and squat. They may have brightly colored patterns, and have distinct eyes with slitlike pupils. Several anatomical features make them distinct among frogs. These include the lack of palatine bones, eight opisto-coelous, presacral vertebrae, articulation of sacrum and urostyle formed by two bony proturusions (bicondylar), a distinct sternum, and the presence of free ribs on Vertebrae II–IV. Also, the ankle bones (tibiale and fibulare, or astragalus and calcaneum) are fused only at their ends.

Distribution

These frogs occur in central and southern Europe, including the Iberian Peninsula and Italy, northwestern Africa, and Israel.

Habitat

Painted frogs occur primarily in wet or moist areas, including the edges of fast moving streams with rocky substrates. Midwife toads prefer slightly drier habitats that include wooded areas and open habitats near ponds and

The male midwife toad (*Alytes obstetricans*) wraps long strings of eggs around his hind feet and protects them until they hatch. (Photo by Nuridsany et Pérennou/Photo Researchers, Inc. Reproduced by permission.)

streams. During the day, these frogs hide beneath cover objects such as rocks and logs.

Behavior

Frogs in this group are primarily active only at night, when they forage for insects near moist areas. During the day, they seek shelter from the sun under cover objects or in tunnels they excavate. Those that dig their own burrows do so digging head-first with their forearms through sandy soils, and they sometimes push their heads against the top of the tunnel to pack the soil tightly. The tunnel systems may be quite complex and elaborate.

Feeding ecology and diet

Larval and adult insects make up the majority of the diets of discoglossids. Stomach contents may include flies, grasshoppers, moth larvae, weevils and other beetles, ants, isopods, spiders, and snails. These frogs actively forage for prey at night.

Reproductive biology

During the breeding season, males call from in or near waters. In some species, the females also call in response to the males, but at a lower frequency. In all species, males mate with females by grabbing them around the waist (inguinal amplexus). Frogs in the genus *Alytes* are known for the parental care given by the males. During reproduction, males fertilize up to 100 egg strings, which they wrap around their legs. The males carry and protect these eggs until they hatch, at which time they return to the water to allow the tadpoles to swim free. In *Discoglossus*, females deposit up to 1,000 eggs attached to aquatic vegetation or on the bottom of the stream. Some tadpoles overwinter and metamorphose the next spring or summer. All tadpoles have keratinized mouth parts and two small fused spiracular tubes with a single anteromedial spiracle.

Conservation status

One species, *Discoglossus nigriventer*, is listed by the IUCN as Extinct, one (*Alytes muletensis*) as Critically Endangered, and two (*Discoglossus montalenti* and *Alytes dickhilleni*) as Vulnerable. Other populations seem to be robust and viable, but may eventually be affected by habitat destruction. Seven species are listed in Appendix II of the Convention on the Conservation of European Wildlife and Natural Habitats.

Significance to humans

None known.

1. Iberian midwife toad (*Alytes cisternasii*); 2. Midwife toad (*Alytes obstetricans*); 3. Tyrrhenian painted frog (*Discoglossus sardus*); 4. Painted frog (*Discoglossus pictus*). (Illustration by Patricia Ferrer)

Species accounts

Iberian midwife toad

Alytes cisternasii

TAXONOMY

Alytes cisternasii Boscá, 1879, Badamoz, Mérida, Spain. No subspecies recognized.

OTHER COMMON NAMES

French: Alyte de cisternas; German: Iberische Geburtshelferkröte; Spanish: Sapo-partero ibérico; Portuguese: Sapo-parteiro ibérico, sapo partero de cisternas.

PHYSICAL CHARACTERISTICS

The average body length ranges between 1.4 in (36 mm) for males and 1.7 in (42 mm) for females. The frogs are small and stocky, and have large eyes with a vertical pupil. Males and females also vary in several relative morphometric variables, including the diameter of external ear, the width of the head, and the length of portions of the hind limb. These frogs display several characteristics commonly found in toads, including: small parotid glands behind the ear, warty skin, and two broad bumps on each hand. Their color pattern can vary, but consists mostly of a brown background with dark spots and red warts. Rows of red warts may be present on the upper eyelids, extending down the side of the body from the ear to the hind limb.

DISTRIBUTION

Endemic to the Iberian Peninsula; inhabits the southwestern and central parts of this region, including southern Portugal and much of Spain.

HABITAT

The Iberian midwife toad is usually found in dry habitats with sandy soils. In the southern part of its distribution, it inhabits wetter environments, particularly near small, temporary streams.

BEHAVIOR

These frogs are nocturnal, spending most daylight hours hiding in burrows they dig in sandy soils.

FEEDING ECOLOGY AND DIET

Iberian midwife toads actively forage at night for small insects and crustaceans, including flies, grasshoppers, moth larvae, weevils and other beetles, and ants. To a lesser extent, they may also prey on isopods, spiders, and snails.

REPRODUCTIVE BIOLOGY

This species is particularly known for the care that males give to their offspring. Between September and March, males call for several hours during the night. Females respond, but with a weaker call. Mating begins when the male grabs the female around the waist. After some time, the female ejects an egg mass. The male then releases his lumbar grip, takes hold of the female around the chest, and inseminates the eggs. After 10–15 minutes, the male uses his legs to wrap the egg mass around his ankles. A male can mate again, and can carry up to four egg clutches around his legs (as many as 180 eggs). Females can breed up to four times per season.

The eggs nearly double in size as they mature, likely because of water absorption. Males prevent drying of the eggs by resting in moist areas or wading in waters. When the eggs are ready to hatch, the male enters shallow water to deposit the larvae. Larvae metamorphose after 110–140 days of development; sexual maturity is not reached for at least two years.

CONSERVATION STATUS

Not threatened. This species is fairly common throughout most of its range. However, in a few areas in Spain, populations are in decline owing to the destruction of forest habitats.

SIGNIFICANCE TO HUMANS

None known. ◆

Midwife toad

Alytes obstetricans

TAXONOMY

Bufo obstetricans Laurenti, 1768, France. Three subspecies are recognized.

OTHER COMMON NAMES

French: Crapaud accroucheur, alyte accoucheur; German: Geburtshelferkröte, Glockenfrosch; Spanish: Sapo partero común.

PHYSICAL CHARACTERISTICS

Midwife toads are generally small and squat, with large heads. The average body size for both males and females is 2.17 in (55 mm). As in other midwife toads, the large eyes have a vertical, slit-shaped pupil. Small parotoid glands are present behind

☐ *Alytes obstetricans*
☐ *Alytes cisternasii*

the eardrum. Warts are present on the skin, which may be spotted with black, brown, olive, or green. A row of large, reddish warts extends from behind the eardrum to the hind limb. The palm of the forelimb has three bumps (metacarpal tubercles). The underside is off-white, and the throat and the chest may be spotted with gray.

DISTRIBUTION
Alytes obstetricans occurs in eight European countries: Portugal, Spain, France, Belgium, the Netherlands, Luxembourg, Germany, and Switzerland. The distribution of the species clearly follows the habitat differences between the mountainous regions of Central Europe, where it is present, and the plains extending to the North Sea, where the species is absent. For this reason, it is absent from the coast of Belgium and most of the Netherlands.

HABITAT
Midwife toads prefer permanent bodies of water, such as ponds and streams, because larvae often overwinter. The type of water may vary by region, but these toads generally avoid fast-moving waters. On land, they are generally found hiding in moist, warm, sandy or loose soils with little vegetation; but they are also found under gravel, stone walls, embankments with small stones, and large stone slabs.

BEHAVIOR
As with other midwife toads, their most interesting behaviors have to do with reproduction and parental care by males. These frogs are nocturnal, and spend most of the daylight hours hiding in burrows.

FEEDING ECOLOGY AND DIET
Midwife toads have a diet similar to that of other toads. They actively forage at night, mostly for insects, arthropods, isopods, and snails.

REPRODUCTIVE BIOLOGY
As with other toads in the genus, males care for the eggs by attaching them to their legs during amplexus and carrying them until they eggs hatch. The mating season varies with climate, males carry eggs from beginning of February in some areas, but not until late March in others.

Generally, males call only at night, but may call from their hiding places during the day. Amplexus and fertilization are similar to that of other midwife toads, but males have been reported to stimulate ovulation by scratching the female with the toes of their hind limbs.

Males carry up to 150 eggs (from three different females) until they hatch, about three to six weeks after breeding. Larvae overwinter and metamorphose the following year.

CONSERVATION STATUS
Although not threatened according to the IUCN, this species is in decline in several areas, including the northernmost portions of its distribution, where several populations have disappeared entirely. The primary cause of population declines seems to be habitat destruction and alteration.

SIGNIFICANCE TO HUMANS
None known. ◆

Painted frog
Discoglossus pictus

TAXONOMY
Discoglossus pictus Otth, 1837, Sicily. Three subspecies are recognized.

OTHER COMMON NAMES
French: Discoglosse peint; German: Gemalter Scheibenzüatngler; Spanish: Sapillo pintojo.

PHYSICAL CHARACTERISTICS
Painted frogs have stout bodies with flat, wide heads, and their pupils are shaped like upside-down teardrops. Average body size is 2.76–3.15 in (70–80 mm). They are quite colorful (as the common name implies), and vary from having large dark spots with bright edges, two dark brown bands, or a band along the back and two along the sides. They may also have longitudinal glands on the back.

DISTRIBUTION
Mediterranean Africa in Tunisia, northern Algeria, and Morocco; Sicily (Italy), Malta, and Gozo (Ghawdax); one subspecies introduced to France and Spain.

HABITAT
Painted frogs seem to prefer human-made habitats, including orchards and vineyards, stone-sided cisterns, irrigation pipes and canals in cultivated areas, campsites, and cattle tracks filled with water. They can also be found near small brooks, as well as in holes they dig under stones. One subspecies lives and breeds in brackish water.

BEHAVIOR
Most of the knowledge of this species has been acquired from studies of the introduced populations and regards their reproductive behavior. They are primarily nocturnal, and excavate small, flat burrows under stones to use as refugia.

Discoglossus pictus
Discoglossus sardus

FEEDING ECOLOGY AND DIET
Painted frogs actively forage at night for insects and other invertebrates.

REPRODUCTIVE BIOLOGY
Mating season occurs from January to early November. Males clasp females in the lumbar region, and after 35 seconds to two hours, depending on the subspecies, females will lay up to 50 eggs. Females mate with several males consecutively, laying up to 1,000 eggs in one night. The eggs have no common jelly coating, and form a loose mass on the water surface, or sink to the bottom. Eggs usually hatch within six days of mating, and in one to three months, tadpoles metamorphose. Adulthood is reached after one year.

CONSERVATION STATUS
Not threatened. However, in Europe, populations that live in or near agricultural habitats appear to be in decline because of the loss of farmlands. Those living near rivers and seasonal ponds seem to be less threatened. Populations in France are protected, and several in northern Africa are endangered.

SIGNIFICANCE TO HUMANS
None known. ◆

Tyrrhenian painted frog
Discoglossus sardus

TAXONOMY
Discoglossus sardus Tschudi, 1837, Sardinia. No subspecies recognized.

OTHER COMMON NAMES
English: Sardinia painted frog; French: Discoglosse sarde; German: Sardischer Scheibenzüatngler.

PHYSICAL CHARACTERISTICS
This is a small, squat frog that can be dark brown, dark gray, reddish, or red-brown, with or without dark brown spots. Average body size is 2.76–3.15 in (70–80 mm).

DISTRIBUTION
These frogs are restricted to Sardinia, Corsica, and several small islands of the Tyrrhenian Sea. They are also found on the Italian mainland on the small peninsula Monte Argentario (Tuscany).

HABITAT
These frogs occur in a variety of habitats, including open, windy, desolate coastlines and coniferous forest streams. They prefer stagnant water or slow-running brooks, but have also been found in slightly brackish waters.

BEHAVIOR
What is known seems to be similar to that of other frogs in the genus.

FEEDING ECOLOGY AND DIET
Insects and other invertebrates make up most of the diet of these frogs.

REPRODUCTIVE BIOLOGY
Reproduction seems to be similar to that of other frogs in the genus. Females lay eggs in small clumps or singularly on the bottom of a stream or on or near aquatic vegetation.

CONSERVATION STATUS
Not threatened. However, because populations are so small, the species may be at greater risk of decline owing to habitat destruction and fragmentation.

SIGNIFICANCE TO HUMANS
None known. ◆

Resources

Books

Crespo, E. G. *Contribuição para o conhecimento da Biologia dos* Alytes ibéricos, Alytes obstetricans *Boscai Latase, 1879 e* Alytes cisternasii *Boscá, 1879 (Amphibia-Salientia):—A Problemática da Especiação de* Alytes cisternasii. Ph.D. Dissertation, Universidade de Lisboa, 1979.

Duellman, William. E., and Trueb, Linda. *Biology of Amphibians.* Baltimore: Johns Hopkins University Press, 1994.

Garcia Paris, Mario. *Los Anfibios de España.* Madrid: Ministerio de Agricultura, Pesca y Alimentacion, 1985.

Gasc, Jean-Pierre, et al., eds. *Atlas of Amphibians and Reptiles in Europe.* Paris: Societas Europea Herpetologica and Muséum National d'Histoire, 1997.

Herrmann, Hans-Joachim. *Terrarien Atlas Band 1.* Melle, Germany: Mergus Verlag GmbH, 2001.

Zug, George R., Laurie J. Vitt, and Janalee P. Caldwell. *Herpetology.* 2nd edition. San Diego: Academic Press, 2001.

Other

Amphibian Species of the World: An Online Reference. 1 September 2000. (12 April 2002) <http://research.amnh.org/Herpetology/amphibia>

AmphibiaWeb: Information on Amphibian Biology and Conservation. (12 April 2002) <http://elib.cs.berkeley.edu>

Anne M. Maglia, PhD

Mesoamerican burrowing toads
(Rhinophrynidae)

Class Amphibia

Order Anura

Family Rhinophrynidae

Thumbnail description
Moderate-sized burrowing frog with rotund body; triangular head with truncate snout and tiny eyes; exceptionally short, powerful limbs; and loose, pustulose skin

Size
1.8–2.6 in (45–65 mm)

Number of genera, species
1 genus; 1 species

Habitat
Seasonally dry forests and savannas in lowland tropics and subtropics

Conservation status
Not threatened

Distribution
Extreme southern Texas and lowlands of Mexico and Central America to Guatemala on the Atlantic slope and from Guerrero, Mexico, to Costa Rica on the Pacific slope

Evolution and systematics

Rhinophrynus dorsalis (the burrowing toad, Mexican burrowing toad, or Mesoamerican burrowing toad) is the only living representative of the anuran family Rhinophrynidae. Despite its common name, the Mesoamerican burrowing toad is not a toad. It is a curious, almost absurd-looking frog that is related most closely to another bizarre group of frogs—the pipids, of which the flat-headed *Pipa pipa* (Surinam toad) and *Xenopus laevis* (African clawed frog) are the most familiar representatives in laboratories and the pet trade. No subfamilies are recognized.

Externally, the Mesoamerican burrowing toad resembles several other burrowing frogs (e.g., the microhylid *Breviceps*, the hemisotid *Hemisus*, and the myobatrachid *Myobatrachus*), but several skeletal features of the adults and characteristics of the larvae indicate that the Mesoamerican burrowing toad is allied with pipid anurans (e.g., the living *Xenopus, Silurana, Pseudhymenochirus, Hymenochirus,* and *Pipa* and many fossil taxa). The tadpoles of both pipids and the Mesoamerican burrowing toad have broad, flat heads with wide, slitlike mouths that lack keratinous mouthparts and bear marginal barbels; there is a pair of spiracles (instead of only one) located on the

underside of the tadpole body rather than on its side, as in most other anurans.

As a group, pipoid frogs (i.e., Rhinophrynidae and Pipidae) have a rather extraordinary fossil record, in terms of both numbers of fossil representatives and their ages. The Mesoamerican burrowing toad is no exception. It is known from the Upper Pleistocene of Mexico in deposits less than one million years old. A related, extinct species, *Rhinophrynus canadensis*, was described from the Lower Oligocene (ca. 32 million years ago) of Saskatchewan, Canada. Older fossils (ca. 40–50 million years old) include *Eorhinophrynus septentrionalis* from the Middle Eocene and the slightly younger *Chelomophrynus bayi*—both from Wyoming in the United States.

The taxonomy of this species is *Rhinophrynus dorsalis* Duméril and Bibron, 1841, Vera Crúz, Mexico.

Physical characteristics

Typical of burrowing frogs, the Mesoamerican burrowing toad has a short head with tiny eyes and a globular body with

Mesoamerican burrowing toad (*Rhinophrynus dorsalis*). (Illustration by Barbara Duperron)

Mesoamerican burrowing toad (*Rhinophrynus dorsalis*) on the forest floor in Costa Rica. (Photo by Animals Animals ©Michael Fogden. Reproduced by permission.)

loose skin that obscures the short, stout limbs, leaving only the immense hands and feet visible when the frog is at rest. The pectoral girdle and forelimbs are located far forward so that the shoulder blades actually wrap around the back end of the skull. Consequently, the head of the Mesoamerican burrowing toad seems to be even shorter than it actually is, and there is no indication of a neck and no room for a tympanum (external ear). The snout of the Mesoamerican burrowing toad is unique. The nostrils are located much closer to the eyes than to the end of the long, narrow snout, which is truncate at the end. The skin covering the snout firmly adheres to the skull beneath and bears cushionlike pads. Each epidermal skin cell in the snout region has a minute keratin spicule, which is not visible to the naked eye; the spicules are pointed on the top of the snout but rounded on the bottom. The lips are thick, and the lower lip is glandular.

The Mesoamerican burrowing toad lacks teeth and has an unusual triangular tongue. Unlike other anurans, in which the tongue is rolled over the edge of the lower jaw or flipped out of the mouth, in the Mesoamerican burrowing toad the tongue protrudes forward through the buccal groove and out the end of the snout for a short distance. Because this frog feeds underground on termites and ants, it is thought that the tongue is a special adaptation for feeding in confined quarters. Thus, having located and broken through to a subterranean ant or termite tunnel, the frog can place the tip of its snout against the hole and simply extend its sticky tongue each time it detects a passing insect and then retract its tongue and the prey into its mouth.

Tadpole of *Rhinophrynus dorsalis*. (Illustration by Barbara Duperron)

Despite its stocky form, this species is an accomplished burrower. The body is highly flexible, and the stout hind limbs are equipped with large feet with short, thick digits and a pair of digging "spades." While pivoting its body in a circle around its forelimbs, the frog shifts soil away from itself by digging with its hind feet and inflating and deflating its body; it soon disappears, rear end first, into soil, which then fills in over the head as the frog disappears down the shaft it is excavating. Typically, the Mesoamerican burrowing toad is dark brown or nearly black dorsally, whereas the venter varies from dark brown to gray and usually has no pattern. There is a bright stripe on the middle of the back of the frog from its head to the vent; the vertebral stripe is flanked by scattered blotches or spots that vary from bright yellow to yellow-orange or reddish orange.

Distribution

Although extinct rhinophrynids occurred in North America, the Mesoamerican burrowing toad is restricted to the southern tip of Texas in the United States and the lowlands of southern Mexico and Central America (Honduras, Guatemala, Belize, El Salvador, Nicaragua, and Costa Rica).

Habitat

The Mesoamerican burrowing toad is found in savanna habitats and seasonally dry forests.

Behavior

Because this species is fossorial (adapted to digging), it is seen above ground only when it emerges to breed during the rainy season. At this time, adults usually are found in flooded pastures, roadside ditches, pools in savannas, and other ephemeral bodies of water. The frogs spend the dry season underground. Thus, virtually nothing is known about their nonbreeding activity patterns and their interactions with one another and other species.

Feeding ecology and diet

No one has reported observing the Mesoamerican burrowing toad feed. Presumably they do so underground and specialize in termites and ants that use subterranean burrows, because these insects have been recovered from stomach contents of the frogs.

Reproductive biology

Adult members of the species emerge from their subterranean burrows at the beginning of the rainy season to breed. Males call from temporary bodies of water at the water surface. When they call, their internal vocal sacs become enormously distended; with each abrupt inflation of the vocal sacs, the frog is rotated and pushed backward in the water. The loud call has been described as an "uooooooooo" that lasts about 1.4 seconds and is repeated 15–20 times a minute. Choruses of these frogs can be heard over great distances. A fe-male Mesoamerican burrowing toad initiates contact with a breeding male by nudging him in the throat or the chest with her snout. The male then grasps the female from above in the inguinal region and fertilizes the single egg or small groups of eggs that she deposits in the water. Each female produces several thousand eggs. Because she expels only a few at a time, it is possible that each female mates with many males during the breeding season. The fertilized eggs sink to the bottom of the temporary pond and hatch into tadpoles in a few days. The developing tadpoles filter-feed on algae and congregate into swimming groups. These groups may be composed of as few as 50 individuals swimming in a coordinated "ball" about 3.9 in (10 cm) in diameter to several thousand tadpoles in a congregation more than 3.3 ft (1 m) in diameter.

Conservation status

The Mesoamerican burrowing toad is not threatened.

Significance to humans

None known.

Resources

Books

Lee, Julian C. *The Amphibians and Reptiles of the Yucatán Peninsula.* Ithaca: Comstock Publishing Associates, Cornell University Press, 1996.

Periodicals

Henrici, Amy C. "*Chelomophrynus bayi* (Amphibia, Anura, Rhinophrynidae), a New Genus and Species from the Middle Eocene of Wyoming: Ontogeny and Relationships." *Annals of the Carnegie Museum* 60 (1991): 97–144.

Trueb, Linda, and David Cannatella. "The Cranial Osteology and Hyolaryngeal Apparatus of *Rhinophrynus dorsalis* (Anura: Rhinophrynidae) with Comparisons to Recent Pipid Frogs." *Journal of Morphology* 171 (1982): 11–40.

Trueb, Linda, and Carl Gans. "Feeding Specializations of the Mexican Burrowing Toad, *Rhinophrynus dorsalis* (Anura: Rhinophrynidae)." *Journal of Zoology, London* 199 (1983): 189–208.

Linda Trueb, PhD

Clawed frogs and Surinam toads
(Pipidae)

Class Amphibia

Order Anura

Family Pipidae

Thumbnail description
Fully aquatic anurans distinguished by having a dorsoventrally depressed body; holding their limbs in a laterally sprawled position; having fully webbed feet and tiny, dorsally placed eyes; lacking a tongue; and retaining the lateral-line system as adults

Size
Small-to-medium sized anurans ranging from 0.8 to 1.2 in (20 to 30 mm) long up to 4.1–6.7 in (104–170 mm) long

Number of genera, species
5 genera; 30 species

Habitat
All pipids are aquatic and leave the water only under duress; in Africa and South America, they are found in almost every kind of water body, including lakes, rivers, swamps, forest ponds, and varieties of human-made bodies of water

Conservation status
Vulnerable: 1 species

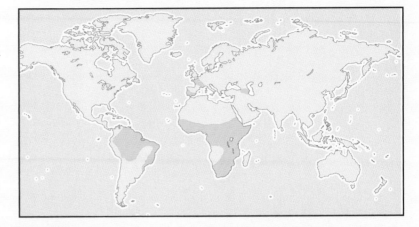

Distribution
Africa south of the Sahara in the Old World; in extreme lower part of Central America, the Amazon Basin of South America, and coastal areas of Venezuela, Guyana, French Guiana, Suriname, and Brazil in the New World

Evolution and systematics

The family Pipidae is distinguished by a few soft anatomical characters, their larvae, and many skeletal features that involve the structure of the skull and the vertebral column. Living representatives are placed into two subfamilies—Xenopodinae, comprising *Xenopus* and *Silurana* in Africa, and Pipinae, comprising *Pipa* in the New World and *Hymenochirus* and *Pseudhymenochirus* in Africa. Pipids are indisputably basal but highly derived anurans; their relationships to other archaeobatrachian frogs are controversial.

The family has an extensive fossil record that spans about 90 million years and two continents—Africa and South America. The most ancient fossil is *Pachybatrachus taqueti*, which lived 84–90 million years ago (Upper Cretaceous) in what is now the Republic of Niger. *Pachybatrachus* is related closely to the living *Hymenochirus* and *Pseudhymenochirus*. A Cretaceous pipid 71–84 million years old, *Saltenia ibanezi*, lived in southern South America (Argentina), and a Paleocene pipid that is 60 million years old, "*Xenopus*" *romeri*, is known from Brazil. (The generic name *Xenopus* is placed inside quotation marks because scientists are not certain that it is a member of the living genus of that name.) There are two slightly younger (34–55 million years old) Eocene fossil pipids from Argentina—*Shelania pascuali* and *S. laurenti*. A contemporaneous Eocene fossil pipid, *Eoxenopoides reuningi*, is known from South Africa.

Two species of *Xenopus*, *X. hasaunas* and *X. arabiensis*, were described from the Lower Oligocene of Libya (30–34 million years ago) and Late Oligocene (26–30 million years ago) of the Republic of Yemen, respectively. The youngest pipid fossils are *Xenopus stromeri*, from the Lower Miocene (16–23 million years ago) of South Africa, and, from the Miocene of Morocco, *Silurana tropicalis*—a fossil that, if correctly identified, is represented by living frogs in central and west Africa today. It seems reasonable to speculate that Pipidae originated on the southern, Gondwanan landmass and was well established before South America completely separated from Africa about 80 million years ago, in the Late Cretaceous. From the Early Triassic through the Jurassic, the southern parts of the incipient continents experienced warm-temperate climatic conditions, not unlike those of eastern temperate Africa today, and during the Cretaceous, tropical conditions prevailed in the northern parts of both Africa and South America, where living pipids are found today.

Physical characteristics

Pipid frogs are medium-sized to large anurans with extraordinarily depressed bodies and flat to wedge-shaped heads with small, dorsally placed eyes. All lack tongues and tympana. The adults retain lateral-line organs, visible on the head

Surinam toad (*Pipa pipa*) uses its highly specialized fingers for underwater feeding. (Illustration by Patricia Ferrer)

and body as a series of "stitches." Usually the forelimbs are small; in all pipids except *Hymenochirus* and *Pseudhymenochirus*, the slender fingers are not webbed. The hind limbs are robust and the feet fully webbed. In all pipids except *Pipa pipa* and *P. snethlageae*, the inner three toes bear keratinized "claws." *Xenopus* species typically have smooth skin, whereas *Pipa* and the hymenochirines have tuberculate skin. The anurans usually are tan to olive brown to gray, with darker spots and mottling dorsally and paler coloring ventrally with darker mottling.

Like other anurans, pipids communicate acoustically, but under water rather than in the air. They lack vocal cords and vocal sacs and have a highly modified laryngeal apparatus to produce the typical "clicking" call. The small, cup-shaped arytenoid cartilages inside the larynges of other anurans are represented by a pair of large, cartilaginous discs, the medial surfaces of which are tightly opposed when the frog is not calling. Contraction of the laryngeal muscles separates the discs and produces the "click"—a sound that is thought to result from implosion of air rushing into the cleft that abruptly opens between the discs. Although a tympanum is absent, there is a large, circular cartilaginous disc located beneath the skin on the side of the head. This is part of the stapes (middle ear bone) that transmits sound vibrations received through the water to the inner ear of the frog.

Distribution

Pipids occur in lowland, tropical South America and sub-Saharan Africa. In Africa they are found from sea level to elevations of 9,000 ft (2,780 m). Introduced populations of

Xenopus laevis occur in the United Kingdom, Europe, South America, and the United States.

Habitat

Pipids are totally aquatic, occupying temporary and permanent bodies of water, including swamps, reservoirs, and slow-moving streams and rivers.

Behavior

Little is known about diel (daily/nightly) activity or inter- and intraspecific territoriality of pipids, because they are exceptionally difficult to observe in nature. Apparently, they remain in a body of water as long as it is suitable. During droughts they burrow into the mud at the bottom of the pond or swamp and estivate up to several months, and during rains they undertake short overland excursions at night, moving from one body of water to another.

Feeding ecology and diet

Because pipids lack a tongue, prey capture is quite different from that in other anurans and is best known in *Xenopus laevis*. Like all other pipids, *Xenopus* species eat anything that they can catch, from aquatic invertebrates to fish, birds, and mammals, as well as their own larvae. Having teeth, *Xenopus* is able to grip its prey. Observations suggest that the frogs bite their prey. While holding it, they use their powerful hind limbs to claw at the prey and shred it and their forelimbs to shove the prey into the mouth. Adult *Xenopus* species have

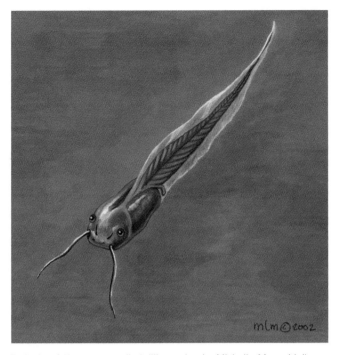

Tadpole of *Xenopus muelleri*. (Illustration by Michelle Meneghini)

Clawed African frogs (*Xenopus laevis*) can spend up to 10 months in an inactive state, buried in the mud, when hot, dry summers deplete their water homes. (Photo by E.R. Degginger. Bruce Coleman Inc. Reproduced by permission.)

been reported to attack prey in groups and collectively tear the body of the prey into fragments that can be ingested. Edentate pipids, those with no teeth, (e.g., *Pipa pipa*) lunge at prey and suck it into the mouth using their forelimbs as *Xenopus* does. Pipids detect prey by chemosensory cues in the water, vision, and vibrations detected by the lateral-line system.

Reproductive biology

Breeding in pipids seems to coincide with the onset of rains and, in this sense, is opportunistic; under appropriate conditions, however, the frogs seem to be capable of breeding

throughout the year. Both sexes vocalize and have repertoires of three to six types of clicking calls that are emitted under water. The advertisement calls of different species are distinguished by their temporal frequencies and the dominant frequency of the call. For example, an isolated male *Xenopus borealis* advertises his presence by two to four single clicks per second. A male of the same species, approaching a female, emits a call of 10 clicks per second, and frogs of both sexes emit a release call consisting of 20 clicks per second.

Tadpole of *Pipa myersi*. (Illustration by Michelle Meneghini)

A Surinam toad (*Pipa pipa*) incubates eggs on its back. The eggs, after fertilization, sink into the spongy skin of the female and remain there 12–20 weeks, until they hatch. (Photo by Tom McHugh/Steinhart Aquarium/Photo Researchers, Inc. Reproduced by permission.)

Old World pipids (*Xenopus, Silurana, Pseudhymenochirus,* and *Hymenochirus*) deposit their eggs in water. Free-swimming tadpoles hatch from the eggs and undergo their development in the water. All larvae lack keratinous mouthparts; those of *Xenopus* and *Silurana* bear sensory barbels at the periphery of the mouth. The larvae of *Xenopus* and *Silurana* are exclusively filter feeders, whereas those of *Pseudhymenochirus* and *Hymenochirus* are predaceous feeders on aquatic insect larvae and ostracods. Among New World pipids, the eggs are deposited on the backs of the females in *Pipa*, although this is not verified in one species, *Pipa myersi*. Two species have free-swimming larvae that lack barbels around the mouth (*P. myersi* and *P. parva*). In the remaining five species, the tadpoles undergo their development on the backs of the females and emerge from her back as miniatures of the adults.

Conservation status

Xenopus gilli (Gill's plantanna) is listed as Vulnerable by the IUCN, and as Endangered according to the South African Red Data Book. Many species in the lowland rainforests of sub-Saharan Africa, however, are threatened by habitat destruction.

Significance to humans

As discussed in the accounts that follow, pipids are tremendously important to humans, because of their use as biomedical experimental animals and their popularity in the pet trade. Medical researchers are investigating several substances found in the skin of some pipids. Among them are magainins (antimicrobial peptides that inhibit the growth of numerous bacteria and fungi) and other peptides, such as xenopsin and caerulein, that cause predators to vomit if they eat these frogs. Pipids are of particular interest to systematic biologists because of their widespread distribution and extensive fossil record.

Species accounts

Common plantanna
Xenopus laevis

SUBFAMILY
Xenopodinae (= Dacylethrinae)

TAXONOMY
Bufo laevis Daudin, 1802, type locality not designated. Five subspecies are recognized.

OTHER COMMON NAMES
English: Plantanna, African clawed frog, clawed toad, clawed frog, upland clawed frog; German: Glatter Krallenfrosch; Spanish: Rana de garras africana.

PHYSICAL CHARACTERISTICS
Xenopus laevis varies in size over the range of the species; males, however, always are smaller than females within a given population. Female frogs range from 2.2 to 5.8 in (57 to 147 mm) and males from 1.8 to 3.8 in (45.6 to 97.5 mm) in snout-vent

Xenopus laevis

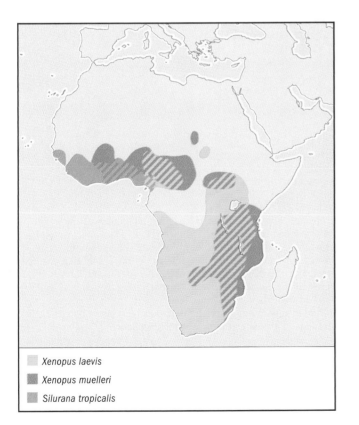

Xenopus laevis
Xenopus muelleri
Silurana tropicalis

length. The body and head are depressed, and the small, round eyes are located on top of the head. The skin is smooth. The hind limbs are long and robust. The three inner toes of the large, fully webbed feet bear small, black claws. The inner metatarsal tubercle is an elevated ridge. The subocular tentacle is minute, less than half the diameter of the eye, and there are 23–31 lateral-line bars between the eye and the vent. Although the dorsal coloration varies, it usually is dark—gray to greenish brown—and marked with darker blotches. The venter is pale and may bear irregular spots. The toe webbing usually is gray and occasionally is tinged with yellow.

DISTRIBUTION
Members of this group occupy mainly savannas of the Republic of South Africa north to Kenya, Uganda, and the Democratic Republic of Congo westward to Cameroon. These frogs are not found in the Congo Basin or the hotter lowlands of eastern Africa. Owing to the pet trade and common use of *Xenopus* as a laboratory animal, the frog (most likely *X. laevis*) has been introduced in Europe, the United States, and South America. Introduced populations thrive in the United Kingdom (Isle of Wight, South Wales, and southeast and southwest England), mainland Europe (Germany and the Netherlands), Chile (specific locality unknown), Ascension Island, and the United States (Tucson, Arizona; and Los Angeles, Orange, Riverside, and San Diego counties in California).

HABITAT
An extraordinary trait of this species is its apparent environmental tolerances and lack of discrimination with regard to its habitat, as long as there is a body of water for the frog to occupy. Doubtless this accounts for the success of the species in the laboratory and as invasive populations around the world. The species occurs in any kind of body of water, including rivers, lakes, reservoirs, swamps, flooded pits, ditches, and wells, and at elevations from sea level to about 9,000 ft (2,780 m) in the Drakensberg Mountains of South Africa. Water quality seemingly is not an issue. The species is common in stagnant, still waters and sluggish streams as well as fast-flowing waters. Similarly, it can be found in water occluded with organic detritus or clear water and in acidic or alkaline waters.

Unlike nearly all other anurans, the common plantanna can tolerate saline waters and has been known to survive in 25% seawater indefinitely and in 40% seawater for a few days. Temperature, as it affects larvae, may be a limiting factor for the species, although they display a remarkable tolerance; in laboratory tests, the critical lethal minimum and maximum temperatures for larvae were shown to be 50°F (10°C) and 95°F (35°C), respectively. Introduced populations survive in ice-covered ponds for several months a year as well as in ponds that are subject to the extreme summer temperatures of southern Arizona. In the introduced southern California populations, when temperatures reach 86°F (30°C), the frogs burrow into the pond bottom to depths of about 12–16 in (30–40 cm), where the temperature is relatively stable at 68°F (20°C).

BEHAVIOR
Under good conditions, members of this group do not leave the water, although they may undertake short nocturnal excursions. Prolonged drought, however, forces them to estivate.

They have been observed to estivate in the laboratory for up to eight months and doubtless are capable of the same in nature, where they burrow backward into drying mud and occupy a vertically oriented chamber. Physiological tests reveal that estivating *X. laevis* have a number of adaptations to survive desiccation and starvation. Rather than excreting ammonia, which is highly toxic, they excrete the less-toxic urea that is accumulated in the blood, liver, and muscle tissue. The frogs can reduce their oxygen consumption by 30%. Frogs maintained in water can survive a year without food and incur as much as a 35–45% loss in body weight. They survive by using stored carbohydrates and lipids for energy for the first four to six months, after which time, the frogs switch to protein catabolism (especially breakdown of body muscle) for energy. If they survive an extended period of drought, they are likely to migrate over land en masse with the onset of torrential rains.

FEEDING ECOLOGY AND DIET
The food preferences of this species are as wide ranging as are the aquatic habitats in which the species is found. Typically, the hatching of tadpoles coincides with algal blooms, and the larvae, which have a highly specialized filter-feeding mechanism, are able to extract algae and the finest suspended organic matter from the water. Recently metamorphosed froglets seem to specialize in small crustaceans and aquatic insect larvae. Adults eat insects and prey on or scavenge other vertebrates—anurans, fish, birds, and small mammals. They have been seen to leap out of the water to capture winged insects. Although the species typically is not found in bodies of water with high natural populations of fishes, it is known to prey occasionally on fish in constrained situations (e.g., isolated pools and hatcheries). Adult frogs feed on other species of anurans and are cannibalistic with respect to their own larvae and young. They will eat small birds and rodents that fall into the water. There is no evidence that the species feeds on land, and in this regard they are quite distinct from most other anurans.

REPRODUCTIVE BIOLOGY
Under laboratory conditions, the common plantanna will breed throughout the year and attains sexual maturity in eight months (i.e., six months after metamorphosis). In nature, when these frogs are exposed to seasonal differences in temperature (e.g., South Africa), development probably is slower. Mating coincides with the onset of heavy rainfall and a temperature of about 68°F (20°C) and thus varies throughout the range of the species. Oogenesis is determined by food supply, and prey abundance correlates positively with rainfall; thus, the onset of rainfall indirectly stimulates egg production. Heavy rains wash fresh sediments into the breeding sites and enrich the nutrients; this, in turn, triggers phytoplankton blooms that provide food for developing larvae.

Both male and female frogs call, and the calls vary in length, pattern, and frequency throughout the range of the species. In South African common plantanna, the male advertisement call is described as a long trill composed of alternating fast-pulsed (43–66 pulses per second) and slow-pulsed (24–42 pulses per second) elements, with the fast element lasting 0.18–0.6 seconds and the slow element lasting 0.34–0.9 seconds. Calling continues for several minutes, and frequencies up to 2.3 kHz are emphasized. When males clasp females in inguinal amplexus, they utter a soft amplectant call. Typically about 1,000 eggs are laid at a time and attached to aquatic vegetation or other underwater objects. The pale brown eggs are about 0.05 in (1.15 mm) in diameter.

CONSERVATION STATUS
Not threatened.

SIGNIFICANCE TO HUMANS
X. laevis is probably one of the most familiar frogs to humans, owing to its long use as a model system for biomedical research and its popularity in the pet trade. Until the 1940s these frogs were used for pregnancy tests; injection of a small amount of urine from a pregnant woman under the frog's skin causes the frog to lay eggs. Biomedical researchers discovered that the species is a convenient experimental organism. It is easy to maintain in aquariums, and it is robust and has a relatively short life cycle. Unlike most other amphibians, it can be induced to provide fertile embryos throughout the year, and the embryos and their cells are large, making them convenient subjects for experimental manipulation and molecular research. In the past, members of this species were used as a food source in Cameroon, Sierra Leone, Central African Republic, Uganda, Rwanda, and the Democratic Republic of the Congo. The indigenous people of these countries used baited wicker baskets or traps to capture the frogs or occasionally drained smaller bodies of water and collected the stranded frogs. ◆

Müller's plantanna
Xenopus muelleri

SUBFAMILY
Xenopodinae (= Dacylethrinae)

TAXONOMY
Dactylethra mülleri Peters, 1844, "Mozambique."

OTHER COMMON NAMES
English: Mueller's clawed frog, Müller's smooth clawed frog, tropical plantanna, northern tropical plantanna, northern plantanna.

PHYSICAL CHARACTERISTICS
Xenopus muelleri is moderately large; females are 2.6–3.5 in (65–90 mm) long, and males are 2.1–2.8 in (52–72 mm) long. The body and head are depressed, and the small, round eyes are located on top of the head; the skin is smooth. The hind

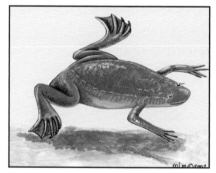

Xenopus muelleri

limbs are long and robust. The three inner toes of the large, fully webbed feet bear small, black claws. The inner metatarsal tubercle is a small, fingerlike projection. The subocular tentacle is long and conspicuous, equal to or more than half the diameter of the eye, and there are 22–27 lateral-line bars between the eye and the vent. Dorsally, the frog is gray and marked with darker blotches. The venter varies from pale gray to darkly marked and may be deep orange-yellow on the belly and legs. The toe webbing is orange-yellow.

DISTRIBUTION
Müller's plantanna has the widest range of any pipid species and occurs in two disjunct populations, usually below 2,625 ft (800 m). In eastern Africa it is found from southeastern Kenya through Tanzania, Zanzibar and the Mafia islands, Zambia,

Malawi, Botswana, Zimbabwe, and Mozambique and in eastern South Africa south to the area of Saint Lucia/Empangeni (ca. lat. 28°S). The western population extends from Burkina Faso and Ghana eastward to southern Sudan and the northeastern part of the Democratic Republic of the Congo.

HABITAT
This frog is found in hot, dry lowlands exclusive of rainforests. It frequents a wide variety of bodies of water and prefers permanent ones, such as reservoirs, ponds, and quiet regions of rivers during the dry season. Only rarely is it found in the same body of water with another species of *Xenopus*.

BEHAVIOR
Members of the western population of Müller's plantanna seem to pass the dry season in the bank zone of rivers, burrowed in the mud of permanent savanna ponds and occasionally beneath humid layers of leaf litter. The frogs migrate short distances between ponds on rainy nights.

FEEDING ECOLOGY AND DIET
Adults eat toad tadpoles and fish.

REPRODUCTIVE BIOLOGY
Little is known about the breeding behavior of Müller's plantanna. The species uses temporary bodies of water for breeding. Females produce small (0.04 in [1.0 mm] in diameter), dark gray eggs, which are attached singly on aquatic plants and rocks. There are many different reported advertisement calls, suggesting that they vary within and between populations. In South Africa the call of Müller's plantanna is a single note that lasts 0.2 seconds and consists of five to seven pulses at a rate of 26–32 pulses per second and an emphasized frequency of 774–1,142 Hz. The western population is reported to have a call consisting of a repetition of two-pulsed notes at a rate of four to eight pulses per second; the emphasized frequency is unknown.

CONSERVATION STATUS
Not threatened.

SIGNIFICANCE TO HUMANS
None known. ◆

Tropical clawed frog
Silurana tropicalis (= *Xenopus tropicalis*)

SUBFAMILY
Xenopodinae (= Dacylethrinae)

TAXONOMY
Silurana tropicalis Gray, 1864, "West Africa, Lagos," Nigeria.

OTHER COMMON NAMES
None known.

PHYSICAL CHARACTERISTICS
Silurana tropicalis is moderately small; females have an average length of 1.7 in (43 mm), and males have an average length of 1.4 in

Silurana tropicalis

(36.6 mm). The body and head are depressed, and the small, round eyes are located on top of the head; the dorsal skin is pustulose, especially on the heads of males. The hind limbs are short and robust. The inner three toes bear small, black claws, and the inner metatarsal tubercle is in the form of a claw. The subocular tentacle is minute, less than half the diameter of the eye, and there are 18–20 lateral-line bars between the eye and the vent. Dorsally, the frog is olive to brown, with fine gray and black marks that never coalesce into larger spots. The venter is white to gray with scattered black mottling.

DISTRIBUTION
The tropical clawed frog is found in western Africa from the Casamance River (Senegal) to the Cross River (Nigeria); the eastern limit of distribution is undetermined.

HABITAT
Confined to lowland tropical forest below 2,297 ft (700 m), the tropical clawed frog is found in still and running waters. Occasionally, it is found in savanna ponds near forests after heavy rainfalls.

BEHAVIOR
During heavy rain, tropical clawed frogs move between ponds at night. In the dry season the species is found along riverbanks under flat stones, in holes in the banks, or under roots by day; at night, it is found in small rock pools along the river. If isolated in pools, the frog burrows into the mud at the bottom of the pool.

FEEDING ECOLOGY AND DIET
Little is known about the diet of the tropical clawed frog. The species apparently is an opportunistic feeder and has been reported to eat arthropods and tadpoles—whether its own or those of other species is unknown.

REPRODUCTIVE BIOLOGY
Male frogs grasp females in the inguinal region and perform mating turnovers in the water before attaching eggs to aquatic plants in forest pools. Apparently, they breed throughout the year when it rains. Although the frogs prefer larger bodies of water, they will use small, water-filled holes in the forest; during the day, adult frogs hide nearby under dead trunks in shallow water. The advertisement call is described as a deep rattling trill that sounds like "roaroaroa" and lasts one to 10.5 seconds, with an emphasized frequency of about 1 kHz.

CONSERVATION STATUS
Not threatened.

SIGNIFICANCE TO HUMANS
None known. ◆

Surinam toad
Pipa pipa

SUBFAMILY
Pipinae

TAXONOMY
Rana pipa Linnaeus, 1758, "Surinami."

OTHER COMMON NAMES
French: Pipa américain; German: Wabenkröte.

PHYSICAL CHARACTERISTICS

Arguably, *Pipa pipa* is the most bizarre-looking frog known. The frog is large, with adult females being 4–7 in (105–171 mm) long and the slightly smaller males being 4–6 in (105–154 mm) long. It has an improbably flat, tri-angular head sur-

Pipa pipa

mounting an extremely depressed, wide body equipped with short, muscular hind limbs and immense, fully webbed feet. The tip of each finger is divided into four lobes, each of which is distally bifurcate. The eyes are minute and sometimes covered with skin. The nostrils are located at the tip of the snout and are valvular and slitlike. The lateral-line organs around the mouth are elaborated into spine-shaped dermal appendages; those at the corner of the mouth are associated with an enlarged, flat, bifurcate flap of skin. The dorsal and lateral surfaces of the body are tuberculate. Although coloration varies from light tan to dark brown with variable mottling, a distinctive T-shaped mark always is present on the venter; the top of the "T" traverses the chest between the forelimbs, and the leg of the "T" runs down the middle of the abdomen.

DISTRIBUTION

The Surinam toad is distributed widely in the Amazon Basin, occurring in eastern Venezuela, Guyana, Surinam, Brazil, Colombia, Ecuador, Peru, and Bolivia. It also is known from the Guianan region and Trinidad.

HABITAT

These odd frogs are found in slow-moving streams and rivers and lowland rainforest ponds and swamps, the bottoms of which are covered with organic detritus.

BEHAVIOR

Because the Surinam toad is extraordinarily difficult to observe in nature, most of its reported "natural history" is based on observations of captive individuals. The frogs usually are found lying immobile amidst the detritus on the bottom of ponds. On rainy nights they move from one pond to another.

FEEDING ECOLOGY AND DIET

In nature the Surinam toad has been observed to eat small fish and aquatic invertebrates. The frog lacks teeth. Typically, it lunges toward its prey, opens its mouth, and inflates its body, creating a vacuum into which the prey is sucked; the frog uses its forelimbs to push food into its mouth.

REPRODUCTIVE BIOLOGY

Preliminary amplexus lasts 24–30 hours, with a male clasping a female around the waist while bobbing and pumping his body;

Pipa pipa

during this time the female's back becomes tumescent, and the lips of her vent swell. Amplectant pairs of Surinam toads conduct a complex, repeated ritual of midwater acrobatic turnovers, with the female leading the male, who clasps her around her waist. The 11–14-second sequence includes ascending to the surface, doing a complete turnover without breaking the surface of the water, and descending to the bottom. When the female is upside down and ascending through the water, she expels three to five eggs that are fertilized and caught on the male's belly. As the rollover is completed, the eggs drop on the female's back and are implanted there by pressure from the male's clasp. The entire oviposition sequence takes about three hours. The skin of the female's back gradually swells up around the individual eggs (about 50); after 10 days only a small portion of the top of the embryo is visible, and the outer membrane of the egg covers it. Fully developed froglets begin to emerge from the female's back within three to four months. The call of the Surinam toad is a metallic clicking noise produced at the rate of four per second for periods of 10–20 seconds.

CONSERVATION STATUS

Not threatened.

SIGNIFICANCE TO HUMANS

The Surinam toad is a source of food for some indigenous Amazonian people. ◆

Resources

Books

Channing, Alan. *Amphibians of Central and Southern Africa.* Ithaca, NY: Comstock Publishing Associates, Cornell University Press, 2001.

Rödel, Mark-Oliver. *Herpetofauna of West Africa.* Vol. 1, *Amphibians of the West African Savanna.* Frankfurt: Edition Chimaira, 2000.

Tinsley, R. C., and H. R. Kobel, eds. *The Biology of Xenopus.* London: Clarendon Press, 1996.

Periodicals

Báez, Ana María, and Lourdes Analía Púgener. "A New Paleogene Pipid Frog from Northwestern Patagonia." *Journal of Vertebrate Paleontology* 18 (1998): 511–524.

Cannatella, David C., and Linda Trueb. "Evolution of Pipoid Frogs: Intergeneric Relationships of the Aquatic Frog Family Pipidae (Anura)." *Zoological Journal of the Linnean Society* 94 (1988): 1–38.

————. "Systematics, Morphology, and Phylogeny of Genus *Pipa* (Anura: Pipidae)." *Herpetologica* 42 (1986): 412–449.

Henrici, Amy C., and Ana María Báez. "First Occurrence of *Xenopus* (Anura: Pipidae) on the Arabian Peninsula: A New Species from the Late Oligocene of the Republic of Yemen." *Journal of Paleontology* 75 (2001): 870–882.

Rabb, George B., and Mary S. Rabb. "On the Mating and Egg-Laying Behavior of the Surinam Toad, *Pipa pipa.*" *Copeia* 4 (1960): 271–276.

————. "Additional Observations on Breeding Behavior of the Surinam Toad, *Pipa pipa.*" *Copeia* 4 (1963): 636–642.

Trueb, Linda, and David Massemin. "The Osteology and Relationships of *Pipa aspera* (Amphibia: Anura: Pipidae), with Notes on Its Natural History in French Guiana." *Amphibia-Reptilia* 22 (2001): 33–54.

Linda Trueb, PhD

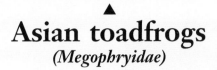

Asian toadfrogs

(Megophryidae)

Class Amphibia

Order Anura

Family Megophryidae

Thumbnail description
Small to large frogs that are exceptionally well camouflaged and often secretive

Size
0.59–5.51 in (15–140 mm)

Number of genera, species
11 genera; 107 species

Habitat
Forest, mountain streams

Conservation status
Not threatened

Distribution
Eastern Asia

Evolution and systematics

Asian toadfrogs are a group of frogs with diverse morphologic features, and there is not one character that easily defines them. For this reason, using anatomy to form an understanding of the natural groupings within the megophryids or their relationship to other frog families has been difficult. There are no fossil megophryids, and thus the age of this group and its ancient distribution cannot be confirmed. Like other organisms with similar distributions and limitations to dispersal, their current ranges are the result of geologic events that took place more than 30 million years ago. In the Late Oligocene, the sea levels were much lower than they are today, creating a continuous landmass from mainland Asia to the Indo-Australian archipelago, including a portion of the Philippine islands. The tropical rainforest climate at that time was similar to the environments where megophryids are found today, and many of the megophryid genera likely were established and distributed across much of this area. Subsequent fluctuations in sea levels and shifting tectonic plates eroded these land-bridge connections, isolating representatives of *Megophrys*, *Xenophrys*, and *Leptobrachium* on many islands, including Borneo, Sumatra, Java, and the Philippines. The radiation of *Scutiger* and *Oreolalax* is intimately tied to the

uplifting of the Tibet plateau that followed the collision of the Indian plate with Eurasia some 50 million years ago.

As of the year 2002, the family Megophryidae included 107 species divided among 11 genera and two subfamilies. The arrangement of tubercles on the hand and two distinct types of tadpoles easily distinguish the subfamilies. The subfamily Leptobrachiinae includes *Leptobrachella*, *Leptobrachium*, *Leptolalax*, *Oreolalax*, *Scutiger*, and *Vibrissaphora*. In this group a large tubercle is present at the base of the first finger. The tadpoles of all of these genera typically have a downward oriented mouth and a robust keratinized horny beak on both the upper and lower jaws. The subfamily Megophryinae includes *Atympanophrys*, *Brachytarsophrys*, *Megophrys*, *Ophryophryne*, and *Xenophrys*. The tubercle on the hand, at the base of the first finger, extends well onto the thumb. The tadpoles have a large umbelliform (funnel-like) mouth that is directed upward. The keratinized beak is reduced and present only on the margin of the lower jaw.

All Asian toadfrogs have eight vertebrae and intervertebral discs that are not fused to adjacent vertebrae at the time of metamorphosis. The sacral diapophyses are dilated, and the pectoral girdle is arciferal, with a long bony sternum. On the roof of the mouth the neopalatines are absent; to compensate,

Burmese spadefoot toad (*Xenophrys parva*) is a diminutive species and is both slow of movement and cryptic of color. (Photo by R. D. Bartlett. Reproduced by permission.)

a palatal process of the maxilla is elongated. Asian toadfrogs share a common ancestor with North American and European spadefoot toads (Pelobatidae) and parsley frogs (Pelodytidae). Some researchers have recognized these three families as the suborder Pelobatoidea. Asian toadfrogs can be distinguished from their sister groups by their paddle-shaped tongue and a hyoid that is simplified and elongated; the hyoid lacks any remnant of a cartilaginous connection to the back of the skull.

Physical characteristics

Asian toadfrogs come in just about every size and shape imaginable. The largest species, the broad-headed toad (*Brachytarsophrys*), attains a maximum length of 6.6 in (168 mm), and the smallest species, the Borneo frog (*Leptobrachella*), is a mere 0.7 in (17.8 mm) long. Females are typically larger than males, except among the moustache toads (*Vibrissaphora*) and two species of alpine toads (*Scutiger*). Other sexually dimorphic characters include keratinized nuptial patches on the chest and fingers of breeding male alpine toads and cat-eyed frogs (*Oreolalax*) and bizarre keratinized spines seen on the upper lip of the male moustache toad during the breeding season.

The group takes its family name from the genus *Megophrys*, derived from Greek words (*meg* + *ophrys*) that mean "large eyebrow." This refers to the species *Megophrys montana*, which has long, fleshy appendages above the eyes. This trait is present in most species of Megophryinae, though it is not as pronounced. Leptobrachiinae includes two genera of warty toadlike species, the cat-eyed frogs and alpine toads. The remainder of the Leptobrachiinae, the slender frogs (*Leptolalax*), leaf litter frogs (*Leptobrachium*), and moustache toads, are not as toadlike, but they have fairly large eyes in proportion to their heads. The Borneo frog has unique toe disks that are swollen and have a pointed tip; all other megophryids have simple rounded toes with no visible disks. The pupils are vertical in all genera except the Borneo frog and the mountain

toad (*Ophryophryne*), in which they are either horizontal or diamond-shaped. The color pattern is generally cryptic, but in some species the males may exhibit breeding coloration in the form of bright highlights on the digits or vocal sac, and even fewer have brilliant spots (red, yellow, or orange) along the flanks or thighs.

Distribution

Asian toadfrogs occur throughout Southeast Asia, as far north as Shanxi, China; south to the island of Java (Indonesia); and east from Bengal, India, to Mindanao in the Philippines. The Asian horned frog (*Megophrys montana*) has been collected at sea level on the beaches of Sarawak, whereas the Nyingchi lazy toad (*Scutiger nyingchiensis*) has adapted to the harsh climate at 16,732 ft (5,100 m) on the southern slopes of the Himalayas. Of the 11 genera of megophryids, only one genus, *Leptobrachella*, occurs exclusively on the island of Borneo (Indonesia and Malaysia) or a nearby oceanic island. The most widely distributed genus and the largest, in terms of number of species, is *Xenophrys*. It occurs throughout much of the range of the family at moderate elevations, between 1,968 and 6,889 ft (600–2,100 m). *Leptobrachium* and *Leptolalax* have a more restricted distribution that covers southern China, Indochina, and the island of Borneo; *Ophryophryne* and *Brachytarsophrys* occur in Vietnam and southern China. The five species of *Vibrissaphora* live on only two widely disjunct mountain ranges in southern China and northern Vietnam. Because of their preference for montane habitats, certain species in each genus are unique to a specific mountain range. Of the 107 species of Asian toadfrogs, 26 are known from only a single location.

Habitat

Asian toadfrogs have an extensive north-south distribution that encompasses temperate to tropical rainforest climates. In their northern distribution (approximately 8–35° north latitude), they are found in seasonal monsoon forests where the dry season may last 2–5 months and where the annual rainfall is 106 in (270 cm) a year. In their southern equatorial limit, they occur in aseasonal tropical rainforests in which the annual rainfall may be as much as 263 in (668 cm) a year. Asian toadfrogs require primary or old-growth secondary montane forests. In either case, the forest canopy is generally dense, and the ground is covered with accumulated leaf litter. The tadpoles need clear mountain streams of varying depths. The slender-bodied tadpoles of *Leptolalax* and *Leptobrachella* live in swift torrents among the small stones that line the streambed. Larger-bodied tadpoles, such as those of *Leptobrachium*, *Vibrissaphora*, *Scutiger*, and *Oreolalax* occupy the deeper splash pools, where they are able to avoid the current. The funnel-mouth tadpoles opt for calmer edges of larger streams or are found in the clear, shallow seeps that are scarcely deeper than the tadpoles' bodies.

Behavior

Asian toadfrogs are nocturnal, coming out at dusk to forage and breed. In seasonal climates, their activity is dictated by the wet and dry seasons, but in the tropics they may be active all year long. In the north, breeding takes place during the wet season; Asian toadfrogs may be very abundant at this time. During the dry season it is difficult to find Asian toadfrogs, and it is thought that they return to the forest, taking refuge under rocks and logs. There is one report of five male moustache toads "hibernating" in a tree hole. Asian toadfrogs are poor jumpers; in fact, the leaf litter frog is more inclined to walk slowly away from a disturbance than it is to hop. The cryptic appearance of most species is their only defense against predation, but if they are threatened, the broad-headed toads will open their large mouths and lunge as if to bite.

Feeding ecology and diet

The broad-headed toads are sit-and-wait predators, consuming fairly large prey that may be moving along the forest floor. Little else is known about the foraging activities of the remaining species. Random examination of stomach contents has found that moths, spiders, crickets, cockroaches, beetles, scorpions, centipedes, and snails are all potential prey of Asian toadfrogs.

Funnel-mouth tadpoles feed on minute particles on the surface of the water. While feeding, the larvae position their upturned lips at the level of the water. Taking advantage of the gentle currents that they prefer, they simply allow water and any small particles on the surface film to flow over the edge of the funnel and into the mouth. Papillae (small fleshy fingerlike projections) around the lips direct the food particles into the mouth. When the papillae come in contact with a particle that is too large, the tadpole quickly dives to avoid the obstruction and resurfaces to resume the feeding process. The non–funnel-mouth tadpoles of Leptobrachiinae all forage on the detritus or algae that accumulate in streams. One study has shown that in the same microhabitat, leptobrachine larvae consume food particles that are on average three times the size of what the funnel-mouth tadpoles eat.

Reproductive biology

In seasonal climates breeding activity occurs during the wet season. In Vietnam, this is typically late fall to early spring, and it may last one to two months. In these climates, it is not uncommon to find peak breeding activity when evening temperatures are 41–44°F (5–7°C). Males also may be heard calling during the day, but these efforts are never made with the enthusiasm that is heard at night. Female leaf-litter frogs from Borneo are full of eggs in January, June, July, and August; these equatorial megophryids may breed all year round.

The males of most species situate themselves along the stream bank, either in the vegetation or under the boulders that are at the sides of the stream. The semi-arboreal *Ophryophryne* calls from vegetation up to 3 ft (1 m) above the ground. Larger species (*Megophrys* and *Brachytarsophrys*) space themselves 162–324 ft (50–100 m) along the stream bank. Smaller species, such as *Leptolalax*, *Ophryophryne*, and *Scutiger*, may form aggregations of five to 10 males along a 75-ft (23 m) stretch of a stream. The calls of *Xenophrys*, *Brachytarsophrys*, and *Lepto-*

The Malayan horned frog (*Megophrys nasuta*), is one of the largest species in the family. It is a terrestrial form that is clad in the hues of dead leaves. (Photo by R. D. Bartlett. Reproduced by permission.)

brachium sound like a guttural bark, "grrrack," whereas *Ophryophryne* produces a series of eight to 10 quick whistles. *Leptolalax* and *Leptobrachella* have calls made up of series of quick pulses that sound like a finger dragging across a comb. Amplexus is either axillary, where males clasp the females at the level of the shoulder, or inguinal, where males clasp females around the waist. Alpine toads (*Scutiger*) engage in inguinal amplexus and are said to use their keratinized chest patches to stimulate the female to lay eggs. The eggs always are deposited on the underside of large boulders at the edges of streams. Parental care has been recorded in the moustache toad—males remain with the clutches until they develop into tadpoles.

Conservation status

Asian toadfrogs vary from being extremely common to exceedingly rare. The slender mud frog (*Leptolalax pelodytoides*) has a vast distribution and during the breeding season can be the most abundant species in a stream habitat. On the other hand, Sung's slender frog (*Leptolalax sungi*) is known only from a 50-yd (50 m) stretch of one mountain stream. Efforts are being made to protect the rapidly disappearing habitat in which Asian toadfrogs live. As of the year 2002, no Asian toadfrogs were listed as endangered or threatened by the IUCN.

Significance to humans

During the winter breeding season, when Asian toadfrogs are most abundant, the larger species, such as the Annam broad-headed toad (*Brachytarsophrys intermedia*), are eaten by local people. The immediate relevance of the smaller species to humans has not been discovered. Given their preference for mature, undisturbed forests and the larval requirement for clean streams, however, they may prove to be a valuable indicator for evaluating the health of Southeast Asian montane forests.

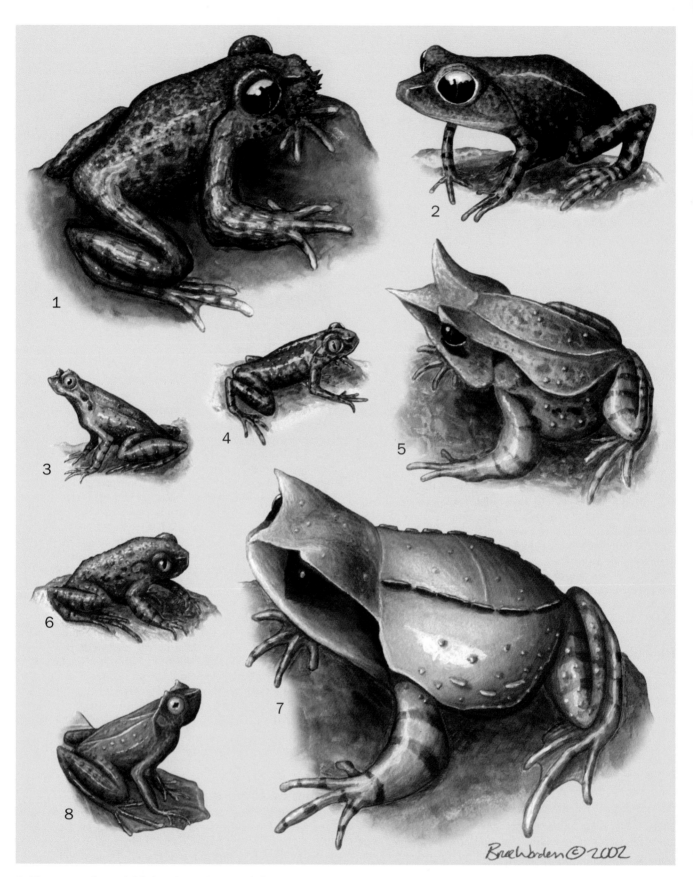

1. Ailao moustache toad (*Vibrissaphora ailaonica*); 2. Bana leaf litter frog (*Leptobrachium banae*); 3. Burmese spadefoot toad (*Xenophrys parva*); 4. Schmidt's lazy toad (*Oreolalax schmidti*); 5. Asian horned frog (*Megophrys montana*); 6. Slender mud frog (*Leptolalax pelodytoides*); 7. Annam broad-headed toad (*Brachytarsophrys intermedia*); 8. Asian mountain toad (*Ophryophryne microstoma*). (Illustration by Bruce Worden)

Species accounts

Bana leaf litter frog
Leptobrachium banae

SUBFAMILY
Leptobrachiinae

TAXONOMY
Leptobrachium banae Lathrop Murphy, Orlov, and Cuc, 1998, Gia Lai Province, Vietnam.

OTHER COMMON NAMES
None known.

PHYSICAL CHARACTERISTICS
Males are 2.3–2.9 in (57.2–73.0 mm) long, and females are 3.1–3.3 in (79.9–84.2 mm) long. This heavy-bodied frog has a head that is broad and flat. The limbs are slender and short and seem disproportionately small for the body. The protruding eyes are dark, except for the upper third, which is white. The pupil is vertical. A narrow white membrane is visible around the margin of the eye. From above, the Bana leaf litter frog is uniformly dark brown with red spots on the flanks and hind limbs; the belly is gray with minute white spots.

DISTRIBUTION
This species inhabits the central highlands of Vietnam.

HABITAT
This species is known only from primary forests at elevations between 2,620 and 3,280 ft (800–1,000 m).

BEHAVIOR
The Bana leaf litter frog is nocturnal and terrestrial; it spends most of its time taking refuge in the leaf litter deep in the forest.

FEEDING ECOLOGY AND DIET
The feeding habits are not known, but the diet likely includes a variety of large to medium-size insects.

REPRODUCTIVE BIOLOGY
Solitary males have been heard calling from burrows or under logs some distance away from the nearest stream.

CONSERVATION STATUS
The species is not listed as threatened, but prevailing habitat destruction for coffee plantations is jeopardizing the only known population of this species.

SIGNIFICANCE TO HUMANS
None known. ◆

Slender mud frog
Leptolalax pelodytoides

SUBFAMILY
Leptobrachiinae

TAXONOMY
Leptobrachium pelodytoides Boulenger, 1893, Karin Hills, Burma (Myanmar). This species has the most extensive distribution of any Asian toadfrog and probably represents a complex of species.

OTHER COMMON NAMES
English: Mountain short-legged toad; German: Schlamm-Schlankfrosh; Vietnamese: Cóc mày buèn.

Megophrys montana
Leptobrachium banae

Vibrissaphora ailaonica
Leptolalax pelodytoides
Oreolalax schmidti

PHYSICAL CHARACTERISTICS

Males grow up to 1.31 in (33.4 mm) and females to 1.62 in (41.2 mm). An elongate frog, it is orange to light brown, with irregular dark brown mottling on the back and head. The chin and belly are creamy white, and the slender limbs have black transverse bars. The upper lip includes several vertical black bars and one cream-colored vertical bar at the apex of the snout. Small tubercles may be scattered along the back. The tadpole is long and slender and has a subterminal mouth. The body and tail are light brown, and the edges of the tail fin are translucent.

DISTRIBUTION

The slender mud frog ranges across Hong Kong, southern China, Myanmar, Thailand, Vietnam, and Malaysia.

HABITAT

This species occurs in the vicinity of montane streams but is tolerant of disturbed habitats.

BEHAVIOR

The frogs are nocturnal and terrestrial.

FEEDING ECOLOGY AND DIET

The diet is unknown, but it presumably consists of small insects.

REPRODUCTIVE BIOLOGY

Males call from rocks in and along the edges of streams. Once approached by a female, the male places his chin on the female's shoulder and guides her to a site to lay eggs. Eggs are deposited on the underside of partially submerged rocks.

CONSERVATION STATUS

This species is common in montane stream habitats and is not considered threatened.

SIGNIFICANCE TO HUMANS

None known. ◆

Schmidt's lazy toad

Oreolalax schmidti

SUBFAMILY

Leptobrachiinae

TAXONOMY

Scutiger schmidti Liu, 1947, Mount O-mei, Szechwan, China.

OTHER COMMON NAMES

English: Webless toothed toad.

PHYSICAL CHARACTERISTICS

Males are 1.7–2.0 in (44–52 mm) in length, and females are 1.8–2.1 in (45–54 mm) long. This species has an overall toad-like appearance, including numerous scattered warts on the back and limbs and a thick layer of skin that hides the tympanum. Males have two large nuptial patches on the chest and many keratinized spines on the first finger. The color of the back is grayish brown; the belly and chin are flesh-colored and partially transparent. The short limbs are covered with dark brown transverse bars. The pupil is vertical, and the iris is golden. The thick-bodied tadpole has a subterminal mouth, and the body and tail are ashy-brown with gold and green flecks.

DISTRIBUTION

This species is distributed throughout the Hengduanshan Mountains in southern Szechwan and Yunnan, China.

HABITAT

The species lives around the headwaters of high mountain streams in desolate high-altitude valleys at elevations of 5,700–7,800 ft (1,740–2,380 m).

BEHAVIOR

Schmidt's lazy toad is nocturnal, terrestrial, and inactive.

FEEDING ECOLOGY AND DIET

The diet is unknown.

REPRODUCTIVE BIOLOGY

The males are extremely persistent callers and will not be deterred even if they are covered with leeches or if someone overturns the stone under which they are calling. Females lay about 120 eggs in balls affixed to the underside of rocks in small mountain streams at a time when the tadpoles from the previous year are on the verge of metamorphosing. Males seem to seek out and court females; individual females may be surrounded by several calling males.

CONSERVATION STATUS

Not threatened.

SIGNIFICANCE TO HUMANS

None known. ◆

Ailao moustache toad

Vibrissaphora ailaonica

SUBFAMILY

Leptobrachiinae

TAXONOMY

Vibrissaphora ailaonica Yang Chen, and Ma, 1983, Jingdong County, Yunnan, China. Some authorities include *Vibrissaphora* in the genus *Leptobrachium.*

OTHER COMMON NAMES

English: Ailao spiny toad.

PHYSICAL CHARACTERISTICS

Males are slightly larger than females. Males can grow to 3.2 in (81.6 mm) and females to 3.1 in (78.5 mm). Both sexes have a fine network of ridges over the body and limbs. Keratinized spines develop on the upper lip of the males; females have white spots that correspond to the spines found on the males. The adults are reddish brown with indistinct dark spots; juveniles are light brown with more conspicuous spotting. The pupil is vertical, and the iris is nearly black except for the top third, which is a striking lime green. Tadpoles have a thick body and a subterminal mouth; the body and tail are brown with a light brown line above that bifurcates into a Y where the body joins the tail.

DISTRIBUTION

This species is distributed across the Ailao Shan and Wuliang Shan mountain ranges; a southern extension of the Tibet plateau, in Yunnan province, China; and probably northern Vietnam.

HABITAT

The Ailao moustache toad prefers closed canopy montane forests at elevations of 7,220–8,200 ft (2,200–2,500 m).

BEHAVIOR
This species is primarily terrestrial, though males become aquatic during the breeding season.

FEEDING ECOLOGY AND DIET
Not known.

REPRODUCTIVE BIOLOGY
The breeding season occurs in late winter and lasts 2–6 weeks. There appears to be a considerable paternal investment; males construct nests underneath large boulders and undergo substantial physical transformations. At the onset of the breeding season, 20–60 keratinized spines develop on the upper lip of the males; the forearms become very thick, and the skin begins to loosen, forming numerous folds on the back and sides of the body. After the female deposits eggs in the nest, the male stays to guard the eggs while attempting to attract additional females. Nesting sites may contain several males and several egg masses. Fertilized eggs take more than a month to develop into tadpoles, and metamorphosis does not take place for two years.

CONSERVATION STATUS
Not threatened.

SIGNIFICANCE TO HUMANS
None known. ◆

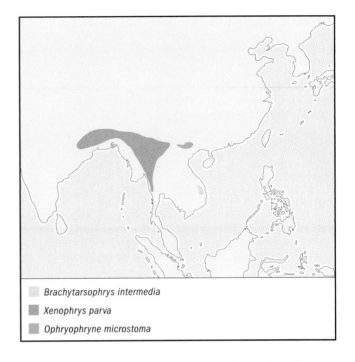

▢ Brachytarsophrys intermedia

▨ Xenophrys parva

▨ Ophryophryne microstoma

Annam broad-headed toad
Brachytarsophrys intermedia

SUBFAMILY
Megophryinae

TAXONOMY
Megalophrys intermedius Smith, 1921, Annam, Vietnam.

OTHER COMMON NAMES
English: Annam spadefoot toad; Vietnamese: Cóc mắt trung gian.

PHYSICAL CHARACTERISTICS
Males grow up to 4.6 in (118.3 mm) and females to 5.5 in (139.5 mm) in length. One of the largest Asian toad frogs, it has a stout body and a broad, flat head. The limbs are short and thick, and webbing is absent on the feet. A fleshy appendage is present above the eyes, and there are irregular folds and ridges on the flanks and back. The color of the back is light brown to reddish brown. The pupil is vertical, and the iris is nearly black. Tadpoles have a funnel-shaped mouth and are brown on the back, with contrasting black and white bars that extend from the belly to the tail.

DISTRIBUTION
The Annam broad-headed toad occupies the central highlands of Vietnam.

HABITAT
The species prefers montane forest and streams at elevations of 2,460–3,940 ft (750–1,200 m).

BEHAVIOR
This toad is solitary and inactive and relies on its cryptic appearance to avoid predation. If disturbed, it will not attempt to escape but rather gape its mouth and threaten to bite.

FEEDING ECOLOGY AND DIET
The Annam broad-headed toad is a sit-and-wait predator that will eat nearly anything that approaches. The diet is known to include smaller frogs, beetles, crickets, spiders, and moths. Small rodents are a likely prey as well.

REPRODUCTIVE BIOLOGY
This species is known to breed in late fall and early spring. Males migrate to streams and call from underneath large boulders, where the eggs are deposited.

CONSERVATION STATUS
Not threatened.

SIGNIFICANCE TO HUMANS
The Annam broad-headed toad is occasionally used as a food source by local people. ◆

Asian horned frog
Megophrys montana

SUBFAMILY
Megophryinae

TAXONOMY
Megophrys montana Kuhl and Van Hasselt, 1822, Java, Indonesia.

OTHER COMMON NAMES
English: Asian spadefoot toad; Dialect (unspecified): Katak bertanduk, Takang.

PHYSICAL CHARACTERISTICS
Males grow to 1.7–3.6 in (44–92 mm) in length and females to 2.6–4.4 in (67–111 mm). This is a stocky, large-bodied frog with a bizarre, elongated "horn" on the upper eyelid and, in some forms, a fleshy appendage projecting off the nose. This skin is smooth, except for one or two pairs of fleshy ridges that extend from behind the head to the groin. The color of the back is light brown to reddish brown, occasionally with a few black tubercles. The flanks bear numerous fleshy tubercles and are slightly darker in color than those on the back. The color

and overall shape of this species is a perfect imitation of a dried leaf. The pupil is vertical, and the iris is dark brown. Tadpoles have a funnel mouth, and the body and tail are brown.

DISTRIBUTION
The species inhabits Thailand, Malaysia, Sumatra, Java, Natuna, Borneo (Indonesia), and the Philippines.

HABITAT
This species prefers dense tropical forests from elevations between sea level and 7,220 ft (2,200 m). On rare occasions it is discovered in agricultural areas.

BEHAVIOR
The Asian horned frog is nocturnal. Its impeccable camouflage makes it extremely difficult to see on the forest floor. If it is discovered, either during the day or at night, it will crouch down further into the leaf litter and wait for the disturbance to go away.

FEEDING ECOLOGY AND DIET
This species consumes relatively large prey, including cockroaches, scorpions up to 3.9 in (10 cm) in length, and snails with diameters greater than 1.6 in (4 cm).

REPRODUCTIVE BIOLOGY
During the breeding season these frogs migrate to streams to breed. Males call individually and do not form choruses. The call sounds like a resonant honk.

CONSERVATION STATUS
Not threatened.

SIGNIFICANCE TO HUMANS
None known. ◆

Asian mountain toad
Ophryophryne microstoma

SUBFAMILY
Megophyrinae

TAXONOMY
Ophryophryne microstoma Boulenger, 1903, Tonkin, Vietnam.

OTHER COMMON NAMES
English: Narrow-mouthed horned toad.

PHYSICAL CHARACTERISTICS
Males grow up to 1.43 in (36.2 mm) in length and females up to 1.79 in (45.4 mm). This bizarre toothless frog has a narrow mouth and an extremely truncated snout. Small, pointy tubercles are present above the eye and leaflike venations are on the back. The color of the back ranges from light to dark brown, with some irregular mottling on the head and back. The pupil is diamond-shaped, and the iris is golden brown.

DISTRIBUTION
The Asian mountain toad lives in Vietnam and southwestern China.

HABITAT
The species prefers montane forests and streams at elevations above 1,300 ft (400 m).

BEHAVIOR
This nocturnal frog is terrestrial to semi-arboreal. It relies on its camouflage to avoid predation.

FEEDING ECOLOGY AND DIET
The species feeds on small insects.

REPRODUCTIVE BIOLOGY
Males typically call from an elevated position above a stream, either in the vegetation or on rocks. The call includes a series of five to 10 quick whistles. Males do not form choruses, but they often are compelled to respond with calls to nearby calling males.

CONSERVATION STATUS
This species is not threatened, but its habitat is disappearing quickly. The habitats in Vietnam (populations in the north and in the central highlands) are becoming urbanized or altered for agricultural purposes.

SIGNIFICANCE TO HUMANS
None known. ◆

Burmese spadefoot toad
Xenophrys parva

SUBFAMILY
Megophryinae

TAXONOMY
Xenophrys monticola Günther, 1864, Khasi Hills, India. A study of karyotypes found *X. parva* to have six large and seven small chromosomes, a pattern shared with other *Xenophrys* species. Preliminary DNA analyses indicate that the Vietnamese populations are distinct from those in Nepal. Further population sampling across the broad distribution of this species undoubtedly will confirm that this is a complex of species.

OTHER COMMON NAMES
English: Concave-crowned horned toad.

PHYSICAL CHARACTERISTICS
Males grow to 1.8 in (44.6 mm) and females to 2.0 in (51.0 mm) in length. The width of the body is slightly less than that of the head, and the snout is flat and shieldlike. A distinct ridge extends from the tip of the nose over the tympanum to the shoulder. There are one or two fleshy ridges on the back and a small cone-shaped tubercle is often present above the eye. The top of the head has a dark brown triangle, and there is a similar brown X on the back. The color of the back ranges from reddish brown to golden brown, and the throat and chest are mottled with dark brown. The pupil is vertical, and the iris is golden brown. Tadpoles have a funnel-shaped mouth, and the body is almost uniformly light brown, except for the translucent extremities of the tail fin.

DISTRIBUTION
The species inhabits eastern and central Nepal, northeastern and eastern India, Bangladesh, Myanmar, northern Thailand, northern Vietnam, and southern China.

HABITAT
This montane frog prefers forests and streams at elevations above 4,950 ft (1,400 m).

BEHAVIOR
The species is terrestrial and relatively active. It hides in leaf litter by day and relies on its cryptic coloration to avoid predators.

FEEDING ECOLOGY AND DIET
The Burmese spadefoot toad actively forages at night, feeding on small insects, including crickets, spiders, and moths.

REPRODUCTIVE BIOLOGY
The species is known to breed in early spring.

CONSERVATION STATUS
Not threatened.

SIGNIFICANCE TO HUMANS
None known. ◆

Resources

Books

Bourret, René. *Les Batraciens de l'Indochine.* Vol. 6. Hanoi: Institut Océanographique de l'Indochine, 1942.

Duellman, William. E., and Linda Trueb. *Biology of Amphibians.* New York: McGraw-Hill Book Co., 1986.

Inger, Robert, F. "Distribution of Amphibians in Southern Asia and Adjacent Islands." In *Patterns of Distribution of Amphibians: A Global Perspective,* edited by William Duellman. Baltimore: Johns Hopkins University Press, 1999.

Zhao, Er-Mi. "Distribution Patterns of Amphibians in Temperate Eastern Asia." In *Patterns of Distribution of Amphibians: A Global Perspective,* edited by William Duellman. Baltimore: Johns Hopkins University Press, 1999.

Zug, George R., Laurie J. Vitt, and Janalee P. Caldwell. *Herpetology: An Introductory Biology of Amphibians and Reptiles.* 2nd ed. San Diego: Academic Press, 2001.

Periodicals

Dring, Julian. "Frogs of the Genus *Leptobrachella* (Pelobatidae)." *Amphibia-Reptilia* 4, no. 2–4 (1983): 89–102.

Dubois, Alain, and Annemarie Ohler. "A New Species of *Leptobrachium (Vibrissaphora)* from Northern Vietnam, with a Review of the Taxonomy of the Genus *Leptobrachium* (Pelobatidae, Megophryinae)." *Dumerilia* 4, no. 1 (1998): 1–32.

Inger, Robert F. "Diets of Tadpoles Living in a Bornean Rain Forest." *Alytes* 5, no. 4 (1986): 153–164.

Lathrop, Amy. "Taxonomic Review of the Megophryid Frogs (Anura: Pelobatoidea)." *Asiatic Herpetological Research* 7 (1997): 68–79.

Xie, Feng, and Zhuwang Wang. "Review of the Systematics of Pelobatids (Anura: Pelobatidae)." *Cultum Herpetologica Sinica* 8 (June 2000): 356–370.

Other

Frost, Darrel R. *Amphibian Species of the World: An Online Reference.* Vers. 2.20. 1 Sept. 2000 (8 May 2002) <http://research.amnh.org/herpetology/amphibia/index.html>.

Orlov, Nikolai, Roman Khalikov, Robert W. Murphy, and Amy Lathrop. *Atlas of Megophryids (Megophryidae: Anura: Amphibia) of Vietnam.* Compact Disc. Saint Petersburg: Zoological Institute of Saint Petersburg, 2000.

Amy Lathrop, MA

Spadefoot toads
(Pelobatidae)

Class Amphibia

Order Anura

Family Pelobatidae

Thumbnail description
Rotund, moderate-size frogs with vertical pupils and a keratinous tubercle on the hind foot

Size
2.0–3.2 in (51–81 mm)

Number of genera, species
3 genera; 11 species

Habitat
Spadefoot toads usually live in arid to semiarid areas, such as fields and woodlands with sandy or loose soil

Conservation status
No species listed by IUCN

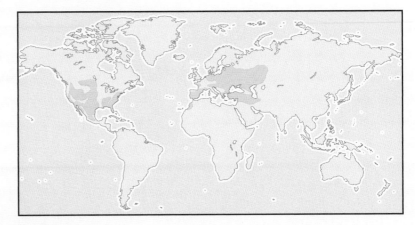

Distribution
North America and Europe to central Asia and northern Africa

Evolution and systematics

The earliest fossil Pelobatidae are from the late Cretaceous of North America, and extend through the middle Eocene of Europe to the Pleistocene. Pelobatidae are characterized by having sacral vertebrae (the vertebrae attached to the hips) fused to their "tail" vertebrae (whereas these are jointed in many frogs); the presence of a metatarsal spade; and bony ornamentation on the bones covering the brain (the frontoparietals). Within the family, the Nearctic genera *Spea* and *Scaphiopus* are each other's closest relative; some researchers recognize only a single genus, *Scaphiopus*. For a long time, members of the family Megophryidae commonly were considered to be a subfamily of Pelobatidae. Research eventually showed, however, that megophryids are not the closest relative to pelobatids, and the megophryids were removed from Pelobatidae family in 1985. The extinct subfamily Eopelobatinae has also been considered a member of Pelobatidae by some authors, but this relationship is not well supported. No subfamilies are recognized.

Physical characteristics

Frogs in this family are often mistaken for toads (exemplified by the common name, "spadefoot toads"). They do not have the warty skin of true toads, however, and they have teeth in the upper jaw (which true toads lack). All pelobatids have teeth on the maxilla and premaxilla. Palatines, bones that support the upper part of the inside of the mouth, are absent from the skull, and there are two frontoparietal bones covering the brain. Some species (*Pelobates* and *Scaphiopus*) have exostosis, or additional bony deposits, on the dorsal surface of the skull.

In members of this family, the facial nerve exits through the anterior acoustic foramen in the auditory capsule. The pupil is vertically elliptical. There are eight vertebrae before

the pelvis, and ribs are absent. The pectoral girdle is arciferal with a distinct sternum and omosternum. The small leg bones, the fibulare and tibiale, are fused only at their ends.

Larvae are aquatic with complete larval mouths (beaks).

Distribution

The family has a discontinuous distribution. The genus *Pelobates* occurs throughout most of western Eurasia and in the northwestern tip of Africa. The genera *Scaphiopus* and *Spea* occur throughout temperate North America, north to southern Canada, and south to southern Mexico.

Habitat

Spadefoot toads normally are found in arid to semiarid areas, such as fields, farmlands, dunes, and woodlands. They prefer rocky or sandy areas or regions where the soil is loose. Spadefoot toads typically inhabit low-lying areas that retain water after heavy rains. In eastern North America, one species occurs in cool, moist areas.

Behavior

All spadefoot toads are adapted to digging (fossorial). They are primarily nocturnal, but males call both day and night. Usually, males call while floating near the surface of shallow waters. They are very secretive and spend most of their time hiding in burrows. Those that live in strictly desert areas are active on the surface for only about two weeks during the year. As an adaptation to living in dry places, all spadefoot toads burrow down far enough so that the moisture content

A spadefoot toad hibernates in a cocoon underground during the dry season in the desert. (Illustration by Patricia Ferrer)

in the soil is the same as in their skin. Some may form a co-coon of dead skin to help protect against desiccation. These behaviors ensure that they can live for long periods without losing much water to the environment. During the rainy sea-son, they burrow only about 2 in (5 cm) below the surface, but during droughts they can be found more than 3 ft (1 m) underground. They burrow feet first, like most other digging frogs, but spadefoot toads have thick, shovel-like, keratinous spades on their feet to help them move dirt quickly. They al-ternate from left to right, pushing dirt forward, while rock-ing their bodies backward into the hole they are excavating.

Like most frogs, spadefoot toads rely on several an-tipredator mechanisms to ward off would-be attackers. If they detect motion, they stay completely still and depend on their camouflaged skin to blend in with the environment. If threat-ened, they inflate their lungs to make themselves appear big-ger. Some toads also produce distasteful skin secretions, which often are accompanied by a strong odor (some smell like gar-lic, and others like peanut butter).

Feeding ecology and diet

During the night, when there is enough moisture in the air to keep them from becoming desiccated, spadefoot toads come out of their burrows to hunt for food. The adult diet generally consists of invertebrates, including beetles, snails, spiders, and caterpillars. The diets of spadefoot tadpoles are much more remarkable. Most anuran larvae eat vegetative matter, but spadefoot larvae include some of the few species that eat aquatic insects and small crustaceans as well. They also eat plant material, filtering particles from the water col-umn. Spadefoot tadpoles sometimes group together in huge schools, which may help stir up settled plant material from the bottom of the pond. Schooling also may help protect against predation by insect larvae.

Because spadefoots breed in relatively shallow, temporary waters, they are under constant stress from drying waters, in-creasing temperatures, reduced food densities, and crowding. If the density of tadpoles reaches a certain point, some of the larvae of certain species eat their fellow tadpoles. The canni-

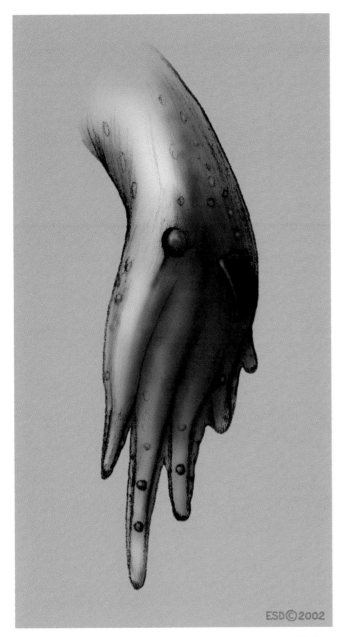

The spade (shown at right) on the hind leg of *Scaphiopus couchii* gives the group its common name (spadefoot toads). (Illustration by Emily Damstra)

Close-up of the spade of the eastern spadefoot toad (*Scaphiopus holbrookii*) that allows the toad to burrow. (Photo by Jeff Lepore/Photo Researchers, Inc. Reproduced by permission.)

stead, males wait in their burrows for optimal weather conditions (triggered by the low-frequency vibrations of rainfall) and then race to any available waters and let out a relatively loud call for their size (much like a deep "bleet" from a sheep). This call can be heard for about a mile (1.6 km), and other males will join in and set up adjoining territories in the water. Females then emerge from their burrows and join the males in their territories.

Spadefoot toads generally breed in shallow, temporary pools, such as cow ponds and drainage ditches. Because these waters may last for only a few weeks, much of the population mates on the first night of heavy rains. Males clasp females in front of the hind legs during amplexus (mating). Hundreds of thousands of small, dark eggs are laid in gelatinous clumps or bands attached to aquatic vegetation. Because the waters in which the eggs are laid may have begun to evaporate, to survive to adulthood these eggs must develop into toadlets that can leave the water in a matter of days. Thus the developmental cycle of most spadefoot toads is fast. Hatching occurs within 24–72 hours in hot weather or up to one week in cooler temperatures. Generally, spadefoot tadpoles metamorphose in about four weeks. Tadpoles range from tan to

bal morphs develop larger heads, sharp beaks, strong jaw muscles, and shortened intestines. Research indicates that cannibalism is adaptive in these species, because it allows for increased caloric intake, which, in turn, accelerates the rate of growth of the tadpoles. This ensures that the larvae reach the necessary size to metamorphose before the pond dries. Research also indicates that the cannibals use chemical cues to recognize related tadpoles and avoid eating their close relatives.

Reproductive biology

Because spadefoot toads live where rains and available water are unpredictable, they do not have a breeding season. In-

This spadefoot toad (*Scaphiopus*), found in Arizona, has a call that is compared to the bleating of sheep. (Photo by Joe McDonald. Bruce Coleman Inc. Reproduced by permission.)

dark brown; some are finely dotted with orange pigmentation and have transparent tail fins (with dark rims) that typically maintain transparency throughout the larval period.

Conservation status

No species are listed by the IUCN. In some parts of their range, several species are not considered threatened or endangered. However, population numbers and geographic ranges of several species are showing signs of rapid decline throughout their ranges. Several species (e.g., *Scaphiopus holbrookii, Spea intermontana*) are of special concern in parts of their range, because they are found only rarely. Other species, such as *Pelobates fuscus*, are listed as protected, threatened, or endangered by some agencies.

Significance to humans

Spadefoot toads are of no special significance to humans, though a few species are found in the pet trade.

1. Plains spadefoot toad (*Spea bombifrons*); 2. Common spadefoot (*Pelobates fuscus*); 3. Couch's spadefoot toad (*Scaphiopus couchii*). (Illustration by Emily Damstra)

Species accounts

Common spadefoot
Pelobates fuscus

TAXONOMY
Bufo fuscus Laurenti, 1768, Vienna, Austria.

OTHER COMMON NAMES
English: Spadefoot toad; French: Craupad brun; German: Kroblauchkröte.

PHYSICAL CHARACTERISTICS
The average length of adults is 3.2 in (81 mm). Despite having stout bodies with short limbs, these frogs are fairly agile on land. The dorsum is mottled green to brown; the skin on the dorsum is smooth.

DISTRIBUTION
Widespread in Europe from France to eastern Siberia; not present on the Iberian Peninsula.

HABITAT
Occurs in areas with loose soil, especially forests and fields used for agriculture. Usually, it is found in low-lying areas near shallow ponds or ephemeral waters.

BEHAVIOR
The frogs of this mostly nocturnal species take refuge in burrows that they excavate with keratinous spades on the hind limbs.

FEEDING ECOLOGY AND DIET
The diet consists primarily of insects, mollusks, and worms.

REPRODUCTIVE BIOLOGY
The breeding season extends from about April to June, during which time males call from shallow waters. About 1,000 small eggs are laid in a short, thick strand.

CONSERVATION STATUS
P. fuscus insubricus is listed by the IUCN as Endangered. *P. fuscus* is listed as Endangered by the Red Data Books of Estonia, Moldavia, and Krasnodar and Middle Urals (Russia). Habitat destruction and pollution seem to be the major causes of its decline.

SIGNIFICANCE TO HUMANS
None known. ◆

Couch's spadefoot toad
Scaphiopus couchii

TAXONOMY
Scaphiopus couchii Baird, 1854, Coahuila and Tamaulipas, Mexico.

OTHER COMMON NAMES
English: Spadefoot toad; French: Pied-en-bêche méridional; German: Südlicher Schaufelfuß; Spanish: Sapo con espuelas.

PHYSICAL CHARACTERISTICS
In this stout species with short limbs, adults are about 3 in (76 mm) long. The dorsum is bright greenish yellow to brown with dark green, brown, or black markings. The ventral surface is white, and the skin is granular.

DISTRIBUTION
Distributed widely in the United States, in parts of California, Arizona, New Mexico, Texas, and Oklahoma. In Mexico it ex-

▢ Pelobates fuscus

▢ *Spea bombifrons*
■ *Scaphiopus couchii*

ists in Baja California and on the coasts south to Nayarit and San Luis Potosi.

HABITAT
Well-drained, sandy areas, and common in deserts, short grass prairies, grasslands, and farmlands.

BEHAVIOR
A fossorial species, generally nocturnal, but recent metamorphs are slightly more active on the surface than adults. Takes refuge in burrows excavated using sickle-shaped spades on the hind limbs.

FEEDING ECOLOGY AND DIET
Primarily beetles, ants, grasshoppers, and termites; these toads often go for months without eating.

REPRODUCTIVE BIOLOGY
The breeding season is concurrent with heavy rains from April to September. Breeding may not occur in years in which rainfall is not sufficient. Males form large choruses. Up to 3,000 eggs are laid in clumps in shallow waters. Eggs hatch in about a day, and tadpoles transform in about six weeks.

CONSERVATION STATUS
Not threatened, though some populations are in decline owing to habitat destruction.

SIGNIFICANCE TO HUMANS
None known. ◆

Plains spadefoot toad
Spea bombifrons

TAXONOMY
Spea bombifrons Cope, 1863, Fort Williams, North Dakota, United States.

OTHER COMMON NAMES
Spanish: Sapo de espuela de los llanos.

PHYSICAL CHARACTERISTICS
A stout species with a prominent boss between the eyes. The snout-vent length, on average, is 1.5–2.5 in (38–64 mm). The dorsum is gray, brown, or cream; some individuals may have dark pigmentation surrounding red or yellow granules or four pale stripes on the dorsum.

DISTRIBUTION
Distributed widely in North America from Manitoba and Alberta, Canada, southward to Chihuahua, Mexico; disjunct populations exist in the United States in southern Texas and New Mexico.

HABITAT
Primarily inhabits dry grassland or farmland with sandy or loose soil.

BEHAVIOR
A fossorial and generally nocturnal toad, rarely present on the surface unless there are heavy rains.

FEEDING ECOLOGY AND DIET
The typical diet consists of invertebrates.

REPRODUCTIVE BIOLOGY
Known as an explosive breeder, this species emerges by the hundreds during warm spring rains. Adults return to burrows until the next heavy rain. Eggs hatch within two days.

CONSERVATION STATUS
Some populations are declining owing to habitat destruction. Although not threatened by IUCN criteria, the species is listed as rare or species of Special Status by states or provinces in parts of its range.

SIGNIFICANCE TO HUMANS
None known. ◆

Resources

Books
Duellman, William E., and Linda Trueb. *Biology of Amphibians.* Baltimore: Johns Hopkins University Press, 1994.

Stebbins, Robert C. *A Field Guide to Western Reptiles and Amphibians.* Boston: Houghton Mifflin, 1985.

Zug, George R., Laurie J. Vitt, and Janalee P. Caldwell. *Herpetology: An Introductory Biology of Amphibians and Reptiles.* 2nd ed. San Diego: Academic Press, 2001.

Periodicals
Pfennig, David W. "Polyphenism in Spadefoot Toad Tadpoles as a Locally Adjusted Evolutionarily Stable Strategy." *Evolution* 46 (1992): 1408–1420.

Anne M. Maglia, PhD

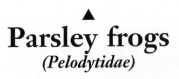

Parsley frogs
(Pelodytidae)

Class Amphibia

Order Anura

Family Pelodytidae

Thumbnail description
Moderately small, primarily nocturnal, terrestrial
Eurasian frogs

Size
1.8–2.2 in (45–55 mm)

Number of genera, species
1 genus; 3 species

Habitat
Moist areas from low elevations to midmountain
regions

Conservation status
Data Deficient: 1 species

Distribution
Iberian Peninsula and southwestern Europe; Caucasus Mountains in Asia

Evolution and systematics

Fossil Pelodytidae (genus *Miopelodytes*) are known from the Middle Miocene of Nevada in the United States, and the Eocene of Germany (genus *Propelodytes*).

The designation Pelodytidae most commonly includes the fossil forms, and is characterized by the fusion of the "ankle bones" (astragalus and calcaneum). Fossils of *Miopelodytes* and the extinct *Propelodytes arevacus* have a fused astragalus and calcaneum, but other *Propelodytes* do not. Therefore, the status of *Propelodytes* as a pelodytid is questionable. No subfamilies are recognized.

Physical characteristics

Parsley frogs (named for their speckled green coloration), are small and gracile, with large, bulging eyes. They are distinguished from all other frogs by a unique set of morphological features that includes the presence of a parahyoid bone, fused Vertebrae I and II, fused astragalus and calcaneum, and three tarsalia bones in the foot. The average body size is 1.57–1.97 in (40–50 mm).

Distribution

The three species have a discontinuous distribution in Europe and western Asia. One species is in the northwestern Caucasus and western Trans-Caucasus, Russia, Georgia, and Turkey. The second is in southern Portugal and southern

Spain, and the third is in Belgium, through France to eastern Spain and northwestern Italy.

Habitat

Parsley frogs are regularly found in deciduous and coniferous forested canyons, valleys drained by streams, and coastal zones. They can be found in or near shallow ponds, streams, and flooded quarries. One species seems prefer small streams with stony areas and/or sandy bottoms. Larval pelodytids are regularly found in brackish waters. These frogs can be found as far as 900 ft (275 m) away from the nearest water source.

Behavior

Pelodytids are generally nocturnal. During the day, they retreat to refugia under rocks or hide among vegetation at the base of large rocks or stone walls. At night, they forage near water sources. Parsley frogs hibernate from September to March, depending on the altitude and weather conditions.

Feeding ecology and diet

Parsley frogs generally forage at night. Their diet consists primarily of invertebrates, including flies, crickets, slugs, and worms.

The parsley frog (*Pelodytes punctatus*) is named for its green coloration—it appears to be garnished with parsley. (Photo by Francesc Muntada/Corbis. Reproduced by permission.)

Reproductive biology

Breeding in these frogs occurs during the spring and summer, with a second breeding season possible in the fall. Mating and egg laying seems to be triggered by rainfall. During the breeding season, males emit a low-volume acoustic sig-

nal, and apparently may call from under water. Amplexus (mating) is inguinal. Although the species are generally terrestrial, they breed in slow-moving to still waters, with eggs and tadpoles normally found in waters with high oxygen content and low plant nutrients. In France, several populations of pelodytid tadpoles have been found inhabiting brackish waters. The tadpoles have denticles, a sinistral spiracle, and well-defined jaw sheaths. Tadpole development can be prolonged, with some tadpoles regularly overwintering and completing their development the following year. Generally, tadpoles are medium-sized, but if they take two years to develop, can be quite large.

Conservation status

Although not listed (with one exception) by the IUCN or CITES, most populations are declining because of habitat destruction. *Pelodytes caucasicus* is categorized as Data Deficient by the IUCN; it is also listed in the Red Data Books of Russia, Georgia, and Azerbaijan and in the Bern Convention (Annex 2). *Pelodytes punctatus* is listed as endangered by the national standards of Belgium, Luxembourg, France, and as vulnerable in the other countries where it is found.

Significance to humans

None known.

Species accounts

Parsley frog
Pelodytes punctatus

TAXONOMY
Rana punctata Daudin, 1803, Beauvoise, Oise, France. No sub-species are recognized.

OTHER COMMON NAMES
English: Common parsley frog; French: Pélodyte ponctué; German: Westlicher Schlammtaucher; Spanish: Sapo moteado.

PHYSICAL CHARACTERISTICS
This species is small, averaging only about 1.6 in (4 cm). It is brown in color, with green flecks on the dorsum. The common name seems to have originated because its coloring makes it appear to be coated with parsley. The parsley frog is similar to other species of pelodytids, but differs by several

Pelodytes punctatus

Pelodytes punctatus

morphological and morphometric characteristics. *P. punctatus* is a smaller frog, with shorter hind legs. Also, the teeth found on the vomer bone (hard palate) are very close to the internal opening of the nares (the small hole connecting the nostrils and the inside of the mouth); this is not the case in other pelodytids.

DISTRIBUTION
P. punctatus is found in several countries in Europe. Its distribution includes Belgium, France, Luxembourg, eastern Spain, and northwestern Italy. Small populations occur in northern Spain.

HABITAT
This species generally inhabits open areas, including agricultural lands, as well as coniferous and deciduous forests. Although a terrestrial species, it can be found near slow, to still waters, such as deep ponds, small pools, flooded quarries, and slow-moving streams. It seems to prefer waters with stony or sandy areas. During the mating season, parsley frogs enter water to breed.

BEHAVIOR
A primarily terrestrial, nocturnal species, these frogs generally hide under stones or in holes in the ground. They emerge only at night, after moderate rainfalls. Parsley frogs migrate to water during breeding season, and both males and females are good swimmers. Depending on the weather, climate, and altitude, the frogs may hibernate during the winter months (November to March).

FEEDING ECOLOGY AND DIET
These frogs actively forage at night, searching for small invertebrates, including crickets and flies.

REPRODUCTIVE BIOLOGY
Depending on the climate, the breeding season begins in early spring (late February to April) and may occur again in fall (November to December). Reproduction seems to be triggered by rainfall. Males emit a low volume call from below the surface of the water. Amplexus (mating) is inguinal. Females lay an average of 50–300 eggs. During extended reproductive seasons, females may produce up to 1,600 eggs. The eggs are laid in small strings attached to aquatic plants. Tadpoles develop for approximately seven to eight months, and before metamorphosis, grow to be nearly 2.5 in (6.5 cm) long, which is larger than the adult frog. Metamorphosis occurs in January or February.

CONSERVATION STATUS
Although not categorized by the IUCN, this species is listed as endangered by the national standards of Belgium, Luxembourg, France, and as vulnerable in the other countries where it is found. The most likely cause for its declines is the alteration and loss of its habitat through drainage of marshlands, canalization of rivers, and destruction of stream habitats. Its range is subsequently highly fragmented, and most populations are in steady decline.

SIGNIFICANCE TO HUMANS
None known. ◆

Resources

Books

Duellman, William E., and Linda Trueb. *Biology of Amphibians.* Baltimore: Johns Hopkins University Press, 1994.

Garcia Paris, Mario. *Los Anfibios de España.* Madrid: Ministerio de Agricultura, Pesca y Alimentacion, 1985.

Gasc, Jean-Pierre, et al., eds. *Atlas of Amphibians and Reptiles in Europe.* Paris: Societas Europaea Herpetologica and Muséum National d'Histoire Naturelle, 1997.

Herrmann, Hans-Joachim. *Terrarien Atlas Band 1.* Melle, Germany: Mergus Verlag GmbH, 2001.

Zug, George R., Laurie J. Vitt, and Janalee P. Caldwell. *Herpetology.* 2nd edition. San Diego: Academic Press, 2001.

Other

Amphibian Species of the World: An Online Reference. September 1, 2000. [cited April 19, 2002] <http://research .amnh.org/herpetology/amphibia>

Anne M. Maglia, PhD

Ghost frogs
(Heleophrynidae)

Class Amphibia

Order Anura

Family Heleophrynidae

Thumbnail description
Medium-sized frogs with triangular discs on the fingers and toes

Size
1.4–2.6 in (35–65 mm)

Number of genera, species
1 genus; 6 species

Habitat
Montane forest

Conservation status
Endangered: 1 species; Vulnerable: 1 species

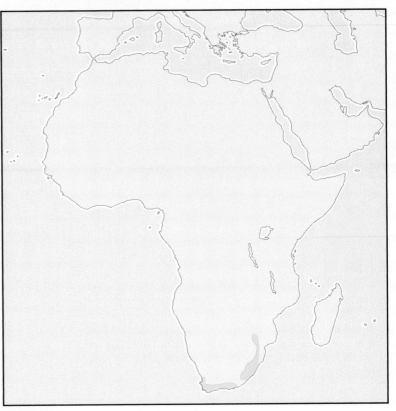

Distribution
South Africa, Lesotho, and Swaziland

Evolution and systematics

No fossils are known. The family has its closest relatives in South America and Australia, which is interesting further evidence of continental drift and the great age of this family. Although it was placed earlier as a subfamily within the Leptodactylidae, it now is recognized as a distinct family. No subfamilies are recognized.

Physical characteristics

The large, triangular discs on the fingers and toes are characteristic, along with a vertical pupil and a dorsal color pattern usually consisting of large spots on a brown or green background. The adult males of the smaller species, such as *Heleophryne orientalis*, do not exceed 1.4 in (35 mm), while the females of larger species grow to more than 2.6 in (65 mm). The body is flattened with protruding eyes, and the limbs are thin and long. The pupil is vertical, the tongue is disc-shaped, and the upper jaw bears teeth. The frogs swim well, with toes that are nearly fully webbed in some species. Most species have large dark spots on a paler background. The background color is typically tan to pale gray, but dark brown, yellowish, or bright green individuals are found. The tadpoles are streamlined and cling to rocks in fast-flowing streams. Most tadpoles have no keratinized jaw sheaths, except *Heleophryne rosei*, which has only a lower jaw sheath.

Distribution

This family is endemic to the high mountains and escarpment of the Drakensberg range and its extensions in southern Africa. Species are found from sea level to 9,843 ft (3,000 m). The recent loss of natural forest has caused streams to dry up, especially in areas where pines have been planted. Two species have very restricted ranges associated with pine plantations.

Habitat

Adults are found in forest or riverine forest. They may move 0.6 mi (1 km) or more from streams outside the breeding season, even into alpine grassland. The larvae are restricted to fast-flowing streams with rocky substrates. They are found attached to rocks in the fast current and also in quiet backwaters.

A male Purcell's ghost frog (*Heleophryne purcelli*) sits on a waterfall ledge. (Photo by Alan Channing. Reproduced by permission.)

Behavior

The frogs congregate after the rains near waterfalls or other fast-flowing water once the rivers subside. After breeding, the adults stay near the stream to feed but will move long distances away from water until the next rainy season heralds a new breeding season. Adults remain concealed in cracks or in holes during the day, emerging at night to feed and breed.

Feeding ecology and diet

These frogs take a range of insects, arthropods, and snails. They readily eat smaller species of frogs.

Reproductive biology

During the breeding season the body skin becomes loose, forming large, slimy folds, with the toes fringing with web. Males move into the streams as sexual activity increases and remain aquatic until the breeding season ends. The loose skin provides additional surface area, so that the males can obtain oxygen from the water. The breeding season is from spring to mid-summer (October to January in southern Africa). The male calls from within the spray zone of a waterfall or concealed in a rock crack or under a large rock. In some species the call is loud, but in others it is quiet, audible only from 10 ft (3 m) or less. Eggs are laid in quiet backwaters, but they also may be laid out of water in seepage zones, singly in slow-flowing areas and small pools. Some species attach their eggs under rocks in a stream. The eggs develop into free-swimming tadpoles. There is no parental care. The tadpoles graze on algae growing on rocks, leaving grazing trails.

Conservation status

The family is endemic to the Drakensberg mountain chain running through South Africa, Lesotho, and Swaziland. Two species are common, with wide distributions, while one is classified as Vulnerable (*Heleophryne rosei*) and another as Endangered (*Heleophryne hewitti*) by the IUCN. *H. hewitti* is known from short sections of only four rivers, all within 6.2 mi (10 km) along the slopes of the Elandsberg Mountains. *H. rosei* is restricted to a few streams on one side of Table Mountain in Cape Town, South Africa.

Significance to humans

These animals are not used for food. Although the skin contains toxins that protect the animal from mammalian predators, these toxins are not significant for humans.

Species accounts

Natal ghost frog
Heleophryne natalensis

TAXONOMY
Heleophryne natalensis Hewitt, 1913, eastern South Africa, Lesotho and Swaziland.

OTHER COMMON NAMES
Spanish: Sapo de espuela de los llanos.

PHYSICAL CHARACTERISTICS
The body is flat-tened, and the eyes are large and pro-truding. The back is brown to black with green or yellowish markings. These frogs have a marbled throat and triangular discs on the fingers and toes that are only slightly wider than the fingers and toes themselves.

Heleophryne natalensis

DISTRIBUTION
This species is known from the eastern mountains of South Africa, including those in Lesotho and Swaziland.

HABITAT
These frogs are found where the streams are fast flowing in natural forest. Adults can be found up to 0.6 mi (1 km) from water, in holes in banks and cliffs.

- Heleophryne natalensis
- Heleophryne rosei

BEHAVIOR
The adults sometimes can be seen during the day as they sit and wait for prey in the splash zone of waterfalls. They mostly hide under rocks in the river during the day, however, and come out after dark.

FEEDING ECOLOGY AND DIET
The Natal ghost frog eats small insects, spiders, and other arthropods.

REPRODUCTIVE BIOLOGY
The males call from vegetation near streams or from rock ledges or under large boulders within the spray zone of small waterfalls. The eggs are deposited under rocks in a stream. Within days they hatch into free-living tadpoles with as many as four upper rows and 17 lower rows of labial teeth.

CONSERVATION STATUS
Not threatened.

SIGNIFICANCE TO HUMANS
None known. ◆

Rose's ghost frog
Heleophryne rosei

TAXONOMY
Heleophryne rosei Hewitt, 1925, Table Mountain above Cape Town, South Africa.

OTHER COMMON NAMES
English: Thumbed ghost frog, Table Mountain ghost frog, Skeleton Gorge ghost frog.

PHYSICAL CHARACTERISTICS
This is a moderately sized frog, with the larger female up to 2.4 in (60 mm) and the smaller male up to 2 in (50 mm). The coloration of adults is striking: of-ten a pale green background with purple to brown blotches. The fingers and toes have large

Heleophryne rosei

triangular terminal discs. A rudimentary thumb is present as a distinct inner metacarpal tubercle. The feet are half webbed, with one phalanx of the fifth toe free of web. The tadpole has neither an upper nor lower jaw sheath but up to 17 rows of posterior labial teeth. The tadpole also has a large oral disc and is able to climb up wet vertical rock faces.

DISTRIBUTION
This species is known only from the eastern side of Table Mountain in Cape Town in a few perennial streams.

HABITAT
The typical habitat of this frog includes moist, forested gorges, with vertical rock faces covered with moss.

BEHAVIOR
The frogs are found on rock ledges or up in vegetation at night, retreating under large rocks and in cracks of rocks during the day.

FEEDING ECOLOGY AND DIET
These frogs eat a range of small insects and other forest arthropods.

REPRODUCTIVE BIOLOGY
Breeding starts in November when the streams are low but the temperature is high. The male's secondary sexual characters in-clude a number of small black spines on the outside surfaces of the forearms, on the back, and on the top of the back legs. The eggs have not been found, but in other species they are deposited under rocks in streams. The tadpoles develop for about 12 months.

CONSERVATION STATUS
This species is listed as Vulnerable by the IUCN and in the South African Red Data Book. The population is small, geographically restricted, and threatened by the plantations of pines on the mountain that cause the streams to dry up.

SIGNIFICANCE TO HUMANS
None known. ◆

Resources

Books

Channing, A. *Amphibians of Central and Southern Africa*. Ithaca: Cornell University Press, 2001.

Alan Channing, PhD

Seychelles frogs

(Sooglossidae)

Class Amphibia

Order Anura

Family Sooglossidae

Thumbnail description
Small, secretive frogs with varying, generally subdued coloration and generalized body form

Size
Snout-vent length of adults ranging from 0.41 to 2.2 in (10.5–55 mm)

Number of genera, species
2 genera; 3 species

Habitat
Tropical rainforest, including both undisturbed and disturbed forest

Conservation status
Endangered: 1 species; Vulnerable: 2 species

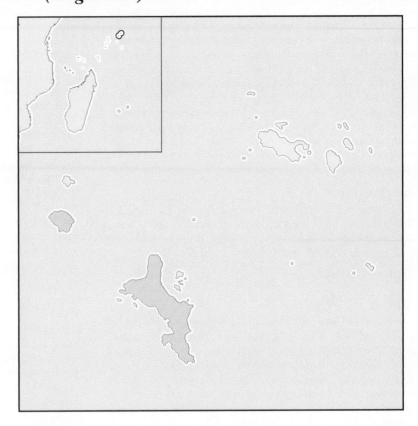

Distribution
Granitic islands of the Seychelles, western Indian Ocean

Evolution and systematics

The three species of sooglossid frogs were discovered at the turn of the nineteenth century and were described from specimens sent to European museums by field biologists. These frogs mistakenly were thought to be bufonids or ranids until 1931, when Noble placed them in a separate subfamily (Sooglossinae), which he thought was a subgroup of the Pelobatidae. The taxonomic history of sooglossids has been tortuous, and even today the phylogenetic history and classification are uncertain. However, all frog systematists today rank them as a full family. Sooglossids have no fossil record, but it is believed that they originated many millions of years ago and may be transitional between the more primitive arciferal frogs (with separate shoulder girdles) and the more advanced firmisternal groups (with fused shoulder girdles). No subfamilies are recognized.

Sooglossids are confined to the high granitic islands of the Seychelles archipelago, which are isolated from major landmasses in the western Indian Ocean. The islands are 1,000 mi (1,600 km) distant from Africa (Mombasa), 580 mi (930 km) northeast of Madagascar, and 1,800 mi (2,900 km)

southwest of India (Bombay). Because amphibians are intolerant of saltwater and have no obvious means of transoceanic dispersal, the presence of endemic frogs in the Seychelles was somewhat of a mystery until the history of these islands was elucidated. The main islands of the Seychelles are composed of granite rocks, which are of a continental nature. The islands are the mountaintop remnants of a partially submerged microcontinent that was left behind as India drifted northward toward Asia during the Cenozoic. The exact date that the Seychelles microcontinent separated from India is unknown, but it probably occurred sometime between 55 and 65 million years ago. The geological history of the Seychelles suggests that the ancestors of the modern sooglossids drifted to their present position and have been isolated for many millions of years. The observation that these frogs have no obvious sister group also suggests they have been isolated for a very long time. These facts indicate that the Seychelles microcontinent has never been submerged fully since it detached from India; otherwise there would be no surviving endemic frogs and other ancient endemic groups, such as the Seychellean caecilians (Amphibia, Gymnophiona).

Physical characteristics

Sooglossid frogs are small to medium-small, ordinary frogs. They have subdued colors that generally make them difficult to see where they live on the forest floor among litter and rocks and on low vegetation. The smallest species, *Sooglossus gardineri*, or Gardiner's frog, is among the smallest frogs in the world, with adults growing to only about 0.39–0.47 in (10–12 mm) in snout-vent length. *Nesomantis thomasseti*, or Thomasset's frog, is much larger at about 1.8 in (45 mm) in snout-vent length. Females are slightly larger than males in all three species. There are no obvious differences in coloration between the sexes, and young are colored nearly the same as adults.

Distribution

Sooglossids are restricted to two granitic islands, Mahé and Silhouette, of the Seychelles in the western Indian Ocean. The islands lie just south of the equator between 4° and 5° south latitude and 55° and 56° east longitude.

Habitat

Sooglossids occur in the rainforests above the 656 ft (200 m) contour line. They have not been observed on coastal plains. Presumably, their ancestral habitat was undisturbed forest, but they obviously survive in disturbed and even highly disturbed forests. The Seychelles frog, *Sooglossus sechellensis*, and Gardiner's frog are not associated with streams, whereas Thomasset's frog usually is found near streams.

Behavior

Sooglossids are secretive frogs, generally hiding in leaf litter, hollow stems, rock crevices, and leaf axils of low vegetation. Generally, they are not active on the surface except during rainy weather.

Feeding ecology and diet

These frogs eat a wide variety of small invertebrates, including mites, fruit flies, moths, mosquitoes, and other forest floor insects. Thomasset's frog often perches on rocks near streams at night and feeds on flying insects.

Reproductive biology

Sooglossids call day or night from hiding places; each species has a distinctive call. Gardiner's frog has a high-pitched "peep" and the Seychelles frog a "wrracck toc toc toc toc"; Thomasett's frog produces a call similar to that of the Seychelles frog, which sounds like "wrracck wrracck wrracck toc toc toc."

These frogs have the primitive form of the mating embrace (inguinal amplexus), in which the male clasps the female just in front of her hind limbs with his forelimbs. The Seychelles frog and Gardiner's frog deposit their eggs in hidden nests on the forest floor. Both species engage in parental care, in which the female remains with the eggs until they hatch. This finding is contrary to statements in the early literature, which claimed that the male Seychelles frog guards the young. In the latter species, the eggs hatch into tadpoles, which climb onto their mother's backs and are carried around until they metamorphose into froglets. The froglets remain on their mother's backs a short time but soon jump off to live an independent life.

The eggs of Gardiner's frog hatch directly into tiny froglets about the size of a grain of rice, which soon leave the nest. There is no post-hatching tadpole stage in this species, and the mothers do not transport the young on their backs. Nothing is known about the reproduction of Thomasset's frog; presumably, they deposit their eggs in hidden nest sites on land, and the females guard the eggs until they hatch directly into small froglets. This is a suggestion based on the reproductive biology of the two other species and the fact that no unidentified tadpoles have been found in aquatic habitats in the Seychelles.

Conservation status

Gardiner's frog and the Seychelles frog are listed as Vulnerable by the IUCN; both occur in dense populations and are distributed widely at higher elevations. Thomasset's frog, however, which is listed as Endangered by the IUCN, is less common and has a more restricted range. All three species occur in a national park on one of the islands. Although there appears to be no immediate threat to their survival, the fact that they are restricted to two tiny islands with expanding human populations is reason for concern.

Significance to humans

None known.

Species accounts

Gardiner's frog
Sooglossus gardineri

TAXONOMY
Nectophryne gardineri Boulenger, 1911, Mahé, Morne Pilot, 2,700 ft (823 m), and Silhouette, highest jungle.

OTHER COMMON NAMES
None known.

PHYSICAL CHARACTERISTICS

These are among the world's smallest frogs. The average snout-vent length of adult males is 0.4 in (10.2 mm), with a maximum of 0.43 in (11 mm); the average for females is 0.47 in (11.9 mm), with a maximum of 0.5 in (13 mm). The coloration varies widely. Some frogs are uniformly reddish brown on the dorsum, and others are tan; some have scattered spots, and others have stripes on the dorsum. The sides of the head and body are usually darker than the dorsal and ventral surfaces.

Sooglossus gardineri

DISTRIBUTION
The species occurs at elevations above 660 ft (200 m) on Mahé and Silhouette, Seychelles Archipelago, in the western Indian Ocean.

HABITAT
They inhabit the forest floor and low vegetation.

BEHAVIOR
The frogs are active night and day during the rainy season.

FEEDING ECOLOGY AND DIET
This species feeds on tiny ground and litter-layer invertebrates.

REPRODUCTIVE BIOLOGY
The call is a high-pitched "peep." The female deposits eight to 15 eggs in hidden nests on the forest floor. The female guards the eggs until they hatch into tiny froglets about 0.12 in (3 mm) long. There is no larval stage.

CONSERVATION STATUS
The IUCN lists the Gardiner's frog as Vulnerable.

SIGNIFICANCE TO HUMANS
None known. ◆

Seychelles frog
Sooglossus sechellensis

TAXONOMY
Arthroleptis sechellensis Boettger, 1896, Auf den Seychellen.

OTHER COMMON NAMES
None known.

PHYSICAL CHARACTERISTICS

This is a medium-sized sooglossid. The average snout-vent length is about 0.59 in (15 mm) in males and about 0.79 in (20 mm) in females. The dorsum is golden brown, and the sides and upper surfaces of the legs have scattered black spots. There is a large, often triangular black spot on top of the head between the eyes.

Sooglossus sechellensis

DISTRIBUTION
This species occurs at elevations above 660 ft (200 m) on Mahé and Silhouette, Seychelles archipelago, in the western Indian Ocean.

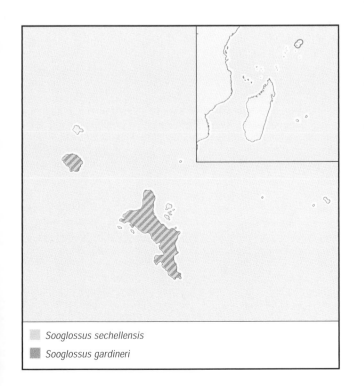

Sooglossus sechellensis
Sooglossus gardineri

HABITAT
The Seychelles frog inhabits leaf litter on the forest floor and at the edges of rainforest.

BEHAVIOR
These secretive frogs are seldom seen at the water surface.

FEEDING ECOLOGY AND DIET
The Seychelles frog feeds on small insects, mites, and other invertebrates that live in forest litter and rotten logs.

REPRODUCTIVE BIOLOGY
Males call day or night from hidden sites on the forest floor during the rainy season: "wrracck toc toc toc toc." Females deposit six to 15 small white eggs in hidden nests. They remain with the eggs until they hatch into tadpoles. Tadpoles are transported on the mother's back until they metamorphose into tiny froglets. There is no aquatic tadpole stage.

CONSERVATION STATUS
The IUCN classifies this species as Vulnerable.

SIGNIFICANCE TO HUMANS
None known. ◆

Resources

Books
Noble, Gladwyn K. *The Biology of the Amphibia.* New York: McGraw-Hill, 1931.

Nussbaum, R. A. "Amphibians of the Seychelles." In *Biogeography and Ecology of the Seychelles Islands,* edited by D. R. Stoddart. Hague: Dr. W. Junk, 1984.

Periodicals
Green, D. M., R. A. Nussbaum, and Y. Datong. "Genetic Divergence and Heterozygosity Among Frogs of the Family Sooglossidae." *Herpetologica* 44, no. 1 (1988): 113–119.

Griffiths, I. "The Phylogenetic Status of the Sooglossine." *Annals and Magazine of Natural History* 2, no. 22 (1959): 626–640.

———. "The Phylogeny of the Salientia." *Biological Review* no. 38 (1963): 241–292.

Noble, Gladwyn K. "An Analysis of the Remarkable Cases of Distribution Among the Amphibia, with Descriptions of New Genera." *American Museum Novitates* no. 212 (1926): 1–24.

Nussbaum, R. A. "Mitotic Chromosomes of Sooglossidae (Amphibia: Anura)." *Caryologia* 32, no. 3 (1979): 279–298.

———. "Phylogenetic Implications of Amplectic Behavior in Sooglossid Frogs." *Herpetologica* 36, no. 1 (1980): 1–5.

———. "Amphibian Fauna of the Seychelles Archipelago." *National Geographic Society Research Reports* no. 18 (1985): 53–62.

Nussbaum, R. A., A. Jaslow, and J. Watson. "Vocalization in Frogs of the Family Sooglossidae." *Journal of Herpetology* 16, no. 3 (1982): 198–203.

Ronald A. Nussbaum, PhD

Australian ground frogs

(Limnodynastidae)

Class Amphibia

Order Anura

Family Limnodynastidae

Thumbnail description
Small to large frogs that range from rotund burrowing forms to terrestrial species with powerful legs

Size
0.9–4.3 in (22–108 mm)

Number of genera, species
9 genera; 48 species

Habitat
These frogs are wide ranging, from arid habitats to the wet/dry tropics to temperate and subtropical zones with summer or winter peaks in rainfall and in all vegetation types found within these climatic areas

Conservation status
Critically Endangered: 1 species; Endangered: 2 species; Vulnerable: 2 species; Lower Risk/Near Threatened: 1 species; Data Deficient: 1 species

Distribution
Australia (including Tasmania) and New Guinea

Evolution and systematics

Based on studies of the ilium from disarticulated material, three extant genera, *Limnodynastes*, *Lechriodus*, and *Kyarranus* have been recorded from the Oligo-Miocene, and *Lechriodus* also has been recorded from the early Eocene. The extant species *Limnodynastes ornatus* has been recorded from a Quaternary site. Other *Limnodynastes* material has been recorded from a Plio-Pleistocene site and *Neobatrachus* from the Miocene/Pliocene boundary. *Lechriodus* is well represented in material from the Riversleigh site in northwestern Queensland, Australia, and its occurrence in the Tertiary helps explain the current disjunct distribution of the genus, with four species in New Guinea and a single representative in southeastern Queensland. No subfamilies are recognized.

The Australopapuan ground frogs have had a checkered taxonomic and phylogenetic history that remains the subject of debate. Early workers, such as Cope, placed individual genera in several families, but the seminal work of Parker in 1940 recognized the frogs as two subfamilies of the Leptodactylidae. The Myobatrachinae and Cycloraninae were defined clearly, on the basis of numerous skeletal and myological features. Within the Cycloraninae were the recognized genera *Adelotus*,

Cyclorana, *Heleioporus*, *Lechriodus*, *Limnodynastes*, *Mixophyes*, *Notaden*, and *Philoria*. *Cyclorana* later was shown to be a hylid genus, and the subfamily became the Limnodynastinae. *Heleioporus* was split into the nominate genus and *Neobatrachus*; and *Kyarranus* was recognized as a new genus with affinities to *Philoria*.

Megistolotis was described in 1979 but has since been made synonymous with *Limnodynastes*. The problematic genus *Rheobatrachus* was described in 1973, and it has been allied variously with the Limnodynastinae, the Myobatrachinae, or its own subfamily, Rheobatrachinae, or family Rheobatrachidae. *Mixophyes* has posed problems in the acceptance of monophyly of the Limnodynastinae. The Myobatrachidae inclusive of all genera formerly assigned to the Limnodynastinae and Myobatrachinae was recognized in 1973 on the basis of geographical distribution. Monophyly of the Myobatrachidae with or without *Rheobatrachus* or *Mixophyes* or both has been challenged, though the data used are old and not always substantiated by later studies. Familial status as recognized here must remain subject to debate and further analysis with new data. *Mixophyes* currently is placed within the Limnodynastidae, but the genus does not share many of the features that unite the lineage. For example, they engage in axillary am-

Spencer's burrowing frog (*Limnodynastes spenceri*) begins its burrowing. (Photo by Margaret Davies. Reproduced by permission.)

plexus, their tongue muscles differ, and the first two vertebrae are not fused.

Physical characteristics

In all Limnodynastidae (except *Mixophyes*) the first two vertebrae are fused. The alary (wing-like) processes of the hyoid are on stalks, though the actual shape of these processes can vary. The superficial jaw muscle, the *M. intermandibularus*, underlies the *M. submentalis*, and the cricoid ring is complete.

Spencer's burrowing frog (*Limnodynastes spenceri*) during its "corkscrew" burrowing. (Photo by Margaret Davies. Reproduced by permission.)

Some genera or species groups (e.g., *Notaden, Neobatrachus, Heleioporus, Limnodynastes dorsalis,* and *L. ornatus* groups) are burrowers and exhibit the burrowing morphotype of short limbs, short heads, and rotund bodies. Others are more streamlined and have powerful hind limbs (e.g., *Mixophyes*), and still others (e.g., the *Limnodynastes tasmaniensis* group) have a body form that is intermediate between these extremes.

Foam-nesting species (except *Heleioporus*) exhibit seasonal development of flanges on the second and third fingers in females. Nuptial excrescences in males vary from heavily spinous structures on the thumb and second finger (e.g., *Heleioporus* and *Limnodynastes lignarius*) to glandular pads that may be restricted to the base of the thumb, found solely on the thumb and second fingers (e.g., *Limnodynastes spenceri*), or extend to three fingers (e.g., *L. ornatus.*).

Many species are dull brown or gray, but others have red or gold marks on the thighs or brilliantly colored labial glands. Still others have spectacular dorsal markings of yellow superimposed with black and red warty markings in the form of a cross (e.g., *Notaden bennettii*).

Distribution

Lymnodynastids occur throughout Australia. *Limnodynastes convexiusculus* also inhabits southern New Guinea, and a single species of *Mixophyes* (*M. hihihorlo*) and four species of *Lechriodus* occur in New Guinea.

Habitat

These frogs inhabit arid desert and seasonally arid grasslands, woodlands, and open forest; they are found along perennial and ephemeral (temporary) streams and around permanent and ephemeral ponds from sea level to elevations above the snow line.

Behavior

Seasonal activity is governed by the availability of moisture. All limnodynastids are crepuscular or nocturnal. Many limnodynastids burrow to avoid dry conditions. All burrow backward, but in one of two ways. Backward-sliding burrowers shuffle with the hind limbs and descend at an angle to the surface, whereas circular burrowers corkscrew vertically downward, turning themselves around as they descend. The different forms of burrowing are associated with differences in muscle mass in the lower hind legs. Of the burrowing species, members of the genera *Neobatrachus* and *Heleioporus* form cocoons and reduce evaporative water loss dramatically while estivating underground. Burrowing species resorb water from the bladder to maintain water balance during their subterranean periods. *Adelotus brevis* engages in male-male combat, but encounter calls are used by other species to maintain calling sites.

Feeding ecology and diet

The diet is arthropod-based, but it is restricted by the gape of the mouth and by seasonal availability of prey items. The form of the tongue governs whether animals capture prey by biting or by tongue flicking.

Reproductive biology

Marked differences in breeding seasons are common. Many species, in particular most of the burrowing species, are explosive breeders. They come to the surface in response to heavy rain and breed in temporary pools. Others are strictly seasonal, and still others (e.g., *Limnodynastes tasmaniensis*) breed continuously if conditions are right.

Calls vary considerably, from clicks (e.g., *Limnodynastes tasmaniensis*) to whoops (e.g., *Heleioporus, Notaden*) to trills (e.g., some *Neobatrachus*). The complex calls of the amazing repertoire emitted by many Australian hylid frogs are lacking in the ground frogs. Calling is nocturnal in the group.

The Limnodynastidae have highly varying forms of egg deposition. Some frogs produce eggs in jelly that are laid in water (e.g., *Neobatrachus* and *Notaden*); others produce eggs in jelly that are deposited out of water. The tadpoles enter the stream on hatching (*Mixophyes*). All species of *Limnodynastes, Lechriodus,* and *Adelotus* produce a foam nest in water, though some *Limnodynastes tasmaniensis* in southern South Australia lay eggs that are not in a foamy mass. *Heleioporus* lays eggs, also in a foamy mass, in a burrow that, on flooding, releases tadpoles into the water. The other foam-nesting species, *Philoria* and *Kyarranus*, lay their eggs either out of water or in shallow water, and the nonfeeding tadpoles develop in the broken-down foam and jelly mass.

Conservation status

Philoria frosti is listed as Critically Endangered by the IUCN; *Mixophyes fleayi* and *M. iteratus* as Endangered, *M. balbus* and *Heleioporus australiacus* as Vulnerable; *Adelotus brevis* as Lower Risk/Near Threatened; and *Notaden weigeli* as Data Deficient. Reasons for declines have not been identified positively, though *Mixophyes* may have been decimated by chytrid fungus. *Philoria* is an alpine species that is threatened by ski-field development as well as unidentified factors that are causing declines elsewhere in the Australian Alps.

Significance to humans

Some burrowing species, such as *Notaden bennettii*, have been recognized as a source of water to aboriginal people living in arid areas. Skin secretions have been investigated for pharmacological activity, and an unidentified toxic substance has been recorded in *Heleioporus*.

Spencer's burrowing frog (*Limnodynastes spenceri*) has burrowed into the sand, and is nearly completely covered. (Photo by Margaret Davies. Reproduced by permission.)

Painted frog (*Neobatrachus pictus*) cocooned. (Photo by Margaret Davies. Reproduced by permission.)

1. Tusked frog (*Adelotus brevis*); 2. Woodworker frog (*Limnodynastes lignarius*); 3. Baw Baw frog (*Philoria frosti*); 4. Giant barred frog (*Mixophyes fasciolatus*); 5. Painted frog (*Neobatrachus pictus*); 6. Northern spadefoot toad (*Notaden melanoscaphus*). (Illustration by John Megahan)

Species accounts

Tusked frog
Adelotus brevis

TAXONOMY
Cryptotis brevis Günther, 1863, Clarence River, New South Wales, Australia.

OTHER COMMON NAMES
None known.

PHYSICAL CHARACTERISTICS
In this medium-size, sexually dimorphic species, males are larger than females, which is unusual. Males are 1.3–1.7 in (34–44 mm), and females are 1.1–1.5 in (29–38 mm). The head of the male is broad and flat with two large tusks on the lower jaw. The head of the female is not as expanded posteriorly, and the tusks, if present, are small. The ventral surface of both males and females is pigmented heavily with strong white marbling; the groin and the back of the hind legs are bright red. Fingers and toes are basally webbed.

DISTRIBUTION
This species occurs on the Great Dividing Range and the coast from central and eastern Queensland to the southern coast of New South Wales, Australia.

HABITAT
This species inhabits temperate rainforest or wet sclerophyll forest floor and sometimes is found in open grasslands.

BEHAVIOR
Male combat occurs at calling sites; the tusks are used to attack rivals.

FEEDING ECOLOGY AND DIET
Males tend to eat more snails and fewer arthropods than do females, a dietary divergence related to habitat separation of male and female frogs. Males usually spend more time in moist habitats, where snails are abundant, whereas females are apt to be found on the drier forest floor, where arthropods are more abundant.

REPRODUCTIVE BIOLOGY
The call is a single repeated "cluck." Females lay unpigmented eggs in foam nests in still water. Males remain with the foam nest after deposition. Tadpoles are ovoid to elliptical in shape, with shallow fins, and they seem to feed on detritus.

CONSERVATION STATUS
Listed as Lower Risk/Near Threatened. Some mortality in Brisbane has been attributed to chytrid fungus. Loss and degradation of habitat may be a threat.

SIGNIFICANCE TO HUMANS
None known. ◆

Woodworker frog
Limnodynastes lignarius

TAXONOMY
Megistolotis lignarius Tyler Martin, and Davies, 1979, 4 mi (6.5 km) north of Lake Argyle Tourist Village on Kununnura and Lake Argyle Road, Kimberley Division, Western Australia.

OTHER COMMON NAMES
English: Carpenter frog.

PHYSICAL CHARACTERISTICS
This moderately large frog is characterized by an extremely large tympanum, a broad head, and an extensive row of vomerine teeth. The males are 1.7–2.4 in (43–62 mm), and the females are 1.9–2.4 in (47–61 mm). Males have muscular forelimbs and spiny nuptial pads on the thumb and second finger.

DISTRIBUTION
This species is confined to escarpment country in the Kimberley Division of Western Australia and the Northern Territory.

HABITAT
The frog inhabits scree slopes and escarpments near ephemeral or perennial streams.

BEHAVIOR
The frogs take shelter in caves and crevices during the dry season and emerge to mate during the wet season.

FEEDING ECOLOGY AND DIET
Nothing is known of the feeding ecology or diet of this species, but it probably feeds on arthropods.

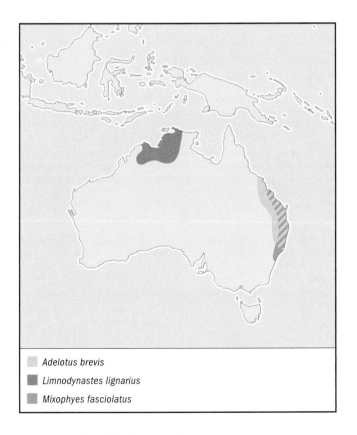

Adelotus brevis

Limnodynastes lignarius

Mixophyes fasciolatus

REPRODUCTIVE BIOLOGY
Males call from concealed locations beneath rocks or vegetation beside streams. The call is a soft tap similar to the sound of wood being struck. Females lay a foam nest of unpigmented eggs beneath vegetation or rocks. The eggs hatch into darkly pigmented tadpoles with ventral suctorial mouths adapted to the fast-flowing streams. Tadpoles actively seek out riffles in the stream and usually are found in the fastest-flowing sections.

CONSERVATION STATUS
Not threatened. The species is secure across its range.

SIGNIFICANCE TO HUMANS
None known. ◆

Giant barred frog
Mixophyes fasciolatus

TAXONOMY
Mixophyes fasciolatus Günther, 1864, Clarence River, New South Wales, Austalia.

OTHER COMMON NAMES
None known.

PHYSICAL CHARACTERISTICS
This large frog has a proportionately large head, powerful hind limbs, and moderately webbed toes. The males are 2.4–2.6 in (60–65 mm), and the females are 2.8–4 in (72–101 mm). The dorsum is brown or gray with well-defined blotches on the body, stripes on the head, and transverse bars on the limbs.

☐ *Notaden melanoscaphus*
■ *Neobatrachus pictus*
▨ *Philoria frosti*

DISTRIBUTION
The species occurs along the Great Dividing Range and eastern coast from Bundaberg in Queensland to the southern coast of New South Wales, Australia.

HABITAT
The frogs inhabit the forest floor adjacent to streams.

BEHAVIOR
A crepuscular and nocturnal species, little is known of its behavior other than its reproductive strategy.

FEEDING ECOLOGY AND DIET
The diet consists of insects and smaller frogs.

REPRODUCTIVE BIOLOGY
Breeding usually takes place along streams but sometimes in unconnected pools near streams and drainages away from streams. Amplectant pairs sit in the water facing the bank. A few eggs are laid and then kicked in a spray of water onto the bank or rock face by the female, where the eggs stick to the surface. Hatching tadpoles fall into the water. The stream-adapted tadpoles have ventral suctorial mouths.

CONSERVATION STATUS
Not threatened. The species seems to be secure, which may be because of a broader use of habitat than its congeners.

SIGNIFICANCE TO HUMANS
None known. ◆

Painted frog
Neobatrachus pictus

TAXONOMY
Neobatrachus pictus Peters, 1863, Loos, 2.8 mi (4.5 km) west of Gawler (Buchsfelde), South Australia.

OTHER COMMON NAMES
English: Trilling frog.

PHYSICAL CHARACTERISTICS
This moderate-size, rotund frog has short limbs and a short head. The males are 1.8–2.3 in (46–58 mm), and the females are 1.9–2.2 in (48–55 mm). This frog is brightly colored, with yellow, gray, or pale brown markings with irregular dark patches and a warty dorsum. In breeding males, these warts are tipped with spines. The inner metatarsal tubercle is large, compressed, keratinized, and black. The toes are almost fully webbed, and the pupil is vertically elliptical when constricted.

DISTRIBUTION
This species lives in southeastern South Australia and probably in adjoining parts of Victoria and New South Wales, Australia.

HABITAT
This frog inhabits clay pans, grassy marshes, roadside pools, and open woodland.

BEHAVIOR
This burrowing species forms a cocoon. Once below the surface of the ground, the outer keratinous layer of skin is lifted from the body, as if to be shed, but remains attached, enclosing the entire animal except for the nostrils, which remain open to the exterior. The cocoon splits and is shed after rain percolates down through the soil to the estivating animal. The

frog digs its way to the surface and breeds and feeds before digging down again as drought sets in.

FEEDING ECOLOGY AND DIET
Nothing is known about the feeding ecology or diet of this species, but it probably feeds on arthropods.

REPRODUCTIVE BIOLOGY
Many animals emerge from the ground after heavy rains. The call is a musical trill. Small, pigmented eggs are laid as loose clumps in vegetation at the edge of pools. Tadpoles, which have massive jaw sheaths, grow as large as 3.5 in (up to 90 mm) before metamorphosing.

CONSERVATION STATUS
Not threatened.

SIGNIFICANCE TO HUMANS
Skin secretions are believed to be harmful if ingested. ◆

Northern spadefoot toad
Notaden melanoscaphus

TAXONOMY
Notaden melanoscaphus Hosmer, 1962, Borroloola, Northern Territory, Australia.

OTHER COMMON NAMES
French: Pied-en-bêche du nord.

PHYSICAL CHARACTERISTICS
This moderate-size, rotund species has short limbs and a short snout. The males are 1.3–1.9 in (34–48 mm), and the females are 1.8–1.9 in (45–49 mm). The dorsum is gray or olive brown, with large, dark, symmetrical blotches. The dorsum is warty, and the warts commonly are tipped with white. Juveniles typically have bright yellow, red, and black spots. The inner metatarsal tubercle is black and keratinized.

DISTRIBUTION
The species is widespread in eastern and northern Kimberley, Western Australia, across to Townsville, Queensland, Australia.

HABITAT
The frog is found in flooded grassland after torrential rains.

BEHAVIOR
This burrowing species exudes a sticky, yellow-orange, gluelike substance when disturbed; the exudate hardens quickly and is difficult to remove. It does not appear to be toxic to other frogs. These frogs run rather than hop; at night they can be mistaken for small rodents.

FEEDING ECOLOGY AND DIET
The mouth is small, and, hence, these frogs are restricted in their diet to small arthropods. They flick the tongue rather than bite at prey.

REPRODUCTIVE BIOLOGY
Males inflate the entire body while lying in shallow water and call, "whoop, whoop, whoop." If disturbed, they deflate, sink,

and remain silent until the threat has passed. Males grasp females in inguinal amplexus, and eggs are laid as a surface film that sinks as the larvae hatch. The tadpoles seem to be filter feeders rather than grazers.

CONSERVATION STATUS
Not threatened. The species is secure across its range.

SIGNIFICANCE TO HUMANS
None known. ◆

Baw Baw frog
Philoria frosti

TAXONOMY
Philoria frosti Spencer, 1901, Mount Baw Baw, Victoria, Australia.

OTHER COMMON NAMES
None known.

PHYSICAL CHARACTERISTICS
Males of this species are 1.7–1.8 in (42–46 mm), and females are 1.9–2.2 in (47–55 mm). This moderate-size frog with an indistinct tympanum has a rather drab, dark brown, warty appearance and well-developed parotoid glands; the ventral surfaces and groin are cream or yellowish.

DISTRIBUTION
This species is found only at elevations above 3,800 ft (1,160 m) on Mount Baw Baw, Victoria, Australia.

HABITAT
This frog inhabits tunnels in sphagnum bogs or lives beneath logs and rocks on the sides of streams.

BEHAVIOR
A reclusive species.

FEEDING ECOLOGY AND DIET
Nothing is known about the feeding ecology or diet of this species, but it probably feeds on arthropods.

REPRODUCTIVE BIOLOGY
Males call in frost hollows. Unpigmented eggs are laid in a foam nest in small, seepage-fed depressions beneath logs or rocks or dense vegetation. The jelly breaks down, and tadpoles lacking mouthparts feed on their yolk supply and develop within the liquefied jelly over a period of five to eight weeks.

CONSERVATION STATUS
This species is listed as Critically Endangered, but the cause, other than habitat threat, has not been identified. Population decline seems to be a widespread phenomenon at high elevations, and ultraviolet light, temperature, and prolonged drought cannot be tied directly to the declines.

SIGNIFICANCE TO HUMANS
None known. ◆

Resources

Books

Anstis, M. *Tadpoles of South-eastern Australia: A Guide with Keys.* Sydney: Reed New Holland, 2002.

Barker, John, Gordon C. Grigg, and Michael J. Tyler. *A Field Guide to Australian Frogs.* Chipping Norton, Australia: Surrey Beatty, 1995.

Campbell, A., ed. *Declines and Disappearances of Australian Frogs.* Canberra, Australia: Environment Australia, 1999.

Cogger, H. G. *Reptiles and Amphibians of Australia.* 6th edition. Sydney: Reed New Holland, 2001.

Cogger, H. G., E. E. Cameron, and H. M. Cogger. *Zoological Catalogue of Australia.* Vol. 1, *Amphibia and Reptilia.* Canberra, Australia: Australian Government Publishing Service, 1983.

Littlejohn, M. J., M. Davies, J. D. Roberts, and G. F. Watson. "Family Myobatrachidae." In *Fauna of Australia.* Vol. 2A, *Amphibia and Reptilia*, edited by C. J. Glasby, G. J. B. Ross, and P. Beesley. Canberra, Australia: AGPS, 1993.

Malone, B. S. "Mortality during the Early Life History Stages of the Baw Baw Frog, *Philoria frosti* (Anura: Myobatrachidae)." In *Biology of Australasian Frogs and Reptiles*, edited by G. Grigg, R. Shine, and H. Ehmann. Chipping Norton, Australia: Surrey Beatty and Sons, 1985.

Periodicals

Davies, M., and G. F. Watson. "Morphology and Reproductive Biology of *Limnodynastes salmini, L. convexiusculus* and *Megistolotis lignarius* (Anura: Leptodactylidae: Limnodynastinae)." *Transactions of the Royal Society of South Australia* 118, no. 3 (1994): 149–169.

Katsikaros, K., and R. Shine. "Sexual Dimorphism in the Tusked Frog, *Adelotus brevis* (Anura: Myobatrachidae): The Roles of Natural and Sexual Selection." *Biological Journal of the Linnean Society* 60, no. 1 (1997): 39–51.

Parker, H. W. "The Australasian Frogs of the Family Leptodactylidae." *Novitates Zoologicae* 42, no. 1 (1940): 1–106.

Schauble, C. S., C. Moritz, and R. W. Slade. "A Molecular Phylogeny for the Frog Genus *Limnodynastes* (Anura: Myobatrachidae)." *Molecular Phylogenetics and Evolution* 16, no. 3 (2000): 379–391.

Tyler, M. J. "*Limnodynastes* Fitzinger (Anura: Leptodactlidae) from the Cainozoic of Queensland." *Memoirs of the Queensland Museum* 28, no. 2 (1990): 779–784.

———. "*Kyarranus* Moore (Anura: Leptodactylidae) from the Tertiary of Queensland." *Proceedings of the Royal Society of Victoria* 103, no. 1 (1991): 47–51.

Tyler, M. J., and H. Godthelp. "A New Species of *Lechriodus* Boulenger (Anura: Leptodactylidae) from the Early Eocene of Queensland." *Transactions of the Royal Society of South Australia* 117, no. 4 (1993): 187–189.

Tyler, M. J., H. Godthelp, and M. Archer. "Frogs from a Plio-Pleistocene Site at Floraville Station, Northwest Queensland." *Records of the South Australian Museum* 27, no. 2 (1994): 169–173.

Tyler, M. J., A. A. Martin, and M. Davies. "Biology and Systematics of a New Limnodynastine Genus (Anura: Leptodactylidae) from Northwestern Australia." *Australian Journal of Zoology* 27, no. 1 (1979): 135–150.

Organizations

Environment Australia. GPO Box 787, Canberra, ACT 2601 Australia. Phone: 61 (2) 6274-1111. Web site: <http://www.ea.gov.au>

Margaret Davies, PhD

▲
Australian toadlets and water frogs
(Myobatrachidae)

Class Amphibia

Order Anura

Family Myobatrachidae

Thumbnail description
Small to large frogs in which the first two presacral vertebrae are not fused, and in which are seen widely varying reproductive modes, ranging from fully aquatic to direct development and including some bizarre forms of parental care

Size
0.6–3.1 in (16–79 mm)

Number of genera, species
21 genera; 121 species

Habitat
Streams, alpine meadows, seasonally arid open forest and grasslands, woodlands, temperate rainforest

Conservation status
Extinct: 3 species; Critically Endangered: 5 species; Endangered: 1 species; Vulnerable: 5 species; Data Deficient: 17 species

Distribution
Australia (including Tasmania) and southern New Guinea

Evolution and systematics

The fossil record is poor and based on studies of the ilium, believed to be the most diagnostic of disarticulated bones. The extant species *Crinia signifera* and *C. georgiana* have been recorded from Pleistocene deposits, whereas *C. remota*, also an extant species, has been recorded from a deposit of unknown age, probably Holocene or late Pleistocene. An extinct species, *C. presignifera*, has been identified from the Oligo-Miocene of Queensland. No subfamilies are recognized.

The relationships of the Australopapuan ground frogs have been the subject of much argument. Originally believed to be part of the widely distributed family Leptodactylidae, they were placed in a single family, Myobatrachidae, in 1973 on the sole basis of distribution. An argument was raised at this stage that knowledge of the subfamilies Myobatrachinae and Limnodynastinae was insufficient to decide whether they were different enough from each other to merit familial status. Composition of the subfamilies has been subject to dispute, with two enigmatic genera, *Mixophyes* and *Rheobatrachus*, being of uncertain affinities. *Mixophyes* has been placed within the Limnodynastinae uncritically, but *Rheobatrachus* has been placed variously in either the Limnodynastinae or Myobatrachinae or even in its own subfamily or family.

Monophyly of the two major subfamilies (leaving aside the two questionable genera) has never been in dispute, but monophyly of Myobatrachidae has yet to be established. Attempts to answer this question have relied on early studies that were shown to be based on spurious data (probably owing to misidentification of the material under examination). Family or subfamily status of these two lineages is likely a semantic argument in the absence of new data, but within the Myobatrachidae as recognized here, relationships of all genera recognized in 2001, except *Rheobatrachus*, have been established using mitochondrial genes. The monotypic genus *Bryobatrachus* has been shown to be the sister taxon to *C. tasmaniensis* (both taxa are endemic to Tasmania) and, pending morphologic investigation of the generic status of this lineage, has been placed in *Crinia*.

The monotypic genus *Spicospina* is the sister taxon to *Uperoleia*, which resolves the enigmatic biogeographical observation of an absence in southwestern Australia of this speciose and widely distributed genus. Classification issues relate to the status of *C. tasmaniensis* and *C. nimbus* and to the familial position of *Rheobatrachus*.

The turtle frog's (*Myobatrachus gouldii*) modifications for forward burrowing include its thickened snout and modified front limbs. (Photo by Margaret Davies. Reproduced by permission.)

Physical characteristics

Apart from the larger aquatic species (*Rheobatrachus*) with fully webbed toes, myobatrachids are small frogs, varying from slender, long-legged froglets to squat, short-legged toadlets. Some are highly modified as frontward burrowers with thickened snouts, broad hands, and reduced digits; some dig backward with two raised, compressed metatarsal tubercles; and many others have an unmodified terrestrial or semi-aquatic body form.

Most frogs vary from dull slate gray to brown, often with bright flash markings in the groin or armpits or both, but others are brilliantly colored with yellow and black, as in the Corroboree frog, *Pseudophryne corroborree*, or with blue, orange, and red, as in *Spicospina flammocaerulea*.

Features of the skeleton and muscles indicative of this family include the lack of fusion of the first two presacral vertebrae, the shape of the alary processes of the hyoid (broad and winglike except in *Rheobatrachus*), and the relationship of the

Sandhill frog (*Arenophryne rotunda*) tracks in the sand. (Photo by Margaret Davies. Reproduced by permission.)

superficial muscles of the throat and of the leg muscles, (except in *Rheobatrachus*). The finger and toe discs are small or absent in members of this family.

Distribution

Species are found throughout Australia and *Uperoleia mimula* and *Crinia remota* have been recorded in southern New Guinea. Myobatrachids occur in sand dunes close to the oceans through intermediate elevations to the alpine meadows of the Australian Alps in the Great Dividing Range.

Habitat

Adults live in habitats ranging from seasonally arid grasslands to sand dunes in which no surface free water is available to temperate rainforest to open woodland to alpine meadows and to rainforest streams. Larvae develop in permanent streams, temporary ponds, nests in damp mossy habitats, underground egg membranes, and the stomachs of female parents as well as the hip pockets of males.

Behavior

Most species are nocturnal or crepuscular, but the day frogs of the genus *Taudactylus* are active during daylight hours. Species found in the wet/dry tropics are strictly seasonal, as are most temperate species. A few species are active and call after rain throughout the year. Members of the genus *Uperoleia* burrow to escape the dry season in seasonally arid areas and remain underground until the rains come, whereas the two frontward-burrowing genera *Myobatrachus* and *Arenophryne* spend daylight hours underground irrespective of the season.

Male *Uperoleia lithomoda* wrestle with intruding males if challenged at their calling sites. Vocalizations are part of the ensuing struggle. Females of *Crinia georgiana* can be clasped simultaneously by as many as five males, resulting in multiple paternity of the eggs but with reduced fertilization success compared with single matings. This has been attributed to intense conflict between males attempting to mate with a single female, and usually the conflict is resolved when extra males clasp the female in suboptimal positions, such as ventrally.

Feeding ecology and diet

Some species (e.g., *Arenophryne*) feed mostly on ants, and *Myobatrachus* feeds almost exclusively on termites. All species are arthropod feeders. Type of prey is restricted by the gape of the mouth and seasonal availability. No species is known to eat other frogs. *Myobatrachus* live in termite mounds; hence, prey items are readily available. Most other frogs feed opportunistically.

Reproductive biology

Most myobatrachids are strictly seasonal breeders, but some frogs breed in all seasons. Those in the wet/dry tropics respond to monsoon rains, though many species of *Uperoleia* call throughout much of the wet season. Some males (e.g., *Assa*) call away from nesting sites, whereas others call beside water and, after females have selected them, move to the water in amplexus. *Arenophryne* and *Myobatrachus* call from underground. *Crinia nimbus* and *Taudactylus* call during the day, but all other species call in the evening and at night. Female *Uperoleia inundata* approach a calling male from behind and wriggle underneath him when he is in full voice. He rapidly deflates his vocal sac and clasps her in the inguinal region, and they move to water to deposit eggs. Female *Assa* follow the male to the nesting site, and female *Crinia georgiana* often are subjected synchronously to attempted matings by multiple males.

Myobatrachids have a wide range of reproductive modes, but they do not include foam nesting. *Uperoleia*, many species of *Crinia*, *Paracrinia haswelli*, and *Taudactylus* all lay aquatic eggs and have free-swimming tadpoles. Others, such as some species of *Geocrinia* and all *Pseudophryne* except *P. douglasi* lay terrestrial eggs that develop out of water for differing periods of time and then hatch at times of rains and are flushed or wriggle into the water.

The eggs of *Arenophryne* and *Myobatrachus* are laid underground, undergo the entire larval period within jelly membranes, and hatch as froglets. One of the most unusual reproductive modes is exhibited by *Assa*, in which the newly hatched larvae wriggle up the flanks of the male, lodge in a pair of inguinal pouches, and undergo their entire development at these sites. *Rheobatrachus* has an equally unusual form of reproduction, in which the larvae develop in the stomach of the female.

Conservation status

Rheobatrachus silus, *R. vitellinus*, and *Taudactylus diurnus* are listed as Extinct. In addition, five species are Critically Endangered; one is Endangered; five are Vulnerable; and 17 are Data Deficient. *Taudactylus acutirostris*, which the IUCN lists as Critically Endangered, is categorized as extinct by the 2002 Environment Australia Threatened Species List.

Many species have restricted distributions, with endemic centers in southwestern Australia, Tasmania, Kimberley, Pilbara, and Cape York, which increases their vulnerability. The presumed extinct species may have been decimated by chytrid fungus; no known anthropogenic influence can be cited. Other factors that may influence frog survival are herbicides, pesticides, urbanization, salinization, and fire.

The sandhill frog (*Arenophryne rotunda*) burrows frontwards into the sand. (Photo by Margaret Davies. Reproduced by permission.)

The sandhill frog (*Arenophryne rotunda*), nearly covered after burrowing into the sand. (Photo by Margaret Davies. Reproduced by permission.)

Significance to humans

Secretions of pharmacological activity have been isolated from the skins of *Uperoleia*, *Taudactylus*, and *Pseudophryne*. None has been developed further. Skin secretions are potentially toxic, though studies of toxicity have not been undertaken.

1. Southern gastric brooding frog (*Rheobatrachus silus*); 2. Hip pocket frog (*Assa darlingtoni*); 3. Moss frog (*Crinia nimbus*); 4. Eungella torrent frog (*Taudactylus eungellensis*); 5. Sandhill frog (*Arenophryne rotunda*). (Illustration by Barbara Duperron)

Species accounts

Sandhill frog
Arenophryne rotunda

TAXONOMY
Arenophryne rotunda Tyler, 1976, False Entrance Well Tank, Edel Land, Shark Bay, Western Australia.

OTHER COMMON NAMES
English: Round frog.

PHYSICAL CHARACTERISTICS
This is a small species; males are 1–1.3 in (26–33 mm), and females are 1.1–1.3 in (28–33 mm). These frogs have short limbs and broad hands with reduced phalanges on the first digit. The dorsal skin is mottled with small warts and ridges and often a pale mid-dorsal stripe.

DISTRIBUTION
The narrow range extends from Kalbarri north to Shark Bay and Dirk Hartog Island, Western Australia.

HABITAT
This frog lives in seasonally arid sand dunes close to the coast with no surface free water.

BEHAVIOR
The frog burrows head first into the sand and takes shelter during the day at the interface of damp and dry sand at the top of the water table at depths that are seasonally variable. At night they walk around on the dunes leaving characteristic tracks.

FEEDING ECOLOGY AND DIET
Probably catholic and opportunistic in their feeding, their diet consists mainly of ants as well as beetles, arachnids, spiders, and true bugs. Individuals have been observed at night with their heads down openings of ant nests, where they presumably are feeding.

REPRODUCTIVE BIOLOGY
Males call from underground from about April until July, and pairs form, also underground, by November, sometimes in aggregations at the same site. About five months later, directly developing eggs are deposited about 31.5 in (80 cm) underground over a period of three months.

CONSERVATION STATUS
Not threatened. Although this species has a restricted distribution, it is locally abundant.

SIGNIFICANCE TO HUMANS
None known. ◆

Hip pocket frog
Assa darlingtoni

TAXONOMY
Crinia darlingtoni Loveridge, 1933, Queensland National Park, McPherson Range, Queensland, Australia.

OTHER COMMON NAMES
English: Pouched frog, marsupial frog; French: Rainettes-à-bourse; German: Beutelfrösche.

PHYSICAL CHARACTERISTICS
In this small species, males are 0.6–0.7 in (15–19 mm), and females are 0.7–0.8 in (18–21 mm). The pointed snout expands into a broad body, which is gray, pale brown, or pinkish brown to red dorsally; a dark stripe commencing anteriorly to the nostrils extends posterolaterally, passing through the eye and terminating midway along the flank. The toes lack fringes and webbing but have slightly expanded tips. Vomerine teeth are absent.

DISTRIBUTION
This small frog inhabits the mountain ranges on the New South Wales/Queensland border and the McPherson Ranges in northeastern Australia.

HABITAT
Found in deeply-shaded leaf litter in montane rainforest, the frogs take shelter under rocks, vegetated soil banks, or overhangs.

BEHAVIOR
Little known. Reproductive behavior more well known.

FEEDING ECOLOGY AND DIET
The feeding ecology of this species is unknown, but it feeds on arthropods.

REPRODUCTIVE BIOLOGY
Males call from leaf litter or under logs; when approached by females, they increase the intensity of their calls and lead the

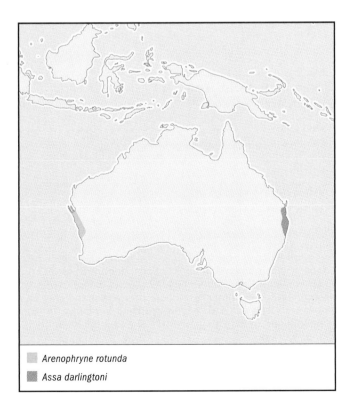

■ *Arenophryne rotunda*
▨ *Assa darlingtoni*

female to a nesting site, where inguinal amplexus occurs. Eggs are laid in two layers on moist, decomposing leaves deep in the leaf litter or on soil under leaf litter. The female attends the nest during this time. After several days (11 in captivity), the male covers the egg mass with the anterior part of his body as the jelly capsules rupture; using tail movements, the hatchling tadpoles move up his flanks and enter one of the bilateral inguinal pouches (up to six in each pouch). Tadpoles lack a spiracle and are supplied with yolk. Fully formed froglets emerge either frontward or backward 48–69 days later.

CONSERVATION STATUS
Not threatened.

SIGNIFICANCE TO HUMANS
None known. ◆

Moss frog
Crinia nimbus

TAXONOMY
Bryobatrachus nimbus Rounsevell Zeigeler, Brown, Davies, and Littlejohn, 1994, 984 ft (300 m) north of Lake Esperance, Herz Mountains National Park, Tasmania.

OTHER COMMON NAMES
None known.

PHYSICAL CHARACTERISTICS
This species is small; males are 0.7–1.1 in (19–27 mm), and females are 1–1.2 in (25–30 mm). The frogs have maxillary teeth and no reduction in the bones of the ear. The last two presacral vertebrae fuse with the sacrum; the dorsum has consistent

□ *Crinia nimbus*
▨ *Taudactylus eungellensis*
▨ *Rheobatrachus silus*

markings of a chevron-shaped patch between the eyes and dark parallel lines extending posteriorly from the shoulder, with a second pair of dark patches laterally on the posterior part of the body. The fingers and toes lack fringes and webbing.

DISTRIBUTION
This frog occurs in mountains in southern Tasmania ranging from sea level to 3,600 ft (1,100 m).

HABITAT
The species is confined to subalpine moorland or implicate rainforest.

BEHAVIOR
This cryptozoic species hides under low vegetation or in breeding chambers, often in cushions of sphagnum moss, lichens, or similar vegetation. The frogs do not appear to aggregate or to use open surface water.

FEEDING ECOLOGY AND DIET
Nothing is known about the feeding ecology or diet of this species, but it probably feeds on arthropods.

REPRODUCTIVE BIOLOGY
The call is a series of "toks"; frogs call diurnally in spring and early summer from the ground in dense vegetation or from breeding chambers. Egg capsules are extremely large, with an ovum diameter of about 0.2 in (4 mm) and a capsule diameter of about 0.6 in (15 mm); clutches of four to 16 eggs are laid in moss or lichens. The jelly capsules break down, and the entire period of larval development takes place in the liquefied jelly. Tadpoles do not feed, and the oral disk lacks keratinous jaws or labial teeth. Tadpoles spend the winter under snow and metamorphose after about 395 days.

CONSERVATION STATUS
Not threatened.

SIGNIFICANCE TO HUMANS
None known. ◆

Eungella torrent frog
Taudactylus eungellensis

TAXONOMY
Taudactylus eungellensis Liem and Hosmer, 1973, Eungella National Park, 33.6 mi (54 km) west of Mackay, Queensland, Australia.

OTHER COMMON NAMES
English: Eungella day frog.

PHYSICAL CHARACTERISTICS
This is a small to medium-size frog; males are 1–1.1 in (25–28 mm), and females are 1.1–1.4 in (28–36 mm). The frogs have relatively powerful hind limbs; the finger and toe discs are clearly expanded, and the terminal phalangeal bones are T-shaped. The dorsal surface is smooth or granular and gray or brown with irregular darker brown patches. Ventrally, the skin is smooth and white with yellow suffusions on the lower abdomen and thighs.

DISTRIBUTION
The frog occurs only in the Clarke Range, mostly in Eungella National Park, in central and eastern Queensland, Australia.

HABITAT
Frogs are found on rocks, boulders, and waterfalls in fast-flowing streams at elevations of 490–3,280 ft (150–1,000 m) in disturbed and undisturbed rainforest.

BEHAVIOR
This species is both diurnal and nocturnal and communicates visually by head bobbing and by waving the hind legs.

FEEDING ECOLOGY AND DIET
Nothing is known, but it probably feeds on arthropods.

REPRODUCTIVE BIOLOGY
The call resembles a gentle rattling sound. Tadpoles have weakly keratinized jaws but lack labial tooth rows. The oral disc is small, almost ventrally positioned, and surrounded by a complete row of papillae—all adaptations to fast-flowing streams.

CONSERVATION STATUS
This species is listed as Endangered by the IUCN. It was one of the stream frogs that partially vanished in the 1980s, but small relict populations appear to be maintaining themselves. The three largest populations are in the same catchment, and no populations are now known from the southern and northern areas of the former distribution.

SIGNIFICANCE TO HUMANS
None known. ◆

Southern gastric brooding frog
Rheobatrachus silus

TAXONOMY
Rheobatrachus silus Liem, 1973, Kondalilla National Park, Queensland, Australia.

OTHER COMMON NAMES
English: Southern platypus frog.

PHYSICAL CHARACTERISTICS
This was a medium-size species; males were 1.3–1.6 in (33–41 mm), and females were 1.8–2.1 in (45–54 mm). The species had a small head with large, dorsally protruding eyes and powerful hind limbs with fully webbed toes. The dorsum was dull gray to slate with obscure darker and paler patches.

DISTRIBUTION
Apparently the range was restricted to the Conondale and Blackall Ranges in southeast Queensland.

HABITAT
This aquatic species usually was found in perennial streams in closed forest.

BEHAVIOR
This frog was a strong swimmer but would drift in the water. The species was observed sitting on exposed rocks and was capable of traveling across land through moist habitat.

FEEDING ECOLOGY AND DIET
Insects were taken either from the stream surface or from surrounding rocks by grabbing with the mouth and using the hands to push in the prey.

REPRODUCTIVE BIOLOGY
Males called with a soft pulsed note of about 33 pulses and a low dominant frequency. Females swallowed either the large, unpigmented eggs or the newly hatched tadpoles (not known which), and the entire development through metamorphosis took place inside the stomach; 18–25 young were brooded in this manner for six to seven weeks. Tadpoles were supplied with yolk, and neither they nor the mother fed throughout this period. The stomach wall became thin and vascular, and gastric acid secretion was switched off in response to prostaglandin E_2, which was secreted by the egg capsules and developing tadpoles. When fully formed, the young were released through the mother's mouth over a period of about six days. The female arched her back and dilated her esophagus, and the young were ejected onto her tongue and climbed out. About four days after birth of the last young, the female resumed feeding, and the stomach converted to its pregestation condition.

CONSERVATION STATUS
This species is listed as Extinct by the IUCN and Environment Australia and has not been seen in the wild since 1981. The reasons for its disappearance remain unknown. The habitat was logged during their persistence in large numbers, and few frogs were collected for scientific purposes, but dead and dying frogs were seen in 1977.

SIGNIFICANCE TO HUMANS
The mechanisms for switching off acid secretion by these frogs are the same used in medicine today for gastric ulcers. ◆

Resources

Books
Anstis, M. *Tadpoles of South-eastern Australia: A Guide with Keys.* Sydney: Reed New Holland, 2002.

Barker, John, Gordon C. Grigg, and Michael J. Tyler. *A Field Guide to Australian Frogs.* Chipping Norton, Australia: Surrey Beatty, 1995.

Campbell, A., ed. *Declines and Disappearances of Australian Frogs.* Canberra, Australia: Environment Australia, 1999.

Cogger, H. G. *Reptiles and Amphibians of Australia.* 6th edition. Sydney: Reed New Holland, 2001.

Cogger, H. G., E. E. Cameron, and H. M. Cogger. *Zoological Catalogue of Australia.* Vol. 1, *Amphibia and Reptilia.* Canberra, Australia: Australian Government Publishing Service, 1983.

Ehmann, H., and G. Swan. "Reproduction and Development in the Marsupial Frog *Assa darlingtoni* (Leptodactylidae: Anura)." In *Biology of Australasian Frogs and Reptiles*, edited by G. Grigg, R. Shine, and H. Ehmann. Chipping Norton, Australia: Surrey Beatty and Sons, 1985.

Littlejohn, M. J., M. Davies, J. D. Roberts, and G. F. Watson. "Family Myobatrachidae." In *Fauna of Australia.* Vol. 2A, *Amphibia and Reptilia*, edited by C. J. Glasby, G. J. B. Ross, and P. Beesley. Canberra, Australia: AGPS, 1993.

Resources

Roberts, J. D. "The Biology of *Arenophryne rotunda* (Anura: Myobatrachidae): A Burrowing Frog from Shark Bay, Western Australia." In *Research in Shark Bay, Report of the France-Australe Bicentennary Expedition Committee*, edited by P. F. Berry, S. D. Bradshaw, and B. R. Wilson. Perth, Australia: West Australian Museum, 1990.

Tyler, M. J., ed. *The Gastric Brooding Frog.* London and Canberra: Croom Helm, 1983.

Periodicals

Byrne, P. G., and J. D. Roberts. "Simultaneous Mating with Multiple Males Reduces Fertilization Success in the Myobatrachid Frog *Crinia georgiana*." *Proceedings of the Royal Society Biological Sciences Series B* 266, no. 1420 (1999): 717–721.

Liem, D. S., and W. Hosmer. "Frogs of the Genus *Taudactylus* with Descriptions of Two New Species (Anura: Leptodactylidae)." *Memoirs of the Queensland Museum* 16, no. 3 (1973): 435–457.

McDonald, K. R. *Rheobatrachus* Liem and *Taudactylus* Straughan and Lee (Anura: Leptodactylidae) in Eungella National Park, Queensland: Distribution and Decline. *Transactions of the Royal Society of South Australia* 114, no. 4 (1990): 187–194.

Mitchell, N., and R. Swain. "Terrestrial Development in the Tasmanian Frog *Bryobatrachus nimbus* (Anura: Myobatrachinae): Larval Development and a Field Staging Table." *Papers and Proceedings of the Royal Society of Tasmania* 130, no. 1 (1996): 75–80.

Mitchell, N., R. Swain, and R. S. Seymour. "Effects of Temperature on Energy Cost and Timing of Embryonic and Larval Development of the Terrestrially Breeding Frog *Bryobatrachus nimbus*." *Physiological and Biochemical Zoology* 73, no. 6 (2000): 829–840.

Read, K., J. S. Keogh, I. A. W. Scott, J. D. Roberts, and P. Doughty. "Molecular Phylogeny of the Australian Frog Genera *Crinia*, *Geocrinia* and Allied Taxa (Anura: Myobatrachidae)." *Molecular Phylogenetics and Evolution* 21, no. 2 (2001): 294–308.

Retallick, R. W. R., and J.-M. Hero. "The Tadpoles of *Taudactylus eungellensis* Liem and Hosmer and *T. liemi* Ingram (Anura: Myobatrachidae) and a Key to the Stream-Dwelling Tadpoles of the Eungella Rainforest in East-central Queensland, Australia." *Journal of Herpetology* 32, no. 2 (1998): 304–309.

Rounsevell, D. E., D. Ziegeler, P. B. Brown, M. Davies, and M. J. Littlejohn. "A New Genus and Species of Frog (Anura: Leptodactylidae: Myobatrachinae) from Southern Tasmania." *Transactions of the Royal Society of South Australia* 118, no. 3 (1994): 171–185.

Tyler, M. J. "*Crinia tchudi* (Anura: Leptodactylidae) from the Cainozoic of Queensland, with the Description of a New Species." *Transactions of the Royal Society of South Australia* 115, no. 2 (1991): 99–101.

Organizations

Environment Australia. GPO Box 787, Canberra, ACT 2601 Australia. Phone: 61 (2) 6274-1111. Web site: <http://www.ea.gov.au>

Margaret Davies, PhD

▲
Leptodactylid frogs
(Leptodactylidae)

Class Amphibia
Order Anura
Family Leptodactylidae

Thumbnail description
Small to large terrestrial and aquatic frogs with arciferal pectoral girdles, usually with teeth on the upper jaw, and no intercalary elements between the penultimate and terminal phalanges of digits

Size
0.4–10 in (10–250 mm)

Number of genera, species
45 genera; 1,124 species

Habitat
Tropical rainforest, temperate rainforest, semiarid grasslands, montane forests, and grassland above tree line

Conservation status
Critically Endangered: 5 species; Endangered: 2 species; Vulnerable: 13 species; Data Deficient: 18 species

Distribution
Leptodactylids occur throughout South America, the West Indies, Central America, and Mexico and also range into the extreme southern United States

Evolution and systematics

Fossils of the subfamily Telmatobiinae are known from Paleocene, Eocene, and Oligocene deposits in Brazil and Argentina. Ceratophryine fossils are known from the Miocene and Pliocene of Argentina, and early Tertiary fossils of *Eleutherodactylus* are known from the West Indies.

Formerly, Leptodactylidae included the South African *Heleophryne*, now placed in its own family (Heleophrynidae), and two subfamilies in Australia, now recognized as Limnodynastidae and Myobatrachidae; Limnodynastidae also included *Cyclorana*, now recognized as a pelodryadine hylid. Most of the features of leptodactylids are primitive for neobatrachians. There is no compelling evidence that the family is monophyletic, and it probably is paraphyletic with respect to several other neotropical families of frogs.

Classification within Leptodactylidae has not been stable; herein seven subfamilies are recognized. Of these, Cycloramphinae, Eleutherodactylinae, and Odontophryinae commonly have been recognized as tribes within Telmatobiinae.

Ceratophryinae

This group consists of medium to large frogs with broad heads, large mouths, robust bodies, and relatively short limbs. The skull is massive and casqued, and the dermal roofing bones are exostosed (with bony outgrowths). The sternum is cartilaginous. The transverse processes on the anterior vertebrae are greatly expanded, and the sacral diapophyses are rounded. The terminal phalanges are knoblike, and dermal glandular pads are absent on the dorsal surfaces of the tips of the digits. The usual karyotype consists of 13 pairs of chromosomes, but some *Ceratophrys* are polyploids with as many 52 pairs of chromosomes. Pigmented aquatic eggs hatch into carnivorous tadpoles. The subfamily is widely distributed in the tropical lowlands of South America from northern Argentina northward to northern Colombia. It contains 3 genera and 12 species: *Ceratophrys* (8 species), *Chacophrys* (1 species), and *Lepidobatrachus* (3 species).

Cycloramphinae

This group contains small to medium-sized frogs with normal heads and limbs. The skull is not casqued, and the dermal roofing bones are not exostosed. The sternum is

Thoropa miliaris, of the subfamily Telmatobiinae. (Photo from Natural History Museum, University of Kansas. Reproduced by permission.)

cartilaginous but has an osseous plate in *Paratelmatobius*. The transverse processes on the anterior vertebrae are short, and the sacral diapophyses are rounded or dilated. The terminal phalanges are knoblike, and dermal glandular pads are absent on the dorsal surfaces of the digital pads. Chromosomes are in 13 pairs. Eggs are deposited in moist situations and hatch as stream-inhabiting tadpoles or as nonfeeding tadpoles that complete their development in terrestrial nests. The subfamily is restricted to southeastern Brazil. It contains 8 genera and 44 species: *Crossodactylodes* (3 species), *Cycloramphus* (25 species), *Paratelmatobius* (6 species), *Rupirana* (1 species), *Scythrodes* (1 species), *Thoropa* (5 species), and *Zachaenus* (3 species).

Eleutherodactylinae

This group consists of small to medium-sized frogs, mostly with normal heads and limbs. The skull is not casqued, and the dermal roofing bones are not exostosed. The sternum is cartilaginous. The transverse processes on the anterior vertebrae are short, and the sacral diapophyses are rounded. The terminal phalanges are knoblike or T-shaped, and paired dermal glandular pads are absent on the dorsal surfaces of the terminal digits. Chromosomes are in 13 pairs in most genera (9 in *Holoaden*) but vary from 9 to 17 pairs in *Eleutherodactylus*. A few large, unpigmented eggs are deposited on land or in bromeliads. Eggs hatch as froglets; there is no aquatic larval stage. At least one species, *Eleutherodactylus jasperi*, gives birth to living young. The subfamily is widely distributed in South America from northern Argentina northward; *Eleutherodactylus* also occurs throughout the West Indies, Central America, Mexico, and southern Texas and Florida in the United States. It contains 12 genera and 745 species: *Adelophryne* (5 species), *Atopophrynus* (1 species), *Barycholos* (2 species), *Dischidodactylus* (2 species), *Eleutherodactylus* (689 species), *Euparkerella* (4 species), *Geobatrachus* (1 species), *Holoaden* (2 species), *Ischnocnema* (5 species), *Phrynopus* (29 species), *Phyllonastes* (6 species), and *Phyzelaphryne* (1 species).

Hylodinae

This subfamily contains small to large frogs with normal heads and limbs. The skull is not casqued, and the dermal

roofing bones are not exostosed. The sternum is cartilaginous but tends to calcify in old adults. The transverse processes on the anterior vertebrae are short, and the sacral diapophyses are rounded. The terminal phalanges are T-shaped, and a pair of dermal glandular pads is present on the dorsal surfaces of the digital pads. Chromosomes are in 3 pairs. Pigmented eggs deposited in ponds or streams hatch into herbivorous tadpoles with two upper and three lower rows of labial teeth. The subfamily is restricted to southeastern Brazil and extreme northeastern Argentina. It contains 3 genera and 34 species: *Crossodactylus* (10 species), *Hylodes* (19 species), and *Megaelosia* (5 species).

Leptodactylinae

This subfamily consists of small to large frogs with normal heads and limbs. The skull is not casqued, and the dermal roofing bones are not exostosed. The sternum consists of a bony style. The transverse processes on the anterior vertebrae are not expanded, and the sacral diapophyses are rounded or slightly dilated. In most genera, the terminal phalanges are knoblike, but they are T-shaped in *Lithodytes*, and dermal glandular pads are absent on the dorsal surfaces of the terminal digits. The chromosome complement consists of 10–13 pairs, except for two tetraploid species of *Pleurodema* that have 22 pairs. Most genera deposit eggs in aquatic foam nests, with tadpoles hatching as free-living herbivrous tadpoles, but eggs are laid in clumps or strings in *Limnomedusa*, *Pseudopaludicola*, and some *Pleurodema*; eggs of *Adenomera* are in terrestrial foam nests and usually hatch as nonfeeding larvae. The subfamily occurs throughout South America and tropical and subtropical Mesoamerica as far north as southern Texas, United States, and also in the West Indies. It contains 9 genera and 152 species: *Adenomera* (7 species), *Edalorhina* (2 species), *Hydrolaetare* (1 species), *Leptodactylus* (66 species), *Limnomedusa* (1 species), *Lithodytes* (1 species), *Physalaemus* (41 species), *Pleurodema* (12 species), and *Pseudopaludicola* (11 species).

Odontophryinae

This group contains medium-sized frogs with robust bodies and relatively large heads. The skull is not casqued, and the dermal bones are not exostosed except in *Proceratophrys*. The sternum is cartilaginous. The transverse processes on the anterior vertebrae are not widely expanded, and the sacral diapophyses are rounded or slightly dilated. The terminal phalanges are knoblike, and dermal glandular pads are absent on the dorsal surfaces of the terminal digits. The known chromosome complement is 11 pairs. Eggs are deposited in ponds and hatch into herbivorous tadpoles. The subfamily ranges from eastern Brazil to central Argentina. It contains 3 genera and 27 species: *Macrogenioglottus* (1 species), *Odontophrynus* (9 species), and *Proceratophrys* (17 species).

Telmatobiinae

This subfamily contains the basal leptodactylids that have normal heads and bodies. The skull is not casqued and dermal roofing bones are not exostosed, except in *Caudiverbera*. The sternum is cartilaginous but tends to calcify in old adults. The transverse processes on the anterior vertebrae are short, and the sacral diapophyses are rounded. The terminal phalanges are knoblike (T-shaped in *Batrachyla*), and dermal glandular pads are absent on the dorsal surfaces of the terminal digits. Chromosomes are in 13 pairs. Pigmented eggs deposited in water hatch into herbivorous tadpoles. The subfamily is restricted to the temperate forests and Patagonian Region of southern Chile and Argentina but extends northward in the Andes to Ecuador. It contains 11 genera and 92 species: *Alsodes* (14 species), *Atelognathus* (8 species), *Batrachophrynus* (2 species), *Batrachyla* (5 species), *Eupsophus* (8 species), *Hylorina* (1 species), *Insuetophrynus* (1 species), *Somuncuria* (1 species), *Telmatobius* (47 species), and *Telmatobufo* (3 species).

Physical characteristics

Leptodactylids range in size from minute species of *Eleutherodactylus* with a snout-vent length of only 0.4 in (10 mm) to large terrestrial species (*Ceratophrys aurita*) and aquatic species (*Telmatobius culeus*) with snout-vent lengths of 10 in (250 mm). Body shape varies from robust toadlike species (e.g., *Odontophrynus*) with extremely large heads (ceratophryines) to dorsoventrally flattened aquatic species (e.g., *Atelognathus* and some *Telmatobius*) with loose flaps of skin. Some long-legged, terrestrial species (e.g., some *Eleutherodactylus* and some *Leptodactylus*) resemble ranids but lack webbing between the toes. Some other arboreal *Eleutherodactylus* have expanded digits.

All members of the family have eight separated presacral vertebrae, except that the first and second are fused in *Telmatobufo*. The two halves of the pectoral girdle overlap midventrally to produce the arciferal condition. Usually the pectoral girdle contains two cartilaginous elements, the sternum and omosternum; in leptodactylines, the sternum has a bony style and in *Paratelmatoibius*, a bony plate. Maxillary and premaxillary bones usually bear teeth. The terminal phalanges of the digits are knoblike or T-shaped. The skin on the dorsum varies from smooth (with or without longitudinal ridges) to pustular or tubercular. Species of *Ceratophrys* and *Proceratophrys* have fleshy eyelid "horns," and *Edalorhina* and many species of *Eleutherodactylus* have elongate tubercles on the snout, eyelids, and/or heels. The constricted pupil on the eye is horizontally elliptical in most leptodactylids, but it is vertically elliptical in some telmatobiines (*Caudiverbera*, *Hylorina*, and *Telmatobufo*), leptodactylines (*Hydrolaetare* and *Limnomedusa*), and one ceratophryine (*Lepidobatrachus*).

Dorsally most leptodactylids are varying shades of gray, brown, or dull green, and the venter usually is dull white or cream. However, many species of *Eleutherodactylus* have pale longitudinal stripes and/or bright flash colors on the flanks or limbs that are not visible when the frog is in a resting position. The striped pattern is most evident in the black *Lithodytes lineatus*, which also has red spots in the groin and on the thighs.

Most leptodactylid tadpoles have a globular body with a single sinistral spiracle and well-developed caudal fins; the oral disc usually has keratinized jaw sheaths and two anterior and three posterior rows of labial teeth. Tadpoles of *Lepidobatrachus* have paired spiracles and lack keratinized mouthparts. Some stream-inhabiting tadpoles (e.g., *Cycloramphus* and *Thoropa*) have long, muscular tails with extremely low fins.

South American leptodactylid frog (*Eleutherodactylus*) in Peru. (Photo by Animals Animals ©Paul Freed. Reproduced by permission.)

Distribution

With the exception of the Atacama Desert, leptodactylids occur throughout South America from the Straits of Magellan northward; they range from sea level to 16,200 ft (5,000 m) in the Andes. In so doing, the family contains the southernmost frog in the world (*Pleurodema bufonina*) and the species reaching the highest elevation in the New World (*Pleurodema marmorata*). *Leptodactylus* ranges northward to southern Texas, United States, and on Hispaniola, Puerto Rico, and the Lesser Antilles in the West Indies. *Eleutherodactylus* occurs throughout the West Indies and Mesoamerica to southwestern United States. The only other genera not confined to South America

The South American bullfrog (*Leptodactylus pentadactylus*) is relatively large and has vibrant colors. (Photo by Animals Animals ©Patti Murray. Reproduced by permission.)

Surinam horned frog (*Ceratophrys cornuta*) digs itself backwards into the leaves and waits for a meal to come by. It may remain motionless for hours at a time. (Photo from Natural History Museum, University of Kansas. Reproduced by permission.)

are *Pleurodema*, which extends into Panama, and *Physalaemus*, which ranges northward into Mexico.

Habitat

Leptodactylids occur wherever moisture is present sometime during the year. Ceratophryines mostly inhabit dry regions, but three species inhabit humid forests. Cycloramphines, hylodines, and eleutherodactylines mostly inhabit humid forests, but some eleutherodactylines occur above the tree line in the Andes; many *Eleutherodactylus* are arboreal in tropical forests. Leptodactylines occur in semiarid regions as well as humid forests, and three species of *Pleurodema* exist above tree line in the Andes; *Hydrolaetare* is aquatic. Odontophryines inhabit humid forests, grasslands, and semiarid regions. Telmatobiines are most diverse in humid temperate forests but also range into semiarid regions; *Telmatobius* and *Batrachophrynus* inhabit lakes and streams in the high Andes, and *Somuncuria* inhabits streams originating from hot springs.

Behavior

Most leptodactylids are nocturnal; daytime retreats are under logs or leaf litter, in burrows, or in bromeliads or other epiphytes. However, hylodines are diurnal in mesic montane environments. Two genera of leptodactylines, *Edalorhina* and *Pseudopaludicola*, also are diurnal. Even at high elevations when nighttime temperatures are only slightly above freezing, many species of *Eleutherodactylus* and *Phrynopus* and three species of *Pleurodema* are active at night.

The large, carnivorous *Ceratophrys* secret themselves in shallow excavations amidst leaf litter with only the tops of their heads visible. During the dry season, *Lepidobatrachus* burrow into the mud in the bottoms of drying ponds; once underground, they shed successive layers of skin that harden into a cocoon that protects them from desiccation.

Escape behavior in most leptodactylids consists of leaping away from potential predators, but some (e.g., *Ceratophrys* and *Edalorhina*) sit still and rely on their cryptic coloration and disruptive outlines to avoid predators; this is accompanied by stretching out the limbs in the cryptically colored *Proceratophrys appendiculata*. When disturbed, *Caudiverbera* and some species of *Leptodactylus* inflate their lungs and thereby increase their size to a potential predator. Many species of *Physalaemus* and *Pleurodema* have a pair of large, elevated, and brightly colored glands on the posterior part of the body. These frogs assume a defensive posture by lowering their heads and elevating the posterior part of the body, thereby presenting the glands to the potential predator. These glands have been interpreted as "eyespots" and can be construed by the predator as representing a much larger organism.

Feeding ecology and diet

Most leptodactylids are sit-and-wait predators on small arthropods. But *Caudiverbera*, *Ceratophrys*, and large species of *Leptodactylus* also feed on other vertebrates, including frogs, lizards, and small snakes, birds, and mammals. Some species of *Eleutherodactylus* feed only on ants, and *Physalaemus* feeds almost exclusively on termites.

Reproductive biology

Leptodactylids living in seasonal environments and at least some living in continuously humid forests have defined breeding seasons usually associated with the beginning of the rainy season. Those species living in continuously humid environments may breed several times a year.

Males of most species of leptodactylids vocalize to attract females. Calls vary from a single "peep" or series of short notes in various species of *Eleutherodactylus* to a loud "baaa" in *Ceratophrys* and a loud "whoorup" in some *Leptodactylus*. At least some species of *Physalaemus* and *Eleutherodactylus* have more complex calls consisting of notes that are territorial and others that are courtship calls. Selection of a mate seems to be mostly by female choice. Once a female approaches a male, he grasps her from above with his hands in her armpits (axillary amplexus), except in the telmatobiine *Batrachyla*, in which amplexus is around the waist.

All ceratophryines, cycloramphines, odontophryines, odontophryines, and telmatobiines deposit their eggs in water or at the edge of water and have aquatic tadpoles. Clutches vary from a few dozen to hundreds of eggs, mostly depending on the size of the frog. Most leptodactylines construct a foam nest by the pair or with only the male kicking the mucous secreted with the eggs and trapping air bubbles within it. The nests float on water or are constructed in depressions that become inundated shortly after the eggs hatch. The small leptodactyline *Adenomera* has terrestrial foam nests, and the eggs hatch as nonfeeding larvae that complete their development in the nest. The foam nests develop a sticky exterior and contain moisture within, thereby protecting the eggs from desiccation. These frogs commonly deposit their eggs earlier than sympatric pond-breeders, and therefore in temporary ponds the tadpoles get an early start before potential competitors and predators.

Insofar as known, all eleutherodactylines deposit their eggs in moist situations on the ground or in epiphytic plants. Clutches usually contain fewer than 50 relatively large eggs that undergo direct development, thereby eliminating the aquatic larval stage. One species, *Eleutherodactylus jasperi*, is known to give birth to one to six living young; fertilization is internal and the eggs are retained in the oviducts.

Developmental time is highly variable. In most species with aquatic eggs, hatching occurs three to five days after deposition and the larval period lasts for four to nine weeks, but in *Lepidobatrachus* and *Odontophrynus* eggs hatch within two days and the larval period is only about three weeks. In contrast, in some telmatobiines, which deposit eggs in cold water, ovarian development may require as long as 20 days and the larval period lasts up to two years. In eleutherodactylines, development from time of fertilization to hatching of froglets usually is only three to four weeks, and the development period in *Eleutherodactylus jasperi* is only 30 days.

Parental care in the form of male attendance of terrestrial or arboreal clutches of eggs is common among *Eleutherodactylus* in the West Indies, but only a few instances of parental care (by females) are known among *Eleutherodactylus* on the mainland. In some species of *Leptodactylus*, the female remains with the foam nest during embryonic development; after hatching the tadpoles remain closely associated with the female. Female *Leptodactylus bolivianus* have been observed to modify the depth of the pond or to guide the school of tadpoles to deeper water, thereby protecting their tadpoles from possible desiccation. Females of *Leptodactylus fallax* remain with the foam nest; when the larvae hatch, the females deposit unfertilized eggs in the foam nest and the larvae feed on the eggs.

Conservation status

Many leptodactylids are threatened by habitat destruction. Possibly several species are extinct, including the large *Eleutherodactylus karlschmidti* and the live-bearing *Eleutherodactylus jasperi* in Puerto Rico. The 2002 IUCN Red List includes 36 species: 5 are categorized as Critically Endangered; 2 as Endangered; 13 as Vulnerable; and 18 as Data Deficient.

Significance to humans

Several large leptodactylids (*Caudiverbera*, *Batrachophrynus*, *Leptodactylus*, and *Telmatobius*) are consumed by humans. In the Andes of Peru and Bolivia, *Telmatobius culeus* are captured by "raneros" in Lago Titicaca, and *Batrachophrynus macrostomus* are likewise taken from Lago Junín. Restaurants in villages near these lakes commonly advertise that frogs are on the menu. *Ceratophrys* have become popular in the pet trade and are bred in captivity for this purpose. The Puerto Rican *Eleutherodactylus johnstonei* has been introduced intentionally into Colombia and Venezuela because people who vacationed in Puerto Rico were enamored by the call. In contrast, the unintentional introduction of *Eleutherodactylus coqui* in Hawaii has caused distress among inhabitants—they are unused to nocturnal vocalization, because there are no frogs native to Hawaii.

1. Warty tree toad (*Hylodes asper*); 2. Túngara frog (*Physalaemus pustulosus*); 3. Gray four-eyed frog (*Pleurodema bufonina*); 4. Gold-striped frog (*Lithodytes lineatus*); 5. South American bullfrog (*Leptodactylus pentadactylus*); 6. Cururu lesser escuerzo (*Odontophrynus occidentalis*); 7. Rock River frog (*Thoropa miliaris*); 8. Titicaca water frog (*Telmatobius culeus*). (Illustration by Dan Erickson)

1. Surinam horned frog (*Ceratophrys cornuta*); 2. Patagonia frog (*Atelognathus patagonicus*); 3. Puerto Rican coqui (*Eleutherodactylus coqui*);
4. Budgett's frog (*Lepidobatrachus laevis*); 5. Golden coqui (*Eleutherodactylus jasperi*); 6. Helmeted water toad (*Caudiverbera caudiverbera*);
7. Perez's snouted frog (*Edalorhina perezi*); 8. Emerald forest frog (*Hylorina sylvatica*). (Illustration by Dan Erickson)

Species accounts

Surinam horned frog
Ceratophrys cornuta

SUBFAMILY
Ceratophryinae

TAXONOMY
Rana cornuta Linnaeus, 1758, "Virginia" (in error).

OTHER COMMON NAMES
English: Horned frog, packman frog.

PHYSICAL CHARACTERISTICS
This large, robust frog has an immense head, the width of which is about one-half of the snout-vent length, which is as great as 3.1 in (80 mm) in males and 4.7 in (120 mm) in females. The skin on the dorsum and flanks is finely rugose with conical tubercles, and the venter is nearly smooth. A distinguishing feature is the presence of a large, triangular, dermal process ("eyelid horn") extending upward on each eyelid. The fingers are unwebbed, and the hind limbs are moderately short with toes that are about one-half webbed. The dorsum is green or brown with brown markings, and the venter is dull cream except for a dark brown or black throat. The iris is creamy tan with brown flecks. Breeding males have tan nuptial excrescences on the thumbs.

■ *Ceratophrys cornuta*
■ *Lithodytes lineatus*
■ *Pleurodema bufonina*

DISTRIBUTION
This species is widely distributed in the Amazon Basin and Guianan Region in South America.

HABITAT
This frog is a denizen of lowland tropical rainforest.

BEHAVIOR
Using its cryptic color pattern as camouflage, this frogs wriggles into the leaf litter on the forest floor, so that only the head is exposed. Individuals may remain in the same place for several days and nights before moving to another site on rainy nights.

FEEDING ECOLOGY AND DIET
A classic sit-and-wait predator, *Ceratophrys cornuta* apparently will eat anything that moves by it and is not too large to swallow. It makes a short lunge at its prey, which consists of ants, spiders, and other small arthropods, but the bulk of its prey are large grasshoppers, frogs, and even snakes, lizards, and mice.

REPRODUCTIVE BIOLOGY
This is an explosive breeder at the time of the first heavy rains of the rainy season. Males call from the edges of ponds or while sitting in shallow water; the call is a low-pitched "baaa." Amplexus is axillary, and clutches of up to 2,000 small, pigmented eggs are deposited in water. Tadpoles attain a total length of about 2.5 in (65 mm). The body is broadly ovoid with a bluntly rounded snout and small eyes directed dorsolaterally. The oral disc is large and directed anteriorly. The jaw sheaths are massive; a long, pointed median process on the lower sheath inserts into a notch on the upper sheath; there are 13 rows of labial teeth on the upper lip and eight rows on the lower lip. The tadpoles are voracious carnivores and feed on other tadpoles in the pond and even are cannibalistic. Feeding is a gape-and-suck process, during which the prey is punctured by the process on the lower jaw sheath and quickly ingested; attack and swallowing takes only about five seconds.

CONSERVATION STATUS
Although *Ceratophrys cornuta* is locally abundant throughout its range, clearing of forest is restricting its habitat. It is not listed by the IUCN.

SIGNIFICANCE TO HUMANS
This species is found in the pet trade. ◆

Budgett's frog
Lepidobatrachus laevis

SUBFAMILY
Ceratophryinae

TAXONOMY
Lepidobatrachus laevis Budgett, 1899, Paraguayan Chaco, South America.

OTHER COMMON NAMES
Guaraní: Kururú chiní.

☐ *Thoropa miliaris*
■ *Lepidobatrachus laevis*

PHYSICAL CHARACTERISTICS
Adults of this broad-headed frog with a flattened body attain snout-vent lengths of 4.5–5.1 in (110–130 mm). The eyes are small and close together on the top of the head. The snout is broad and sloping. The fingers are unwebbed, and the toes are nearly fully webbed. A large, spade-like, black inner metatarsal tubercle is present on the base of each hind foot. The dorsal skin is glandular, and the skin on the venter is granular. The dorsum is dull brown or gray with faintly darker blotches or paler streaks; the belly is white. The iris is pale cream, and the pupil is round.

DISTRIBUTION
Budgett's frog occurs only in the dry Chaco Region in northern Argentina and southern Paraguay.

HABITAT
This species inhabits dry scrub forest.

BEHAVIOR
This frog is active only during the short rainy season, November through January, when individuals swim in temporary ponds. As the ponds dry up toward the end of the rainy season, the frogs burrow backwards, using the spade-like tubercles on the hind feet, deep in the mud in the bottoms of ponds. Once below the surface, they shed the outer layers of skin several times; this skin forms an impermeable cocoon that protects the frog from desiccation during the long dry season. With the advent of following rainy season, moisture softens the cocoon, and the frogs emerge into the water, eat the shed skin, and begin a new season of activity. Budgett's frog is aggressive and opens its large mouth as a defensive posture.

FEEDING ECOLOGY AND DIET
Apparently most feeding takes place in the water. These frogs eat snails and smaller frogs; in captivity they also will eat fish.

REPRODUCTIVE BIOLOGY
Males call while floating on the water; the call is a loud "eeee." Amplexus is axillary. As many as 1,200 small pigmented eggs are laid in water, and these sink to the bottom, where they hatch in about 18 hours. The tadpoles, which are carnivores, metamorphose about 20 days after hatching. The tadpoles, which reach a total length of about 2 in (50 mm), have broad, depressed bodies, paired spiracles, and large mouths with weak labial teeth and no horny jaw sheaths. The tadpoles feed on smaller tadpoles, which they swallow whole.

CONSERVATION STATUS
Populations seem to be stable, and the species is not listed by the IUCN.

SIGNIFICANCE TO HUMANS
None known. ◆

Rock river frog
Thoropa miliaris

SUBFAMILY
Cyclorampinae

TAXONOMY
Rana miliaris Spix, 1824, "Amazon River" (in error).

OTHER COMMON NAMES
None known.

PHYSICAL CHARACTERISTICS
Males attain a maximum snout-vent length of 2.8 in (71 mm) and females, 3.2 in (81 mm). The head is broad with a rounded snout and large, distinct tympanum. The skin of the dorsum is smooth to weakly granular with scattered tubercles; the venter is smooth. The fingers and toes lack webbing and have slightly swollen tips. Breeding males lack vocal slits but have small nuptial spines on the thumb and first and second fingers. The dorsum is tan or brown, and the groin is dull yellow; the throat and belly are gray and the anterior and posterior surfaces of the thighs are dull yellow with dark brown bars. The iris is reddish copper with black reticulations.

DISTRIBUTION
Thoropa miliaris ranges in the Atlantic Coast Forest from Espírito Santa to São Paulo in southeastern Brazil.

HABITAT
This species inhabits humid tropical and subtropical forests.

BEHAVIOR
This species is nocturnal and terrestrial and is most common along streams.

FEEDING ECOLOGY AND DIET
Presumably the diet includes small arthropods.

REPRODUCTIVE BIOLOGY
Males call from rock faces along streams; the call is a short, pulsed, low-pitched note. Eggs are deposited in streams. Tadpoles wriggle onto wet rocks faces. They have depressed bodies, long and muscular tails without noticeable fins, and ventrally directed oral discs with slender jaw sheaths and two anterior and three posterior rows of labial teeth.

CONSERVATION STATUS
Although not listed by the IUCN, this species is threatened by habitat destruction.

SIGNIFICANCE TO HUMANS
None known. ◆

Puerto Rican coqui
Eleutherodactylus coqui

SUBFAMILY
Eleutherodactylinae

TAXONOMY
Eleutherodactylus coqui Thomas, 1966, 7.3 mi (11.8 km) south of Palmer, Puerto Rico.

OTHER COMMON NAMES
Spanish: Coquí.

PHYSICAL CHARACTERISTICS
Males attain a snout-vent length of 2 in (50 mm) and females, 2.5 in (63 mm). The dorsum is shagreen with scattered small tubercles, and the venter is areolate. The snout is subacuninate, and the tympanum is distinct. The fingers and toes are long, unwebbed, and bear terminal, expanded, truncate discs. The dorsum is various shades of brown, commonly with a middorsal or pair of dorsolateral creamy tan stripes. A distinct dark brown bar extends from the nostrils through the reddish bronze eye to a point above the tympanum. The venter is grayish white.

Eleutherodactylus coqui
Physalaemus pustulosus
Edalorhina perezi

DISTRIBUTION
This frog occurs throughout Puerto Rico to elevations of 3,900 ft (1,200 m). It has been introduced on St. Thomas and St. Croix in the U.S. Virgin Islands, and into southern Florida, Louisiana, and Hawaii, United States.

HABITAT
The Puerto Rican coqui lives in nearly all regions of Puerto Rico; it inhabits humid montane forest, dry forest, gardens, and houses.

BEHAVIOR
This strictly nocturnal species takes refuge under objects, in axils of palms, and especially in bromeliads. At night it is active on the ground but usually on vegetation to heights of more than 50 ft (15 m). Individuals seldom move more than 20 ft (6.5 m) from their diurnal retreats, and when feeding at night they move no more than about 2 in (50 mm). Males establish territories by vocalization and are aggressive toward other males that enter their territories.

FEEDING ECOLOGY AND DIET
Feeding occurs on vegetation at night; the frogs consume vast quantities of insects, principally ants, crickets, and roaches, as well as spiders, snails, and even small frogs.

REPRODUCTIVE BIOLOGY
Breeding occurs throughout the year but is reduced in the driest times of the year (January through March). Males call at night. The call is a multiple note, "co-qui." The "co" solicits response from females, whereas the "qui" is a territorial call, which is repeated rapidly upon the intrusion of another male. Amplexus involves the male sitting on the body of the female with his arms around her body; fertilization is internal via cloacal apposition. Clutches of about 26 eggs are deposited on leaves of bromeliads or other plants. The female abandons the eggs, which are attended by the male, who commonly places his body over the eggs. Development is direct into a froglet within the egg capsule and requires 17–26 days. Late embryos develop a tubercle on the tip of the snout ("egg-tooth") that is used to rip open the capsule. Hatchings are about 0.23 in (6 mm) long. The frogs reach sexual maturity in less than one year and have a life span of four to five years. Females can breed as often as every 58 days.

CONSERVATION STATUS
This ubiquitous species is common throughout its range.

SIGNIFICANCE TO HUMANS
Exportation of plants, especially bromeliads, from Puerto Rico has resulted in the accidental introduction of the Puerto Rican coqui on the U.S. Virgin Islands, into Florida and Louisiana, and Hawaii, where there are no native frogs. People in Hawaii complain about the nocturnal "noise" made by the coqui. Because of its abundance and ease for study in Puerto Rico, this species has been investigated more thoroughly than any other tropical anuran. ◆

Golden coqui
Eleutherodactylus jasperi

SUBFAMILY
Eleutherodactylinae

TAXONOMY
Eleutherodactylus jasperi Drewry and Jones, 1976, 3.7 mi (6 km) southeast of Cayey, Puerto Rico.

Eleutherodactylus jasperi

Leptodactylus pentadactylus

Telmatobius culeus

OTHER COMMON NAMES
Spanish: Coquí dorado.

PHYSICAL CHARACTERISTICS
This small frog attains a maximum snout-vent length of 8.5 in (21.5 mm). The dorsum is shagreen, and the venter is areolate. The snout is bluntly rounded and nearly truncate in dorsal view; the tympanum is about one-half of the diameter of the eye. The fingers and toes are moderately long, unwebbed, and have rounded terminal discs. The dorsum is golden yellow to orange yellow, and the venter is pale yellow, except that the skin covering the abdomen is transparent. The iris is pale gray with black flecks.

DISTRIBUTION
This species has been known only from elevations of 2,100–2,750 ft (650–850 m) in the Sierra de Cayey, Puerto Rico.

HABITAT
This strictly nocturnal frog inhabits arboreal bromeliads in subhumid forest.

BEHAVIOR
This small nocturnal species seeks shelter in bromeliads by day.

FEEDING ECOLOGY AND DIET
Presumably the diet includes small arthropods.

REPRODUCTIVE BIOLOGY
Males call from bromeliads at night; the call consists of a series of six to eight notes, "tuit-tuit-tuit-tuit." *Eleutherodactylus jasperi* is the only member of the family that is known to have internal fertilization and give birth to living young. The species is ovoviparous, in that the eggs are retained in the oviduct and the yolk within the egg capsule supplies all nutrition. The eggs

are up to 0.2 in (5 mm) in diameter and require about 30 days to develop into froglets. The number of young is three to five; upon birth they are 0.3 in (7 mm) long and contain a large amount of yolk in the abdomen.

CONSERVATION STATUS
Although listed as Data Deficient by the IUCN, this species is presumed to be extinct; it was last observed in 1981.

SIGNIFICANCE TO HUMANS
Because of its nearly unique reproductive mode, this small frog was of immense interest to biologists, but only limited data were obtained before it disappeared. ◆

Warty tree toad
Hylodes asper

SUBFAMILY
Hylodinae

TAXONOMY
Elosia aspera Müller, 1924, Barreria, Rio de Janeiro, Brazil.

OTHER COMMON NAMES
None known.

PHYSICAL CHARACTERISTICS
Males attain a maximum snout-vent length of 1.7 in (43 mm) and females, 2 in (50 mm). The snout is rounded, and a tympanum is present but not always distinct. The fingers are unwebbed but fringed, and the toes are unwebbed; terminal

Hylodes asper

Odontophrynus occidentalis

Caudiverbera caudiverbera

segments of all digits are expanded, truncate, and have a pair of scutes on the dorsal surfaces. Males have vocal sacs that are expanded laterally but lack nuptial excrescences. The dorsum is dull brown with irregular darker brown to black markings, but the upper surfaces of the truncate digits are white. The upper lip is white with narrow brown bars. The venter is pale tan with darker mottling or spots, and the iris is pale bronze.

DISTRIBUTION
This species is distributed in the coastal mountain ranges from Rio de Janeiro to Santa Catarina in southeastern Brazil.

HABITAT
Hylodes asper inhabits humid montane forest.

BEHAVIOR
This diurnal terrestrial species is most commonly seen on rocks and low vegetation along mountain streams. Calling males also display by waving their hind feet one at a time.

FEEDING ECOLOGY AND DIET
Presumably the diet consists of small arthropods.

REPRODUCTIVE BIOLOGY
Males call by day from rocks at the edges of streams. The call is a long high-pitched whistling trill. Amplexus is axillary, and eggs are deposited in water, where they hatch into herbivorous tadpoles. The tadpoles have rather slender bodies and long tails with moderately low fins; the oral disc is directed ventrally and has heavy, coarsely serrate jaw sheaths and two anterior and three posterior rows of labial teeth.

CONSERVATION STATUS
Although not listed by the IUCN, this species is threatened by the great reduction in habitat that also affects other inhabitants of the Atlantic Coastal Forest in southeastern Brazil.

SIGNIFICANCE TO HUMANS
None known. ◆

Perez's snouted frog
Edalorhina perezi

SUBFAMILY
Leptodactylinae

TAXONOMY
Edalorhina perezi Jiménez de la Espada, 1870, Napo, Ecuador.

OTHER COMMON NAMES
None known.

PHYSICAL CHARACTERISTICS
Males of this small frog attain a snout-vent length of 1.2 in (30 mm) and females, 1.4 in (35 mm). The snout is short and truncate; prominent, pointed tubercles are present on the upper eyelid, and a distinct dorsolateral fold extends from the orbit to the groin. The dorsum may be tuberculate, smooth with a few scattered tubercles, or having several longitudinal ridges between the dorsolateral folds; the venter is smooth. The dorsum is gray or brown with or without reddish brown streaks, the flanks are black, and the venter is white with extensive black markings. The iris is grayish tan with a reddish copper ring around the pupil.

DISTRIBUTION
This species is distributed in the upper Amazon Basin from southern Colombia to northern Bolivia.

HABITAT
Edalorhina perezi inhabits lowland tropical rainforest.

BEHAVIOR
This diurnal species is active on the forest floor, where its cryptic coloration blends well with the leaf litter.

FEEDING ECOLOGY AND DIET
A great variety of small arthropods, including spiders, flies, crickets, and roaches are eaten while the frogs forage in the leaf litter.

REPRODUCTIVE BIOLOGY
Males call solitarily from the leaf litter by day; the call consists of three to five low whistles with two pulses per note. Amplexus is axillary, and the pair moves to a small body of water, usually temporary ponds, where 78–98 eggs are deposited in a foam nest constructed by the pair kicking the eggs, secretions, and water into a small, spherical mound that floats on the surface of the water. The eggs hatch in four to six days, and the tadpoles develop in water. Tadpoles attain a maximum total length of about 0.8 in (20 mm). The body is ovoid with a bluntly rounded snout and dorsally positioned eyes. The oral disc is directed anteroventrally; the jaw sheaths are finely serrate, and there are two rows of labial teeth on the anterior lip and three on the posterior lip. The body and caudal musculature are tan, and the belly is greenish yellow.

CONSERVATION STATUS
Not listed by the IUCN. However, as in the case of all inhabitants of the Amazonian rainforest, the continuous range of this species is being fragmented by clearing of the forest.

SIGNIFICANCE TO HUMANS
None known. ◆

South American bullfrog
Leptodactylus pentadactylus

SUBFAMILY
Leptodactylinae

TAXONOMY
Rana pentadactyla Laurenti, 1768, "Indiis" (Surinam).

OTHER COMMON NAMES
None known.

PHYSICAL CHARACTERISTICS
Males of this large, robust frog are slightly larger than females; they attain a maximum snout-vent length of 7.3 in (180 mm), whereas the maximum length in females is 6.9 in (176 mm). The body is robust; the head is large with an acutely rounded snout and prominent tympanum. The skin on the dorsum and venter is smooth, and a prominent dorsolateral dermal fold extends from the orbit to the groin. The fingers and toes are long with slender tips and lack webbing. Breeding males have greatly swollen forelimbs and one large, pointed, black spine on the inner surface of the thumb and two black spines on each side of the chest. The dorsum is tan to reddish brown with broad, reddish brown marks on the body between the yellowish tan dorsolateral folds. The dorsal surfaces of the limbs are tan to reddish brown with narrow transverse brown bars. The upper lip is tan with a brown margin and dark brown triangular spots. The venter is cream with bold dark brown to black mottling, especially on the belly and hind limbs. The iris is bronze.

DISTRIBUTION

This frog ranges in lowlands (below 3,800 ft or 1,200 m) from northern Honduras to the Pacific lowlands of Ecuador and throughout the Guianas and northern two-thirds of the Amazon Basin in South America.

HABITAT

Principally a denizen of tropical rainforest, this species also invades dry forest and lower montane forests.

BEHAVIOR

This nocturnal species spends its days in burrows, under logs, or hidden in leaf litter. Defensive mechanisms include noxious skin secretions and posturing by inflating the lungs and elevating the body on all four limbs. When grasped, these frogs usually emit a high-pitched scream.

FEEDING ECOLOGY AND DIET

Juveniles feed on small arthropods, but large adults feed on large arthropods, frogs, lizards, snakes, and small birds and mammals. Tadpoles are omnivorous, feeding on vegetation, tadpoles, and eggs, even of their own species.

REPRODUCTIVE BIOLOGY

Males call solitarily from margins of ponds and backwaters of streams; the call is a loud "whoorup" repeated at intervals of five to 10 seconds. An attracted female is grasped by the male by axillary amplexus and held firmly by the muscular forearms and nuptial spines on the thumbs and chest. About 1,000 eggs are deposited in a large foam nest by backward and forward motions of the male's hind limbs that mix air, water, eggs, and secretions into the nest, which usually is deposited in a depression adjacent to water. The eggs hatch in two to three days; subsequent rains flood the nest site, and the tadpoles move into the pond or slow-moving stream. Development is rapid, and metamorphosis occurs about four weeks after hatching. Tadpoles attain a maximum total length of about 3.3 in (83 mm). The body is ovoid with a rounded snout with large eyes directed dorsolaterally. The oral disc is nearly terminal and bears finely serrate jaw sheaths and two anterior and three posterior rows of labial teeth. The body and caudal musculature are brown.

CONSERVATION STATUS

Populations of this species seem to be stable. It is not listed by the IUCN.

SIGNIFICANCE TO HUMANS

Some indigenous people eat these frogs, after they have been washed thoroughly. ◆

Gold-striped frog

Lithodytes lineatus

SUBFAMILY

Leptodactylinae

TAXONOMY

Rana lineata Schneider, 1799, Guyana.

OTHER COMMON NAMES

None known.

PHYSICAL CHARACTERISTICS

Males attain a maximum snout-vent length of 1.8 in (45 mm) and females, 2.2 in (56 mm). The body is slender; the snout is rounded, and a distinct tympanum is present. The fingers and toes are unwebbed and slender with slightly dilated tips. The skin on the dorsum is finely spiculate, and the venter is smooth. The dorsum and flanks are black; a pair of broad yellow stripes extends from the tip of the snout to the groin. A large red spot is present in the groin, and a smaller red spot is present on the posterior surface of each thigh. The throat and chest are grayish brown, and the undersurfaces of the hind limbs are gray. The iris is coppery bronze. Males lack nuptial excrescences.

DISTRIBUTION

The species is widely distributed in the upper and middle Amazon Basin and in the Guianan region in northeastern South America.

HABITAT

This species is restricted to humid tropical lowland rainforest.

BEHAVIOR

Juveniles are active on the ground by day and night, whereas adults are primarily nocturnal. Adults have been found in association with the large earthen nest of leaf-cutting ants (*Atta*), and males are known to call from subterranean tunnels in these nests.

FEEDING ECOLOGY AND DIET

This species feeds on a variety of small arthropods and also earthworms on the forest floor.

REPRODUCTIVE BIOLOGY

Males call from mouths of burrows or other partially concealed sites; the call is a series of melodious notes. About 200 unpigmented eggs are deposited in a foam nest constructed at the edge of water. The tadpoles remain in the foam nest for seven to 15 days after hatching and then disperse into the water. Tadpoles metamorphose about nine weeks after hatching. Tadpoles attain a maximum total length of about 2 in (50 mm). The body is elongately ovoid with a truncate snout and dorsally situated eyes. The oral disc is directed anteroventrally and bears slender, finely serrate jaw sheaths and up to two rows of labial denticles on the anterior lip and up to three rows on the posterior lip. The tadpoles are bright pink with a short mid-dorsal white stripe.

CONSERVATION STATUS

Clearing of rainforest threatens to limit the distribution of this species. It is not listed by the IUCN.

SIGNIFICANCE TO HUMANS

Lithodytes is not known to have toxic skin secretions like the poison frogs of the family Dendrobatidae; however, the color pattern of *Lithodytes* closely resembles that of the sympatric poison frog *Epipedobates femoralis* and thus may be a case of mimicry. ◆

Túngara frog

Physalaemus pustulosus

SUBFAMILY

Leptodactylinae

TAXONOMY

Paludicola pustulosa Cope, 1864, New Grenada and Truando River (Colombia).

OTHER COMMON NAMES
None known.

PHYSICAL CHARACTERISTICS
This small, toadlike anuran attains a maximum snout-vent length of 1.4 in (35 mm). The head is relatively small with a subacuminate snout and no distinct tympanum. The dorsum is tuberculate, and the venter is smooth. A well-defined elongate gland is present on the flank. The first finger is longer than the second, and the fingers and toes lack webbing. The dorsum is dull brown with or without irregular darker brown spots; the venter is grayish white with black spots, and the iris is tan with black flecks. Breeding males have brown nuptial excrescences on the thumbs.

DISTRIBUTION
The species is widely distributed in the lowlands of southern Mexico, Central America, northern Colombia, the coastal region and llanos of Venezuela eastward to Guyana, and the islands of Trinidad and Tobago.

HABITAT
Physalaemus inhabits grasslands, pastures, and open forest formations.

BEHAVIOR
This small species is nocturnal and active only in the rainy season. During the day they are hidden in leaf litter or under objects; in the dry season they burrow in the ground and may remain inactive for many months.

FEEDING ECOLOGY AND DIET
The diet consists of a variety of small arthropods.

REPRODUCTIVE BIOLOGY
Breeding takes place primarily at the beginning of the rainy season when males congregate in small bodies of water, even puddles in roads, and commence calling while floating on the surface of the water. The call consists of a whine followed or not by one or more short notes, "chuck." Females swim to males and preferentially select males with low-pitched "chucks." Amplexus is axillary; as the eggs are extruded, they are kicked into a foam nest by the feet of both individuals. Foam nests float on the water and contain 80–450 small eggs that hatch in two to three days into tiny larvae, which may remain in or under the foam nest for up to five days if the water level has dropped. Tadpoles grow to a length of 0.8 in (20 mm) and metamorphose in five to nine weeks. The larval body is ovoid; the eyes are directed dorsolaterally, and the caudal fins are shallow. The oral disc is directed anteroventrally; the jaw sheaths are moderately massive, and there are two anterior and three posterior rows of labial teeth. In the laboratory, the frogs reach breeding condition two to three months after metamorphosis.

CONSERVATION STATUS
Inasmuch as this small frog is not an inhabitant of dense forest, clearing of forests probably has enhanced its abundance and distribution. It is not listed by the IUCN.

SIGNIFICANCE TO HUMANS
Physalaemus pustulosus has contributed to knowledge of amphibian biology by being the object of studies on vocal communication, sexual selection, and avoidance of predation. ◆

Gray four-eyed frog
Pleurodema bufonina

SUBFAMILY
Leptodactylinae

TAXONOMY
Pleurodema bufonina Bell, 1843, Puerto Deseado and Río Santa Cruz, Patagonia, Argentina.

OTHER COMMON NAMES
None known.

PHYSICAL CHARACTERISTICS
Males of this toadlike species attain a snout-vent length of 1.8 in (45 mm), and females reach 2.2 in (56 mm). The skin on the dorsum is shagreen and glandular; the belly is smooth. The snout is bluntly rounded, and a distinct but small tympanum is present. The fingers are unwebbed, and the toes are basally webbed. A distinct feature is the pair of large, ovoid, lumbar glands that are about one-third of the length of the body. The dorsum is dull brown with or without darker brown spots and/or a tan middorsal stripe; the venter is creamy tan. The iris is pale bronze with black flecks.

DISTRIBUTION
This is the southernmost frog in the world. Its distribution extends from the Straits of Magellan northward to 36° south latitude in Patagonian Argentina and adjacent Chile; the elevational range is from sea level to 7,500 ft (2,300 m).

HABITAT
This small frog inhabits the harsh semiarid Patagonian scrub and steppe, where it is most common in arroyos and margins of lakes.

BEHAVIOR
Pleurodema bufonina is active by day and night, especially after rains. It seeks shelter under stones and in crevices.

FEEDING ECOLOGY AND DIET
Nothing is known; presumably it feeds on small arthropods.

REPRODUCTIVE BIOLOGY
Breeding takes place in shallow water in the austral spring. Males do not call; amplexus is inguinal. Eggs are laid in irregular strings in shallow water. Tadpoles attain a maximum length of about 1.4 in (35 mm); the body is ovoid, and the eyes are small and directed dorsolaterally. The caudal musculature is moderately robust, and the dorsal fin does not extend onto the body. The oral disc is small and directed anteroventrally; the jaw sheaths are broadly arched and finely serrate, and there are two anterior and three posterior rows of labial teeth. The body and caudal musculature are grayish brown, and the belly is gray.

CONSERVATION STATUS
Not threatened.

SIGNIFICANCE TO HUMANS
None known. ◆

Cururu lesser escuerzo
Odontophrynus occidentalis

SUBFAMILY
Odontophryinae

TAXONOMY
Ceratophrys occidentalis Berg, 1896, Arroyo Agrio, Neuquén, Argentina.

OTHER COMMON NAMES
None known.

PHYSICAL CHARACTERISTICS
Males attain a maximum snout-vent length of 2.4 in (60 mm) and females, 2.6 in (65 mm). The body is robust and toadlike. The head is broad with a rounded snout and small tympanum. The skin on the dorsum is pustular with enlarged glands on the eyelids, posterior to the eyes, arms, and legs; the venter is coarsely areolate. The fingers and toes have narrow lateral fringes, and the toes are about one-third webbed. A large, shovel-shaped tubercle is present at the base of the foot. The dorsum is various shades of brown with a middorsal tan stripe; the venter is cream. The iris is dull bronze with black flecks. Breeding males have dark brown nuptial excrescenses on the inner surface of the thumb and dorsal surface of the first finger.

DISTRIBUTION
This species is widely distributed at elevations from near sea level to 6,500 ft (2,000 m) in central and western Argentina.

HABITAT
This frog inhabits arid and semiarid sandy areas; usually individuals are near streams.

BEHAVIOR
This nocturnal species is active above ground only during the rainy season. Using their hind feet for digging, individuals spend the dry season underground. In exceptionally dry years, the frogs do not emerge and are known to spend two years in their underground retreats.

FEEDING ECOLOGY AND DIET
For its size, this frog eats relatively large prey—large arthropods and small mice.

REPRODUCTIVE BIOLOGY
Males call day and night while floating in water; the call is a long series of low-pitched notes. Heavily pigmented eggs are deposited in deep, natural pools at the edges of streams. Tadpoles grow to a maximum length of 4.6 in (117 mm); they have globular bodies with dorsally directed eyes, high caudal fins, and oral discs that have two anterior and three posterior rows of labial teeth.

CONSERVATION STATUS
Not threatened.

SIGNIFICANCE TO HUMANS
None known. ◆

Patagonia frog
Atelognathus patagonicus

SUBFAMILY
Telmatobiinae

TAXONOMY
Batrachophrynus patagonicus Gallardo, 1962, Laguna Blanca, Neuquén Province, Argentina.

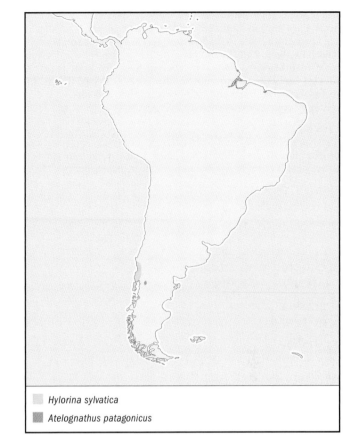

☐ *Hylorina sylvatica*
▨ *Atelognathus patagonicus*

OTHER COMMON NAMES
None known.

PHYSICAL CHARACTERISTICS
Adults attain a maximum snout-vent length of 2 in (50 mm). The snout is acutely rounded in dorsal and lateral views. The eyes are small and directed anterolaterally; the tympanum is obscured by a dermal fold. The skin on the dorsum and venter is smooth; in aquatic adults, loose flaps of skin are present on the sides of the body and on the thighs, but these are absent in terrestrial subadults. The fingers are unwebbed, and the toes are fully webbed. The dorsum is dull brown to live brown with faint darker spots or flecks, and the venter is pale orange; the iris is pale bronze-brown. Breeding males have smooth gray nuptial excrescences.

DISTRIBUTION
This species is known only from the basaltic Laguna Blanca and nearby small lakes in northern Patagonia, Argentina.

HABITAT
Adults inhabit cold lakes with rocky bottoms; subadults are terrestrial in grassy pampas, where they take refuge under stones.

BEHAVIOR
Adults are aquatic and swim among submerged rocks on the bottoms of shallow lakes. Upon metamorphosis, young move onto land and subsequently enter lakes, where they develop loose, baggy skin, which provides additional surface area for integumentary respiration in the cold water.

FEEDING ECOLOGY AND DIET
The diet consists of aquatic arthropods, especially amphipods.

REPRODUCTIVE BIOLOGY
Small eggs are randomly attached to aquatic plants. Tadpoles are bottom-dwellers in shallow water. They reach a total length of about 2 in (50 mm) and have a golden brown dorsum with small brown spots and translucent fins. The body is depressed, and the eyes and nostrils are dorsal; the oral disc is directed anteroventrally and bears two anterior and three posterior rows of labial teeth.

CONSERVATION STATUS
This species is not listed by the IUCN. However, introduction of trout into Laguna Blanca has resulted in a decline in abundance of this species, which might be near extinction.

SIGNIFICANCE TO HUMANS
None known. ◆

Helmeted water toad
Caudiverbera caudiverbera

SUBFAMILY
Telmatobiinae

TAXONOMY
Lacerta caudiverbera Linnaeus, 1758, Peru (in error).

OTHER COMMON NAMES
None known.

PHYSICAL CHARACTERISTICS
Males of this large frog attain a maximum snout-vent length of 4.8 in (120 mm); females as large as 12.8 in (320 mm) have been reported. The body is robust, and the head is large with short, rounded snout. The eyes are small with a vertical pupil and are directed anterolaterally; the tympanum is large and distinct. The skin is smooth with elongate pustules on the dorsum. The fingers are moderately short and unwebbed, and the toes are about one-half webbed. The dorsum is dull brown with faint, paler, irregular markings, and the venter is grayish white. The iris is dull bronze; breeding males have black nuptial excrescences on the inner surfaces of the thumbs.

DISTRIBUTION
This species ranges throughout the lowlands of Chile between 30° and 42° south latitude.

HABITAT
This species is primarily aquatic in ponds, lakes, and rivers.

BEHAVIOR
These large frogs are active by day and night. They are aggressive toward potential predators. The frogs inflate the lungs, elevate the body, open the mouth, lunge, and bite.

FEEDING ECOLOGY AND DIET
Caudiverbera is a voracious carnivore. Adults eat aquatic insect larvae, fishes, frogs, and even small birds and mammals.

REPRODUCTIVE BIOLOGY
Breeding occurs in September and October when males call from shallow water. The call is a loud "oouü." Amplexus is axillary. Eggs are laid in clumps in shallow water; clutches consist of 1,000–10,000 eggs about 0.10–0.12 in (2.7–3.1 mm) in diameter. The eggs hatch about 20 days after deposition, and the larval duration is about two years. The tadpoles reach a maximum length of about 6 in (150 mm). The body is ovoid,

slightly wider than high, with an angular snout; the eyes are directed dorsolaterally. The caudal musculature is moderately robust, and the dorsal fin originates on the posterior part of the body. The oral disc is directed anteroventrally; there are three anterior and three posterior rows of labial teeth. The body and anterior two-thirds of the tail are grayish brown, and the posterior part of the tail is dark brown to black.

CONSERVATION STATUS
This species is negatively affected by habitat degradation and hunting pressure. It is listed as Data Deficient by the IUCN.

SIGNIFICANCE TO HUMANS
Caudiverbera is subjected to human consumption. ◆

Emerald forest frog
Hylorina sylvatica

SUBFAMILY
Telmatobiinae

TAXONOMY
Hylorina sylvatica Bell, 1843, Chonos Island, Chile.

OTHER COMMON NAMES
None known.

PHYSICAL CHARACTERISTICS
This frog attains a maximum snout-vent length of 2.5 in (62 mm). The skin on the dorsum is slightly tubercular, and the belly is smooth. The snout is bluntly rounded; the eyes are large and prominent with vertical pupils, and the tympanum is distinct and about one-half of the diameter of the eye. The fingers and toes are long, slender, and unwebbed. The dorsum is pale green with coppery brown markings; the venter is pale cream. The iris is brown; breeding males have smooth, gray nuptial excrescences on the thumbs.

DISTRIBUTION
This species is restricted to the austral humid forests of southern Chile and adjacent Argentina.

HABITAT
Hylorina inhabits humid forests.

BEHAVIOR
This nocturnal species spends its days under logs; at night it perches in bushes.

FEEDING ECOLOGY AND DIET
Presumably this species feeds on small arthropods.

REPRODUCTIVE BIOLOGY
Most reproductive activity occurs in January, when males call from the edges of ponds. The call is a series of low-pitched notes. Amplexus is axillary. Clumps of 400–500 eggs (ova about 0.08 in or 2 mm in diameter) are deposited at the bases of plants in shallow water. In about 10 days, tadpoles hatch in developmental Stage 21. Tadpoles attain a maximum size of about 2.4 in (60 mm) and require about one year to develop to metamorphosis. Tadpoles have a broad, slightly depressed body with dorsolateral eyes. The oral disc is directed anteroventrally and has two anterior and two posterior rows of labial teeth.

CONSERVATION STATUS
Although this frog is not listed by the IUCN, extensive deforestation is restricting its habitat.

SIGNIFICANCE TO HUMANS
None known. ◆

Titicaca water frog
Telmatobius culeus

SUBFAMILY
Telmatobiinae

TAXONOMY
Cycloramphus culeus Garman, 1875, Lake Titicaca, Peru and Bolivia.

OTHER COMMON NAMES
None known.

PHYSICAL CHARACTERISTICS
This large, aquatic frog attains a snout-vent length of about 6 in (150 mm). The snout is acutely rounded, and the eyes are relatively small and protuberant dorsally; a tympanum is not evident. The skin is nearly smooth and tends to be loose and somewhat baggy. The digits are long with narrowly rounded tips; fingers are unwebbed, and the toes are about one-half webbed. The dorsum is dull olive green or dark brown, with or without paler or darker spots. The venter is creamy gray, and the iris is dull bronze.

DISTRIBUTION
This species is known only from Lake Titicaca and nearby lakes in the Titicaca Basin at elevations of about 12,300 ft (3,800 m) in the Andes in southern Peru and adjacent Bolivia.

HABITAT
This strictly aquatic frog inhabits shallower parts of lakes where the water temperature is about 50°F (10°C).

BEHAVIOR
Living in cold, well-oxygenated water, *Telmatobius culeus* has a low metabolic rate; the lungs are relatively small, and apparently all respiration occurs through the skin.

FEEDING ECOLOGY AND DIET
Nothing is known.

REPRODUCTIVE BIOLOGY
Eggs are laid in water and hatch into feeding tadpoles, which attain maximum total lengths of about 3.1 in (80 mm). The tadpoles have a large, globular body with a round snout. The oral disc is directed anteroventrally and bears keratinized jaw sheaths and two anterior and three posterior rows of labial teeth. The body is dark gray with white flecks, and the caudal fins are tan.

CONSERVATION STATUS
This frog is not listed by the IUCN. However, in Lake Titicaca, *Telmatobius culeus* is threatened by pollution and hunting.

SIGNIFICANCE TO HUMANS
This frog is a staple for residents in the vicinity of Lake Titicaca; the frogs are collected with seines (nets) and sold in local markets. ◆

Resources

Books
Cei, José M. *Batracios de Chile.* Santiago: Ediciones se la Universidad de Chile, 1962.

Joglar, Rafael L. *Los Coquíes de Puerto Rico.* San Juan: Editorial de la Universidad de Puerto Rico, 1998.

Ryan, Michael J. *The Túngara Frog.* Chicago: University of Chicago Press, 1985.

Periodicals
Cei, José M. "The Amphibians of Argentina." *Monitore Zoologico Italiano Monografia* 2 (1980): 1–609.

Heyer, W. Ronald. "A Preliminary Analysis of the Intergeneric Relationships of the Frog Family Leptodactylidae." *Smithsonian Contributions from Zoolology* 31 (1975): 1–55.

Heyer, W. Ronald, A. Stanley Rand, Carlos A. G. da Cruz, Oswaldo L. Peixoto, and Craig E. Nelson. "Frogs of Boracéia." *Arquivos de Zoología Univ. São Paulo* 31 (1990): 231–410.

Lynch, John D. "Evolutionary Relationships, Osteology, and Zoogeography of the Frog Family Leptodactylidae." *Miscellaneous Publications of the Museum of Natural History, University of Kansas* 53 (1971): 1–238.

Lynch, John D. "A Re-assessment of the Telmatobiine Leptodactylid Frogs of Patagonia." *Occasional Papers of the Museum of Natural History, University of Kansas*, no. 72 (1978): 1–57.

Lynch, John D., and William E. Duellman. "Frogs of the Genus *Eleutherodactylus* (Leptodactylidae) in Western Ecuador: Systematics, Ecology, and Biogeography." *Natural History Museum, University of Kansas Special Publication* 23 (1997): 1–236.

William E. Duellman, PhD

▲
Vocal sac-brooding frogs
(Rhinodermatidae)

Class Amphibia
Order Anura
Family Rhinodermatidae

Thumbnail description
Small frogs; green, tan, or brown (or a combination of these colors) with a distinctive fleshy proboscis at the tip of the snout

Size
Snout-vent length to 1.3 in (33.0 mm)

Number of genera, species
1 genus; 2 species

Habitat
Forest and open areas; often near streams

Conservation status
Data Deficient: 2 species

Distribution
Chile, Argentina

Evolution and systematics

No fossils have been described for the family.

Frogs belonging to the family Rhinodermatidae have been included in the families Brachycephalidae, Dendrobatidae, and Leptodactylidae at various times. Since 1971 they have been recognized in their own family.

Rhinoderma rufum was originally named *Heminectes rufus*. These frogs were later considered to be a local variant of *Rhinoderma darwinii*, rather than a valid species. Subsequent work suggested that *Heminectes* is a synonym of *Rhinoderma*. The new combination of *Rhinoderma rufum* was proposed based on differences in the mating call, karyotype, larval development, and male parental care between the two species. No subfamilies are recognized.

Physical characteristics

Rhinodermatids are small frogs; males range from 0.9–1.2 in (22–31 mm), females from 1–1.3 in (25–33 mm). The tympana (external eardrums) are indistinct. The most distinctive external characteristic is a fleshy proboscis,

found in both sexes and all age classes. The forelimbs and hind limbs are rather long and slender. These frogs are extremely variable in color. Dorsally, they may be uniformly tan, brown, or reddish brown; uniformly pale green or dark green; or a combination of brown and green, in variable patterns. The underside has blotches of black and white. Brooding males are easily distinguished by their enlarged vocal sacs.

Distribution

Vocal sac-brooding frogs are found in central to southern Chile, and in Argentina near the Chilean border. Many populations have declined or disappeared from their historical sites during the past 15 years.

Habitat

These terrestrial frogs are found in wet temperate southern beech forest (*Nothofagus*), often near slowly running streams or in swampy areas, and in open areas around human habitation.

Cross-section of male Darwin's frog carrying developing young in his vocal pouch. The young feed off their yolks in the pouch. As froglets, they emerge from his mouth and swim away. (Illustration by Wendy Baker)

A male Darwin's frog (*Rhinoderma darwinii*) brooding tadpoles. (Photo by Martha L. Crump. Reproduced by permission.)

Behavior

Both species exhibit seasonal patterns of activity. They take refuge during the colder months, presumably under moss or logs on the ground, and breed during the warmer months. Both species are primarily diurnal. Territoriality has not been reported.

Feeding ecology and diet

The feeding ecology and diet of these frogs have not been studied. Anecdotal field observations suggest they oppor-

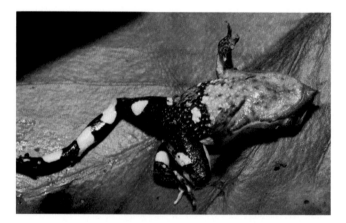

Darwin's frog (*Rhinoderma darwinii*) demonstrates its anti-predator behavior—the frog flips over onto its back and "plays dead." (Photo by Martha L. Crump. Reproduced by permission.)

tunistically eat insects and other small invertebrates. In captivity they eat fruit flies, aphids, and juvenile crickets. Both species are sit-and-wait predators, that is, they sit in one place and snap up prey that come within striking distance.

Reproductive biology

Breeding is seasonal. Males call from land to attract females. Eggs are fertilized on moist ground, and males attend the eggs. Just before the eggs hatch, the males take the eggs into their mouths, where they slide into the vocal sacs. In *Rhinoderma darwinii*, the tadpoles develop within the vocal sac until they metamorphose 50–70 days later. In *Rhinoderma rufum*, the male releases the tadpoles into water, where they continue to develop for an unknown period of time.

Conservation status

Both species are listed as Data Deficient by the IUCN. However, *Rhinoderma rufum* is listed as Endangered and *R. darwinii* as Vulnerable by CITES. Possible reasons for population declines and disappearances include habitat destruction and modification, climate change, and detrimental effects from increased levels of ultraviolet radiation. No specific efforts are known to be underway to protect these species.

Significance to humans

None known.

Species accounts

Darwin's frog
Rhinoderma darwinii

TAXONOMY
Rhinoderma darwinii Duméril and Bibron, 1841, Valdivia, Chile. No subspecies are recognized.

OTHER COMMON NAMES
French: Le rhinoderme de Darwin; German: Darwin-Nasenfroschs; Spanish: Ranita de Darwin, sapito de Darwin, sapito vaquero.

PHYSICAL CHARACTERISTICS
These are small frogs; males are up to 0.9–1.1 in (22–28 mm), females are 1–1.2 in (25–31 mm), with moderately developed membranes between the first and second toes and between the second and third toes. The membrane between the third and fourth toes is smaller, and there is no membrane between the fourth and fifth toes. The metatarsal tubercle is evident, but less prominent than in *R. rufum*.

Rhinoderma darwinii

DISTRIBUTION
This species occurs in central and southern Chile, from the province of Maule south to the province of Aisén, from 0–4,921 ft (0–1,500 m) elevation. In Argentina, the frogs occur near the border with Chile, in the provinces of Neuquén and Río Negro.

HABITAT
The frogs are found both in primary and in disturbed forest. They are also commonly found in open areas around human habitation, and in open wooded or grassy areas. Most individuals are found in or near swampy areas or slowly running water.

BEHAVIOR
This species is primarily diurnal, but males also call at night. Some individuals display an unusual behavior when disturbed. They flip over onto their backs, revealing their contrasting black and white undersides. If a frog near a stream is frightened, it may jump into the water and float downstream on its back.

FEEDING ECOLOGY AND DIET
Darwin's frogs are sit-and-wait predators. By day, they sit in one place and snap up moving insects and other small invertebrates that come within striking distance.

REPRODUCTIVE BIOLOGY
The male mating call is a rapidly repeated "piiiip, piiiiip, piii-iip, piiiiip." Calling is most prevalent beginning in the spring

Rhinoderma darwinii
Rhinoderma rufum

and continuing through the breeding season (November through March).

Observations made in captivity reveal that a male leads a female to a sheltered place that serves as the site for egg deposition. After considerable courtship movements by both frogs, the female crawls underneath the male. He holds onto her very loosely, in contrast to the typical strong amplectant hold of most frogs.

Darwin's frogs deposit and fertilize large eggs (about 0.16 in/4 mm in diameter) on land. In a population studied from the far south of the range, clutch size was estimated to be three to seven eggs. The male stays near the eggs for about 20 days, until the eggs are nearly ready to hatch. At that point, the male takes the eggs into his mouth where they enter his vocal sac and soon hatch. The tadpoles develop within the vocal sac for the next 50–70 days. After the young metamorphose, they crawl back into the father's mouth. The father opens his mouth and the froglets hop out onto land.

The tadpoles lack the typical morphology of free-swimming tadpoles. They do not have external gills, spiracle, beak, or keratinized teeth, and their caudal fins are poorly developed.

Studies of the lining of the vocal sacs of brooding males suggest that the epithelial cells secrete a substance that is taken up by the tadpoles through their skin. Tracers experimentally introduced into the lymphatic sacs moved into tissues of the tadpoles, further supporting the idea that tadpoles receive nutrients from the lining of the vocal sac.

CONSERVATION STATUS
Listed as Data Deficient by IUCN but Vulnerable by CITES. Although the frogs are locally common in some areas (particularly at low elevations), populations are declining or disappearing in other areas (especially at high elevations). The causes of these declines and disappearances are unknown, but habitat destruction is a major threat. Some areas that previously supported dense populations of Darwin's frogs are now planted in non-native pine or eucalyptus, or have been converted to pastures or human residential areas. Climatic change may also be affecting the species, as the climate throughout much of the range is warmer and drier than it was 15–20 years ago. The frogs may also be affected by increased levels of ultraviolet radiation, as the frogs are diurnal and often bask in sunlight.

SIGNIFICANCE TO HUMANS
None known. ◆

Chile Darwin's frog
Rhinoderma rufum

TAXONOMY
Heminectes rufus Philippi, 1902, Vichuquén, Chile. No subspecies are recognized.

OTHER COMMON NAMES
Spanish: Ranita de Darwin de Chile, sapito de Darwin de Chile.

PHYSICAL CHARACTERISTICS
Small frogs (males to 1.2 in [31 mm], females to 1.3 in [33 mm]), with membranes between each of the toes; the membranes between the first and second and the second and third toes are especially well devel-

Rhinoderma rufum

oped. The metatarsal tubercle is more prominent than in *R. darwinii*.

DISTRIBUTION
Central Chile, from the province of Bío-Bío north to the province of Maule, between 164 and 1,640 ft (50 and 500 m) elevation.

HABITAT
These frogs are found on the ground, in southern beech (*Nothofagus*) forest, usually near slowly running water.

BEHAVIOR
No study of behavior under natural field conditions has been published.

FEEDING ECOLOGY AND DIET
Presumably these frogs are sit-and-wait predators that feed on small insects and other small invertebrates.

REPRODUCTIVE BIOLOGY
The male mating call is a rapid "pip, pip, pip, pip," with long pauses between repetitions. These frogs deposit and fertilize their eggs on moist ground. The eggs are smaller than those of *R. darwinii*, about 0.10 in (2.4 mm) in diameter on average. Clutch size is estimated to be 12–24 eggs. After about eight days, the male takes the eggs into his vocal sac. The eggs hatch there, and the tadpoles remain in the vocal sac until they have developed horny jaws and the digestive tract has elongated and spiraled. At that point, the male releases the tadpoles into water. The tadpoles undergo free-swimming aquatic development for an unknown number of days until metamorphosis.

CONSERVATION STATUS
Listed as Data Deficient by IUCN and listed on CITES. Investigators have been unable to find any individuals within the past decade. Historically, they occurred in a very restricted area. Much of their known habitat is currently planted in non-native pine or eucalyptus, or has been converted to pasture or human residential areas.

SIGNIFICANCE TO HUMANS
None known. ◆

Resources

Periodicals

Busse, Klaus. "Bemerkungen zum Fortpflanzungsverhalten und zur Zucht von *Rhinoderma darwinii*." *Herpetofauna* 13 (1991): 11–21.

Crump, Martha L. "Natural History of Darwin's Frog, *Rhinoderma darwinii*." *Herpetological Natural History* 9, no. 1 (2002): 21–30.

Formas, Ramón, Emilio Pugin, and Boris Jorquera. "La identidad del batracio Chileno *Heminectes rufus* Philippi, 1902." *Physis Sección C. Buenos Aires* 34 (1975): 147–157.

Goicoechea, Oscar, Orlando Garrido, and Boris Jorquera. "Evidence for a Trophic Paternal-Larval Relationship in the Frog *Rhinoderma darwinii*." *Journal of Herpetology* 20 (1986): 168–178.

Jorquera, Boris. "Biologia de la reproducción del genero *Rhinoderma*." *Anales del Museo de Historia Natural Valparaíso* 17 (1986): 53–62.

Jorquera, Boris, Emilio Pugin, and Oscar Goicoechea. "Tabla de desarrollo normal de *Rhinoderma darwini*." *Archivos de Medicina Veterinaria* 4 (1972): 1–15.

Jorquera, Boris, Emilio Pugin, Orlando Garrido, Oscar Goicoechea, and Ramón Formas. "Procedimiento de desarrollo en dos especies del genero *Rhinoderma*." *Medio Ambiente* 5 (1981): 58–71.

Martha Lynn Crump, PhD

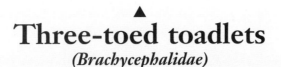

Three-toed toadlets
(*Brachycephalidae*)

Class Amphibia

Order Anura

Family Brachycephalidae

Thumbnail description
Small toad-like anurans with reduced number of segments in their digits and a fully ossified pectoral girdle lacking a sternum

Size
0.3–0.8 in (8.5–20.0 mm) snout-vent length

Number of genera, species
2 genera; 6 species

Habitat
Humid forest

Conservation status
Not threatened

Distribution
Atlantic coastal forest of eastern Brazil

Evolution and systematics

No fossils are known. These small anurans were formerly placed in Bufonidae, from which they differ by the absence of a Bidder's organ (a growth of ovarian tissue on the testis). The relationships of the family are unknown, but it has been suggested that brachycephalids are related to *Euparkerella*, a telmatobiine (tribe Eleutherodactylini) leptodactylid. No subfamilies are recognized.

Physical characteristics

These small toad-like anurans reach a maximum snout-vent length of 0.8 in (20 mm). The head is short, and the body is robust. The limbs are short to moderately long. The digits are reduced, so there are only two or three functional fingers and three or four functional toes. The dorsum is orange to greenish yellow or brown. The two halves of the pectoral girdle overlap midventrally (arciferal condition) and are fully ossified; a sternum is absent. Teeth are absent on the maxillaries and premaxillaries. The phalanges are short and reduced in number; the terminal phalanges are T-shaped. In *Brachycephalus ephippium*, a dermal bony shield ossifies dorsal to the vertebral column.

Distribution

All members of the family have restricted distributions in the coastal mountains to elevations of approximately 2,240 ft (750 m) from Espírito Santo southward to Paraná in eastern Brazil.

Habitat

Terrestrial amidst leaf litter on the forest floor.

Behavior

During the rainy season, brachycephalids are active by day and slowly walk about on the leaf litter. Males are territorial and advertise vocally and visually; male-male encounters have been observed. During the dry season, the toadlets seek shelter beneath leaf litter or under logs.

Feeding ecology and diet

Small arthropods, principally springtails and mites.

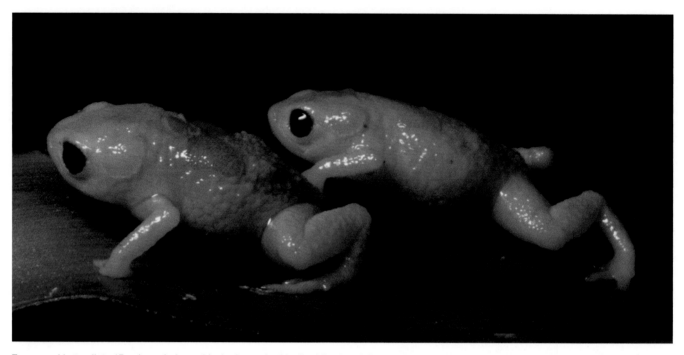

Two pumpkin toadlets (*Brachycephalus ephippium*) on a leaf in the Atlantic rainforest. (Photo by Kevin Schafer/Corbis. Reproduced by permission.)

Reproductive biology

The advertisement call is a long, low-pitched buzz in *Brachycephalus*. Males grasp females around the waist (inguinal amplexus). Relatively large, unpigmented eggs are deposited terrestrially and undergo direct development into miniatures of the adults. *Brachycephalus ephippium* deposits up to five eggs per clutch but clutches of *B. didactyla* consist of a single egg.

Conservation status

Although not officially listed as threatened, these toadlets are restricted to the Atlantic coastal forest, much of which has been cleared.

Significance to humans

The skin secretions of *Brachycephalus ephippium* contain extremely strong toxins, tetradotoxin, and analogues.

Species accounts

Pumpkin toadlet
Brachycephalus ephippium

TAXONOMY
Bufo ephippium Spix, 1824, Ilhéus, Bahia, Brazil (probably erroneous).

OTHER COMMON NAMES
English: Spix's saddleback toad; Portuguese: Botão de ouro, sapinho dourado.

PHYSICAL CHARACTERISTICS
This robust, short-legged toadlet is bright yellow to orange; the iris is black. Adults attain a snout-vent length of 0.5–0.8 in (12.5–19.7 mm). A bony shield ossifies dorsal to the vertebral column.

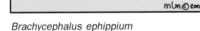

Brachycephalus ephippium

DISTRIBUTION
Serra do Mar and Serra da Mantiqueira in southeastern Brazil.

HABITAT
Terrestrial on and amid leaf litter on the forest floor in the Atlantic coastal forest.

BEHAVIOR
Individuals actively walk slowly on the leaf litter by day during the rainy season. When the relative humidity approaches 100%, the toadlets often ascend low perches. Pumpkin toadlets commonly clean themselves by wiping the head and body with their limbs. During the rainy season, males are territorial and advertise their presence vocally. On approach by an intruder, the male toadlet moves an arm up and down in front of its eye. This movement may be derived from the wiping behavior. If an intruder does not retreat, a resident male may embrace or push the intruder until it departs.

FEEDING ECOLOGY AND DIET
Toadlets actively forage on the leaf litter and consume a variety of small arthropods, of which collembolans make up 54% of the diet, mites, 8%; and insect larvae, 6%.

REPRODUCTIVE BIOLOGY
Reproductive activity occurs throughout most of the rainy season. Males call while in a high posture allowing for expansion of the large subgular vocal sac. The call consists of a continuous series of buzzes lasting two to six minutes with emphasized frequencies at 3.4–5.3 kHz. The first notes in the series are shortest with five or six pulses. Succeeding notes increase in length to as many as 15 pulses, but most of the notes have 10 pulses and a nearly constant pitch. Initial amplexus is inguinal as the male walks behind the female as she selects an oviposi-

Brachycephalus ephippium

Brachycephalus pernix

tion site in the leaf litter or under a log. Before oviposition the male moves forward and grasps the female nearly in an axillary position. This shift in position results in juxtaposition of the vents of both toadlets, maximizing fertilization. During a period of approximately 30 minutes, five large (0.2 in [5.1–5.3 mm] diameter), yellowish white eggs are deposited. The male leaves the site, but the female uses her hind feet to press and roll the eggs in the soil, particles of which adhere to the eggs and camouflage them. Then the eggs remain unattended. Embryos have a large yolk sac. The mouth is differentiated at 25 days of age, and a small tail is evident. By 41 days, fingers and toes are fully formed, and two egg teeth are present on the snout. By 54 days the tail is reduced in size, only one egg tooth is present, and the body is pigmented. Hatching occurs in 64 days. The miniature reddish brown toadlets retain a vestigial tail but no egg tooth.

CONSERVATION STATUS
Not threatened. It exists within various protected areas in the Atlantic coastal forest.

SIGNIFICANCE TO HUMANS
Dermal glands secrete extremely strong toxins, tetradotoxin and analogues, which have biomedical importance. ◆

Southern three-toed toadlet

Brachycephalus pernix

TAXONOMY

Brachycephalus pernix Pombal, Wistuba, and Bornschein, 1998, Morro do Anhagava, Serra da Baitaca, Paraná, Brazil.

OTHER COMMON NAMES

None known.

PHYSICAL CHARACTERISTICS

This robust, short-legged toadlet has a bright orange body, but the flanks, vent region, limbs, and area around the eye are black. Adults attain a snout-vent length of 0.5–0.6 in (12.0–15.8 mm). Ossified warts and a dermal shield are absent.

Brachycephalus pernix

DISTRIBUTION

Southern part of the Serra do Mar, Paraná, Brazil.

HABITAT

Leaf litter in humid forest.

BEHAVIOR

Diurnal; visual and vocal communication similar to that of *B. ephippium*.

FEEDING ECOLOGY AND DIET

These toadlets feed on small arthropods in the leaf litter by day; mites and insect larvae are the most common prey.

REPRODUCTIVE BIOLOGY

The advertisement call is a low buzz. Reproductive activity occurs throughout the rainy season. Terrestrial eggs undergo direct development into miniature toadlets.

CONSERVATION STATUS

Not threatened.

SIGNIFICANCE TO HUMANS

None known. ◆

Resources

Periodicals

Pires Jr., O. R., A. Sebben, E. F. Schwartz, S. W. R. Larguna. C. Bloch Jr., R. A. V. Morales, and C. A. Schwartz. "Occurrence of Tetradotoxin and Its Analogues in the Brazilian Frog *Brachycephalus ephippium* (Anura: Brachycephalidae)." *Toxicon* 40 (2002): 761–766.

Pombal Jr., J. P. "Oviposition and Development of the Pumpkin Toadlet, *Brachycephalus ephppium* (Anura: Brachycephalidae)." *Reuta Bras. Zool.* 16 (1999): 967–976.

———. "A New Species of *Brachycephalus* (Anura: Brachycephalidae) from Atlantic Rain Forest of Southeastern Brazil." *Amphibia-Reptilia* 22 (2001): 179–185.

Pombal Jr., J. P., I. Sazima, and C. F. B. Haddad. "Breeding Behavior of the Pumpkin Toadlet, *Brachycephalus ephippium* (Brachycephalidae)." *Journal of Herpetology* 28 (1994): 516–519.

Pombal Jr., J. P., E. M. Wistuba, and M. Bornschein. "A new species of Brachycephalid (Anura) from the Atlantic rainforest of Brazil." *Journal of Herpetology* 32 (1998): 70–74.

José P. Pombal, Jr, PhD

True toads, harlequin frogs, and relatives
(Bufonidae)

Class Amphibia

Order Anura

Family Bufonidae

Thumbnail description
These are tiny to very large, generally warty or dry-skinned frogs, with usually unspecialized digital tips; most have parotoid glands

Size
0.6–9.8 in (15–250 mm)

Number of genera, species
33 genera; 344 species

Habitat
Deserts, savanna, dry and humid forests, from sea level to 16,404 ft (5,000 m)

Conservation status
Critically Endangered: 1 species; Endangered: 6 species; Vulnerable: 6 species; Lower Risk/Near Threatened: 2 species; Data Deficient: 3 species

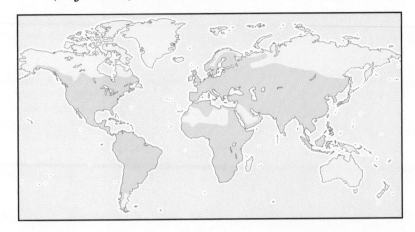

Distribution
Worldwide, except for Madagascar, Australia (introduced), and New Guinea; bufonids just barely cross Wallace's Line to the east, and are present on the Indonesian island of Sulawesi

Evolution and systematics

Bufonidae currently contains 33 genera. Subfamily names have been proposed, but these were based on geographic distribution rather than evolutionary relationships and have not been widely accepted.

Scientists have not determined which are the closest relatives of Bufonidae. Although many data from DNA sequences have been accumulated recently, no other group of frogs has emerged as a close relative to Bufonidae. Although not known with certainty, South America is generally believed to be the continent of origin for the group.

Bufo is known as far back as the Oligocene (Whitneyan) of North America, the Miocene of Europe, western Asia, and North Africa, and questionably from the Middle Paleocene (with certainty from the Miocene) of South America.

Evolutionary novelties that unite the species of Bufonidae include the presence of Bidder's organ; a unique pattern of insertion of the rectractor muscle of the tongue (hyoglossus); the loss of the posterior constrictor muscles of the larynx; the absence of teeth; and the presence of the "otic element," an independent bone in the temporal region that fuses indistinguishably to the posterior arm of the squamosal bone.

Physical characteristics

The term *toad* is usually applied to frogs in the family Bufonidae. Sometimes "toad" is used for any frog that is rough-skinned, regardless of its evolutionary relationships. More often, toad is used to describe any member of the family Bufonidae. English is not the only language to recognize frogs and toads; the distinction is made in languages as diverse as French, German, Quechua, and Bahasa Indonesia.

Toads in the family Bufonidae are relatively diverse in their appearance. Yet all of them share certain structural characteristics that unite them into this taxon. Most frogs have teeth on the upper jaws, but all bufonids lack them. At the turn of the twentieth century, the presence or absence of teeth was considered to be a significant character for classification, and several species that herpetologists now know are unrelated were grouped into Bufonidae simply because they lacked teeth. Although several other groups of frogs have independently lost teeth, the absence of teeth in bufonids remains a diagnostic feature of the group.

Only the Bufonidae among frogs have a Bidder's organ. In amphibians, the testis and ovary develop from an undifferentiated mass of gonadal tissue. During larval development, the gonadal tissues of future males secrete testosterone, which causes the animal to develop as a male. In the absence of testosterone production, the animal will become a female. In other words, being female is the default sex. The organ of Bidder is a bit of gonadal tissue that apparently retains its female attributes in male toads, which also develop normal testes. If the testes of adult toads are removed surgically, Bidder's organ will transform into a functional ovary. Thus it seems that the presence of a functional testis, which produces male hormones, suppresses the development of the ovarian tissue of the bidder's organ. It is not known whether Bidder's organ has an adaptive or functional role in the natural life of toads.

Bufonids are basically hoppers and walkers, never leapers. The joint between the hip bones (pelvic girdle) and vertebral column (at the sacrum) is modified such that the range of motion is not in the longitudinal vertical plane, as in leapers such as *Rana*, but rather movement is from side to side. Also, several genera of bufonids apparently have evolved skeletal mod-

An American toad (*Bufo americanus*) swallows an earthworm in Pennsylvania, USA. (Photo by Joe McDonald. Bruce Coleman Inc. Reproduced by permission.)

Harlequin frogs (*Atelopus varius*) inhabit rainforests from Costa Rica to northwestern South America. (Photo by Michael Fogden. Bruce Coleman Inc. Reproduced by permission.)

ifications that perhaps reflect the reduced locomotor abilities of these toads. They have seven or fewer vertebrae (rather than eight as in most frogs). The coccyx is fused to the sacrum, rather than having a flexible joint. The left and right halves of the shoulder girdle are fused to each other, rather than having a flexible joint at midline. The number of bones in the hands and feet is reduced; the lengths of the fingers and toes are correspondingly shorter; and the hands and feet look more like a mitten rather than a glove (the name *Atelopus* means "incomplete foot"). All of these modifications suggest a reduced ability to jump. In fact, this is true. But also, these toads are all rather small, and the skeletal modifications might also result from a smaller size.

Skin glands are present in almost all amphibians and are generally widely distributed throughout the skin as small structures that are not obvious. In contrast, the parotoid gland consists of closely spaced skin glands concentrated into a prominent organ behind the ear. Within bufonids, one can distinguish two groups, those with parotoid glands and those without. Parotoid glands are found in all species of *Bufo*, as well as members of several of the non-*Bufo* genera. However, several non-*Bufo* genera lack parotoid glands. In some species the glands are difficult to distinguish without a close examination of the skin in cross section. When disturbed, the toad can discharge a milky venom from the glands, sometimes through the air. The secretions of large toads have been known to kill predators such as dogs.

Distribution

The genus *Bufo* has a world-wide distribution, with radiations in North America, Central America, South America, the West Indies, Africa (but not Madagascar), Europe, and all of Asia including Japan, the Philippines, Southeast Asia, and Sulawesi east of Wallace's Line.

The non-*Bufo* genera in the Neotropics include *Crepidophryne* and *Atelophryniscus*, which are endemic to Central America; *Atelopus* in Central and South America; and *Andinophryne, Atelopus, Dendrophryniscus, Frostius, Melanophryniscus, Metaphryniscus, Osornophryne, Oreophrynella, Truebella,* and *Rhamphophryne* in South America. The other genera endemic to Africa are *Altiphrynoides, Capensibufo, Didynamipus, Laurentophryne, Mertensophryne, Nectophrynoides, Nectophryne, Nimbaphrynoides, Schismaderma, Spinophrynoides, Stephopaedes, Werneria,* and *Wolterstorffina*. Some genera are found in Southeast Asia: *Ansonia, Leptophryne, Pedostibes, Pelophryne,* and *Pseudobufo*. Genera endemic to the Indian subcontinent include *Adenomus* and *Bufoides*.

Habitat

It is difficult to make general statements for a group with as many species and with as broad a geographic range as Bufonidae. Species may be found in near-desert to primary tropical rainforest habitats, from sea level to 16,400 ft (5,000 m) in treeless alpine environments. The genus *Bufo* occupies the greatest range of latitudes and altitudes of any frog. Most bufonids would be termed terrestrial; very few are fully aquatic or arboreal.

Golden toads (*Bufo periglenes*) mating. They are secretive most of the year and are seen only during breeding season. (Photo by Michael Fogden. Bruce Coleman Inc. Reproduced by permission.)

Behavior

Behavior in this diverse group of toads varies. Very little is known about some species, while others are more well-studied. *Atelopus varius* is known to have exceptional homing ability.

Feeding ecology and diet

Like most frogs, bufonids feed mainly on a diet of arthropods. Ants form a large part of the diet in tropical areas. Despite their size, large toads such as *Bufo marinus* are not decidedly carnivorous, although they are capable of eating small mammals such as mice.

Reproductive biology

The mating call of most species is a trilled call emitted at a rather steady pitch, rather than an untrilled or pure tone that might rise or drop in pitch. Most bufonids lay numerous, small pigmented eggs that are enclosed single file in strings of jelly, rather than in a discoid or globular egg mass. These egg masses are usually laid in temporary ponds rather than large bodies of water or streams. Typically, the eggs develop quickly, with tiny tadpoles hatching out in large numbers. Most of these die, and the few that make it through metamorphosis do so as very tiny toadlets, regardless of the ultimate size of the adult. Even *Bufo marinus* has small toadlets; therefore the total weight increase over the life of the animal may be several orders of magnitude. Tiny toadlets are notoriously difficult to identify to the exact species. A few species of bufonids deviate from this general pattern and are presumed to have direct development, because the eggs observed in dissected specimens are large, few in number, and not pigmented. Few toads are known to exhibit parental care.

Many species of toads are known to hybridize in nature, and hybrids have been produced in the laboratory between species that are very distantly related, even from different continents.

Conservation status

The IUCN lists 1 species as Critically Endangered (*Bufo periglenes*), 6 as Endangered, 6 as Vulnerable, 2 as Lower Risk/Near Threatened, and 3 as Data Deficient. Although many species are not threatened and some are so common in human settlements as to be considered pests, several species appear to have suffered dramatic and mysterious population declines since the 1980s, along with many other amphibians.

Significance to humans

Because of their ubiquity, toads have been the subject of myth and lore.

1. Yungas redbelly toad (*Melanophryniscus rubriventris*); 2. Roraima bush toad (*Oreophrynella quelchii*); 3. Malcolm's Ethiopia toad (*Altiphrynoides malcolmi*); 4. Golden toad (*Bufo periglenes*); 5. Aquatic swamp toad (*Pseudobufo subasper*); 6. Harlequin frog (*Atelopus varius*); 7. Marine toad (*Bufo marinus*). (Illustration by Brian Cressman)

1. Green toad (*Bufo viridis*); 2. Chirinda toad (*Stephopaedes anotis*); 3. Houston toad (*Bufo houstonensis*); 4. Long-fingered slender toad (*Ansonia longidigita*); 5. Brown tree toad (*Pedostibes hosii*); 6. Common Sunda toad (*Bufo melanostictus*). (Illustration by Brian Cressman)

Species accounts

Malcolm's Ethiopian toad
Altiphrynoides malcolmi

TAXONOMY
Nectophrynoides malcolmi Grandison, 1978, 3.7–4.97 mi (6–8 km) south of Goba, Balé Province, Ethiopia. This species was formerly included in *Nectophrynoides*, a group of montane toads with specialized, but variable, modes of development.

OTHER COMMON NAMES
None known.

PHYSICAL CHARACTERISTICS
Adult males are 0.59–0.75 in (15–19 mm) and females 0.9–1.1 in (23–28 mm) in snout-vent length. The parotoid glands are very small, if present at all, and cranial crests are absent.

DISTRIBUTION
Balé Mountains, Ethiopia.

HABITAT
These toads are inhabitants of high mountains, at 10,500–13,100 ft (3,200–4,000 m). The normal ambient temperature in which the larvae are known to develop is 41°F (5°C).

BEHAVIOR
Nothing is known except the reproductive behavior.

FEEDING ECOLOGY AND DIET
Nothing is known.

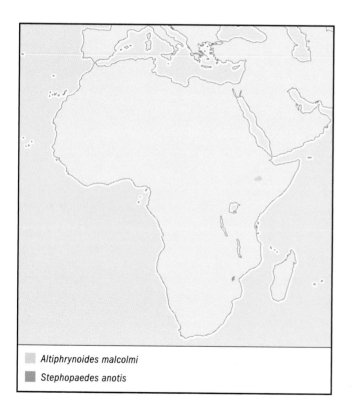

Altiphrynoides malcolmi
Stephopaedes anotis

REPRODUCTIVE BIOLOGY
Fertilization is internal in these toads. During mating, the male grasps the female just in front of the hind limbs (inguinal amplexus), but in contrast to the usual positions, amplexus occurs belly to belly, rather than with the male behind the female. The eggs are retained until the early neurula stage, when the embryo is beginning to develop a spinal chord. Then the eggs are laid and continue their development in the egg capsule, without active feeding. The embryos lack the mouth parts needed to feed, and they also have a short gut, indicating that it does not function in digestion. Thus, the embryo derives all of its nutrition from the yolk. Females have eggs that are huge relative to the body; the average clutch size is 18 eggs, with an egg diameter of 0.1 in (2.73 mm). The terrestrial egg clutches are thought to be communal, laid by as many as 20 females that are attracted by the chorus of calling males.

CONSERVATION STATUS
Not listed by the IUCN.

SIGNIFICANCE TO HUMANS
None known. ◆

Long-fingered slender toad
Ansonia longidigita

TAXONOMY
Ansonia longidigita Inger, 1960, Mount Kina Balu, Borneo, Malaysia.

OTHER COMMON NAMES
English: Long-fingered stream toad.

PHYSICAL CHARACTERISTICS
In general, *Ansonia* are small, slender toads that lack parotoid glands. The snout protrudes over the tip of the lower jaw. The leg and digits are slender and the eyes relatively large. The males of *Ansonia longidigita* are 1.4–1.97 in (35–50 mm) and the females 1.77–2.75 in (45–70 mm) in snout-vent length. This toad is dark brown, with a few darker crossbars on the hindlimbs.

DISTRIBUTION
This toad is known only from Borneo.

HABITAT
Long-fingered slender toads live in lower and upper montane regions (up to 7,220 ft or 2,200 m) with intact forest.

BEHAVIOR
Little is known.

FEEDING ECOLOGY AND DIET
As with many toads, ants are an important part of the diet.

REPRODUCTIVE BIOLOGY
The call is a high trill. Breeding occurs near swift rocky streams, where males gather to call. The tadpoles are small and stream-adapted, with large ventral suctorial mouths that occupy part of the belly. However, like *Atelopus*, the number of

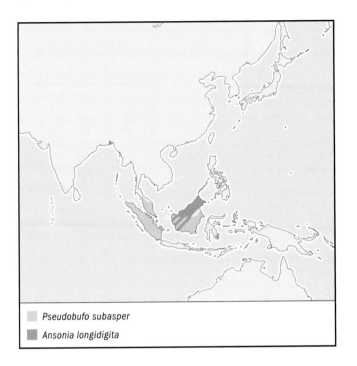

☐ *Pseudobufo subasper*

■ *Ansonia longidigita*

denticle rows is only two above and three below the mouth. In many ways, *Ansonia* appear to be ecological equivalents of *Atelopus* in the New World. However, research on the phylogenetic relationships of toads using DNA data has shown that the similarities between *Ansonia* and *Atelopus* result from evolutionary convergence.

CONSERVATION STATUS
Not listed by the IUCN.

SIGNIFICANCE TO HUMANS
None known. ◆

Harlequin frog
Atelopus varius

TAXONOMY
Phrynidium varium Lichtenstein and von Martens, 1856, Veragua, Panama.

OTHER COMMON NAMES
English: Harlequin toad.

PHYSICAL CHARACTERISTICS
These are often called harlequin frogs or toads because many of them are so brightly colored as to appear to be in a jester's costume. The coloration is usually a combination of markings of black and some starkly contrasting color such as yellow, green, orange, or red. In some populations the males and females are colored differently, in others they are similar. Males are about 1.06–1.57 in (27–40 mm), females 1.34–1.9 in (34–48 mm) in snout-vent length. Like most species of *Atelopus*, this one lacks a tympanum and cranial crests.

DISTRIBUTION
This toad is known from Costa Rica and Panama.

HABITAT
These toads inhabit humid lowland and lower montane forests.

BEHAVIOR
These diurnal toads may be seen actively moving across open areas as if impervious to predators; adults sleep at night on large flat leaves of vegetation over montane streams. The author has encountered as many as 50 individuals in an hour. Toads of the genus *Melanophryniscus* behave similarly. Harlequin frogs have exceptional homing ability. Field experiments showed that 31 of 44 individuals that were displaced 32.8 ft (10 m) from their point of capture returned to within 3 ft (1 m) of that spot in a week. Some individuals were faithful to a particular boulder for two years. Male harlequin frogs have pronounced aggressive encounters. One may chase and pounce on another male and use his body to squash his opponent. Males may also signal each other by raising a front foot and waving it in a circular motion in the air, either before or after a battle. Interestingly, males will tolerate each other more in the dry season, when limited wet areas necessitates that they crowd together. When the rains come, aggression is more pronounced.

In the species *Atelopus zeteki* and *Atelopus varius*, the extremely potent toxin tetrodotoxin has been found in the skin. This compound was named for the fugu or pufferfish (*Tetraodon*), from which it was isolated. At the least, tetrodotoxin makes the toad bad-tasting; at worst, it is lethal to the predator. There is a link between being diurnal, brightly colored, and toxic. A predator can easily spot this gaudy toad during the day, but the experience of grabbing this prey will be unpleasant. The predator learns to avoid potential prey that have these visual characteristics.

FEEDING ECOLOGY AND DIET
They are known to feed on a diverse set of arthropods, including flies, wasps, ants, caterpillars, and spiders.

REPRODUCTIVE BIOLOGY
There is apparently no courtship; the short buzzlike call serves a territorial function. Amplexus, as in other species of *Atelopus*, may last several days, with the female carrying the male around on her back. The eggs are completely cream-colored and are laid in strings, presumably in the streams in which the tadpoles are found.

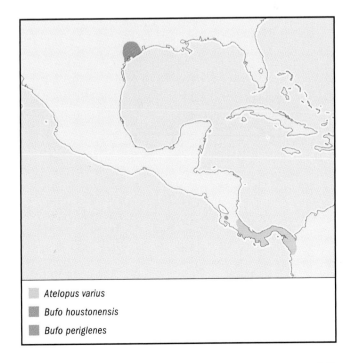

☐ *Atelopus varius*

■ *Bufo houstonensis*

■ *Bufo periglenes*

The tadpole has an enlarged mouth and sucker that extends onto the belly. The tadpoles adhere to the undersides of rocks in swiftly flowing mountain streams.

CONSERVATION STATUS
Their diurnal habits make these toads easy to observe. But this has led to the unsettling realization that well-established populations of the Costa Rican *Atelopus varius* have mysteriously disappeared since the mid-1980s. However, the species is not listed on the IUCN Red List.

SIGNIFICANCE TO HUMANS
In the past, this animal was exported in large numbers for the pet trade. It apparently has been one of the victims of the worldwide decline of amphibians. Most populations in Costa Rica appear to be extinct. ◆

Houston toad
Bufo houstonensis

TAXONOMY
Bufo houstonensis Sanders, 1953, Fairbanks, Harris County, Texas, United States.

OTHER COMMON NAMES
None known.

PHYSICAL CHARACTERISTICS
Males range from 1.92–2.6 in (49–66 mm) and females from 2.24–3.15 in (57–80 mm). This toad resembles others in the

Oreophrynella quelchii

Melanophryniscus rubriventris

Bufo marinus

Bufo americanus group: The dorsal surfaces are very warty, with obvious larger warts and many smaller warts between. Some dark dorsal spots surround the larger warts. Cranial crests are moderately developed.

DISTRIBUTION
This toad is known from a few counties in southeast Texas in the United States.

HABITAT
This toad is usually associated with sandy soils in loblolly pine forests.

BEHAVIOR
The Houston toad is one of the first frogs to call in the spring, in January or February, when the 24-hour minimum air temperature reaches 57°F (14°C).

FEEDING ECOLOGY AND DIET
Nothing is known.

REPRODUCTIVE BIOLOGY
The call is a high-pitched, pleasant musical trill lasting four to 11 seconds. Males often begin calling at sunset from burrows and then move out to occupy the highest sites around a pond to continue calling. Amplexus lasts a minimum of six hours before oviposition. Choruses last three to five nights, unless cold weather intervenes. Pigmented eggs are laid in strings and hatch in as little as seven days. The time from oviposition to metamorphosis is relatively constant, from 60–65 days. Metamorphic young are 0.27–0.35 in (7–9 mm) in length. The species is known to hybridize in the wild with *Bufo woodhousii* and *Bufo valliceps*.

CONSERVATION STATUS
The IUCN categorizes this species as Endangered. A captive breeding program was begun at the Houston Zoo.

SIGNIFICANCE TO HUMANS
Continued survival of this toad depends on effective management. Expansion of local golf courses and parks threatens to remove some of the species' critical habitat. ◆

Marine toad
Bufo marinus

TAXONOMY
Rana marina Linnaeus, 1758, America.

OTHER COMMON NAMES
English: Cane toad; French: Bufo géant; German: Aga-Kröte; Spanish: Sapo grande.

PHYSICAL CHARACTERISTICS
This is a very large toad with a broad U-shaped furrow between the eyes. The parotoid glands are large and triangular. Adults may reach 9 in (230 mm) and weigh 3.3 lb (1.5 kg). A close relative, *Bufo paracnemis*, is even larger and may be as big as a dinner plate.

DISTRIBUTION
This large, rather plain toad is native to South and Central America, Mexico, and the south of Texas. Its closest relatives are in South America, so its presence in Central America and northward represents a gradual migration across the Isthmus of

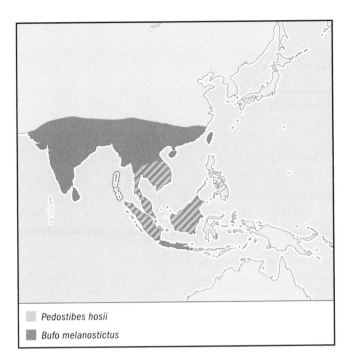

Pedostibes hosii
Bufo melanostictus

Hawaii, the Philippines, and Australia in the 1930s. The common name "cane toad" came from its intended use to control insect pests of sugar cane. The toad did quite well in its new home. In Australia, especially, it spread rapidly, becoming a pest and outcompeting many local animals. It has also caused economic damage by fouling water supplies used by cattle. Its notoriety in Australia has been recognized with a movie and a book with the title *Cane Toads: An Unnatural History*. ◆

Common sunda toad

Bufo melanostictus

TAXONOMY

Bufo melanostictus Schneider, 1799, Orient.

OTHER COMMON NAMES

English: Asian common toad, Asian toad, black-lipped toad, black-spined toad, common Asian toad, common Indian toad, Indian toad, keeled-nosed toad, Maharashtra stream toad, Southeast Asian broad-skulled toad.

PHYSICAL CHARACTERISTICS

This is a rather typical-looking moderate-sized toad. Males are 2.24–3.27 in (57–83 mm) and females 2.56–3.34 in (65–85 mm). The distinctive features are the bony crests that border the eyes and extend from behind the eye to the parotoid gland, which is moderately large and oval. Like many *Bufo*, the body is generally warty, but the bony crest and warts are tipped with many small black spines of keratin; hence the name *melanostictus*.

DISTRIBUTION

Southwestern and southern China, Taiwan, Hainan; from Pakistan and Nepal through India to Sri Lanka; Andaman Islands, Sumatra, Java, Borneo, and Bali. It has apparently invaded Borneo recently.

HABITAT

The most common place to find these toads is in association with human dwellings. This toad seems to be at home in cities, as long as there is some temporary water for breeding.

BEHAVIOR

Little is known, except for reproductive behavior.

FEEDING ECOLOGY AND DIET

Like many toads, it eats arthropods, especially ants.

REPRODUCTIVE BIOLOGY

The mating call of this toad is a moderately pitched trill, sounding somewhat like a rattle. The tadpoles of these toads are typical *Bufo* tadpoles, small (0.47–0.63 in or 12–16 mm long) and black, without obvious modifications of the mouthparts; in other words, they have no expanded lips or extra rows of denticles.

CONSERVATION STATUS

Not threatened.

SIGNIFICANCE TO HUMANS

Its main significance to humans is that it thrives in human habitats and is actively expanding its range. How this might affect local species is unknown at present. ◆

Panama. It is one of the few frog species found on both sides of the Andes in northern South America.

HABITAT

In its natural habitat, this toad prefers secondary forests and open areas in lowland and foothill areas.

BEHAVIOR

These toads breed opportunistically when there is rain, and the breeding may occur over several months. Both temporary and permanent ponds and edges of lakes are used.

FEEDING ECOLOGY AND DIET

In natural settings the marine toad eats a variety of arthropods, from large roaches to ants. The species does quite well around human populations. Adult toads will gather under streetlamps to prey on insects that gather there; the same toad may return to the same lamp night after night. They are well-known for eating from the food dishes of pet dogs and cats. A biologist saw one sit for hours nabbing flies around a large pile of excrement. Sometimes large, common pests are ideal study organisms. The first experimental studies of how a frog projects its tongue were done on *Bufo marinus*.

REPRODUCTIVE BIOLOGY

The call is a very low-pitched trill, lasting for 10–20 seconds. The females produce up to 25,000 eggs during one spawning. In south Florida these toads will breed in swimming pools, depositing long gelatinous strings consisting of thousands of eggs, to the chagrin of homeowners. The tadpoles are small and black and often form large schools.

CONSERVATION STATUS

This species is not listed by the IUCN. Informal but active extirpation efforts are underway in several areas where the species have been introduced.

SIGNIFICANCE TO HUMANS

The marine toad is also quite common in many tropical climates because of human introductions. It was introduced to the West Indies as early as the mid-nineteenth century, and to

Golden toad
Bufo periglenes

TAXONOMY

Bufo periglenes Savage, 1966, 2 mi (3.22 km) east-northeast of Monteverde, Puntarenas Province, Costa Rica.

OTHER COMMON NAMES

English: Alajuela toad, Monteverde toad; French: Crapaud doré; Spanish: Sapo dorado de Monteverde.

PHYSICAL CHARACTERISTICS

This species is spectacular in that both sexes are brightly colored. Males are a uniform bright orange, and females are blackish green with red spots. The coloration of this toad makes it impossible to confuse with anything else. The males are 1.53–1.89 in (39–48 mm) in snout-vent length and the females 1.65–2.2 in (42–56 mm). The cranial crests are not well developed, and the tympanum and middle ear are absent.

DISTRIBUTION

The golden toad is known from two localities at elevations of 4,920–5,250 ft (1,500–1,600 m) along the continental divide of northwestern Costa Rica.

HABITAT

This toad is a denizen of the elfin, windswept montane rainforests along the crest of the cordillera. The areas where it is known are part of the Monteverde Cloud Forest Preserve.

BEHAVIOR

Although such bright colors are usually associated with skin toxins, this species has not been examined for these.

FEEDING ECOLOGY AND DIET

Not known.

REPRODUCTIVE BIOLOGY

It is not clear whether this species has an advertisement call. The breeding of these toads is explosive and coincides with heavy thunderstorms. As many as several hundred males

emerge, but only about 100 females at most. Several males may battle, attempting to dislodge a male already in amplexus with a female. The eggs of this species are about 0.11 in (3 mm) in diameter, which is a little larger than the eggs of most *Bufo*, but they are laid in the typical strings. The tadpoles are about 1.18 in (30 mm) in length, which is larger than tadpoles of most species of *Bufo*. In addition, the larvae can develop in the absence of food. Mostly likely, the larger size of the eggs provides sufficient yolk supply for them to survive if food is not present.

CONSERVATION STATUS

The golden toad is one of many frog species whose recent disappearance has caused much concern. After their discovery, these toads bred regularly each year until about 1988, when only a few emerged. In 1989 only one was observed, and thereafter none. The IUCN lists the species as Critically Endangered, although most experts believe it is extinct.

SIGNIFICANCE TO HUMANS

This beautiful toad is a reminder of the fragility with which some species pass their existence and a symbol for amphibian conservation. ◆

Green toad
Bufo viridis

TAXONOMY

Bufo viridis Laurenti, 1768, Vienna, Austria. The exact taxonomic status of many populations assigned to this species is not clear because there are diploid, triploid, and tetraploid populations in various parts of the range that have been named as subspecies or full species. These populations differ in call characteristics as well as the number of sets of chromosomes.

OTHER COMMON NAMES

English: European green toad; French: Crapaud vert; German: Wechselkröte; Spanish: Sapo verde.

PHYSICAL CHARACTERISTICS

Males reach about 2.44–3.22 in (62–82 mm) in snout-vent length, and females may reach 3.9 in (100 mm); however, local populations in parts of the range are extremely variable in size. This toad has well-defined marbled green dorsal markings, usually with darker edges, against a tan background. The paratoid glands are oval and parallel to each other rather than diverging. Cranial crests are absent.

DISTRIBUTION

The green toad occurs in Europe east of the Rhine River, including the southern tip of Sweden; the Balearic Islands, Corsica, and Sardinia; western Asia, including Iran, Mongolia, and China; southwestern Asia and the Arabian Peninsula; and northern Africa, from Morocco to Libya.

HABITAT

The green toad is usually found in open, drier lowland areas in Europe and more mountainous regions in Asia.

BEHAVIOR

The green toad is nocturnal; in cities these toads may congregate around street lamps and eat insects. Some physiological color change may occur in the intensity of the green marbling.

Bufo viridis

FEEDING ECOLOGY AND DIET
This toads eats arthropods and insects of all kinds; anecdotes report that it will starve rather than eat earthworms.

REPRODUCTIVE BIOLOGY
The call is a high-pitched trill that lasts for about 10 seconds; it is said to resemble a bird more than a toad. Small (0.04 in or 1.2 mm diameter) blackish eggs (1,000–2,000) are laid in strings. Like most *Bufo* tadpoles, these are small (0.47–0.59 in or 12–15 mm) and blackish.

CONSERVATION STATUS
Not listed by the IUCN.

SIGNIFICANCE TO HUMANS
None known. ◆

Yungas redbelly toad
Melanophryniscus rubriventris

TAXONOMY
Atelopus rubriventris Vellard, 1947, San Andrés, Salta Provice, Argentina.

OTHER COMMON NAMES
None known.

PHYSICAL CHARACTERISTICS
Individuals are about 1.57–1.77 in (40–45 mm) in snout-vent length. This is a moderately warty toad, but the warts appear to be glandular swellings. There are no parotoid glands. The head and snout are relatively short compared to *Atelopus*. The dorsum may be mostly black or may have some yellow spots. The belly and palms and soles are uniformly red-orange.

DISTRIBUTION
This species is known from the subtropical valleys of northwestern Argentina.

HABITAT
These toads live in humid hilly regions along small streams.

BEHAVIOR
Like *Atelopus*, these toads are diurnal; they are also toxic. *Melanophryniscus* exhibit the same sort of unken reflex when disturbed as do the European fire-bellied toads, *Bombina*.

FEEDING ECOLOGY AND DIET
Nothing is known.

REPRODUCTIVE BIOLOGY
The call of males is a soft, short trill. Although this species is often found near streams, the eggs are attached to vegetation in small bodies of standing water. The tadpoles are bottom dwellers in ponds.

CONSERVATION STATUS
Not listed by the IUCN.

SIGNIFICANCE TO HUMANS
None known. ◆

Roraima bush toad
Oreophrynella quelchii

TAXONOMY
Oreophryne quelchii Boulenger, 1895, summit of Mount Roraima, Guyana.

OTHER COMMON NAMES
None known.

PHYSICAL CHARACTERISTICS
These small toads are about 0.78 in (20 mm) in snout-vent length. This species lacks cranial crests. The venter has brown and yellow blotches, and the dorsum is dark brown and warty. Species of *Oreophrynella* are distinctive in the morphology of the foot, in which the first digit is opposed to the others and the second toe is distinctly shorter than the remaining ones. Originally, it was thought that this foot functioned as a branch grasper, much as in the treefrog *Phyllomedusa* (Hylidae). However, observations of these toads in life indicate that the foot is used for clambering across rocks. The toads are basically walkers and rarely if ever hop. The number of vertebrae is reduced to six, which is probably related to the mode of locomotion.

DISTRIBUTION
This toad is found on Mount Roraima on the border of Venezuela and Guyana, one of the highlands of the Guianan shield region of South America consisting of flattened tabletop mountains with steep sides, known as *tepuis*.

HABITAT
The few available data indicate that this species lives among boulders in dense vegetation on the mountaintops.

BEHAVIOR
These toads do not leap or hop; rather, they walk slowly across rocks. When disturbed, they tuck their head, hands, and feet under the body and roll off of the rock face like a dislodged stone.

FEEDING ECOLOGY AND DIET
Nothing is known.

REPRODUCTIVE BIOLOGY
The call of this species is not known with certainty, but that of a related species (*O. huberi*) was described as a shrill "pi, pi, pi." Development of the eggs is apparently direct; no tadpoles have ever been found. Females have been observed attending terrestrial egg clutches.

CONSERVATION STATUS
Not listed by the IUCN.

SIGNIFICANCE TO HUMANS
None known. ◆

Brown tree toad
Pedostibes hosii

TAXONOMY
Nectophryne hosii Boulenger, 1892, Mt. Dulit, Sarawak, Borneo, Malaysia.

OTHER COMMON NAMES
English: Boulenger's Asian tree toad, common tree toad.

PHYSICAL CHARACTERISTICS
This is a moderately large toad, with males 2.09–3.07 in (53–78 mm) and females 3.5–4.1 in (89–105 mm). The dorsum is only moderately warty. A small parotoid gland is present; there is a slight bony ridge just behind the eye, but otherwise cranial crests are not obvious. Consistent with this toad's arboreal

habits, the digital tips are slightly widened, but these are not true digital discs as found in species of the families Hylidae or Rhacophoridae. Female brown tree toads are often featured in pet enthusiast magazines because some of them are dark purple with yellow spots. The significance of this coloration is not known.

DISTRIBUTION
This apparently widespread species is found in Borneo, Sumatra, peninsular Malaysia, and Thailand.

HABITAT
Species of *Pedostibes* are perhaps the only truly arboreal toads. They are found in lowland primary forests and not in open areas.

BEHAVIOR
Unlike many toads, brown tree toads apparently do not form breeding aggregations with large numbers of males calling.

FEEDING ECOLOGY AND DIET
Ants form the major part of the diet.

REPRODUCTIVE BIOLOGY
The males make a call, which has been described as a slurred squawk. Adults breed at clear streams. The tadpoles are similar but not identical to those of *Bufo*. The color is dark brown, but not quite black as in *Bufo*. The tadpoles live in the side pools of streams.

CONSERVATION STATUS
Not listed by the IUCN.

SIGNIFICANCE TO HUMANS
None known. ◆

Aquatic swamp toad
Pseudobufo subasper

TAXONOMY
Pseudobufo subasper Tschudi, 1838, Borneo.

OTHER COMMON NAMES
None known.

PHYSICAL CHARACTERISTICS
This is a large toad, with males 3.03–3.7 in (77–94 mm) and females 3.62–6.1 in (92–155 mm) in snout-vent length. In general, most toads have relatively unremarkable body shapes and are terrestrial. They have not really invaded the aquatic or arboreal niches. *Pseudobufo subasper* is an exception. This toad is basically a *Bufo* that has become an aquatic specialist. The feet are fully webbed, and the webbing is thin, in contrast to the rather thick webbing found in most toads. The nostrils are placed dorsally, and the fingers are slender and unwebbed. The vertebral column exhibits a reduction in ossification that is consistent with it being an aquatic species. Paratoid glands are present.

DISTRIBUTION
This toad is found in the Indonesian province of Kalimantan on the island of Borneo, Sumatra, peninsular Malaysia.

HABITAT
These toads are found associated with pools in swampy areas near large rivers.

BEHAVIOR
Little information is available for these toads; it is not known when they call or breed.

FEEDING ECOLOGY AND DIET
Not known.

REPRODUCTIVE BIOLOGY
Not known.

CONSERVATION STATUS
Not listed by the IUCN.

SIGNIFICANCE TO HUMANS
None known. ◆

Chirinda toad
Stephopaedes anotis

TAXONOMY
Bufo anotis Boulenger, 1907, southeast Mashonaland, Zimbabwe.

OTHER COMMON NAMES
English: Boulenger's earless toad, Chirinda forest toad, Mashonaland toad.

PHYSICAL CHARACTERISTICS
These are moderately small toads about 1.57–1.77 in (40–45 mm) in snout-vent length. The cranial crests are poorly developed, and the tympanum is absent. The parotoid glands are large. The dorsum is not so much warty as granular, and the brown coloration renders this animal cryptic against dead leaves.

DISTRIBUTION
The Chirinda toad is known only from the Chirinda Forest in Zimbabwe and from forest in adjacent Mozambique.

HABITAT
This toad dwells in the leaf litter on the forest floor.

BEHAVIOR
Little is known, except for reproductive behavior.

FEEDING ECOLOGY AND DIET
The diet consists of leaf-litter arthropods, mainly ants.

REPRODUCTIVE BIOLOGY
It is questionable whether this species of *Stephopaedes* calls; no direct observations of vocalizations are known. Unlike most bufonids, this species breeds in restricted pools of water and in holes in the trunks of a particular species of tree. The eggs are about 0.1 in (2.5 mm) in diameter and are laid in short strings that quickly fall apart. The tadpoles are remarkable in having a crown of epithelial tissues forming a closed circle around the eyes and nostrils. It may function as an additional respiratory surface for the restricted habitats in which these tadpoles live.

CONSERVATION STATUS
Although not listed by the IUCN, this species is considered to be vulnerable owing to its restricted distribution.

SIGNIFICANCE TO HUMANS
None known. ◆

Resources

Books

Crump, Martha L. *In Search of the Golden Frog.* Chicago: University of Chicago Press, 2000.

Duellman, William E., and Linda Trueb. *Biology of Amphibians.* New York: McGraw-Hill, 1986.

Inger, Robert F., and Robert B. Stuebing. *A Field Guide to the Frogs of Borneo.* Kota Kinabalu: Natural History Publications, 1997.

Sanchíz, Borja. *Salientia. Part 4: Encyclopedia of Paleoherpetology.* Munich: Verlag Dr. Friedrich Pfeil, 1998.

Savage, Jay M. *The Amphibians and Reptiles of Costa Rica.* Chicago: University of Chicago Press, 2002.

Periodicals

Crump, Martha L. "Homing and Site Fidelity in a Neotropical Frog, *Atelopus varius* (Bufonidae)." *Copeia* 1986 (1986): 438–444.

Daly, John W., Robert J. Highet, and Charles W. Myers. "Occurrence of Skin Alkaloids in Non-dendrobatid Frogs from Brazil (Bufonidae), Australia (Myobatrachidae) and Madagascar (Mantellinae)." *Toxicon* 22 (1984): 905–919

Ford, Linda S., and David C. Cannatella. "The Major Clades of Frogs." *Herpetological Monographs* 7 (1993): 94–117.

Graybeal, Anna. "Phylogenetic Relationships of Bufonid Frogs and Tests of Alternate Macroevolutionary Hypotheses Characterizing their Radiation." *Zoological Journal of the Linnean Society* 119 (1997): 297–338.

Graybeal, Anna, and David C. Cannatella. "A New Taxon of Bufonidae from Peru, with Descriptions of Two New Species and a Review of the Phylogenetic Status of Supraspecific Bufonid Taxa." *Herpetologica* 51 (1995): 105–131.

McDiarmid, Roy W. "Comparative Morphology and Evolution of Frogs of the Neotropical Genera *Atelopus, Dendrophryniscus, Melanophryniscus,* and *Oreophrynella.*" *Science Bulletin of the Los Angeles County Museum of Natural History* 12 (1971): 1–66.

McDiarmid, Roy W., and Stefan Gorzula. "Aspects of the Reproductive Ecology and Behavior of the Tepui Toads, Genus *Oreophrynella* (Anura, Bufonidae)." *Copeia* 1989 (1989): 445–451.

Roessler, Martha K. P., Hobart M. Smith, and David Chiszar. "Bidder's Organ: A Bufonid By-product of the Evolutionary Loss of Hyperfecundity." *Amphibia-Reptilia* 11 (1990): 225–235.

Wake, Marvalee H. "The Reproductive Biology of *Nectophrynoides malcolmi* (Amphibia: Bufonidae), with Comments on the Evolution of Reproductive Modes in the Genus *Nectophrynoides.*" *Copeia* 1980 (1980): 193–209.

David Cannatella, PhD

Poison frogs

(Dendrobatidae)

Class Amphibia

Order Anura

Family Dendrobatidae

Thumbnail description
Small, agile frogs that occur in rainforests of the New World tropics; more primitive genera are cryptically colored with nontoxic skin, whereas the derived genera are brightly colored poison frogs

Size
Most species are 0.75–1.5 in (15–35 mm) in length; a few species reach 2.5 in (62 mm)

Number of genera, species
9 genera; 207 species

Habitat
Rain and cloud forest

Conservation status
No species listed by the IUCN

Distribution
Southern Central America through tropical South America

Evolution and systematics

The relationship of the poison frog family to other frogs remains the subject of controversy. Dendrobatidae lies within a clade of frogs, the neobatrachians, that diverged in the early Cretaceous or late Jurassic. Within this clade, two major groups, the hyloids and ranoids, diverged. Since 1959 numerous studies have placed dendrobatids inconsistently within these two groups. Most recent studies have shown that placement within the hyloids, specifically the leptodactylid/bufonid clade, is most likely.

The largest dendrobatid genus, *Colostethus*, comprises about 100 species, whereas the more derived genera, including *Dendrobates* and *Phyllobates*, contain about 36 and five species, respectively. Relationships at the generic level are fairly well understood. In 1991 the most primitive dendrobatid known, *Aromobates*, was described. *Colostethus* and *Mannophryne* are basal or primitive groups closely related to *Aromobates*, whereas *Dendrobates* is the most derived, or advanced, genus.

Relationships of species within each genus are not well worked out. Partly this is because many species have small ranges and occur in areas that cannot be reached easily; thus, many species have not been studied in detail. The genus *Phyllobates* is best understood. This genus contains five species that occur from Colombia to southern Nicaragua and are united by being the only species that have a unique alkaloid (batrachotoxin) in the skin. Other genera, such as *Colostethus* and *Epipedobates*, may be composed of groups of species that are not related closely; thus, it is probable that these genera will be subdivided into smaller genera in the future. Many species in the poison frog family have yet to be discovered. At least three or four new species in this group are described each year. No subfamilies are recognized.

Physical characteristics

Poison frogs are typically small frogs, less than 1 in (2.5 mm) in body length. The name of one genus, *Dendrobates*, is derived from the Greek *dendro* (tree) and *bates* (walker), an allusion to the fact that they can walk or hop up trees. Dendrobatids have short but powerful hind limbs and are agile jumpers and, in some cases, climbers. They are characterized by the presence of divided scutes (thick pads of skin) on the upper surfaces of the fingers and toes.

Bright colors and distinctive patterns appear on many species of poison frogs. (Photo by Michael & Patricia Fogden/Corbis. Reproduced by permission.)

Poison frogs can be found in a variety of colors and patterns including combinations of red, orange, green, blue, purple, and black. (Photo by Buddy Mays/Corbis. Reproduced by permission.)

Poison frogs derive their common name from the fact that the more derived, or advanced, frogs in the group are brightly colored and have toxic skin. All frogs have glands in the skin that produce a variety of noxious substances serving to protect them from predators. However, poison frogs are unique, because their skin glands contain a large array of alkaloids, which are especially toxic nitrogen-containing chemicals once believed to be produced only by plants. Investigations of these chemicals have found hundreds of different alkaloids in the four derived genera of dendrobatids. Individual species or populations of frogs have different alkaloids, and within a population individual frogs may have different combinations of alkaloids. Some of these alkaloids are encountered in only one or a few species of frogs, whereas others may be present in numerous species. Examples are batrachotoxin, found in only four of the five species of the genus *Phyllobates*, and epibatidine, found in just one species in the genus *Epipedobates*. It was once thought that the frogs produced these chemicals, but it is now believed that they are obtained from the frogs' diet.

The derived poisonous species of dendrobatids are brightly colored. Some species of *Dendrobates* are bright yellow with black spots and black and blue legs, whereas others are a brilliant green with black markings. At least one species, *Dendrobates pumilio*, has numerous differently colored individuals throughout its range. Some populations are bright red, others are yellow with black spots, and still others are blue. The more primitive species, such as those in the genera *Aromobates*, *Colostethus*, and *Mannophryne*, are cryptically colored, generally brown frogs that lack alkaloids in their skin.

Distribution

Poison frogs occur in the neotropics from Nicaragua south through Costa Rica, Panama, and northern South America to southern Brazil and Bolivia. Within their range, most species are found only in undisturbed primary rainforest or cloud forest, although a few species occur in converted pastureland, *cerrado* (a savanna-like habitat) in southern Brazil, or cacao plantations in Central America.

No dendrobatids are known to be extinct at present, but many dendrobatids have small distributions within the overall range of the family, so information on the status of these species is difficult to obtain. Some species seem to be on the verge of extinction (e.g., *D. mysteriosus* in the Cordillera del Condor of Peru), and others have not been seen in the wild for years and may be extinct (e.g., *D. speciosus* from Panama). Only one species, *Dendrobates auratus*, has become established outside its natural range. This species was introduced into Hawaii.

Habitat

Throughout the range, most species of dendrobatids occur in primary rainforest. The more primitive species, such as those in the genus *Colostethus* and *Mannophryne*, occur strictly in leaf litter on the forest floor. Frogs in the genus *Dendrobates* are partially or completely arboreal, depending on the species. For example, some species of *Dendrobates*, such as *D. auratus*, *D. pumilio*, and *D. castaneoticus*, largely inhabit the forest floor leaf litter but frequently climb trees and vines. *Dendrobates vanzolinii*, a species in western Brazil, lives in the lower canopy of the forest and avoids the forest floor. Other species, such as *D. arboreus* in Panama, seem to have an almost entirely arboreal existence.

Nearly all species of dendrobatids deposit eggs on land. Among the more primitive species, tadpoles are transported to small, slow-moving streams or pools on the forest floor. The more derived species transport their tadpoles to container habitats, such as bromeliads, tree holes, *Heliconia* bracts, Brazil nut capsules, or other types of small phytotelmata (water-holding plants). Although most of these container habitats are arboreal, some, such as Brazil nut capsules and other seed husks, are on the forest floor.

Behavior

All species of dendrobatids, except the most primitive, *Aromobates*, are diurnal. They tend to be most active in early morning at first light and in late afternoon, particularly on rainy days. Dendrobatids are also most active and conspicuous during the rainy season. Their collective range encompasses a large area of the neotropics, and consequently the rainy season in any one area may be longer or shorter or occur earlier or later in the year compared with other areas. Dendrobatids may be found during the dry season, but they are generally less active during this time.

Many species of dendrobatids seem to be territorial. Territoriality typically is associated with reproduction. In most species, males are territorial, and females are not. However, in several species of *Colostethus* females are territorial.

Feeding ecology and diet

Dendrobatids usually feed on tiny arthropods, ranging in average size from 0.03 to 0.07 in (0.8 to 1.7 mm) in length. Prey include ants, mites, small beetles, small flies, springtails, and tiny spiders, among others.

Evidence is accumulating that toxic dendrobatids obtain alkaloids from their diet. When frogs collected as tadpoles were raised under similar conditions and fed either leaf-litter arthropods or fruit flies, only those fed on arthropods developed alkaloids in the skin. Poison frogs fed fruit flies dusted with alkaloids have been shown to absorb those toxins from the diet and secrete them from the skin glands. Some species of ants have the same alkaloids as those in dendrobatid skin; thus, ants may be one of the main sources of alkaloids. Further work on species of dendrobatid frogs representing both primitive and derived clades showed that three derived species (which are also toxic) had diets consisting 50–73% of ants, whereas the diets of five nontoxic species contained only 6–16% ants. Thus, the diet of the derived, toxic, brightly colored dendrobatids is composed of a much higher percentage of ants than the primitive, cryptically colored, nontoxic species.

Epipedobates trivittatus with tadpoles on its back, from Brazil. The eggs are deposited on land, and when they have developed into tadpoles, the male parent crouches by them, and they wiggle onto his back. He then transports them to small pools in the rainforest when they undergo the remainder of their development and tranform into small frogs. (Photo by Janalee P. Caldwell. Reproduced by permission.)

Reproductive biology

Dendrobatid frogs breed primarily during the rainy season. Males typically call intensely at first light for a period of several hours and then are quiet during midday, unless rainfall occurs. They may call again in late afternoon, especially on wet days. Like other frogs, each species has a unique call. Call characteristics are important in distinguishing closely related species that may be very similar in morphologic features.

A red-backed poison dart frog (*Dendrobates reticulatus*) carries its tadpoles on its back. (Photo by Michael Fogden. Bruce Coleman Inc. Reproduced by permission.)

All species of dendrobatids (with the possible exception of *Aromobates nocturnus*) deposit eggs on land. Depending on the species, eggs may be deposited on leaves in leaf litter on the forest floor, or they may be attached above the waterline to the inside of a tree hole or a bromeliad tank.

The eggs are attended by the male or female parent, depending on the species. When the eggs have developed into tadpoles, the transporting parent sits among the tadpoles in the nest, and they wiggle up the parent's leg and onto the back. The tadpoles are attached to the body of the parent by mucopolysacchrides, chemicals that dissolve easily in water. The parent frog then carries the tadpoles to a body of water, such as a small stream or pool or a bromeliad tank. The parent submerges the back half of its body in the water, and the tadpoles gradually dislodge and swim away, thus ending the period of parental care for most species. In some species, parents spend substantial amounts of time searching for suitable pools.

In terms of reproduction, differences exist between the primitive and the advanced dendrobatids and even within these groups. Among the primitive species, clutch size is larger, and all tadpoles are transported at once. In some species of *Colostethus*, for example, clutch size may be from 25 to 35 offspring. Clutch size in *Dendrobates* is much smaller, generally from three to six eggs. In these species, depending on aspects of the biology, each tadpole may be transported singly by the parent to a different aquatic site, usually a tree hole or some other type of phytotelmata.

Parental care is more complex in some species of *Dendrobates*. In at least one species, *D. vanzolinii*, males and females may remain together as pairs and care for their offspring together. Eggs are deposited on the inside of tiny tree holes or vine holes above the water level. After the tadpoles develop, the male parent carries each one to another tiny tree hole, in which it completes its development. About every five days, the male and female court, which appears to trigger ovulation in the female. However, instead of depositing fertilized eggs above the waterline, the pair returns to the site of their tadpole. The tadpole goes through a stereotypic movement in which the body stiffens and vibrates. The female parent appears to respond to this movement by the tadpole by backing into the water and depositing one or two unfertilized eggs for the tadpole to consume. This type of parental care presumably evolved in response to the lack of nutrients in the tree holes; the tadpoles are dependent on the nutritive eggs produced by the parent to survive.

In another species of *Dendrobates*, *D. pumilio*, the female rather than the male transports the tadpoles from the nest in the leaf litter to individual leaf axils that contain a tiny bit of water. The female then returns about every five days to deposit eggs for the tadpole to consume. Although the male parent appears not to be involved, as is the case for *D. vanzolinii*, there is some suggestion that the female seeks the calling male and remains near him for a period of time, possibly to stimulate ovulation.

Tadpoles of all species of dendrobatids, except those in *Dendrobates*, are typical herbivores that graze on algae and detritus. In contrast, those of *Dendrobates* are predaceous. This trait may have evolved in response to confinement of the tadpoles in small, unproductive habitats, where the ability to kill and eat small macroinvertebrates, such as mosquito larvae, would have been advantageous. In some species, tadpoles of *Dendrobates* readily kill and eat smaller tadpoles of the same and different species; thus, as discussed earlier, parent frogs in some species transport only one tadpole at a time and place it singly at an isolated site for development so that it will not be consumed by one of its larger siblings.

Conservation status

Most species of dendrobatid frogs occur in rainforest habitats that are vulnerable to deforestation. In addition, the extent of the distribution ranges of many species is unknown, because areas where they may occur are unexplored. These facts make determination of the conservation status difficult. In Ecuador five species of *Colostethus* and one species of *Dendrobates* are declining in numbers; all are Andean species that occur above 3,940 ft (1,200 m). At present, no species are cited as threatened on the IUCN Red List. In contrast, all species, except those in the genera *Colostethus*, *Mannophryne*, and *Nephelobates*, are listed on CITES Appendix II.

Significance to humans

Derived species of dendrobatids produce large numbers of alkaloids in the skin. The toxins in most species have not been studied thoroughly, and little is known about the potential pharmacological uses. An alkaloid produced by one species of *Epipedobates* is a painkiller 200 times more potent than morphine. Some Indian tribes in Colombia have used skin toxins of three species in the genus *Phyllobates* as poison for their blowgun darts; blowguns were used in hunting small game.

1. Trinidad poison frog (*Mannophryne trinitatis*); 2. Venezuelan skunk frog (*Aromobates nocturnus*); 3. Blue-toed rocket frog (*Colostethus caeruleo-dactylus*); 4. Blue-bellied poison frog (*Dendrobates minutus*); 5. Stephen's rocket frog (*Colostethus stepheni*). (Illustration by Joseph E. Trumpey)

1. Phantasmal poison frog (*Epipedobates tricolor*); 2. Green poison frog (*Dendrobates auratus*); 3. Amazonian poison frog (*Dendrobates ventri-maculatus*); 4. Golden dart-poison frog (*Phyllobates terribilis*); 5. Brazilian poison frog (*Dendrobates vanzolinii*); 6. Harlequin poison frog (*Dendrobates histrionicus*); 7. Strawberry poison frog (*Dendrobates pumilio*); 8. Brazil nut poison frog (*Dendrobates castaneoticus*); 9. Imitating poison frog (*Dendrobates imitator*). (Illustration by Joseph E. Trumpey)

Species accounts

Venezuelan skunk frog
Aromobates nocturnus

TAXONOMY
Aromobates nocturnus Myers Paolillo, and Daly, 1991, about 1.2 mi (2 km) airline east-southeast of Agua de Obispos, Trujillo, Venezuela.

OTHER COMMON NAMES
None known.

PHYSICAL CHARACTERISTICS
The Venezuelan skunk frog is large compared with other dendrobatids; females may reach 2.5 in (62 mm) in snout-vent length. This species derives its generic name from the production of a particularly noxious substance that has a skunklike odor but which has defied analysis. The substance is not toxic and is not an alkaloid, like the chemicals found in the skin of advanced dendrobatids, but the noxious odor, released by the frog upon being handled, is sufficient to protect it from predators.

DISTRIBUTION
The Venezuelan skunk frog is known only from the type locality in northwestern Venezuela.

HABITAT
This species occurs in small streams and rivulets in dense Andean cloud forest at an elevation of 7,382 ft (2,250 m).

BEHAVIOR
The species is entirely nocturnal, in contrast to all other species of dendrobatids. Also in contrast to all other dendrobatids, the Venezuelan skunk frog is strictly aquatic, found only by small streams, usually sitting or swimming in water.

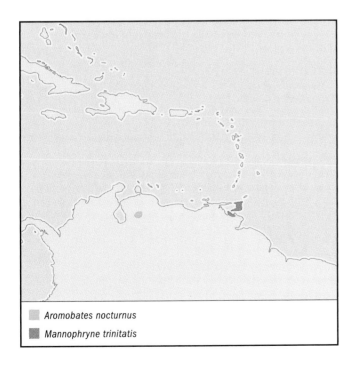

◻ *Aromobates nocturnus*
◼ *Mannophryne trinitatis*

FEEDING ECOLOGY AND DIET
No information is available on the diet or feeding, but individuals sitting out at night readily took insects tossed to them. They probably feed strictly on small insects and arthropods, like most other frogs.

REPRODUCTIVE BIOLOGY
Individuals have not been observed calling, and no information is available on reproduction.

CONSERVATION STATUS
The range of this species probably is restricted to a small area; thus, disturbance of the area could have a severe impact on populations of these frogs.

SIGNIFICANCE TO HUMANS
This frog, discovered only in the early 1980s, is significant because of its basal position in the poison frog family. It may offer clues to the relationship of poison frogs to other families of frogs. Its discovery illustrates how much remains to be discovered about tropical frogs. ◆

Blue-toed rocket frog
Colostethus caeruleodactylus

TAXONOMY
Colostethus caeruleodactylus Lima and Caldwell, 2001, about 25 mi (40 km) south of Manaus, Amazonas, Brazil.

OTHER COMMON NAMES
None known.

PHYSICAL CHARACTERISTICS
The snout-vent length is 0.60–0.67 in (15.4–17.4 mm) for females and 0.58–0.63 in (14.9–16.3 mm) for males. These small frogs are brown on the dorsum, with white chins and bellies. Males have sky-blue fingers and blue discs on the toes during the breeding season. Females have blue discs on the fingers and toes.

DISTRIBUTION
This species is known only from the type locality.

HABITAT
The frogs occur in leaf litter in an isolated patch of slightly disturbed lowland *igapó* (flooded) forest intersected with small hills and valleys. During the rainy season, rising rivers overflow into small streams in the valleys, creating a system of deep, interconnected, meandering pools. The frogs occur on the slopes above the streams, and their tadpoles develop in the seasonal pools that form in the streams.

BEHAVIOR
Males are territorial, defending small areas of forest approximately 1,000 ft² (10 m²) in size. Short, loud encounter calls are produced by the resident male when an intruding male approaches.

FEEDING ECOLOGY AND DIET
This species feeds on a variety of small insects and other arthropods.

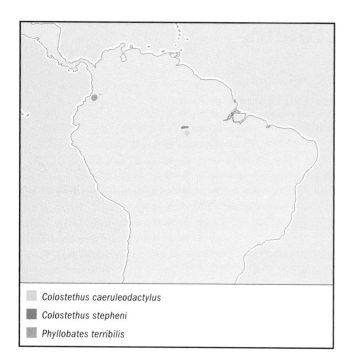

Colostethus caeruleodactylus

Colostethus stepheni

Phyllobates terribilis

REPRODUCTIVE BIOLOGY
Reproduction takes place during the rainy season, from January through April. Courtship lasts all of one day and part of the following morning, after which an average of 19 eggs are deposited in rolled or folded leaves on the forest floor. Males attend the clutches and transport all the developing tadpoles near the end of the rainy season, when *igapó* pools are at their maximum depth.

CONSERVATION STATUS
This species is known only from the type locality. Should the forest in this area be removed, the species would become extinct. No special protection is provided the forest at present; it is under the control of private landowners.

SIGNIFICANCE TO HUMANS
None known. ◆

Stephen's rocket frog
Colostethus stepheni

TAXONOMY
Colostethus stepheni Martins, 1989, proveniente da bica da vila residencial da Usina Hidroeléctrica de Balbin, Município de Presidente Figueiredo, Amazonas, Brazil.

OTHER COMMON NAMES
None known.

PHYSICAL CHARACTERISTICS
The snout-vent length is 0.66–0.70 in (17.0–18.0 mm) in females and 0.59–0.64 in (15.2–16.5 mm) in males. This small frog has a brown dorsum with a white oblique lateral stripe.

DISTRIBUTION
This species is known only from the region of the type locality.

HABITAT
Individuals occur in the leaf litter of lowland tropical forest.

BEHAVIOR
Males produce three types of vocalizations: an advertisement call to attract females, an encounter call to signal that another male is invading the caller's territory, and a courtship call to communicate at close range, particularly with a gravid female.

FEEDING ECOLOGY AND DIET
This species feeds on tiny arthropods found in leaf litter.

REPRODUCTIVE BIOLOGY
Males call during the rainy season, from November to April. Peak calling times are at dawn and dusk, although males may call anytime during the day before or after heavy rainfall. Unlike most other dendrobatids, tadpoles of this species develop entirely in small terrestrial nests in cuplike leaves on the forest floor. Clutch size varies from three to six eggs; males remain with the clutches and guard them from potential predators, such as small lizards or large spiders.

CONSERVATION STATUS
Not listed by IUCN.

SIGNIFICANCE TO HUMANS
None known. ◆

Green poison frog
Dendrobates auratus

TAXONOMY
Dendrobates auratus Girard, 1855, Taboga Island, Panama.

OTHER COMMON NAMES
German: Goldbaumsteiger.

PHYSICAL CHARACTERISTICS
The snout-vent length is 1.06–1.65 in (27.0–42.0 mm) in females and 0.98–1.56 in (25.0–39.5 mm) in males. This relatively large dendrobatid typically has calligraphic brilliant green markings on a black background. There is substantial variation among populations in both hue (ranging from white to blue-green) and especially pattern (from thick stripes to dots).

DISTRIBUTION
This species occurs from Nicaragua through Costa Rica and Panama to Colombia.

HABITAT
The green poison frog is found in lowland primary rainforest.

BEHAVIOR
Males are territorial at high population densities but may not be at low population densities. Males attempt to attract and mate with many females and can care for offspring of several different females simultaneously. This behavior increases male reproductive success but imposes a cost on the survival probability of each offspring. Females do not defend territories, but some females guard particular males and will attack other females to prevent them from approaching their mates. This species provides an excellent example of sexual conflict.

FEEDING ECOLOGY AND DIET
Like other species of *Dendrobates*, this one feeds primarily on tiny ants and mites. Other prey include tiny beetles, flies, and springtails.

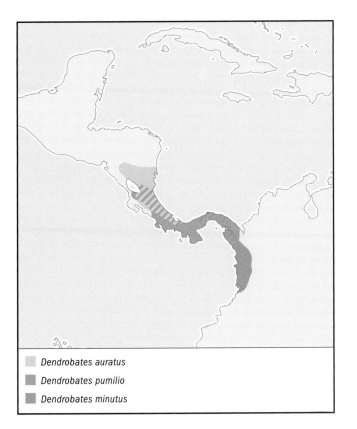

Dendrobates auratus
Dendrobates pumilio
Dendrobates minutus

Dendrobates histrionicus
Dendrobates ventrimaculatus
Dendrobates castaneoticus

REPRODUCTIVE BIOLOGY
Eggs are laid in leaf litter. The male visits the eggs periodically over the two weeks of development, shedding water on them, removing fungus, and rotating the eggs. The male then transports the tadpoles on his back, usually one at a time, to small pools of water, typically in tree holes.

CONSERVATION STATUS
Not threatened.

SIGNIFICANCE TO HUMANS
This species is popular in the pet trade, and most individuals are raised in captivity. ◆

Brazil nut poison frog
Dendrobates castaneoticus

TAXONOMY
Dendrobates castaneoticus Caldwell and Myers, 1990, near the Rio Xingu, Pará, Brazil.

OTHER COMMON NAMES
None known.

PHYSICAL CHARACTERISTICS
The snout-vent length is 0.83–0.88 in (21.5–22.7 mm) in females and 0.70–0.79 in (17.9–20.3) in males. The body is black with white spots; the arms and legs are black with brilliant orange spots.

DISTRIBUTION
This species is known from the type locality and two other localities within 155 mi (250 km) in Pará, Brazil.

HABITAT
This species is found in primary rainforest.

BEHAVIOR
This species is diurnal and commonly is seen hopping through leaf litter or climbing vines and trees in the forest. One individual hopped straight up the trunk of a large tree and disappeared into the canopy.

FEEDING ECOLOGY AND DIET
Like other species of *Dendrobates*, this one feeds primarily on tiny ants and mites. Other prey include tiny beetles, flies, and springtails.

REPRODUCTIVE BIOLOGY
Neither eggs nor calling males have been observed. During the rainy season, males transport tadpoles singly to fallen Brazil nut capsules on the forest floor.

CONSERVATION STATUS
The extent of the range and the number of populations in this species are unknown. Like many species of dendrobatids, the size of the distribution range is small.

SIGNIFICANCE TO HUMANS
None known. ◆

Harlequin poison frog
Dendrobates histrionicus

TAXONOMY
Dendrobates histrionicus Berthold, 1845, Pacific versant of northwestern Colombia, probably the upper Río San Juan drainage in the present-day Departamento Risaralda.

OTHER COMMON NAMES
None known.

PHYSICAL CHARACTERISTICS
The snout-vent length is 1.1–1.5 in (28.0–38.0 mm) in females and 0.95–1.5 in (24–38 mm) in males. This large dendrobatid has extensive variation in color and pattern among populations. The color is typically red with yellow and orange variants.

DISTRIBUTION
The species inhabits Chocó of western Colombia and northwestern Ecuador.

HABITAT
This species occurs in lowland rainforest.

BEHAVIOR
Resident males establish small territories. They respond aggressively when the call of another male is played on a tape recorder near them.

FEEDING ECOLOGY AND DIET
This species feeds on small insects and arthropods, particularly ants and mites.

REPRODUCTIVE BIOLOGY
If a female approaches a calling male, the male continues calling until the female begins to follow him. He leads the female under the leaf litter, where deposition of eggs occurs. After the eggs develop into tadpoles, the female transports them on her back to small pools of water in the axils of plants such as *Heliconia*, where the tadpoles undergo the remainder of their development. This species has female parental care, as in the strawberry poison frog, but unlike the strawberry poison frog, the male does not tend the eggs.

CONSERVATION STATUS
Not threatened.

SIGNIFICANCE TO HUMANS
None known. ◆

Dendrobates imitator

Dendrobates vanzolinii

Epipedobates tricolor

Imitating poison frog
Dendrobates imitator

TAXONOMY
Dendrobates imitator Schulte, 1986, Tarapoto, Peru.

OTHER COMMON NAMES
German: Zweipunkt-Baumsteiger.

PHYSICAL CHARACTERISTICS
The snout-vent length is 0.67–0.87 in (17.0–22.0 mm). This small dendrobatid has considerable variation in color but generally is black with yellow stripes on the dorsum.

DISTRIBUTION
This species is known from the eastern foothills of the Andes in Departamentos San Martín and Huánuco, Peru.

HABITAT
The imitating poison frog occurs below 3,300 ft (1,000 m) in primary rainforest.

BEHAVIOR
Little is known of this frog's behavior.

FEEDING ECOLOGY AND DIET
Like other species of *Dendrobates*, this one feeds primarily on tiny ants and mites. Other prey include tiny beetles, flies, and springtails.

REPRODUCTIVE BIOLOGY
Not known.

CONSERVATION STATUS
Not threatened.

SIGNIFICANCE TO HUMANS
This species represents the only known example of mimetic radiation (in which different populations of a single species mimic different species) in amphibians. Three populations of this frog occur in sympatry with one of three other species of poison frogs, *D. variabilis*, *D. ventrimaculatus*, and *D. fantasticus*, none of which is related closely to the imitating poison frog. These three species differ dramatically with respect to color pattern. Each population of the imitating poison frog looks virtually identical to the species with which it occurs in sympatry. Molecular phylogenetic analysis has confirmed that the separate populations of imitating poison frog are all closely related members of a single species. The mimicry is likely to be Müllerian in nature, because all involved species are highly toxic. ◆

Blue-bellied poison frog
Dendrobates minutus

TAXONOMY
Dendrobates minutus Shreve, 1935, Barro Colorado Island, Panama.

OTHER COMMON NAMES
German: Zwergbaumsteiger.

PHYSICAL CHARACTERISTICS
The snout-vent length is 0.47–0.61 in (12.0–15.5 mm) in females and 0.47–0.59 in (12.0–15.0 mm) in males. This tiny dendrobatid typically is bronze on the dorsum, with a black-and-white or black-and-blue marbled venter.

DISTRIBUTION
This species occurs on the Pacific coast from Panama to central Colombia.

HABITAT
Individuals are found in rainforest below 3,300 ft (1,000 m).

BEHAVIOR
Adult males transport tadpoles on their backs to bromeliad tanks. Tadpoles are predaceous and feed on mosquito larvae.

FEEDING ECOLOGY AND DIET
Adults feed on small insects and other arthropods. Like other dendrobatids, this species is an active, diurnal forager.

REPRODUCTIVE BIOLOGY
Males are territorial. Clutches of two eggs are laid in leaf litter. The male attends the eggs periodically and carries the tadpoles to small pools of water in the leaf axils of plants.

CONSERVATION STATUS
Not threatened.

SIGNIFICANCE TO HUMANS
None known. ◆

Strawberry poison frog
Dendrobates pumilio

TAXONOMY
Dendrobates pumilio O. Schmidt, 1857, between Bocas del Toro and Volcán Chiriqui, Panama.

OTHER COMMON NAMES
German: Erdbeerfröschchen.

PHYSICAL CHARACTERISTICS
The snout-vent length is 0.69–0.95 in (17.5–24.0 mm) in females and 0.71–0.95 in (18.0–24.0 mm) in males. This relatively small dendrobatid typically is red with blue legs, although populations from the Bocas del Toro archipelago in Panama are among the most variable on earth. Populations there vary in color from blue to green and from yellow to red or orange and have patterns with speckles, spots, stripes, or solid colors. Sexual dimorphism is absent, except that males typically have darker throats than females. Genetic and geologic analysis shows that these populations diverged from each other very recently.

DISTRIBUTION
This species occurs in Nicaragua, Costa Rica, and Panama.

HABITAT
This species generally inhabits rainforest but also frequently occurs in cacao and banana groves.

BEHAVIOR
Males are territorial and incessantly produce a nonmusical chirp during the wet season. Field studies have shown that males with larger territories with more three-dimensional structures are more likely to attract mates, possibly because they can advertise more effectively. Mate-choice experiments in the laboratory suggest that females from the Bocas del Toro archipelago prefer to mate with males of the same color.

FEEDING ECOLOGY AND DIET
This species feeds primarily on tiny ants and mites.

REPRODUCTIVE BIOLOGY
The small clutch (two to six eggs) is laid in leaf litter. The male visits the eggs periodically over the two weeks of development, shedding water on them, removing fungus, and rotating the eggs. Females transport tadpoles singly to small pools of water, typically those in the leaf axils of bromeliads or large leafy plants like *Philodendron*. Females return to the pool, on average, every five days for several months to deposit infertile eggs that tadpoles rely on for food in their nutrient-poor pools.

CONSERVATION STATUS
Not threatened.

SIGNIFICANCE TO HUMANS
None known. ◆

Brazilian poison frog
Dendrobates vanzolinii

TAXONOMY
Dendrobates vanzolinii Myers, 1982, Porto Walter on the Rio Juruá, Acre, Brazil.

OTHER COMMON NAMES
None known.

PHYSICAL CHARACTERISTICS
The snout-vent length is 0.67–0.77 in (17.4–19.9 mm) in females and 0.62–0.70 in (16.1–18.1 mm) in males. This small frog has black spots and bars on a bright yellow background and a pattern of blue mesh on the legs.

DISTRIBUTION
This frog is known from western Brazil in the state of Acre and the adjacent Amazonian region in Peru.

HABITAT
The species inhabits lowland rainforest. Individuals avoid leaf litter; instead, they inhabit small trees or shrubs in the lower canopy.

BEHAVIOR
Males are territorial and interact vocally with males in adjacent territories to establish the limits of their territories.

FEEDING ECOLOGY AND DIET
Adults forage continually during the day. They feed on tiny insects and other arthropods, primarily ants and mites.

REPRODUCTIVE BIOLOGY
Pairs of frogs remain together and care for their offspring. Small clutches of two to three eggs are deposited in tiny tree holes above the waterline, and tadpoles develop individually in these nutrient-poor habitats. Pairs undergo courtship about every five days, but instead of depositing fertilized eggs, the female deposits eggs in the water for the tadpoles to consume.

CONSERVATION STATUS
Not threatened. As in many other species of dendrobatids, little is known about the extent of the distribution range. Continual deforestation in the area around Porto Walter, Brazil, has caused the demise of some populations in that area.

SIGNIFICANCE TO HUMANS
This species is unusual among frogs, in that pairs remain together to care for their offspring. Loss of this species would

prevent gaining a better understanding of the evolution of this reproductive mode and those in closely related species. ◆

Amazonian poison frog
Dendrobates ventrimaculatus

TAXONOMY
Dendrobates ventrimaculatus Shreve, 1935, Sarayacu, Ecuador.

OTHER COMMON NAMES
None known.

PHYSICAL CHARACTERISTICS
The snout-vent length is 0.59–0.85 in (15–21.5 mm) in females and 0.57–0.79 in (14.5–20.0 mm) in males. This relatively small dendrobatid typically has linear yellow stripes on a black background, with a bright blue mesh pattern on the legs and venter.

DISTRIBUTION
This species occurs in the Amazon lowlands in Ecuador, Peru, Colombia, Brazil, and French Guiana.

HABITAT
Individuals inhabit lowland forest, where they live in leaf litter and climb into the forest canopy.

BEHAVIOR
These frogs are active during the day in rainforest. They frequently climb into the canopy to feed, court, and deposit eggs in small pools of water held in leaf axils.

FEEDING ECOLOGY AND DIET
These frogs are diurnal, active foragers. They consume small insects, primarily ants, and other arthropods.

REPRODUCTIVE BIOLOGY
Eggs are deposited in the stem axils of such plants as *Heliconia*, near the surface of the water. The male parent transports tadpoles to new pools, or they may slide into the pool below. Cannibalism of small tadpoles by older conspecifics may occur.

CONSERVATION STATUS
Not threatened.

SIGNIFICANCE TO HUMANS
None known. ◆

Phantasmal poison frog
Epipedobates tricolor

TAXONOMY
Epipedobates tricolor Boulenger, 1899, Porvenir, Provincia Bolívar, Ecuador.

OTHER COMMON NAMES
None known.

PHYSICAL CHARACTERISTICS
The snout-vent length is 0.83–1.04 in (21.0–26.5 mm) in females and 0.75–0.97 in (19.0–24.5 mm) in males. This medium-size frog is dark brown to dull red, with wide yellow or whitish stripes along the sides and down the middle of the back.

DISTRIBUTION
This frog is known from southwestern Ecuador and northwestern Peru west of the Andes.

HABITAT
The species inhabits wet and dry habitats but generally occurs near streams in mountain valleys.

BEHAVIOR
Little is known, except for reproductive behavior.

FEEDING ECOLOGY AND DIET
This species feeds on small insects and other arthropods.

REPRODUCTIVE BIOLOGY
Cephalic amplexus, in which the male sits atop the female and clasps her with his forelimbs around her head, has been observed in captive individuals of this species. Large clutches of 15–40 terrestrial eggs are tended by the male. The male carries all the tadpoles at one time on his back to a stream or small pool, where the tadpoles complete their development.

CONSERVATION STATUS
Not threatened.

SIGNIFICANCE TO HUMANS
The phantasmal poison frog has a toxin called epibatidine, an alkaloid that binds to nicotine receptors and acts as an analgesic (painkiller). Remarkably, this painkiller is 200 times more powerful than morphine. Although epibatidine itself is too toxic to use as a painkiller, its discovery led to the synthesis of other drugs that bind to the same receptors and are also highly effective painkillers without the toxic effects of epibatidine. ◆

Trinidad poison frog
Mannophryne trinitatis

TAXONOMY
Mannophryne trinitatis Garman, 1888, Trinidad. Containing 10 species, the genus *Mannophryne* is composed of small frogs that have a throat collar.

OTHER COMMON NAMES
None known.

PHYSICAL CHARACTERISTICS
The snout-vent length is 0.85–1.00 in (22.0–26.0 mm) in females and 0.74–0.85 in (19.0–22.0 mm) in males. Females and males are brown on the dorsum. Females have a bright yellow throat with a black collar and a white venter, whereas males have a gray throat and a black collar.

DISTRIBUTION
The species occurs in Trinidad and northern Venezuela.

HABITAT
The frogs occur around large boulders in intermittent and permanent streams in mountain ranges. They may wander a short distance from the streams during rainy periods.

BEHAVIOR
In contrast to other species of dendrobatids, females (but not males) are territorial. Their small territories are usually 11 ft² (1 m²) or less. Females defend their territories by sitting upright on top of a boulder and pulsating their bright yellow throats at intruders.

FEEDING ECOLOGY AND DIET
This species feeds on small insects and other arthropods.

REPRODUCTIVE BIOLOGY
Courting males use a visual advertisement display in addition to calling, presumably so females will not mistake them for intruders. The visual displays include running forward and jumping up with the front feet off the ground and moving quickly side to side in a crablike motion. Small clutches of eggs are deposited in rock crevices or under leaf litter during the rainy season. Males attend the eggs and transport an average of eight tadpoles to small pools along the streams.

CONSERVATION STATUS
Not threatened.

SIGNIFICANCE TO HUMANS
None known. ◆

Golden dart-poison frog
Phyllobates terribilis

TAXONOMY
Phyllobates terribilis Myers Daly, and Malkin, 1978, Quebrada Guangui, Departamento Cauca, Colombia.

OTHER COMMON NAMES
English: Golden poison frog; German: Schrecklicher Pfeilgiftfrosch, Goldener Giftfrosch.

PHYSICAL CHARACTERISTICS
Females of this large, brilliant yellow dendrobatid are 1.59–1.83 in (40.3–46.5 mm) in length, and males are 1.47–1.76 in (37.3–44.6 mm).

DISTRIBUTION
The species is known from the region of the type locality in Cauca, Colombia.

HABITAT
Individuals are found in lowland rainforest.

BEHAVIOR
This species is diurnal and terrestrial, like most other dendrobatids. It does not have arboreal tendencies, like many species of *Dendrobates*.

FEEDING ECOLOGY AND DIET
The golden dart-poison frog feeds on small insects and other arthropods.

REPRODUCTIVE BIOLOGY
This species has male parental care similar to that of the green poison frog, although the type of pools used for tadpole deposition in nature is not known.

CONSERVATION STATUS
This species is known only from the vicinity of the type locality. Thus, any disturbance of this area could threaten the existence of populations of this species.

SIGNIFICANCE TO HUMANS
This species is the most toxic amphibian and one of the most toxic animals on Earth. Its skin contains the alkaloid batrachotoxin, which is a potent neurotoxin, a type of toxin that affects the nervous system. A microscopic amount is lethal if it reaches the bloodstream. Batrachotoxin acts by forcing sodium channels to remain open. It has become an extremely useful tool for investigating the physiology of sodium channels. Recent research shows that this toxin also is found in certain species of birds from New Guinea and in a North American insect. The toxin probably is produced by plants, though this has not yet been confirmed. The golden dart-poison frog and two closely related species in the Chocó region of Colombia are the only frogs known to be used to make dart poison. ◆

Resources

Books
Heselhaus, Ralf. *Poison-Arrow Frogs: Their Natural History and Care in Captivity.* London: Blandford, 1992.

Schulte, Rainer. *Pfeilgiftfrösche, "Arteneil Peru."* INIBICO, Waiblingen, 1999.

Walls, Jerry G. *Poison Frogs of the Family Dendrobatidae: Jewels of the Rainforest.* Neptune City, NJ: TFH Publications, 1994.

Periodicals
Caldwell, Janalee P. "The Evolution of Myrmecophagy and Its Correlates in Dendrobatid Frogs (Anura: Dendrobatidae)." *Journal of Zoology, London* 240 (1996): 75–101.

———. "Pair Bonding in the Spotted Poison Frog." *Nature* 385 (1997): 211.

Caldwell, Janalee P., and Maria Carmozina Araújo. "Cannibalistic Interactions Resulting from Indiscriminate Predatory Behavior in Tadpoles of Poison Frogs (Anura: Dendrobatidae)." *Biotropica* 30 (1998): 92–103.

Caldwell, Janalee P., and Verônica L. Oliveira. "Determinants of Biparental Care in the Spotted Poison Frog, *Dendrobates vanzolinii* (Anura: Dendrobatidae)." *Copeia* 1999 (1999): 565–575.

Clough, M., and Kyle Summers. "Phylogenetic Systematics and Biogeography of the Poison Frogs: Evidence from Mitochondrial DNA Sequences." *Biological Journal of the Linnaean Society* 70, no. 3 (2000): 515–540.

Daly, John W., H. Martin Garraffo, and Charles W. Myers. "The Origin of Frog Skin Alkaloids: An Enigma." *Pharmaceutical News* 4, no. 4 (1997): 9–14.

Daly, John W., Charles W. Myers, and Noel Whittaker. "Further Classification of Skin Alkaloids from Neotropical Poison Frogs (Dendrobatidae), with a General Survey of Toxic/Noxious Substances in the Amphibia." *Toxicon* 25, no. 10 (1987): 1023–1095.

Myers, Charles W., and John W. Daly. "Dart-Poison Frogs." *Scientific American* 248 (1983): 120–133.

Myers, Charles W., John W. Daly, and Borys Malkin. "A Dangerously Toxic New Frog (*Phyllobates*) Used by Embera

Resources

Indians of Western Colombia, with Discussion of Blowgun Fabrication and Dart Poisoning." *Bulletin of the American Museum Natural History* 161 (1978): 307–366.

Myers, Charles W., Alfredo Paolillo O., and John W. Daly. "Discovery of a Malodorous and Nocturnal Frog in the Family Dendrobatidae: Phylogenetic Significance of a New Genus and Species from the Venezuelan Andes." *American Museum Novitates* 3002 (1991): 1–20.

Summers, Kyle, and W. Amos. "Behavioral, Ecological and Molecular Genetic Analyses of Reproductive Strategies in the Amazonian Dart-Poison Frog, *Dendrobates ventrimaculatus.*" *Behavioral Ecology* 8 (1997): 260–267.

Summers, Kyle, E. Bermingham, L. Weigt, S. McCafferty, and L. Dahlstrom. "Phenotypic and Mitochondrial DNA Divergence in Three Species of Dart-Poison Frogs with Contrasting Parental Care Behavior." *Journal of Heredity* 88 (1997): 8–13.

Summers, Kyle, and M. Clough. "The Evolution of Coloration and Toxicity in the Poison Frog Family (Dendrobatidae)." *Proceedings of the National Academy of Science U S A* 98, no. 11 (2001): 6227–6232.

Symula, Rebecca, Rainer Schulte, and Kyle Summers. "Molecular Phylogenetic Evidence for a Mimetic Radiation in Peruvian Poison Frogs Supports a Müllerian Mimicry Hypothesis." *Proceedings of the Royal Society of London B* 268 (2001): 2415–2421.

Vences, M., J. Kosuch, S. Lötters, A. Widmer, J. Köhler, K.-H. Jungfer, and M. Veith. "Phylogeny and Classification of Poison Frogs (Amphibia: Dendrobatidae), Based on Mitochondrial 16S and 12S Ribosomal RNA Gene Sequences." *Molecular Phylogenetics and Evolution* 15 (2000): 34–40.

Wells, K. D. "Behavioral Ecology and Social Organization of a Dendrobatid Frog (*Colostethus inguinalis*)." *Behavioral Ecology and Sociobiology* 6 (1980): 199–209.

Weygoldt, P. "Evolution of Parental Care in Dart Poison Frogs (Amphibia: Dendrobatidae)." *Zeitschrift für Zoologische Systematik und Evolutionsforschung* 25 (1987): 51–67.

Janalee P. Caldwell, PhD
Kyle B. Summers, PhD

Ruthven's frogs

(Allophrynidae)

Class Amphibia

Order Anura

Family Allophrynidae

Thumbnail description
Small frog that dwells in trees around streams and rivers in lowland tropical forest; body is elongate and covered dorsally with small spicules

Size
Females: 0.85–1.20 in (22–31 mm); males: 0.80–0.95 in (20.6–24.6 mm)

Number of genera, species
1 genus; 1 species

Habitat
Lowland tropical rainforest, particularly around streams and rivers

Conservation status
Not threatened

Distribution
South America

Evolution and systematics

Allophryne ruthveni was described by Helen Gaige in 1926. The type locality is at Tukeit Hill below Kaiteur Falls in British Guiana. Since its description, the relationship of this frog to other frogs has been an enigma. In the original description, Gaige placed it in the toad family, primarily because it lacks teeth. The name *Allophryne* comes from the Greek *allos*, meaning "other," and *phrynos*, meaning "toad," presumably because the author considered the species to be another kind of toad. Other authors considered this frog to be a treefrog or a glass frog, because of the nature of the bones supporting the toe discs. Later researchers examined the internal structure of the toe discs and determined that this character in *Allophryne* is different from that in treefrogs or glass frogs. Those researchers supported placing *Allophryne* in a separate group. No subfamilies are recognized.

Physical characteristics

The dorsal color of this small frog varies from bronze or grayish brown to gold with darker mottling; gold or yellowish brown narrow dorsolateral stripes are present. A variable amount of spotting occurs on the throat of both males and females, although the vocal sac in males is always dark without spots. The vocal sac expands greatly when an individual is calling; it can exceed the size of the head. Sharp spicules, larger and denser in males, are embedded in the skin. The tips of the toes are expanded into discs. The body is elongate, and the head slopes in lateral profile.

Distribution

The frog inhabits the Amazon and Guianan forests from Venezuela through Guyana, Suriname, and French Guiana to Amapá, Brazil, south of the Amazon River in Pará, and west to Rondônia.

Habitat

Individuals generally are found near streams or rivers in lowland forests. Breeding congregations occur around flooded pools, which may form in depressions in the forest or as a result of rising rivers or heavy rains.

Ruthven's frog (*Allophryne ruthveni*). (Illustration by Dan Erickson)

Behavior

During the rainy season, when they are not breeding, individual frogs sit out at night on leaves several feet above the ground, in the general vicinity of streams or rivers. An individual found in Amapá, Brazil, was taken from a terrestrial bromeliad. A gravid female was taken from the stomach of the snake *Leimadophis reginae;* the snake had been collected on the bank of a river in Surinam.

Feeding ecology and diet

No information is available on feeding or diet of this species.

Ruthven's frog (*Allophryne ruthveni*) is the only known member of its family and resides in Guianan area of South America. (Photo by Beat Akeret. Reproduced by permission.)

Reproductive biology

Breeding in this species has been observed in the months of March, May, and July and generally is associated with the rainy season. Males call from the edges of small temporary ponds in the forest or from the flooded edges of rising rivers. In northern Brazil a few individuals have called from positions on the leaves of small trees several meters from the edge of a small pond after heavy rainfall. In southern Venezuela, individuals have called from small trees and bushes near a flooded depression in the forest. Perhaps the most dramatic observation was the explosive breeding event witnessed in March in Pará, Brazil, when several hundred individuals appeared on one evening after only a few individuals had been found in the area during the preceding two months. This congregation was found in a flooded area created by the rising of the Rio Xingu, a large tributary of the Amazon River. The chorus occurred near the end of the rainy season and seemed to be triggered when the river reached a critical stage of flooding.

The call of this species has been described as a series of short notes or of low, raspy trills. A recording of the call from an individual in Roraima, Brazil, disclosed that the calls are given at a rate of 18 notes per minute. During a breeding event observed in the La Escalera region of Venezuela, a pair of frogs in axillary amplexus was found on a branch about 5 ft (1.5 m) above water. This pair was placed in a plastic bag, where the female subsequently deposited approximately 300 pigmented eggs. The eggs did not survive to hatching.

Conservation status

Although not threatened according to the IUCN, the conservation status of this frog is largely unknown, primarily because the frog is rarely encountered. Large areas in the Amazon region and in other places where the frogs occur are being deforested continually; thus, populations of this secretive frog are likely being destroyed before they are discovered.

Significance to humans

This frog has been an enigma since its discovery in 1926, because it cannot be placed within any other known frog clade. Until 1984 it was thought to occur only in the Guianan forests well north of the Amazon River. In 1984, 1985, and 1987 specimens were found considerable distances south of the Amazon River in Amazon rainforest. The discovery of this frog in the Amazon region as recently as the 1980s indicates how much remains to be learned not only about this frog but also about the vast, largely biologically unexplored Amazon region. By 2002 the tadpoles of this species, which could hold clues to the relationship of this species to other frogs, still had not been described.

Resources

Books

Caldwell, Janalee P. "Diversity of Amazonian Anurans: The Role of Systematics and Phylogeny in Identifying Macroecological and Evolutionary Patterns." In *Neotropical Biodiversity and Conservation*, edited by A. C. Gibson. Los Angeles: Mildred E. Mathias Botanical Garden Miscellaneous Publications, 1996: 73–88.

Periodicals

Caldwell, Janalee P., and M. S. Hoogmoed. "Allophrynidae." *Catalogue of American Amphibians and Reptiles* (1998): 666.1–666.3.

Gaige, H. T. "A New Frog from British Guiana." *University of Michigan, Occasional Papers of the Museum of Zoology* 176 (1926): 1–3.

Hoogmoed, M. S. "Notes on the Herpetofauna of Surinam. II. On the Occurrence of *Allophryne ruthveni* Gaige (Amphibia, Salientia, Hylidae) in Surinam." *Zoologische Mededelingen Leiden* 44, no. 5 (1969): 751–781.

Lynch, J. D., and H. L. Freeman. "Systematic Status of a South American Frog, *Allophryne ruthveni* Gaige." *Miscellaneous Publications, Museum of Natural History of the University of Kansas* 17, no. 10 (1966): 493–502.

Janalee P. Caldwell, PhD

Glass frogs
(Centrolenidae)

Class Amphibia

Order Anura

Family Centrolenidae

Thumbnail description
Mostly diminutive green frogs with toe pads, large eyes directed forward, and transparent ventral skin

Size
0.7–3.2 in (18–81 mm)

Number of genera, species
3 genera; 133 species

Habitat
Humid tropical forest

Conservation status
Not threatened

Distribution
Mexico, Central America, and South America

Evolution and systematics

There are no known fossils of these small, fragile frogs; thus, all systematic studies have relied solely on extant species. Centrolenidae is a monophyletic lineage, and as early as 1973, Lynch considered the Centrolenidae to be phylogenetically close to the neobatrachian families of Pseudidae and Hylidae based on the presence of intercalary elements in the digits. In 1993 Linda Ford and David Cannatella noted that this feature is found in other families as well and that further study would be needed to determine if this condition arose independently in centrolenids. Molecular genetic data have yet to be brought to bear on the phylogenetic position of the Centrolenidae. No subfamilies are recognized.

Three genera are recognized within Centrolenidae. *Centrolene*, with 39 species, is characterized by the presence of a humeral spine in males. *Hyalinobatrachium*, literally translated as "glass frog," with 33 species, is characterized by the possession of a bulbous liver. *Cochranella*, with 61 species, lacks the diagnostic features of the other genera. It is thought that many new glass frogs will be discovered as poorly studied areas are explored and new molecular genetic techniques are used in the systematic study of these frogs.

Physical characteristics

The family is characterized by having the two elongate ankle bones (astragulus and calcaneum) fused, a medial process on the third metacarpal bone in the hand, T-shaped terminal phalanges, an intercalary element between the penultimate and terminal phalanges, and deposition of eggs on leaves or rocks above streams. Most centrolids are small, at just 0.7–1.2 in (18–30 mm) in length, but one species, *Centrolene geckoideum*, is a relative giant, reaching 3.2 in (81 mm). Aside from this exceptional species, glass frogs have slender, fragile bodies with long, thin limbs and webbed feet. The digits of these excellent climbers terminate in enlarged toe pads resembling those of tree frogs (Hylidae). In dorsal view, the head is round, with large, protruding eyes set more dorsally than in most frogs.

Most glass frogs are a shade of green, varying from pale lime green to dark forest green. A few species, such as *Cochranella igonta*, are brown or tan. Species may lack any pattern at all, but most possess pale yellow or white spots or even multicolored dots, termed ocelli. The ventral surfaces and frequently the peritoneum (membrane enclosing internal body cavities) of these frogs are transparent, so that the internal organs and

This tan-brown *Chochranella ignota* is one of the few species of glass frogs that is not green. This adult provides parental care in the form of egg attendance in the Andes of Valle de Cauca, Colombia, at 6,360–6,700 ft (1,940–2,050 m). (Photo by Erik R. Wild. Reproduced by permission.)

Fleischmann's glass frogs (*Centrolenella fleischmanni*) in amplexus (mating) in Monteverde Cloud Forest Preserve, Costa Rica. (Photo by Gregory G. Dimijian/Photo Researchers, Inc. Reproduced by permission.)

bones can easily be seen in the living frogs. It is through this transparent skin that one also can see that the bones of centrolenid frogs are either white or green. Unique pigment cells in the skin reflect infrared radiation, the same wavelength of light reflected by plants but invisible to the human eye. This pigmentation is thought to camouflage glass frogs when they sit on leaves, thus protecting them from pit vipers and birds.

Distribution

Glass frogs occur from southern Mexico to Bolivia, east to northeastern Argentina and southeastern Brazil. The genus

Male Fleischmann's glass frog (*Centrolenella fleischmanni*) calling in the rainforest of Central America. (Photo by M.P.L. Fogden. Bruce Coleman Inc. Reproduced by permission.)

Hyalinobatrachium is most diverse in Central America, whereas *Cochranella* and *Centrolene* are more speciose in South America. Many species of glass frogs are endemic to the Andean valleys and thus have restricted distributions.

Habitat

Glass frogs are associated almost exclusively with vegetation along and above streams, predominantly in montane cloud forest. Tadpoles inhabit slow-moving portions of forested streams after hatching from eggs deposited on leafy vegetation above the water.

Behavior

Glass frogs are nocturnally active with diurnal retreats. Males of some species are known to engage in combat for calling or egg-laying sites by wrestling for prime breeding spots. Several species exhibit parental care in the form of egg attendance by one or the other parent, who sits near or directly on the clutch of eggs.

Feeding ecology and diet

Little is known of the diet of glass frogs. They eat primarily insects, although the large *Centrolene geckoideum* has been known to consume small vertebrates.

Reproductive biology

In habitats without seasonal variations, glass frog breeding is continuous throughout the year, whereas in seasonal climates breeding is tied closely to the rainy season. Breeding occurs under the darkness of night either during rains or, in cloud forest species, during evening mists from clouds. Males call from selected sites on leaves overhanging forest streams. Most calls are a high-pitched series of whistles. Females are attracted to calling males, who have secured suitable sites for calling and egg deposition. Males of some species are known to defend these sites by aggressively posturing with stiffened limbs to challenge intruding males. In some species, such as *Cochranella griffithsi*, males physically battle with other males for these calling and egg-deposition sites. Such wrestling matches involve grappling until one individual looses his grip and falls. The humeral spines on males of the genus *Centrolene* are thought to be used as weapons in such territorial disputes, because males sometimes are found with scars on the head and body.

In nearly all species for which reproduction is known, axillary amplexus and egg deposition take place at or near these won breeding sites. The eggs are deposited on leaf surfaces, either above or below, depending on the species. One remarkable exception is the giant *Centrolene geckoideum*, which deposits masses of eggs on rock faces in splash zones of rapids and waterfalls. Adults, either male or female, are known to attend these eggs, often sitting directly on top of them. By night or day, this behavior is thought to protect eggs from invertebrate predators and desiccation. Occasionally, adult *Centrolene geckoideum* attend multiple clutches at varying degrees of developmental maturity at the same location. Larvae hatch and fall into the streams, which, for Andean species, may be rushing torrents. Tadpoles settle in the substrate of eddies or other slow-moving portions of the stream. Tadpoles in oxygen-poor microhabitats often are colored bright red as a result of blood flowing close to the surface of their unpigmented skin. There is no known parental care of the larvae.

Conservation status

There is no official conservation status for any species of glass frog, but many species have restricted distributions that make them vulnerable to extinction.

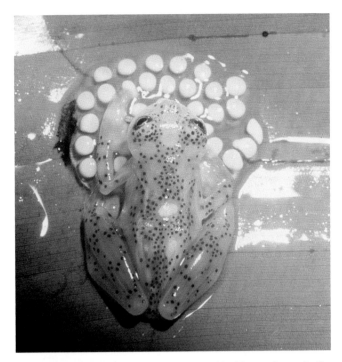

Adult *Hyalinobatrachium aureoguttatum* provides parental care in the form of egg attendance on the underside of a *Heliconia* leaf 16 ft (5 m) above a stream in the Andes of Chocó, Colombia, at 3,080–3,150 ft (940–960 m). (Photo by Erik R. Wild. Reproduced by permission.)

Significance to humans

Glass frogs have had little direct use by humans; however, they have aesthetic value and are potential indicators of overall ecosystem health, especially in tropical montane stream ecosystems.

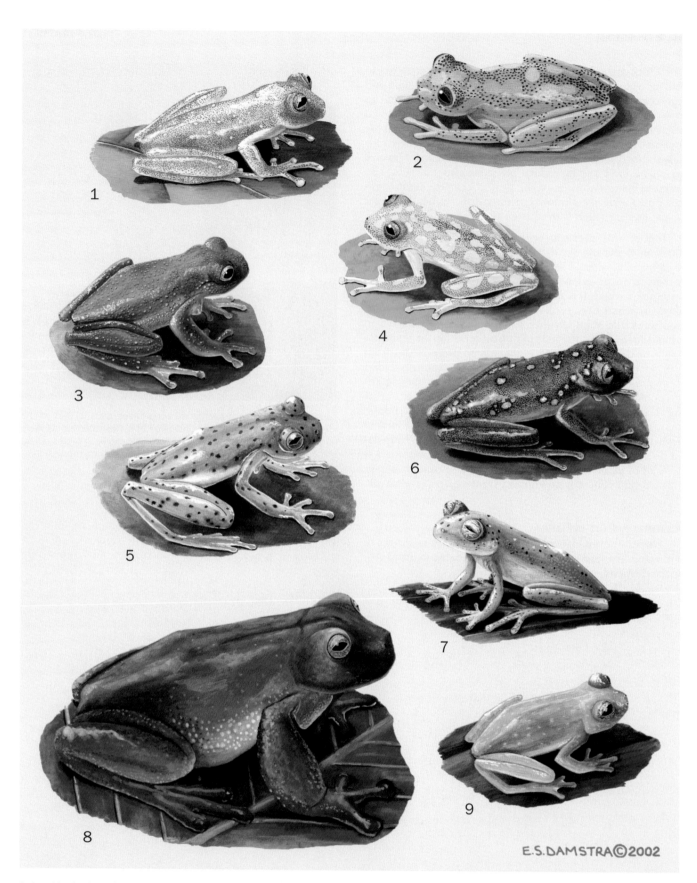

1. Lynch's Cochran frog (*Cochranella ignota*); 2. Atrato glass frog (*Hyalinobatrachium aureoguttatum*); 3. Pichincha glass frog (*Centrolene helo-derma*); 4. La Palma glass frog (*Hyalinobatrachium valerioi*); 5. Nicaragua glass frog (*Centrolene prosoblepon*); 6. Spotted Cochran frog (*Cochranella ocellata*); 7. Ecuador Cochran frog (*Cochranella griffithsi*); 8. Pacific giant glass frog (*Centrolene geckoideum*); 9. Fleischmann's glass frog (*Hyalinobatrachium fleischmanni*). (Illustration by Emily Damstra)

Species accounts

Pacific giant glass frog
Centrolene geckoideum

TAXONOMY
Centrolene geckoideum Jiménez de la Espada, 1872, las riberas del rio Napo en el Ecuador.

OTHER COMMON NAMES
None known.

PHYSICAL CHARACTERISTICS
In this species the males are larger than the females. Males grow to 2.8–3.2 in (70.2–80.7 mm) and females to 2.4–2.9 in (60.7–72.9 mm) in snout-vent length. This largest centrolenid has relatively small eyes, heavily webbed digits, and large, rectangular-shaped toe pads. Males have large, muscular forearms and a long, sharply pointed bony spine on the humerus. The dorsum is lime green to dark forest green. The skin is tuberculate, with some small, scattered white flecks; in males the tubercles are finely spiculate. The bones are green.

DISTRIBUTION
This species ranges through Andean Ecuador and Colombia at elevations of 5,740–9,840 ft (1,750–3,000 m).

HABITAT
The habitat of the Pacific giant glass frog is upper montane cloud forest along swiftly flowing, shaded streams with numerous waterfalls.

BEHAVIOR
The Pacific giant glass frog is nocturnal and uses rock faces or leaves as diurnal retreats. At night, males call from splash zones behind waterfalls or on boulders in torrents. There have been no direct observations, but it is hypothesized that these frogs, like some other centrolids, may be territorial, battling for prime calling and oviposition (egg-laying) sites. Adult males found in the field in Colombia had numerous scars on the face, head, and flanks, which may have been the result of battles between males using their sharp, bony humeral spines.

FEEDING ECOLOGY AND DIET
This large centrolid feeds on a variety of insects and also ingests frogs and fish.

REPRODUCTIVE BIOLOGY
Male Pacific giant glass frogs call at night throughout the year within splash zones behind waterfalls or on boulders in torrents. The call is a loud, high-pitched, trilled whistle, 155–373 milliseconds in duration and produced at intervals of 1.48–5.05 min, with emphasized frequencies of 3,468–4,187 Hz. The calls lack consistent amplitude modulation; this may be related to the din of the rushing water, which would obliterate any subtle characteristics in the calls. Tadpoles are elongate and slender with low caudal fins and eyes positioned dorsally. The oral disc has thin jaw sheaths.

CONSERVATION STATUS
Not threatened.

SIGNIFICANCE TO HUMANS
None known. ◆

Pichincha glass frog
Centrolene heloderma

TAXONOMY
Centrolenella heloderma Duellman, 1981, Quebrada Zapadores, 3.1 mi (5 km) east-southeast of Chiriboga, Provincia de Pichincha, Ecuador.

OTHER COMMON NAMES
None known.

PHYSICAL CHARACTERISTICS
Males are 1.1–1.2 in (26.8–31.5 mm), and females are 1.3 in (32.3 mm) in snout-vent length. This moderately large centrolenid has small eyes. Males have a blunt humeral spine. The toes are about four-fifths webbed, and the digits have expanded toe pads. This species has unique tuberculate skin on the dorsum. The dorsum is dark forest green with bluish white tubercles and a yellow margin on the lip. The bones are green.

DISTRIBUTION
This species inhabits cloud forest on the Pacific slopes of the Andes in Colombia and Ecuador at elevations of 6,430–7,870 ft (1,960–2,400 m).

HABITAT
The Pichincha glass frog inhabits streams in the upper limits of montane cloud forest.

BEHAVIOR
Not known.

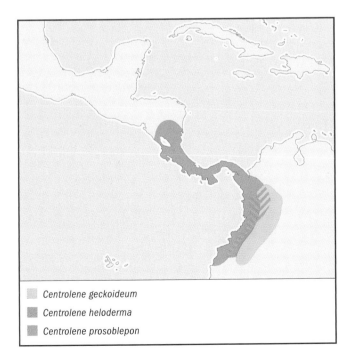

■ *Centrolene geckoideum*
■ *Centrolene heloderma*
■ *Centrolene prosoblepon*

FEEDING ECOLOGY AND DIET
Not known.

REPRODUCTIVE BIOLOGY
Little is known of the reproductive biology of this frog. The call is a harsh peep made from the upper surfaces of leaves and ferns 3.3–13.1 ft (1–4 m) above streams on cliff faces below seepages.

CONSERVATION STATUS
Not threatened.

SIGNIFICANCE TO HUMANS
None known. ◆

Nicaragua glass frog
Centrolene prosoblepon

TAXONOMY
Hyla prosoblepon Boettger, 1892, Plantage Cairo (La Junta) near Limon, Costa Rica.

OTHER COMMON NAMES
None known.

PHYSICAL CHARACTERISTICS
Males are 0.9–1.1 in (21.7–28.1 mm), and females are 1.0–1.1 in (25.4–27.8 mm) in snout-vent length. The dorsum is green with or without black dots. The tips of the digits are pale yellow, and the chest is white. The skin is shagreen on the dorsum, and the belly and thighs are granular. Males possess a pointed humeral spine. The bones are green.

DISTRIBUTION
This species occurs in Nicaragua, Costa Rica, Panama, and the Pacific slopes of Colombia and Ecuador at elevations of 328–4,921 ft (100–1,500 m).

HABITAT
The Nicaragua glass frog inhabits vegetation associated with cascading streams. Tadpoles occupy the bottom of silt-bottomed pools in streams.

BEHAVIOR
Aggressive behavior takes place between calling males. One or both frogs dangle upside down while holding vegetation with their hind legs. The males grapple with each other until one drops off or flattens his body against the leaf.

FEEDING ECOLOGY AND DIET
Not known.

REPRODUCTIVE BIOLOGY
Breeding is coincident with significant rainfall. The call consists of three short beeps with a pitch of 5,300–6,000 Hz at a frequency of one to 43 calls per hour. Calls are made from the tops of leaves over streams. Males are not territorial and initiate amplexus with the female. Egg deposition can occur some distance from the calling site at heights of 0–10 ft (0–3 m) above the ground, usually on the top side of leaves but also on moss-covered rocks and branches. The average clutch of 20 black eggs is attended during the first night by the female, who lies motionless on top of the clutch. Males call vigorously during amplexus and immediately after egg deposition. Tadpoles are elongate and slender, with low caudal fins and eyes posi-

tioned dorsally. The oral disc has thin jaw sheaths and a labial tooth row formula of 2(1)/3.

CONSERVATION STATUS
Not threatened.

SIGNIFICANCE TO HUMANS
None known. ◆

Ecuador Cochran frog
Cochranella griffithsi

TAXONOMY
Cochranella griffithsi Goin, 1961, Río Saloya, Ecuador.

OTHER COMMON NAMES
None known.

PHYSICAL CHARACTERISTICS
Males are 0.8–1.9 in (19.7–26.1 mm), and females are 0.8–1.0 in (20.0–24.8 mm) in snout-vent length. The dorsum is pale yellowish green, with or without dark green flecks. The tips of the digits are pale yellow, and the chest is white. The skin is shargreen on the dorsal surfaces, granular on the belly and posterior surfaces of the thighs, and smooth on other areas. The bones are pale green.

DISTRIBUTION
The Ecuador Cochran frog occurs on the Pacific slopes of the Andes in southern Colombia and adjacent Ecuador at elevations of 3,940–8,700 ft (1,200–2,650 m).

Cochranella griffithsi
Cochranella ignota
Cochranella ocellata

HABITAT
The species inhabits cloud forest.

BEHAVIOR
Males call from leaves of herbs and bushes over cascading streams by night and seek out retreats in such places as the axils of elephant ear (*Colocasia esculenta*) plants by day. Aggressive behavior among males is associated with breeding and territoriality. Competing males grasp each other in a belly-to-belly fashion. While hanging from vegetation by the hind limbs, the combatants wrap their forelimbs about each other's neck. In this position the frogs repeatedly flex and extend their outstretched hind limbs so as to move their bodies up and down while swinging laterally.

FEEDING ECOLOGY AND DIET
Not known.

REPRODUCTIVE BIOLOGY
Males make calls from vegetation over streams. Eggs are laid on the tips of leaves overhanging streams, into which hatchling tadpoles drop and complete their development.

CONSERVATION STATUS
Not threatened.

SIGNIFICANCE TO HUMANS
None known. ◆

Lynch's Cochran frog
Cochranella ignota

TAXONOMY
Centrolenella ignota Lynch, 1990, Peñas Blancas, Farallones de Cali, 3.7 mi (6 km) by the road southwest of Pichindé, Valle de Cauca, Colombia.

OTHER COMMON NAMES
None known.

PHYSICAL CHARACTERISTICS
Males are 0.9–1.0 in (22.3–25.4 mm), and females are 1 in (24.2–24.4 mm) in snout-vent length. The dorsum is tan-brown to pale olive, with black ocelli with orange or yellow centers. The skin is smooth and has elevated flat, white warts. The head is rounded in dorsal view, and the protruding eyes are directed anteriorly. The toes are about one-half webbed, with enlarged toe pads. The bones are pale green.

DISTRIBUTION
This species occurs in the western Andes of Colombia at elevations of 6,230–6,430 ft (1,900–1,960 m).

HABITAT
Lynch's Cochran frog inhabits montane cloud forest streams.

BEHAVIOR
Not known.

FEEDING ECOLOGY AND DIET
Not known.

REPRODUCTIVE BIOLOGY
The advertisement call is a series of chirps, which males make from vegetation over streams.

CONSERVATION STATUS
Not threatened.

SIGNIFICANCE TO HUMANS
None known. ◆

Spotted Cochran frog
Cochranella ocellata

TAXONOMY
Hylella ocellata Boulenger, 1918, Huancabamba, eastern Peru.

OTHER COMMON NAMES
None known.

PHYSICAL CHARACTERISTICS
Males are 0.8–1.0 in (21.0–25.1 mm), and females are 1.1 in (29 mm) in snout-vent length. The dorsum is dark green with large, dark-edged, pale bluish white ocelli. The dorsal skin is shagreen. The bones are green.

DISTRIBUTION
This species lives on the Amazonian slopes of the Andes in Peru at elevations of 5,350–5,580 ft (1,630–1,700 m).

HABITAT
The spotted Cochran frog inhabits cloud forest.

BEHAVIOR
Not known.

FEEDING ECOLOGY AND DIET
Not known.

REPRODUCTIVE BIOLOGY
Not known.

CONSERVATION STATUS
Not threatened.

SIGNIFICANCE TO HUMANS
None known. ◆

Atrato glass frog
Hyalinobatrachium aureoguttatum

TAXONOMY
Centrolenella aureoguttatum Barrera-Rodrigues and Ruiz-Carranza, 1989, Chocó, Colombia, 14 mi (23 km) carretera El Carmen-Quibdo.

OTHER COMMON NAMES
None known.

PHYSICAL CHARACTERISTICS
Males are 0.8–0.9 in (20.4–23.3 mm), and females are 0.9 in (22.9–23.9 mm) in snout-vent length. The dorsum is yellow-green with scattered large brown chromatophores (pigment cells). Two to five large yellow spots free of the brown chromatophores are prominent dorsally. The bones are white.

DISTRIBUTION
This species ranges across the western slopes of the Andes in Colombia at elevations of 150–5,120 ft (45–1,560 m).

Hyalinobatrachium aureoguttatum

Hyalinobatrachium fleischmanni

Hyalinobatrachium valeroi

HABITAT
These frogs are active at night on vegetation 3.3–22.9 ft (1–7 m) above rapidly flowing streams with abundant canopy and high local humidity.

BEHAVIOR
Not known.

FEEDING ECOLOGY AND DIET
Not known.

REPRODUCTIVE BIOLOGY
The call is unknown, but these frogs engage in axillary amplexus. Clutches of 25–35 clear green eggs are deposited in a translucent gelatinous mass on the undersides of leaves, usually *Heliconia*. Parental care is provided in the form of egg attendance within 2 in (5 cm) of the clutch or directly upon it.

CONSERVATION STATUS
Not threatened.

SIGNIFICANCE TO HUMANS
None known. ◆

Fleischmann's glass frog
Hyalinobatrachium fleischmanni

TAXONOMY
Hylella fleischmanni Boettger, 1893, San José, Costa Rica.

OTHER COMMON NAMES
English: Northern glass frog; Spanish: Ranita de vientre transparente.

PHYSICAL CHARACTERISTICS
Males are 0.8–1.0 in (19.2–25.5 mm) in snout-vent length. The dorsum is pale green with pale yellow or yellowish green spots and darker green reticulations. The belly is white, and the tips of the digits are yellow. The bones are white.

DISTRIBUTION
This is the most widespread species of glass frog; it ranges from Guerrero and Veracruz, Mexico, through Central America to Colombia, Venezuela, Guyana, and Surinam at elevations of 200–4,790 ft (60–1,460 m).

HABITAT
This species inhabits vegetation near moderate to fast-flowing streams at low elevations.

BEHAVIOR
Males exhibit territorial behavior by aggressively defending calling and oviposition sites. Both males and females are known to attend the developing egg clutches by sitting on the eggs during the night and sleeping near but not directly on the eggs during the day.

FEEDING ECOLOGY AND DIET
Not known.

REPRODUCTIVE BIOLOGY
The advertisement call of males, made from either the upper or lower surfaces of leaves overhanging streams, is a single untrilled "wheet" that is repeated after a short pause. Males aggressively defend calling and oviposition sites, and successful males may engage in many matings. Females choose a mate and initiate amplexus. Clutches of 18–30 eggs are deposited on the undersides of leaves directly over streams. Sometimes females, but usually males, attend clutches by sitting on eggs at

night and near them but not on them by day. Fruit flies (*Drosophila melanogaster*) deposit eggs on clutches of Fleischmann's glass frog, and the maggots of the fruit fly develop in the clutches and consume the eggs and embryos, resulting in extremely high mortality rates. As many as 80% of clutches may be destroyed by biotic or abiotic factors. Tadpoles are elongate and slender, with low caudal fins and eyes positioned dorsally. The oral disc has thin jaw sheaths and a labial tooth row formula of 2(1)/3. The tadpoles appear bright red as a result of blood flowing beneath the skin.

CONSERVATION STATUS
Not threatened.

SIGNIFICANCE TO HUMANS
None known. ◆

La Palma glass frog
Hyalinobatrachium valerioi

TAXONOMY
Centrolene valerioi Dunn, 1931, La Palma, Costa Rica.

OTHER COMMON NAMES
None known.

PHYSICAL CHARACTERISTICS
Males are 0.8–1.0 in (20.8–26.0 mm) in snout-vent length. The yellowish background with a bold reticulated pattern of green and dark flecks gives the appearance of prominent large yellow

spots on the dorsum of this frog. The texture of the dorsum is smooth and that of the belly and thighs is rugose (wrinkled). The bones are white.

DISTRIBUTION
This species ranges across central Costa Rica to the Pacific slopes of Ecuador.

HABITAT
Not known.

BEHAVIOR
Little is known aside from its reproductive biology.

FEEDING ECOLOGY AND DIET
Not known.

REPRODUCTIVE BIOLOGY
The advertisement call is a single short "seet" that is repeated after a pause. Males of this species provide 24-hour parental care to clutches on leaves, more than that known for any other glass frog. Although diurnal attendance increases survivorship of eggs and unhatched larvae, it exposes the males to predation. The uncanny resemblance between the color pattern of the adult male frog and the appearance of an egg clutch on a leaf led to the suggestion that the patterning is a co-evolutionary adaptation to this increased diurnal predation risk.

CONSERVATION STATUS
Not threatened.

SIGNIFICANCE TO HUMANS
None known. ◆

Resources

Books
McDiarmid, Roy W. "Evolution of Parental Care in Frogs." In *The Development of Behavior: Comparative and Evolutionary Aspects,* edited by G. M. Burghardt and M. Bekoff. New York: Garland Press, 1978.

Periodicals
Grant, Taran, Wilmar Bolivar, and Fernando Castro. "The Advertisement Call of *Centrolene geckoideum*." *Journal of Herpetology* 32, no. 3 (1998): 452–455.

Greer, Beverly J., and Kentwood D. Wells. "Territorial and Reproductive Behavior of the Tropical American Frog *Centrolenella fleischmanni*." *Herpetologica* 36, no. 4 (1980): 318–326.

Jacobson, Susan K. "Reproductive Behavior and Male Mating Success in Two Species of Glass Frogs (Centrolenidae)." *Herpetologica* 41, no. 4 (1985): 396–404.

Lynch, John D., and William E. Duellman. "A Review of the Centrolenid Frogs of Ecuador, with Descriptions of New Species." *Occasional Papers of the Museum of Natural History, University of Kansas* 16 (1973): 1–66.

Rueda-Almonacid, José Vicente. "Estudio Anatomico y Relaciones Sistematicas de *Centrolene geckoideum* (Salienta: Anura: Centrolenidae)." *Trianea* 5 (1994): 133–187.

Ruiz-Carranza, Pedro M., and John D. Lynch. "Ranas Centrolenidae de Colombia. I. Propuesta de Una Nueva Clasificación Genérica." *Lozania* 57 (1991): 1–30.

———. "Ranas Centrolenidae de Colombia. IX. Dos Nuevas Especies del Suroeste de Colombia." *Lozania* 68 (1996): 1–11.

Villa, Jaime. "Biology of a Neotropical Glass Frog *Centrolenella fleischmanni* (Boettger), with Special Reference to Its Frogfly Associates." *Milwaukee Public Museum Contributions in Biology and Geology* 55 (1984): 1–60.

Erik R. Wild, PhD

Amero-Australian treefrogs

(Hylidae)

Class Amphibia

Order Anura

Family Hylidae

Thumbnail description
Small to large primarily arboreal frogs with expanded, adhesive discs on the digits

Size
0.8–4.8 in (20–120 mm)

Number of genera, species
42 genera; 854 species

Habitat
Primarily tropical and subtropical forests, some savannas, grasslands, and deserts; a few species inhabit elevations above the tree line

Conservation status
Critically Endangered: 6 species; Endangered: 5 species; Vulnerable: 5 species; Lower Risk/New Threatened: 5 species; Data Deficient: 8 species

Distribution
Most of the New World, Australia, and New Guinea, and discontinuously in Eurasia

Evolution and systematics

The earliest fossil Hylidae are from the Paleocene of Brazil; elsewhere, fossil hylids are known from as early as the Miocene in Australia, the Oligocene in North America, the Miocene in Europe, and the Pleistocene in Japan. The meager fossil data are consistent with a Gondwanan origin of the family, presumably in South America after its separation from Africa. Independent dispersals from South America were to Australia via Antarctica and to North America and subsequently to Eurasia.

Treefrogs of the family Hylidae presumably are related most closely to other families of frogs in the New World that also have the two halves of pectoral girdle overlapping ventrally (arciferal conditions and intercalary elements between the penultimate and terminal claw-shaped phalanges). These families include Centrolenidae, which differ by having T-shaped terminal phalanges, tarsal elements fused throughout their lengths (fused proximally and distally in Hylidae), and 10 pairs of chromosomes (11 or more in Hylidae). The monotypic Allophrynidae differs by lacking teeth on the maxillaries and premaxillaries and intercalary elements in the digits, but the T-shaped terminal phalange is offset ventrally, as in Hylidae.

Five subfamilies are recognized:

Hemiphractinae

The eggs are brooded on the back of, or in a dorsal pouch of females; the embryos have large, sheet-like gills that at least partially envelop them. Most species have direct development. In those in which the eggs hatch as tadpoles, the spiracle is ventrolateral in position but moves to a lateral position in *Gastrotheca*. The intercalary elements are cartilaginous, and the terminal phalange is offset ventrally. The constricted pupil of the eye is horizontally elliptical. The diplod number of chromosomes is 26 (28 and 30 in some species of *Flectonotus*).

The subfamily contains five genera and 71 species; it is distributed principally in montane regions from Costa Rica to northwestern Argentina, the Guiana Highlands, and eastern Brazil.

Hylinae

The eggs are deposited in water, on vegetation above water, or in subterranean nests near water; all eggs hatch as free-swimming tadpoles, which have a lateral spiracle. The intercalary elements are cartilaginous, and the terminal phalange is offset ventrally. The constricted pupil of the eye is horizontally elliptical. The diploid number of chromosomes is 24, but this number is reduced to 22 in *Acris* and increased to 30 in many species of *Hyla* and to 34 in *Osteopilus brunneus*. The subfamily contains 26 genera with more than 500 species; it has the same distribution as the family, except that it is absent in the Australo-Papuan region.

Pelodryadinae

The eggs are deposited in water or, in a few species, on vegetation above water. The free-swimming tadpoles have filamentous gills and a lateral spiracle. The intercalary elements are cartilaginous, and the terminal phalange is offset ventrally. The constricted pupil of the eye is horizontally elliptical (vertically elliptical in *Nyctimystes*). The diploid number of chromosomes is 26 (24 in *Litoria infrafrenata* and 30 in *Litoria angiana*). The subfamily contains three genera with about 150 species; it is widespread in Australia and New Guinea. Two species are endemic to the Bismarck Archipelago and Solomon Islands, and three Australian species have been introduced into New Zealand and New Caledonia.

A horned treefrog (*Hemiphractus helioi*) is camouflaged on the rainforest floor in Peru. (Photo by Michael Fogden. Bruce Coleman Inc. Reproduced by permission.)

Phyllomedusinae

The eggs are deposited on vegetation above water; the embryos have large, branched gills; and the free-swimming tadpoles have a ventrolateral spiracle. The intercalary elements are cartilaginous, and the terminal phalange is offset ventrally. The constricted pupil of the eye is vertically elliptical. The diploid number of chromosomes is 26. The subfamily contains five genera with 70 species; it is widely distributed in tropical parts of Mexico, Central America, and South America.

Pseudinae

These aquatic frogs deposit eggs in water; the free-swimming tadpoles have feathery gills and a lateral spiracle. The intercalary elements are elongated and ossified; the terminal phalange is essentially in the same plane as the other phalanges. The constricted pupil of the eye is horizontally elliptical. The diploid number of chromosomes is 24. Two genera contain seven species; the subfamily is distributed widely east of the Andes in tropical South America and on the island of Trinidad.

Physical characteristics

The family is characterized by a suite of internal features that distinguish Hylidae from other families. The two halves of the pectoral girdle overlap (arciferal condition); the vertebrae are procoelous, and the first two presacral vertebrae are not fused. The coccyx has a bicondylar (two-headed) articulation with the expanded sacrum. Neopalatine and quadratojugal bones usually are present. An intercalary element (usually cartilaginous) is present between the terminal and penultimate phalanges in each digit of the hand and foot, and, except for Pseudinae, the terminal segment of each digit is offset ventrally. The terminal phalanges are claw-shaped, and the terminal segment of each digit typically is expanded into an adhesive disc.

Several hylid frogs have casque heads, in which the skin on the head is co-ossified with expanded underlying dermal bones in the skull. Casque heads are especially well developed in *Aparasphenodon*, *Corythomantis*, *Hemiphractus*, *Trachycephalus*, and *Triprion*. In some of these frogs (e.g., *Aparasphenodon* and *Triprion*), the upper lips are flared outward, and an additional bone, the prenasal, is present; a different bone, the internasal, is present in *Pternohyla*. Some species of *Gastrotheca* and *Osteocephalus* have bony ridges on the skull, and *Anotheca spinosa* has bony spines directed upward on the back of the skull. All hylids have teeth on the maxillae and premaxillae, and most have teeth on the vomers. *Gastrotheca guentheri* is the only frog known to have true teeth on the dentary bones in the lower jaw, but *Hemiphractus* and *Phyllodytes* have bony projections (odontoids) on the anterior ends of the dentaries in the lower jaw.

Prominent glands are present on the top of the head in some species of *Litoria*, especially *Litoria splendida*, and parotoid glands in the shoulder region are present in many species of *Phyllomedusa*. The dermis of the dorsal skin in some arboreal hylids (e.g., *Gastrotheca weinlandii*, *Phyllomedusa bicolor*, and *P. vaillanti*) contains small, vascularized bony plates (osteoderms) from which small lamellar spines protrude into the epidermis. Presumably, these structures impede water loss through the skin on the body.

Most treefrogs have rather slender bodies and long limbs. The terrestrial and fossorial *Cyclorana* in Australia and the carnivorous *Hemiphractus* in South America, however, have robust bodies and proportionately shorter limbs. All hylid frogs, except pseudines, have ventrally offset terminal discs on their digits; these discs are expanded and adhesive in arboreal species. With the exception of most phyllomedusines, the feet are at least one-half webbed. The fingers may be webbed or not. The fingers are fully webbed in several arboreal species, some of which (e.g., *Agalychnis craspedopus* and *Hyla miliaria*) bear thin flaps of skin at the outer edges of the limbs. Other dermal modifications include a fleshy proboscis in *Hemiphractus* and eyelid "horns" in some species of *Gastrotheca*.

Most hylids have a prominent tympanum (eardrum). Males of most hylid frogs have a single, subgular vocal sac, which is inflated while they perch on the ground or vegetation. *Osteocephalus*, *Phrynohyas*, and *Trachycephalus* call while floating on water. In these frogs the vocal sacs are paired and located behind the angles of the jaws; when inflated, they form balloon-like structures that extend above the head and thus do not inhibit the floating frog. Males of most species develop nuptial excrescences in the breeding season. Commonly, they are keratinized, and in some stream-breeding species (e.g., *Ptychohyla*) they take the form of clusters of spines; a cluster of spines also is present on the humerus in *Hyla armata*. Male *Plectrohyla* and gladiator frogs (*Hyla boans* group) have a sharp spine at the base of the thumb. Burrowing hylids (*Cyclorana* and *Pternohyla*) have enlarged, spade-like inner metatarsal tubercles.

Treefrogs vary tremendously in size and coloration. With a few exceptions, females are larger than males. Several species have snout-vent lengths of less than 1 in (25 mm). The smallest is *Litoria microbelos* in northern Australia; adults of both sexes attain snout-vent lengths of only 0.65 in (16 mm). The largest species is *Hyla vasta* on the island of Hispaniola in the West Indies; females are known to exceed 4.8 in (142 mm). The exceptions are the gladiator frogs and relatives in South and Central America, males of which aggressively defend their nesting sites from other males.

In most hylid frogs the dorsum is brown or green, usually with darker markings. Others have a yellow or gray dorsum, and some, such as *Hyla picturata* with a gaudy red-and-yellow dorsum, are more boldly marked. The ventral surfaces typically are white or pale yellow, but many species have brown or black spots or mottling on the belly; males of many species have bright yellow or dark gray vocal sacs. The most striking

A water-holding frog (*Cyclorana platycephala*) inflates its flexible body full of water after floods on the arid floodplain of the Paroo River, Australia. As the water recedes, the frog burrows underground and lives on its stored water. (Photo by Wayne Lawler/Photo Researchers, Inc. Reproduced by permission.)

aspects of coloration are the so-called flash colors on the flanks and surfaces of the hind limbs, which are not visible when the frogs are in a resting position. These flash colors are especially colorful in some species of *Agalychnis* and *Phyllomedusa*, in which the flanks are marked variously with black, blue, yellow, and white bars. Others, such as several species of *Scinax*, have bright yellow or red bars or spots on the posterior surfaces of the thighs.

With the exception of some Hemiphractinae, all hylid frogs have aquatic, free-swimming, feeding tadpoles, which have a sinistral spiracle and keratinized jaw sheaths. The oral disc usually is directed anteroventrally and lacks marginal papillae on the median part of the upper lip; elsewhere the lips typically bear one or two rows of marginal papillae. Tadpoles of many species that develop in streams have enlarged suctoral oral discs, but the tadpoles of *Duellmanohyla* and *Phasmahyla* have upturned, umbelliform oral discs in the form of an inverted umbrella. Labial teeth generally are present, but they are absent in one group of South American *Hyla*; the labial teeth are reduced greatly in oophagous (egg-eating) tadpoles. Most tadpoles that develop in ponds have two upper rows and three lower rows of labial teeth; the number of rows is increased greatly in many tadpoles that develop in torrential streams. The maximum is 17 upper rows and 21 lower rows of labial teeth.

Most hylid tadpoles have total lengths of less than 2 in (50 mm); in those that develop in ponds, the body is about one-

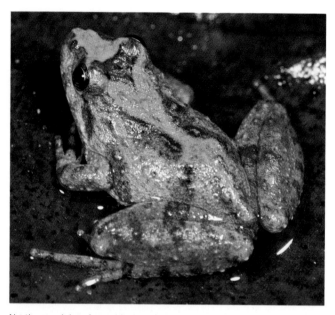

Northern cricket frogs (*Acris crepitans*) can leap three feet in a single jump. They prefer to live along water's edge and leap into the water when frightened, but circle back, as they prefer land to deep water. (Photo by Animals Animals ©Breck P. Kent. Reproduced by permission.)

third of the total length. The largest known tadpole is that of *Pseudis paradoxa*, which reaches a total length of more than 10 in (25 cm). Tadpoles that develop in streams have proportionally longer, more muscular tails with lower fins than those that mature in ponds.

Distribution

The family has a continuous distribution throughout most of the New World, including the West Indies but excluding Arctic North America and southern South America. It is distributed widely in Australia and New Guinea, and two species also inhabit the Bismarck Archipelago and the Solomon Islands. A few species of *Hyla* occur in Europe, southwestern Asia, discontinuously in eastern Asia (including the Japanese Archipelago), and Mediterranean North Africa as well as on the Azores, Madeira, and the Canary Islands in the Atlantic Ocean. Some Australian species have been introduced into New Caledonia and New Zealand, and some North American species have been introduced onto islands in the West Indies. One West Indian species has been introduced into Florida.

Habitat

Hylid frogs are most diverse in tropical and subtropical humid forests, especially in the Amazonian rainforest, where as many as 40 species may occur together. Hylids also are numerous in montane cloud forests, especially in Mexico, Central America, and New Guinea, as well as in the coastal forests of southeastern Brazil and the lowland forests of northern Australia and New Guinea. In Australia some species of *Litoria* and all species of *Cyclorana*, most of which inhabit grass-

lands and deserts, are terrestrial or even fossorial, a habit also characteristic of *Pternohyla* in Mexico. Members of the subfamily Pseudinae are aquatic. By day most hylids secrete themselves in arboreal situations, such as under the loose bark of trees, on the undersides of leaves, and in bromeliads. A few that breed in mountain streams seek diurnal shelter under rocks at the edges of streams or in rock crevices.

Behavior

Nearly all species are nocturnal; *Acris* in North America also is active by day, and some montane species are active by day. In the latter category are the Andean *Hyla labialis* and the Guatemalan *Plectrohyla glandulosa*, both of which bask on bushes or rocks. Thus, hylids are encountered mostly at night, especially after rains, when they feed and breed. Although adults may spend the day in seclusion, most treefrogs perch on branches, leaves, or grasses at night. Aside from natural diurnal retreats, treefrogs also utilize human-made structures, including window shutters, thatch roofs, water tanks, and cisterns.

Some hylids that live in arid regions survive long dry periods by special behaviors to prevent desiccation. In the Australian deserts *Cyclorana* dig burrows with spadelike tubercles on the hind feet; they remain underground for many months. Dehydration is prevented by shedding layer upon layer of skin, which hardens into an impermeable cocoon. Some *Phyllomedusa* in dry regions of South America have lipid glands in the skin. The secretion from these glands is wiped by the hands and feet over the entire body so as to provide an almost impermeable covering that allows the frogs to remain exposed to air for long periods of time. Some of the casque-headed treefrogs (e.g., *Gastrotheca*, *Trachycephalus*, and *Triprion*) back into bromeliads, where water exists at the bases of the leaf axils, or tree holes with water inside; they plug the openings with their bony heads. Treefrogs living in temperate regions hibernate below ground. At least two species (*Hyla versicolor* and *Pseudacris crucifer*) have large quantities of glycerol in their tissues, which acts as an antifreeze; these frogs can tolerate temperatures well below freezing.

Territorial behavior in hylids mostly is acoustic; males of many species are known to emit territorial or aggressive calls in the presence of conspecific males. Such calls usually define a given calling site; in cases where calling fails, males have been observed to grapple or even bite one another. Male gladiator frogs in the American tropics defend their excavated nests by attacking intruders with the sharp spines at the bases of their thumbs. Such attacks may result in deep cuts, punctured eardrums, or even death. At least some of the stream-breeding *Plectrohyla* in Central America presumably also incur damage with their thumb spines, inasmuch as some males have scarred bodies. Captive *Anotheca spinosa* have been seen to puncture the body of another individual in the same tree hole with the sharp spines on their heads.

Hylid frogs are prey for many kinds of animals, especially snakes. Avoidance of predation principally is by the escape behavior of leaping to another branch or leaf; this is carried to an extreme by some species (e.g., *Agalychnis moreletii* and

Giant, or white-lipped, tree frogs (*Litoria infrafrenata*) can reach over 5.5 in (130 mm) in length, not including their legs. (Photo by Joe McDonald. Bruce Coleman Inc. Reproduced by permission.)

Anotheca spinosa) by "parachuting" for a long distance from a high limb. The terrestrial *Litoria nasuta* in Australia escapes by a series of long leaps. Some small species with fully webbed feet are capable of skittering across the surface of the water. *Acris crepitans* skitters after an initial leap from land, and *Scarthyla goinorum* is capable of leaping off a low bush to skitter on the water and then jump up onto another bush.

In an encounter with a potential predator, some *Hyla* and *Phyllomedusa* feign death by tucking their limbs close to the body and remaining motionless on their backs. In contrast, *Hemiphractus* turn their heads up, open their mouths so as to display an orange tongue, and even snap at a potential predator. The volatile, alkaline skin secretions of *Phrynohyas* are insoluble in water and have a deleterious effect on mucous membranes of the eyes and mouth; consequently, most predators avoid these frogs.

Tadpoles of most species seem to exist independently from conspecifics, but tadpoles of *Hyla geographica* and *Phyllomedusa vaillanti* form schools of hundreds of individuals. This behavior may result in less predation. Otherwise, tadpoles avoid predation by either remaining motionless or rapidly hiding amidst aquatic vegetation.

Feeding ecology and diet

All hylids seem to be sit-and-wait predators that feed on a wide variety of arthropods; the selection of food depends primarily on the size of the prey. A few species are specialists on

certain kinds of insects. *Sphaenorhynchus lacteus* feeds almost exclusively on ants, and *Hyla leucophyllata* feeds mostly on moths. The large-headed, broad-mouthed *Hemiphractus* eat large insects and other frogs.

Reproductive biology

Throughout temperate regions and the lowland tropics, hylid frogs respond to rains by moving to breeding sites, either temporary or, less frequently, permanent ponds. The length of the breeding season is determined by the period of rainfall; some northern species (e.g., *Pseudacris crucifer*) even call from the edges of ponds with ice on the water and snow on the banks. Species in dry regions tend to be explosive breeders that are active for only a day or so after heavy rains form temporary ponds. In contrast, hylids inhabiting humid rainforests and montane cloud forests may breed throughout the year in streams and ponds.

In those species that breed in ponds and streams, males congregate for breeding; after a heavy rain in tropical regions, breeding sites may have hundreds of individuals of several species calling at the same time. The calls vary from soft "peeps" to loud "growls." The calls of some species consist of only one note repeated at intervals of a few seconds to several minutes; other calls are a series of notes. In those species that call from bromeliads or tree holes, males usually are solitary. Females are attracted to the breeding site by the calls. Amplexus is axillary.

A pair of tiger-leg monkey frogs (*Phyllomedusa hypochondrialis*) in amplexus. (Photo by Danté Fenolio/Photo Researchers, Inc. Reproduced by permission.)

Diverse reproductive modes are employed by hylid frogs:

- Eggs are deposited in water (ponds or streams), and tadpoles develop in water: most Hylinae and Pelodryadinae and all Pseudinae.

- Eggs are deposited, and early-stage tadpoles develop in natural or constructed basins; subsequent flooding releases tadpoles into ponds or streams: *Hyla boans* group.

- Eggs are deposited in a foam nest floating on water in a pond; tadpoles develop in the pond: *Scinax rizibilis*.

- Eggs are deposited, and tadpoles develop in subterranean nests near ponds; subsequent flooding releases feeding tadpoles into ponds: *Hyla leucopygia*.

- Eggs are deposited on vegetation above water; feeding tadpoles develop in ponds or streams: all Phyllomedusinae and a few Hylinae and Pelodryadinae.

- Eggs are deposited, and tadpoles develop in bromeliads or cavities in trees: several species of Hylinae.

- Eggs are deposited in a pouch on the dorsum of the female; feeding tadpoles live in ponds: some *Gastrotheca*.

- Eggs are deposited in the dorsal pouch or on the back of the female; nonfeeding tadpoles live in bromeliads or tree holes: *Flectonotus*.

- Eggs are deposited in the dorsal pouch or on the back of a female; eggs hatch as froglets: *Cryptobatrachus, Hemiphractus, Stefania*, and some *Gastrotheca*.

At high latitudes and high elevations, as well as in arid environments, females usually deposit only one clutch of eggs per year, but at lower latitudes, especially in the lowland humid tropics, females may lay several clutches per year. Clutch size correlates with female body size within a given repro-

ductive mode. Females of large species, such as *Hyla rosenbergi* and *Phrynohyas venulosa*, that deposit small eggs in water have clutches in excess of 2,000 eggs, whereas in small species, such as *Pseudacris ocularis*, clutches consist of only about 100 eggs. Species that deposit eggs on vegetation over water have smaller clutches, ranging from 10 in the small *Hyla thorectes* to more than 250 in the large *Phyllomedusa bicolor*. Clutch size is less than 100 eggs in those species of *Gastrotheca* that transport eggs that hatch as tadpoles, whereas in those hemiphractines that carry eggs that hatch as froglets, clutches typically contain fewer than 15 proportionately much larger eggs.

No parental care exists among most hylid frogs, but female hemiphractines carry their eggs several weeks or months, depending on the stage at which the eggs hatch. The ultimate in parental care exists in several species that deposit their eggs in bromeliads or tree holes, where food is scarce. After deposition of a clutch of fertilized eggs, the female, accompanied or not by the male, returns to the breeding site and deposits additional fertilized or unfertilized eggs, which are eaten by the tadpoles. This behavior is known only in a few hylines (*Anotheca spinosa, Osteocephalus oophagus, Osteopilus brunneus,* and *Phrynohyas resinifictrix*) in tropical America.

Conservation status

According to the IUCN, six species are Critically Endangered; five are Endangered; five are Vulnerable; five are Lower Risk/Near Threatened; and eight are Data Deficient.

Habitat destruction imperils many species of hylid frogs. This is especially evident in montane regions, where many species have limited distributions. Some of the species of *Hyla, Plectrohyla,* and *Ptychohyla* have not been seen in recent years in areas where they were common before logging and stream pollution. Likewise, the conversion of dry tropical forests to agriculture seems to have limited greatly the distributions of such species as *Triprion spatulatus*. Chytrid fungus may be responsible for drastic declines or the extinction of many species, such as *Hyla calypsa* and *H. xanthosticta* in Central America and *Nyctimystes dayi* and at least three species of *Litoria* in northeastern Australia.

Significance to humans

Hylid frogs are not among those species commonly eaten by Europeans and North Americans, but many indigenous peoples in the American tropics and in the Australo-Papuan region catch and eat a variety of larger hylids, especially *Hyla boans* and *Osteocephalus taurinus* in the Americas and *Nyctimystes* in New Guinea. Indigenous people in New Guinea also eat tadpoles of *Litoria* and *Nyctimystes*, and the large tadpoles of *Pseudis paradoxa* are eaten in South America. Australian Aborigines unearth estivating *Cyclorana platycephala* and squeeze water out of them before replacing the frog in its burrow. Before going on a hunt, some indigenous people in the upper Amazon Basin lick the skin secretions of *Phyllomedusa bicolor*; this has a hallucinogenic effect.

1. Yucatecan shovel-headed treefrog (*Triprion petasatus*); 2. White-lined treefrog (*Phyllomedusa vaillantii*); 3. Manaus long-legged treefrog (*Osteocephalus taurinus*); 4. Rocket frog (*Litoria nasuta*); 5. Green treefrog (*Litoria caerulea*); 6. Cuban treefrog (*Osteopilus septentrionalis*); 7. Chorus frog (*Pseudacris triseriata*); 8. Amazonian skittering frog (*Scarthyla goinorum*); 9. Hartweg's spike-thumb frog (*Plectrohyla hartwegi*); 10.Paradox frog (*Pseudis paradoxa*). (Illustration by Brian Cressman)

1. Cope's gray treefrog (*Hyla chrysoscelis*); 2. European treefrog (*Hyla arborea*); 3. Spiny-headed treefrog (*Anotheca spinosa*); 4. Sumaco horned treefrog (*Hemiphractus proboscideus*); 5. Northern cricket frog (*Acris crepitans*); 6. Water-holding frog (*Cyclorana platycephala*); 7. Hourglass treefrog (*Hyla leucophyllata*); 8. Rosenberg's treefrog (*Hyla rosenbergi*); 9. Riobamba marsupial frog (*Gastrotheca riobambae*); 10. Red-eyed treefrog (*Agalychnis callidryas*). (Illustration by Amanda Humphrey)

Species accounts

Riobamba marsupial frog
Gastrotheca riobambae

SUBFAMILY
Hemiphractinae

TAXONOMY
Hyla riobambae Fowler, 1913, Riobamba, Chimborazo, Ecuador. Before 1972, the frog was referred to as *Gastrotheca marsupiata*, a name now restricted to a species in Peru and Bolivia.

OTHER COMMON NAMES
English: Ecuadorian marsupial frog; Spanish: Rana marsupial.

PHYSICAL CHARACTERISTICS
Males are 1.4–2.3 in (34–57 mm) long, and females are 1.4–2.7 in (34–66 mm) long. A stout-bodied frog with a rounded snout. The skin on the dorsum is smooth or areolate, and the skin on the venter is granular. The limbs are moderately short, and the terminal disks on the digits are only slightly wider than the rest of the digit. Females have a dorsal pouch with an aperture placed posteriorly. The dorsum is tan or various shades of green, with or without darker green or brown markings; the venter is cream with or without gray or brown spots.

DISTRIBUTION
Found at elevations of 5,150–10,400 ft (1,590–3,220 m) in the Andes of northern and central Ecuador.

■ *Hemiphractus proboscideus*
■ *Gastrotheca riobambae*

HABITAT
Montane grasslands, cultivated fields, and gardens in cities.

BEHAVIOR
Terrestrial and primarily nocturnal; it finds diurnal refuges in crevices in stone walls, rock piles, terrestrial bromeliads, and agave plants.

FEEDING ECOLOGY AND DIET
A variety of arthropods, especially beetles.

REPRODUCTIVE BIOLOGY
Males call from the ground or rocks; the call is "wraaack-ack-ack." Although a given male may breed several times a year, females breed only once per year. Females approach males, and mating takes place on land. Once in amplexus, the female elevates the cloaca, and the male exudes seminal fluid and, with his feet, spreads the fluid between the female's cloaca and the opening of the brood pouch. As the female exudes eggs, the male pushes the eggs with his feet over the back of the female and into the pouch. The incubation period of the 64–166 eggs in the pouch is 70–108 days. Tadpoles hatch in the pouch; their wriggling results in the female's sitting in shallow water. Parturition is partly spontaneous and partly assisted by the female, who distends the opening of the pouch with her hind feet and inserts one or both feet into the pouch and scoops out tadpoles. Newly released tadpoles have small hind-limb buds and begin feeding in the water in shallow grassy ponds or irrigation ditches within one to two days after release. Metamorphosis occurs 4–12 months after parturition.

CONSERVATION STATUS
Although not listed by the IUCN, the Riobamba marsupial frog is threatened by pesticides that accumulate in the water where tadpoles develop and by chytrid fungus.

SIGNIFICANCE TO HUMANS
None known. ◆

Sumaco horned treefrog
Hemiphractus proboscideus

SUBFAMILY
Hemiphractinae

TAXONOMY
Cerathyla proboscidea Jiménez de la Espada, 1871, Sumaco, Napo, Ecuador.

OTHER COMMON NAMES
None known.

PHYSICAL CHARACTERISTICS
Males are 1.8–2.0 in (43–50 mm) long; females are 2.3–2.7 in (57–66 mm) long. The Sumaco horned treefrog is a bizarre frog with a triangular head; fleshy, pointed snout; prominent tubercles on the upper eyelids; depressed body; and neural spines of vertebrae evident on the back. The dorsum is brown or tan with green, brown, or gray marks on the body and bars on the limbs; the venter is brown with tan or orange spots.

DISTRIBUTION
Upper Amazon Basin and lower slopes of the Andes in southern Colombia, Ecuador, and Peru.

HABITAT
Humid lowland and lower montane forest.

BEHAVIOR
Nocturnal and arboreal. Defensive display consists of opening the mouth and exhibiting a bright yellow tongue.

FEEDING ECOLOGY AND DIET
Large arthropods, small lizards, and other frogs.

REPRODUCTIVE BIOLOGY
Females carry up to 26 large eggs that adhere to her back; eggs undergo direct development into small frogs.

CONSERVATION STATUS
Not threatened.

SIGNIFICANCE TO HUMANS
None known. ◆

Northern cricket frog
Acris crepitans

SUBFAMILY
Hylinae

TAXONOMY
Acris crepitans Baird, 1854, northeastern United States. Two subspecies are recognized.

OTHER COMMON NAMES
German: Grillenfrosch; French: Acris-grillon; Spanish: Rana grillo.

Acris crepitans
Pseudacris triseriata

PHYSICAL CHARACTERISTICS
Males are 0.7–1.2 in (17–28 mm) long; females are 1.0–1.5 in (25–38 mm) long. The dorsum is weakly tubercular, and the venter is smooth. The snout is acutely rounded, and the discs are not expanded; the toes are about four-fifths webbed. The dorsum is tan with brown or green markings, including a triangle on the head, a mid-dorsal stripe, and bars on the upper lips. The venter is white, and the posterior surfaces of the thighs are white with a longitudinal dark brown stripe.

DISTRIBUTION
Eastern North America.

HABITAT
The vicinity of permanent ponds, marshes, and slow-moving streams.

BEHAVIOR
Nocturnal and diurnal; they are terrestrial and semiaquatic.

FEEDING ECOLOGY AND DIET
Feed under and on the surface of the water on a variety of small arthropods.

REPRODUCTIVE BIOLOGY
Throughout the warm season of the year, males call from shallow water or floating vegetation; the call is a series of clicks. Amplectant pairs deposit up to 400 eggs singly or in clutches of two to seven eggs in shallow water. The eggs hatch in three to four days into small, solitary tadpoles that require five to seven weeks before metamorphosis.

CONSERVATION STATUS
Not threatened.

SIGNIFICANCE TO HUMANS
None known. ◆

Spiny-headed treefrog
Anotheca spinosa

SUBFAMILY
Hylinae

TAXONOMY
Hyla spinosa Steindachner, 1864, Brazil (in error). From 1939 to 1968 the frog was known as *Anotheca coronata* (Stejneger).

OTHER COMMON NAMES
Spanish: Rana de coronata.

PHYSICAL CHARACTERISTICS
Males are 2.5–2.7 in (60–65 mm) long; females are 2.4–3.0 in (58–73 mm) long. The head is casqued with sharp, upwardly pointing spines. The dorsum is brown with darker brown markings; the venter is black.

DISTRIBUTION
Discontinuous from central Veracruz, Mexico, to central Panama.

HABITAT
Humid forests at elevations of 300–5,800 ft (90–1,800 m).

BEHAVIOR
Nocturnal and arboreal.

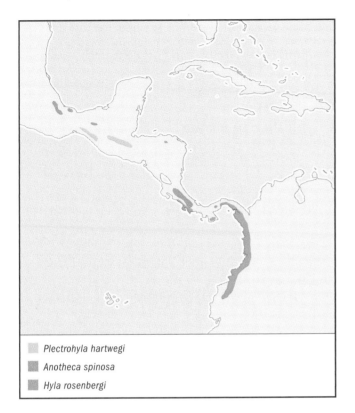

- ▨ *Plectrohyla hartwegi*
- ▨ *Anotheca spinosa*
- ▨ *Hyla rosenbergi*

- ▨ *Hyla arborea*

FEEDING ECOLOGY AND DIET
Feed on arthropods; tadpoles feed on frog eggs and mosquito larvae.

REPRODUCTIVE BIOLOGY
Males call solitarily from bromeliads and tree holes; the call is a series of notes, "boop-boop-boop." Clutches of 48–322 eggs are deposited just above water level on the leaves of bromeliads or on walls of the cavities in trees; only a small percentage of eggs hatch. Hatching tadpoles wriggle into the water and feed on the remaining eggs; the female returns to the site and deposits more eggs, on which tadpoles feed. Tadpoles that are not in crowded containers and that are supplied with sufficient nutritive eggs metamorphose in 60 days.

CONSERVATION STATUS
Not threatened.

SIGNIFICANCE TO HUMANS
None known. ◆

European treefrog
Hyla arborea

SUBFAMILY
Hylinae

TAXONOMY
Rana arborea Linnaeus, 1758, Europe. Five subspecies are recognized.

OTHER COMMON NAMES
French: Rainette verte; German: Laubfrosche; Spanish: Ranita de San Antonio; Russian: Obyknovennaya kvaksha.

PHYSICAL CHARACTERISTICS
Males are 1.3–1.8 in (32–43 mm) long; females are 1.6–2.0 in (40–50 mm) long. A moderately slender treefrog with long legs. The skin on the dorsum is smooth, and the skin on the venter is granular. The dorsum is green or tan with a dark brown stripe from the eye to the groin, bordered above by a narrow white line; the upper lip and venter are creamy white.

DISTRIBUTION
Most of Europe, exclusive of the British Isles and Scandinavia, eastward to the Ural Mountains and northern Turkey. Present on several Mediterranean islands, including Corsica, Crete, Elba, Rhodes, and Sardinia.

HABITAT
Humid and dry forests.

BEHAVIOR
Nocturnal and primarily arboreal.

FEEDING ECOLOGY AND DIET
Feed on a variety of small arthropods.

REPRODUCTIVE BIOLOGY
After spring rains, males call from low vegetation or shallow water in ponds; the call is a rapid "krak-krak-krak." Females deposit 800–1,000 eggs in small clumps in ponds. The eggs hatch in 12–15 days into free-swimming tadpoles that metamorphose in three or more months.

CONSERVATION STATUS
Lower Risk/Near Threatened. The European treefrog is threatened throughout most of its range by habitat destruction and pollution.

SIGNIFICANCE TO HUMANS
None known. ◆

Cope's gray treefrog
Hyla chrysoscelis

SUBFAMILY
Hylinae

TAXONOMY
Hyla femoralis chrysoscelis Cope, 1880, Dallas, Texas, United States.

OTHER COMMON NAMES
None known.

PHYSICAL CHARACTERISTICS
Males are 1.2–1.8 in (30–45 mm) long; females are 1.6–2.2 in (40–53 mm) long. This is a moderately robust treefrog with lightly tuberculate skin on the dorsum, which is green or gray with darker blotches. There is a white spot below the eye. Hidden surfaces of the hind limbs are yellow; the belly is white.

DISTRIBUTION
Eastern North America; the exact range is unknown, because it overlaps with the morphologically identical gray treefrog (*Hyla versicolor*), which differs in call and chromosome number.

HABITAT
Primarily hardwood but also coniferous forest.

BEHAVIOR
Nocturnal and arboreal.

FEEDING ECOLOGY AND DIET
Feeds on a variety of small arthropods.

REPRODUCTIVE BIOLOGY
Breeding takes place after warm spring rains. Males call from bushes and trees bordering ponds; females approach and nudge calling males. Amplexus may last several hours, during which time the female ovulates. In the course of amplexus females carry males to water. The ovarian complement is 485–3,840 eggs, which are laid in small packets of five to 31 eggs attached to aquatic vegetation. Eggs hatch in four to five days into free-swimming tadpoles that require seven to eight weeks to metamorphose. Females may deposit three decreasingly smaller clutches at intervals of eight to 35 days.

CONSERVATION STATUS
Not threatened.

SIGNIFICANCE TO HUMANS
None known. ◆

Hourglass treefrog
Hyla leucophyllata

SUBFAMILY
Hylinae

TAXONOMY
Rana leucophyllata Bereis, 1783, Suriname.

OTHER COMMON NAMES
English: Bereis' treefrog; French: Rainette à bandeau.

PHYSICAL CHARACTERISTICS
Males are 1.3–1.5 in (33–36 mm) long; females are 1.6–1.8 in (40–44 mm) long. This is a slender treefrog with a truncate snout and smooth skin on the dorsum. The toes are about two-thirds webbed, and there is an extensive axillary membrane. The dorsum is creamy yellow with a brown, hourglass-shaped mark on the back; hidden surfaces of the limbs and the

Scarthyla goinorum

Osteocephalus taurinus

Hyla leucophyllata

Hyla chrysoscelis

webbing are orange. The dorsum in some individuals is brown with cream reticulations.

DISTRIBUTION
The Amazon Basin and the Guiana region in South America.

HABITAT
Lowland tropical rainforest.

BEHAVIOR
Nocturnal and arboreal.

FEEDING ECOLOGY AND DIET
Eats mostly moths but also other small insects.

REPRODUCTIVE BIOLOGY
Males call from vegetation around ponds; the call is a ratcheting primary note followed by two to seven shorter secondary notes. While in axillary amplexus, females deposit clutches of about 600 eggs on vegetation over water. The eggs hatch in five to seven days. Tadpoles drop into water; they are macrophagous and feed on the bottom of shallow ponds.

CONSERVATION STATUS
Not threatened.

SIGNIFICANCE TO HUMANS
None known. ◆

Rosenberg's treefrog
Hyla rosenbergi

SUBFAMILY
Hylinae

TAXONOMY
Hyla rosenbergi Boulenger, 1898, Cachabe, Esmeraldas, Ecuador.

OTHER COMMON NAMES
English: Rosenberg's gladiator frog.

PHYSICAL CHARACTERISTICS
Males are 2.8–3.7 in (77–91 mm) long; females are 3.4–3.8 in (82–93 mm) long. The head is broad and flat. The limbs are long, and the fingers and toes are more than three-fourths webbed. The dorsum is tan with faintly darker mottling; the venter is pale bluish green. Males have an elongated spine on the base of the thumb.

DISTRIBUTION
Pacific lowlands from Costa Rica to Ecuador.

HABITAT
Humid lowland rainforest.

BEHAVIOR
Nocturnal and arboreal.

FEEDING ECOLOGY AND DIET
Eats a variety of arthropods.

REPRODUCTIVE BIOLOGY
Males excavate shallow basins into which water seeps on mudflats adjacent to ponds or slow-flowing streams. Males call from basins and defend them from other males; the call is a short series of low-pitched notes, "tonk-tonk-tonk." Attracted by the calls, the female enters and inspects the basin; once in amplexus, the female renovates the basin and deposits 1,700–3,000 eggs. Males remain at the basins until the eggs hatch, in 40–66 hours; at subsequent flooding of the basin, the tadpoles enter open water and metamorphose at an age of about 40 days.

CONSERVATION STATUS
Not threatened.

SIGNIFICANCE TO HUMANS
None known. ◆

Manaus slender-legged treefrog
Osteocephalus taurinus

SUBFAMILY
Hylinae

TAXONOMY
Osteocephalus taurinus Steindachner, 1862, Manaus, Amazonas, Brazil.

OTHER COMMON NAMES
English: Bony-headed treefrog; French: Ostéocéphale taurin.

PHYSICAL CHARACTERISTICS
Males are 2.7–3.5 in (66–85 mm) long; females are 3.1–4.2 in (76–104 mm) long. Long-legged, with toes about three-fourths webbed. The dorsal skin in females is smooth and bears spiny tubercles in males. In large individuals, the skin on top of the head is co-ossified with underlying bones, which form a pair of longitudinal ridges on the top of the head. The dorsum is tan to reddish brown, with brown irregular markings on the back and bars on the limbs; the venter is cream with brown spots or mottling on the chest. The iris is bronze with radiating black lines. Males have paired lateral vocal sacs.

DISTRIBUTION
The Amazon Basin and Guianan region of South America.

HABITAT
Humid lowland rainforest.

BEHAVIOR
Nocturnal and arboreal.

FEEDING ECOLOGY AND DIET
Eats a variety of arthropods, especially orthopterans.

REPRODUCTIVE BIOLOGY
After the initial heavy rains of the season, males congregate at ponds and mostly call while floating on the surface of the water; the call is a loud "boop-boop-boop," followed or not by a "worrr." Females deposit 500–600 small pigmented eggs as a surface film on the water. The eggs hatch in about 24 hours; the free-swimming tadpoles require about 86 days to reach metamorphosis.

CONSERVATION STATUS
Not threatened.

SIGNIFICANCE TO HUMANS
Some indigenous people eat this species. ◆

Cuban treefrog
Osteopilus septentrionalis

SUBFAMILY
Hylinae

TAXONOMY
Hyla septentrionalis Duméril and Bibron, 1841, Cuba.

OTHER COMMON NAMES
German: Kuba-Laubfrosch; Spanish: Rana platanera.

PHYSICAL CHARACTERISTICS
Males are 2.5–3.6 in (60–89 mm) long; females are 3.7–5.7 in (90–140 mm) long. The head is broad and flat; the skin is co-ossified with the skull in large individuals. The dorsum has scattered tubercles, and the venter is granular. The toes are about two-thirds webbed. The dorsum is gray to olive green, with bold darker mottling or elongated blotches; the venter is creamy white.

DISTRIBUTION
Cuba, the Cayman Islands, and the Bahamas. They have been introduced into Puerto Rico, various islands in the Lesser Antilles, and Florida in the United States.

HABITAT
Mesic and dry forest.

BEHAVIOR
These frogs are nocturnal and arboreal. Diurnal retreats include banana plants, burrows, cisterns, and secluded areas in buildings; they are tolerant of brackish water.

FEEDING ECOLOGY AND DIET
Large individuals have a voracious appetite and feed on a variety of insects, small crustaceans, and frogs.

REPRODUCTIVE BIOLOGY
Males call from vegetation near water and from vertical walls adjacent to pools of rainwater; the call consists of a series of loud, low-pitched notes. Eggs are deposited as a surface film on water and hatch in 27–30 hours into free-swimming tadpoles.

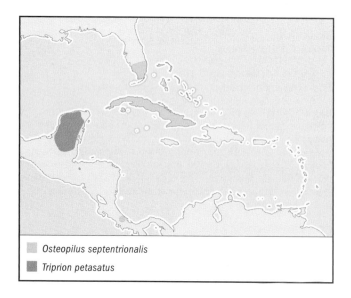

■ *Osteopilus septentrionalis*
■ *Triprion petasatus*

CONSERVATION STATUS
Not threatened.

SIGNIFICANCE TO HUMANS
None known. ◆

Hartweg's spike-thumbed frog
Plectrohyla hartwegi

SUBFAMILY
Hylinae

TAXONOMY
Plectrohyla hartwegi Duellman, 1968, Barrejonel, Chiápas, Mexico.

OTHER COMMON NAMES
Spanish: Ranita de dedos delgados.

PHYSICAL CHARACTERISTICS
Males are 2.2–2.9 in (54–72 mm) long; females are 2.2–3.1 in (54–77 mm) long. The body is robust, with finely tuberculate skin and a short snout. The arms are robust in males, with a bifid spine at the base of the thumb; the toes are about four-fifths webbed. The dorsum is olive tan to green, and the venter is pale gray. The anterior and posterior surfaces of the thighs are mottled boldly with pale cream and black or dark green.

DISTRIBUTION
Moderate to high elevations of the Pacific slopes of Chiápas, Mexico, and the Atlantic slopes of Guatemala and northwestern Honduras.

HABITAT
Montane cloud forest.

BEHAVIOR
Active at night on rocks at the edges of streams and in trees bordering streams.

FEEDING ECOLOGY AND DIET
Eats a variety of arthropods.

REPRODUCTIVE BIOLOGY
Males are not known to vocalize. Females deposit 191–352 eggs in streams. Tadpoles have large, ventral mouths and adhere to stones in pools in streams.

CONSERVATION STATUS
Not threatened.

SIGNIFICANCE TO HUMANS
None known. ◆

Chorus frog
Pseudacris triseriata

SUBFAMILY
Hylinae

TAXONOMY
Hyla triseriata Wied, 1839, Mount Vernon, Ohio River, Indiana, United States.

OTHER COMMON NAMES
French: Rainette faux criquet, rainette faux-grillon de l'Ouest.

PHYSICAL CHARACTERISTICS
Males are 0.8–1.2 in (19–29 mm) long; females are 1.2–1.5 in (29–37 mm) long. The skin on the dorsum is slightly tubercular, and on the venter it is granular. The snout is acutely rounded. The toes are about one-third webbed. The dorsum is grayish tan, with brown mid-dorsal and dorsolateral stripes or rows of spots. There is a broad dark brown or black stripe from the snout through the eye and tympanum to the groin; the venter is white.

DISTRIBUTION
Eastern North America.

HABITAT
Grassland, pastures, cropland, and moist forest.

BEHAVIOR
Nocturnal and terrestrial.

FEEDING ECOLOGY AND DIET
Eats small arthropods, including beetles, grubs, ants, and spiders.

REPRODUCTIVE BIOLOGY
Breeding occurs after the first spring rains. Males call from grasses in the water or at the edge of water in ponds, marshes, and roadside ditches. The call consists of a vibrant, pulsed "cr-reeck." Females deposit 100–1,500 eggs in small clutches of five to 300 eggs attached to vegetation in shallow water. Eggs require about two weeks to hatch into small, free-swimming tadpoles that metamorphose in about two months.

CONSERVATION STATUS
Not threatened.

SIGNIFICANCE TO HUMANS
None known. ◆

Amazonian skittering frog
Scarthyla goinorum

SUBFAMILY
Hylinae

TAXONOMY
Hyla goinorum Bokermann, 1962, Tarauacá, Acre, Brazil.

OTHER COMMON NAMES
None known.

PHYSICAL CHARACTERISTICS
Males are 0.6–0.8 in (15–20 mm) long; females are 0.7–0,9 in (18–23 mm) long. Slender body with a pointed snout and long limbs with fully webbed toes. The dorsum is green with brown and white lateral stripes; the venter is white.

DISTRIBUTION
Upper Amazon Basin from southern Colombia to northeastern Bolivia.

HABITAT
Swampy regions in lowland tropical rainforest.

BEHAVIOR
Nocturnal and arboreal; they perch on leaves just above the surface of the water and are capable of skittering across the surface.

FEEDING ECOLOGY AND DIET
Eats a variety of small arthropods; spiders make up more than 50% of their diet.

REPRODUCTIVE BIOLOGY
Males call from low vegetation above water; the call consists of eight to 10 short, whistle-like notes. Clutches of 130–202 small, pigmented eggs are deposited in ponds. Elongate tadpoles have muscular tails with low fins. Macrophagous tadpoles swim just below the surface of the water and can propel themselves out of water for distances of 8–12 in (20–30 cm).

CONSERVATION STATUS
Not threatened.

SIGNIFICANCE TO HUMANS
None known. ◆

Yucatecan shovel-headed treefrog
Triprion petasatus

SUBFAMILY
Hylinae

TAXONOMY
Pharyngodon petasatus Cope, 1865, Cenote Tamanché, Yucatán, Mexico.

OTHER COMMON NAMES
Spanish: Ranita de casco yucateca.

PHYSICAL CHARACTERISTICS
Males are 2–2.5 in (48–61 mm) long; females are 2.6–3 in (65–75 mm) long. The head is casque-shaped, with a large, upturned prenasal bone and expanded maxillaries forming a broad labial shelf. The dorsum is olive green or tan with dark brown markings. The belly is white, and the undersides of the limbs are tan.

DISTRIBUTION
Yucatán Peninsula of Mexico, northern Guatemala, and Belize; a disjunct population exists in northwestern Honduras.

HABITAT
Semiarid scrub forest and savannas.

BEHAVIOR
A nocturnal species, found on the ground, bushes, and low trees.

FEEDING ECOLOGY AND DIET
Eats a variety of small arthropods and small frogs.

REPRODUCTIVE BIOLOGY
An explosive breeder after heavy rains. Males call from trees, bushes, and ground around temporary pools; the call consists of quickly repeated, low-pitched notes resembling the quacking of a duck. Eggs are laid in clumps in water, where they hatch into free-swimming tadpoles.

CONSERVATION STATUS
Not threatened.

SIGNIFICANCE TO HUMANS
None known. ◆

Water-holding frog
Cyclorana platycephala

SUBFAMILY
Pelodryadinae

TAXONOMY
Chiroleptis platycephalus Günther, 1873, Fort Bourke, New South Wales, Australia.

OTHER COMMON NAMES
None known.

PHYSICAL CHARACTERISTICS
Males are 1.7–2.6 in (42–64 mm) long; females are 2.0–2.9 in (50–72 mm) long. A robust frog with a flat head, small eyes, muscular limbs, a spadelike tubercle on the foot, and extensively webbed toes. The dorsum is dull gray, brown, or green with irregular darker blotches; the venter is dull white.

DISTRIBUTION
Found discontinuously in the interior of Australia.

HABITAT
Dry grassland and desert.

BEHAVIOR
Nocturnal and terrestrial. Using their hind feet, the frogs burrow into soil and shed multiple layers of skin that form a nearly impermeable cocoon, to prevent water loss during months of estivation.

FEEDING ECOLOGY AND DIET
Eats a variety of arthropods.

REPRODUCTIVE BIOLOGY
After rains create temporary ponds, males congregate and call; the call is a long snoring sound, "maw-w-w-w-maw-w-w-w." Eggs are laid in clumps in shallow water; free-swimming tadpoles metamorphose in as few as 30 days.

CONSERVATION STATUS
Not threatened.

SIGNIFICANCE TO HUMANS
Aborigines dig up estivating frogs and squeeze water from them. ◆

Green treefrog
Litoria caerulea

SUBFAMILY
Pelodryadinae

TAXONOMY
Rana caerulea White, 1790, New South Wales, Australia. Some authors place this species in the genus *Pelodryas*.

OTHER COMMON NAMES
English: White's treefrog.

PHYSICAL CHARACTERISTICS
Males are 2.7–3.1 in (66–77 mm) long; females are 2.9–4.5 in (70–110 mm) long. The green treefrog is a robust-bodied treefrog with large, diffuse glands on the back of the head and extensively webbed toes. The dorsum is green; the venter is white.

DISTRIBUTION
Found from northern and eastern Australia to southern New South Wales and in southern New Guinea. It has been introduced into New Zealand.

HABITAT
Dry and humid forests.

BEHAVIOR
Nocturnal and arboreal.

FEEDING ECOLOGY AND DIET
Eats a variety of arthropods, other frogs, and small mammals.

REPRODUCTIVE BIOLOGY
After rains males call from trees, rocks, and ground near swamps and slow-moving streams; the call is a continuously repeated "crawk." Clutches of 200–2,000 eggs are deposited in still water from November to February. Free-swimming tadpoles metamorphose in about six weeks.

CONSERVATION STATUS
Not threatened.

SIGNIFICANCE TO HUMANS
Caerulin, a drug used to control hypertension, was discovered in the skin secretions of this species. Now the compounds have been synthesized, and the drug is produced artificially. This is a common species in the pet trade. ◆

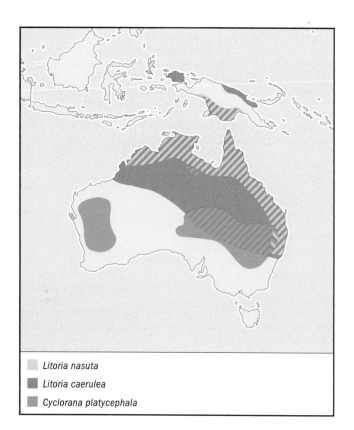

Litoria nasuta

Litoria caerulea

Cyclorana platycephala

Rocket frog
Litoria nasuta

SUBFAMILY
Pelodryadinae

TAXONOMY
Pelodytes nasutus Gray, 1842, Port Essington, Northern Territory, Australia.

OTHER COMMON NAMES
None known.

PHYSICAL CHARACTERISTICS
Males are 1.3–1.8 in (33–45 mm) long; females are 1.5–2.3 in (36–55 mm) long. A streamlined frog with extremely long legs and a pointed snout. The dorsum is colored in shades of brown, with darker longitudinal skin folds or rows of pustules; the venter is white.

DISTRIBUTION
Found in coastal and adjacent areas of northern and eastern Australia, from northern Western Australia to central New South Wales; they also live in southern New Guinea.

HABITAT
Dry and humid forests.

BEHAVIOR
Nocturnal and terrestrial and capable of making a series of long leaps.

FEEDING ECOLOGY AND DIET
Feeds on a variety of arthropods.

REPRODUCTIVE BIOLOGY
Males call from the edges of ponds in November through February; the call is a series of notes, "wick-wick-wick-wick." Batches of 50–100 eggs are laid as a surface film on water; free-swimming tadpoles metamorphose in about one month.

CONSERVATION STATUS
Not threatened.

SIGNIFICANCE TO HUMANS
None known. ◆

Phyllomedusa vaillantii

Agalychnis callidryas

Pseudis paradoxa

Red-eyed treefrog
Agalychnis callidryas

SUBFAMILY
Phyllomedusinae

TAXONOMY
Hyla callidryas Cope, 1862, Darién, Panama.

OTHER COMMON NAMES
English: Red-eyed leaf frog; Spanish: Ninfa de bosque, rana borracha, rana-de àrbol ojos rojos.

PHYSICAL CHARACTERISTICS
Males are 1.6–2.5 in (39–59 mm) long; females are 2.2–2.9 in (51–71 mm) long. Slender, long-legged treefrog with a green dorsum, blue flanks with white vertical bars, a creamy white venter, vertical pupil, bright red iris, and lower eyelid reticulated with white or pale yellow.

DISTRIBUTION
Found at elevations from sea level to 3,100 ft (960 m), from southeastern Mexico to extreme northwestern Colombia.

HABITAT
Humid lowland rainforest.

BEHAVIOR
The frog is nocturnal and arboreal; by day, the limbs are tucked closely against the body, and they sleep on the undersides of leaves.

FEEDING ECOLOGY AND DIET
Eats a variety of small arthropods, especially orthopterns.

REPRODUCTIVE BIOLOGY
Males call from branches and leaves of trees above ponds in the rainy season; the call is a soft single or double note, "cluck." Females approach calling males; once in amplexus, the female descends to the pond, where she absorbs water and then climbs to a leaf above the water. Clutches of 11–78 eggs are deposited on the leaf, which usually is folded around the egg clutch. Females deposit only part of their ovarian complement in one clutch. Hatchling tadpoles drop into the water. Tadpoles are mid-water filter feeders and orient themselves in a head-up position.

CONSERVATION STATUS
Not threatened.

SIGNIFICANCE TO HUMANS
This frog, which is common in the pet trade, has become the "poster frog" for many conservation organizations. ◆

White-lined treefrog
Phyllomedusa vaillanti

SUBFAMILY
Phyllomedusinae

TAXONOMY
Phyllomedusa vaillanti Boulenger, 1882, Santarem, Pará, Brazil.

OTHER COMMON NAMES
French: Phylloméduse de Vaillant.

PHYSICAL CHARACTERISTICS
Males are 2.0–2.4 in (50–58 mm) long; females are 1.8–2.4 in (68–84 mm) long. A large treefrog with a truncate snout and a pair of elevated, longitudinal parotoid glands posterior to the eye and extending to the mid-body. The innermost fingers and toes are longer than the adjacent ones; webbing is absent. The dorsum and side of the head are green, with a row of white granules along the angle of the parotoid gland; the flanks are green above and reddish brown below, with row of elliptical cream or orange spots. The venter is brownish gray, with a pair of cream spots on the throat and a large green spot on the chest. The pupil is vertical, with a pale gray iris.

DISTRIBUTION
Amazon Basin and Guiana region of South America.

HABITAT
Lowland tropical rainforest.

BEHAVIOR
Nocturnal and arboreal; the frog methodically walks on branches of trees and bushes.

FEEDING ECOLOGY AND DIET
Feeds on a variety of arthropods.

REPRODUCTIVE BIOLOGY
After rains males call from vegetation above permanent ponds; the call is a short, harsh "cluck." Females approach males. Once in amplexus, the female carries the male to a pond, where she absorbs water, and then to a large leaf, where clutches of 415–645 eggs are deposited. Eggs hatch in about four days, and tadpoles drop into the water, where they are free-swimming, usually in schools of more than 50 individuals. The tadpoles are midwater filter feeders that orient themselves in a head-up position.

CONSERVATION STATUS
Not threatened.

SIGNIFICANCE TO HUMANS
None known. ◆

Paradox frog
Pseudis paradoxa

SUBFAMILY
Pseudinae

TAXONOMY
Rana paradoxa Linnaeus, 1758, Suriname. Six subspecies are recognized.

OTHER COMMON NAMES
English: Paradoxical frog; French: Grenouille paradoxale; Spanish: Rana boyadera.

PHYSICAL CHARACTERISTICS
Males are 1.6–2.7 in (38–65 mm) long; females are 1.7–3.2 in (40–65 mm) long. The snout is acutely rounded, and the eyes are large and protuberant dorsally. The limbs are long and muscular, with fully webbed toes. The dorsum is greenish tan; the venter is white. The posterior surfaces of the thighs are cream with brown longitudinal stripes.

DISTRIBUTION
Distributed disjunctly in South America—the lower Río Magdalena Valley in Colombia, the Llanos of Colombia and Venezuela, Trinidad, central and southern Brazil, southern Peru, eastern Bolivia, Paraguay, and northern Argentina.

HABITAT
Marshes and permanent ponds in savannas and open forest in tropical lowlands.

BEHAVIOR
Nocturnal and aquatic, but during the breeding season males call by day and night; they float in water with only their eyes above the surface.

FEEDING ECOLOGY AND DIET
Feeds on a variety of aquatic arthropods and small frogs.

REPRODUCTIVE BIOLOGY
Breeding takes place in water. Females are attracted to males by their call, a single loud croak. Frothy masses of eggs are laid amidst aquatic vegetation. Free-swimming tadpoles grow to total lengths of 11 in (270 mm).

CONSERVATION STATUS
Not threatened.

SIGNIFICANCE TO HUMANS
Some indigenous peoples eat the tadpoles. ◆

Resources

Books

Barker, John, Gordon Grigg, and Michael J. Tyler. *A Field Guide to Australian Frogs.* Chipping Norton, Australia: Surrey Beatty and Sons, 1995.

Duellman, William E. *Hylid Frogs of Middle America.* Ithaca, NY: Society for the Study of Amphibians and Reptiles, 2001.

Lescure, Jean, and Christian Marty. *Atlas des Amphibiens de Guyane.* Paris: Muséum National d'Histoire Naturelle, 2001.

Periodicals

Duellman, William E. "The Biology of an Equatorial Herpetofauna in Amazonian Ecuador." *Miscellaneous Publications, Museum of Natural History of the University of Kansas* 65 (1978): 1–352.

Rodríguez, Lily O., and William E. Duellman. "Guide to the Frogs of the Iquitos Region, Amazonian Peru." *Special Publications, Museum of Natural History of the University of Kansas* 22 (1994): 1–80.

Trueb, Linda. "Evolutionary Relationships of Casque-headed Tree Frogs with Co-ossifed Skulls (Family Hylidae)." *Miscellaneous Publications, Museum of Natural History of the University of Kansas* 18, no. 7 (1970): 547–716.

William E. Duellman, PhD

True frogs

(Ranidae)

Class Amphibia

Order Anura

Family Ranidae

Thumbnail description
Small to large firmisternal aquatic or terrestrial frogs with a toothed upper jaw and cylindrical sacral diapophyses, without intercalary elements between penultimate and terminal phalanges of the digits

Size
0.4–12.6 in (10–320 mm)

Number of genera, species
51 genera; 686 species

Habitat
Ranids live in a variety of habitats, including tropical, subtropical, and temperate forests; savannas; grasslands; deserts; and high-elevation sites

Conservation status
Extinct: 3 species; Critically Endangered: 7 species; Endangered: 6 species; Vulnerable: 14 species; Lower Risk/Near Threatened: 4 species; Data Deficient: 12 species

Distribution
Ranids occur throughout the Old World (Eurasia, Africa), on most western Pacific islands, and in northern Australia, North America, and the northern part of South America; absent in the Pacific east of Fiji, in Madagascar (except as introduced), and on numerous isolated oceanic islands

Evolution and systematics

Ranid fossils are known reliably only since the Eocene of Europe, but the fossil record is of little help for the reconstruction of the early history of the group. Although it seems clear that the group is Gondwanan, current evidence does not allow unambiguous distinction between an Asian or an African origin. Molecular data from mitochondrial genes and a few nuclear genes suggest the existence of a monophyletic group (called by some researchers the "epifamily Ranoidae" of the "superfamily Ranoidea") made up of two monophyletic subgroups generally treated as families (Mantellidae and Rhacophoridae) and a third subgroup, the Ranidae, the monophyly of which is highly questionable. The latter subgroup contains several groups that are treated herein as subfamilies or tribes.

Five of these groups (Cacosterninae, Conrauini, Petropedetinae, Ptychadeninae, Pyxicephalinae) seem to be endemic to tropical and southern Africa, and nine (Ceratobatrachini, Limnonectini, Paini, Lankanectinae, Micrixalinae, Nyctibatrachinae, Occidozyginae, Amolopini, Ranixalinae) are endemic to the Oriental region. A fifteenth group (Dicroglossini) is distributed in both regions, and a sixteenth (Ranini) is present in those two regions as well as the Holarctic and the northern parts of South America and Australia.

Most ranids have 13 pairs of chromosomes, but various species have different numbers of chromosomes. Besides the generalized development in water through a tadpole stage,

various kinds of direct development (endotrophy) have evolved independently in several clades. Much variation exists in the labial tooth-row formula besides the generalized and probably plesiomorphic formula of 2/3; a few groups, especially those with rheophilic tadpoles, have up to 14 rows on the anterior lip and 12 on the posterior lip; in a few genera, labial teeth are absent on one or both lips.

The phylogenetic data currently available support the provisional recognition of eleven subfamilies, within some of which distinct tribes can be recognized.

Cacosterninae

This subfamily includes six to eight genera in sub-Saharan Africa: *Anhydrophryne* (one species), *Arthroleptella* (three species), *Cacosternum* (seven species), *Microbatrachella* (one species), *Nothophryne* (one species), *Poyntonia* (one species), and possibly also *Strongylopus* (six species), and *Tomopterna* (eight species). These are minute to medium-size frogs, including some of the smallest known anurans (0.4 in, or 10 mm, in Microbatrachella). Most of these genera have a partially or entirely cartilaginous omosternal style and procoracoid clavicular bar; the latter is sometimes incomplete. Known chromosome numbers are 26 (*Anhydrophryne*, diploid *Tomopterna*), 24 (*Poyntonia*), and 52 (tetraploid *Tomopterna*). Development is exotrophic in most genera; tadpoles have 1–4/2–3 rows of labial teeth. *Anhydrophryne* and *Arthroleptella* lay five to 40 eggs, 0.09–0.2 in (2.2–4.5 mm) in diameter, under shelters, where they undergo direct development.

Dicroglossinae

Despite anatomical heterogeneity, monophyly of this group is established firmly by molecular data. Many members share several characters, including a peculiar scapular girdle with slightly overlapping coracoids, the lowest one having a slight concavity medially; a basally forked omosternum; large nasals in contact medially; no dorsolateral folds; and small numbers of large eggs. Five subgroups are treated provisionally as tribes:

- Ceratobatrachini: This tribe includes five genera on western Pacific islands and in southern Indochina. These genera are *Ceratobatrachus* (one species), *Discodeles* (five species), *Ingerana* (five species), *Palmatorappia* (one species), and *Platymantis* (about 50 species). There are 20–26 known chromosome numbers in the latter genus. Most species in this tribe have dilated digital tips, commonly with a dorsoterminal groove. Females lay four to 47 eggs, 0.09–0.2 in (2.2–5.0 mm) in diameter, under objects or in burrows, where they undergo direct development.

- Conrauini: The unique genus *Conraua* (six species), from tropical sub-Saharan Africa, includes *Conraua goliath*, the largest living anuran (reaching 12.6 in, or 320 mm, in snout-vent length). This genus is characterized by an exceptional development of the procoracoid cartilage and by the medial divergence of the coracoids and clavicles. The diploid chromosome number of *Conraua crassipes* is 26. Males of this genus lack vocal sacs, but they have a unique mode of calling, a strident whistling emitted with an open mouth. Rheophilic tadpoles have 7–14/6–12 labial tooth rows. The current assignment of this tribe to the Dicroglossinae is only tentative.

- Dicroglossini: This group is composed of six genera distributed in southern Asia (India and neighboring countries). Three genera (*Minervarya*, one species; *Nannophrys*, three species; *Sphaerotheca*, three species) are restricted to this region. The distributions of the other three genera include other regions: *Fejervarya* (about 20 species) also occurs in most of the Oriental region, *Euphlyctis* (three species) occurs in the Near East and the Arabian peninsula, and *Hoplobatrachus* (six species) is found in China, Indochina, and tropical Africa. The chromosome numbers are 26 and 52 (tetraploid *Hoplobatrachus*). *Hoplobatrachus* is one of the few well-supported genus-group taxa for which there is substantial of a distribution that includes tropical Asia and tropical Africa but not the Near East. Besides molecular and morphologic data, the monophyly of this genus is supported by its unique tadpole, which has strong jaw sheaths and double tooth rows, a unique character in the Ranoidea. All species of this tribe are exotrophic, tadpoles have 1–5/2–6 rows of labial teeth. Tadpoles of the strange, crevice-dwelling genus *Nannophrys* are semiterrestrial, with elongated bodies and low tail fins, a condition that is strikingly convergent with that of the genus *Indirana*, a member of a distinct ranid lineage. Adults of the aquatic *Euphlyctis* genus retain the larval lateral-line system.

- Limnonectini: This tribe encompasses two genera from Indochina, southern China, and the western Pacific islands: *Limnonectes* (about 50 species) and *Taylorana* (six species). Known chromosome numbers are 22–26. Many species have dilated digital tips with a dorsoterminal groove and various combinations of unusual male secondary sex characters, including the absence of nuptial pads and vocal sacs or an advertisement call, fanglike odontoids on the lower jaw, an enlarged head, and a knob on the posterodorsal part of the head. Most *Limnonectes* have free-swimming tadpoles with 1–3/1–3 rows of labial teeth; some species exhibit parental care, and some might undergo endotrophic development in the female genital tract. Males of the genus *Taylorana* dig nests in the mud, where they call; females deposit five to 13 large eggs (0.12 in, or 3 mm, in diameter) that undergo direct development.

- Paini: This group includes two genera in south-central Asia (through the Himalayas from Afghanistan to eastern China): *Chaparana* (six species) and *Paa* (about 30 species). Despite the fact that the omosternum is not forked at the base and the coracoids do not overlap but are connected by epicoracoid cartilage, several other morphologic characters as well as molecular data suggest that these frogs are members of the Dicroglossinae. Known chromosome numbers in *Paa* are 26 and 64; the latter most likely is a polyploid. Most species in this tribe have various combinations of unusual male secondary sex characters, including keratinized spines on the first three fingers, chest, belly, and forearms; hypertrophied forearms (*Paa*); or differentiated skin, sometimes bearing spines, around the vent (some *Chaparana*). These characters presumably facilitate breeding in fast-flowing waters up to elevations of more than 13,123 ft (4,000 m) in the Himalayas. Tadpoles have three to nine rows of labial teeth on the anterior lip but only three rows on the posterior lip.

Lankanectinae

The aquatic *Lankanectes corrugatus* from Sri Lanka has an unusual combination of characters, such as a forked omosternum; skin folds on the head, body and limbs; retention of the lateral-line system in adults; and vocal sacs and fanglike odontoids on the lower jaw of males but absence of nuptial pads. The tadpole has 2/3 rows of labial teeth.

Micrixalinae

The genus *Micrixalus* (about 10 species) is endemic to the Western Ghats in southern India. These small frogs have nonoverlapping coracoids, an unforked omosternum, T-shaped terminal phalanges, no vomerine teeth or femoral glands, a tongue that usually has a median process, dorsolateral folds, smooth ventral skin, and tadpoles with 1/0 rows of labial teeth.

Nyctibatrachinae

The genus *Nyctibatrachus* (about 12 species) is found in the Western Ghats of India. These are the only known ranids with a vertical pupil. Other notable characters are the presence of femoral glands, T-shaped terminal phalanges, a forked omosternum, and, in tadpoles, jaw sheaths and numerous papillae on the labia but no labial teeth.

Occidozyginae

This subfamily includes *Occidozyga* (one species) and *Phrynoglossus* (10 species) in southeastern Asia and the western Pacific islands. This group is a well-characterized clade in most molecular phylogenetic analyses, some of which suggest that it could be the sister group to all other Ranoidae (including the Mantellidae and Rhacophoridae). The oral disc of the tadpole is reduced to a fleshy rim, without papillae, labial teeth, or an upper jaw sheath; the lower jaw sheath is horseshoe-shaped. The omosternum is forked basally, and vomerine teeth are absent. The chromosome number is 26. Adult *Occidozyga* retain the lateral-line system and have axillary amplexus. Adults of *Phrynoglossus* do not have lateral lines and engage in inguinal amplexus.

Petropedetinae

This group includes seven genera in sub-Saharan Africa: *Arthroleptides* (two species), *Dimorphognathus* (one species), *Ericabatrachus* (one species), *Natalobatrachus* (one species), *Petropedetes* (seven species), *Phrynobatrachus* (about 70 species), and *Phrynodon* (three species). This subfamily of frogs includes some of the smallest known anurans (0.4 in, or 10 mm, among some *Phrynobatrachus*). The terminal phalanges of the digits are T-shaped, and femoral glands may be present. Known chromosome numbers are 26 (*Petropedetes*), 24 (*Dimorphognathus*), and 16–20 (*Phrynobatrachus*). Development is exotrophic in most genera; tadpoles have 1–7/2–6 rows of labial teeth. *Phrynodon sandersoni* frogs lay 12–17 eggs, 0.09 in (2.3 mm) in diameter, on vegetation above ground, where they undergo direct development with the mother remaining in the vicinity.

Ptychadeninae

Three genera from sub-Saharan Africa, *Hildebrandtia* (three species), *Lanzarana* (one species), and *Ptychadena* (about 50 species), share several apomorphic characters, including the loss of the neopalatines; reduced clavicles that usually are fused with the anterior borders of the coracoids; a short, compact, bony metasternal style, fusing of the eighth presacral and sacral vertebrae, and reduction or absence of the otic plate of the squamosal. Frogs of the genus *Ptychadena* have 24 chromosomes. All species are exotrophic. Tadpoles of *Hildebrandtia* have 0/2 rows of labial teeth, and those of *Ptychadena* have 1–3/2 rows.

Pyxicephalinae

This group consists of two genera with widely different habitats and modes of life: *Aubria* (two species) is aquatic and lives in humid tropical forests in central Africa, whereas *Pyxicephalus* (three species), which is adapted to burrowing during dry seasons, occurs in savannas and semiarid to arid habitats in eastern and southern Africa. These two genera share several skeletal apomorphies, femoral glands, and tad-

The crawfish frog (*Rana areolata*) is secretive and resides underground most of the year, frequently using crawfish burrows. It emerges in early spring to breed in temporary ponds. (Photo by Janalee P. Caldwell. Reproduced by permission.)

poles swimming in compact schools, sometimes under the protection of adults that will attack potential predators. *Pyxicephalus adspersus* has 26 chromosomes. Tadpoles have 5/3 rows of labial teeth.

Raninae

This is the largest group of Ranidae. The coracoids do not overlap, the omosternum is not forked at the base, and the nasals usually are small and not in contact medially. These frogs typically have dorsolateral folds and deposit large numbers of small, pigmented eggs. The group includes two provisional tribes:

- Amolopini: This group contains several taxa that are treated by different authors as distinct genera, subgenera, or species groups. The two major ones are the *Amolops* group (about 40 species in four subgenera or genera: *Amo, Amolops, Huia,* and *Meristogenys*) and the *Odorrana* group (about 30 species in at least three subgenera or genera: *Chalcorana, Eburana,* and *Odorrana*). These frogs live in or along swift torrents in the Himalayas and mountains of southern and eastern China, in Indochina, and on the western Pacific islands. They have long hind limbs, smooth venters, and large digital discs with ventrolateral grooves. Tadpoles of the *Odorrana* group have generalized mouthparts and 4–5/3–5 rows of labial teeth. The gastromyzophorous tadpoles of the *Amolops* group are adapted even better to torrent life. They have 4–12/3–10 rows of labial teeth and a sucker on the anterior part of the belly; additionally, they usually have integumentary glands on the body and tail and keratinized spinules in the skin. The known chromosome numbers are 26 and 27 (sexually dimorphic in some *Amolops*).

- Ranini: This is an unresolved, partly catch-all tribe that includes *Batrachylodes* (eight species) on the Solomon Islands, *Nanorana* (three species) in Tibet and surrounding high mountains, *Staurois* (three species) in the western Pacific, and a heterogeneous, nearly cosmopolitan assemblage traditionally called *Rana* (about 200 species), among which more than 35 subgenera and species groups are recognized. These groups include all the well-known Palearctic Ranidae (subgenera *Pelophylax* and *Rana* sensu stricto) and American Ranidae (subgenera *Amerana*, *Aquarana*, *Lithobates*, *Pantherana*, and related groups). They also encompass a few African groups (subgenera *Afrana* and *Amietia*) and numerous Asiatic groups (subgenera *Babina*, *Glandirana*, *Nasirana*, *Nidirana*, *Pseudoamolops*, *Pseudorana*, *Pterorana*, *Rugosa*, and several other poorly known taxa) as well as a few groups present in both regions (*Hylarana* and related groups). Chromosome numbers are 24, 26, and 27 as well as 39 and 52 in the triploid or tetraploid European *Pelophylax*. There are 1–10/2–9 labial tooth rows. Several groups in this tribe have T-shaped terminal phalanges with enlarged digit tips or even fully differentiated discs, but they always have ventrolateral grooves. Many adult males of these species have dorsolateral folds or particular macroglands or other secondary sex characters, such as humeral glands in *Hylarana* and related groups; suprabrachial glands in *Babina* and *Nidirana*; dagger-like prepollex (both sexes) in *Babina*; and vocal sacs, nuptial pads, and advertisement calls. Some *Babina* and *Nidirana* that lay eggs in mud nests engage in simple parental care. In *Batrachylodes*, there is probably direct development of terrestrial eggs.

Ranixalinae

The genus *Indirana* (about 10 species) occurs in the Western Ghats in southern India. These frogs are characterized by their unusual Y-shaped terminal phalanges, digital discs, femoral glands, and semiterrestrial tadpoles with 3–5/3–4 rows of labial teeth, elongated bodies, and low tail fins, which can make long jumps on the ground to escape predators.

Physical characteristics

Few derived characters are common to all groups currently included in the Ranidae. These frogs usually have a firmisternal pectoral girdle, in which the coracoids do not overlap and are connected by an epicoracoid cartilage; however, some groups have a pseudofirmisternal pectoral girdle, in which the coracoids do overlap and are fused to each other. The omosternum usually is ossified and may be forked or unforked basally. The metasternum generally is ossified. There are eight procelous presacral vertebrae, and the last two presacrals are not fused. The sacral diapophyses are cylindrical or slightly dilated. The carpal bones are composed of six elements: the first and second carpals and the first centrale are free, and the third carpal is fused with the fourth and fifth carpals and with the second centrale. Intercalary elements are absent between the penultimate and terminal phalanges of the digits. The terminal phalanges may be simple, slightly dilated, and T-shaped or Y-shaped. All species have teeth along the upper jaw, and most of them have vomerine teeth or ridges.

The musculus sartorius is distinct from the musculus semitendinosus, and the tendon of the latter passes dorsal to the musculus gracilis. The musculus cutaneus pectoris usually is present. Most other characters vary.

Most species have 26 chromosomes, but 16–27 are known in nonpolyploid taxa. Polyploidy is not uncommon; tetraploidy (52 chromosomes) has been reported in several groups (*Hoplobatrachus*, *Pelophylax*, and *Tomopterna*), and some members of *Paa* have up to 64 chromosomes. Triploidy (39 chromosomes) is common in European members of *Pelophylax*.

The snout-vent length varies from 0.4 in (10 mm) in several minute African frogs (*Arthroleptella* and *Phrynobatrachus*) to more than 12.2 in (310 mm) in the African *Conraua* and the large *Discodeles* in the Solomons and neighboring islands. Many aquatic or terrestrial species are 1.6–3.3 in (41–84 mm) in length and have elongated bodies and long limbs. Some stream-adapted forms (e.g., *Amolops* and *Odorrana*) have particularly long hind limbs, whereas a few burrowing taxa (*Pyxicephalus*, *Tomopterna*, *Sphaerotheca*, and *Hildebrandtia*) have short, toadlike bodies with short limbs, spadelike metatarsal tubercles, and, sometimes, strongly ossified skulls. On the other hand, a few high-altitude taxa (*Nanorana*) have partly uncalcified skeletons.

Most ranids have moderate to extensive webbing that extends proximally between the outer metatarsals, but primarily terrestrial or semiarboreal and endotrophic taxa (e.g., *Platymantis*) have reduced webbing, and their outer metatarsals are not separated from each other. Many semiarboreal species or ones that live along or in running water have dilated digital tips or even differentiated terminal discs with grooves. Several kinds of digital discs, which are probably not homologous, exist in the Ranidae: dicroglossines have dorsoterminal grooves, and ranines have ventrolateral grooves. The latter condition includes a completely closed ventral "cell" bordered by a groove below the extremity of the digit; this feature occurs in some genera or subgenera of Raninae (*Amo* and *Staurois*) as well as in the Rhacophoridae and other arboreal families.

Different kinds of glandular structures may be present in the skin (e.g., various warts and folds on the dorsum and flanks, supratympanic folds, dorsolateral folds on the dorsum, rictal glands at the mouth commissure, humeral glands on the upper arm and suprabrachial glands above the arm insertion in males, and femoral glands on the posterior surfaces of the thighs in males or in both sexes). Taxa that climb on rocks (e.g., *Ingerana* and *Staurois*) have a granular venter similar to that in other arboreal families (Hylidae, Hyperoliidae, and Rhacophoridae). Various keratinized structures also arise on the skin, at least seasonally (e.g., nuptial pads bearing a layer of minute spinules on the thumb, or on the first two or three fingers, and on the prepollex, which is dagger-like in both sexes in *Babina*; larger spines in the same places and on the arms, chest, and belly in some stream-breeding species; and spines around the vent).

Males of many species are smaller than the females and have longer hind limbs and more extensive webbing; they also have enlarged forelimbs and internal or external (i.e., protruding

Male African bullfrogs (*Pyxicephalus adspersus*) gather in groups, or leks, during the day to establish territories and attract females. (Photo by Alan Channing. Reproduced by permission.)

through slits during calling) vocal sacs. In a few groups, males display special secondary sex characters, such as fanglike odontoids at the extremity of the lower jaw (e.g., *Lankanectes*, *Limnonectes*, and *Taylorana*), enlarged heads (*Limnonectes*), and knobs on the posterodorsal part of head (*Elachyglossa*).

Coloration varies widely, though rarely (e.g., some *Amolops*, *Odorrana*, and *Pulchrana*) is it as extraordinarily bright as in some other families, such as the Mantellidae or Dendrobatidae. Frogs living or breeding in lentic aquatic habitats tend to be a shade of green, whereas most terrestrial species that live primarily on the forest floor, especially in temperate regions with deciduous trees, commonly are shades of brown, like the color of dead leaves. Species living in savannas and grasslands usually have longitudinal marks or spots on the dorsum, whereas some aquatic species (*Euphlyctis* or *Occidozyga*) have black and white bars at the posterior surfaces of the thighs. Frogs that spend a large part of their lives on rocks in streams (e.g., *Amolops*, *Ingerana*, *Odorrana*, and *Petropedetes*) usually have variegated coloration. The iris coloration varies; generally there is a horizontal dark line continuous with a dark line on the canthus rostralis and another such line on or below the supratympanic fold. This contributes to camouflage of the eye, a particularly visible feature for many vertebrate predators. A dark vertical line also may be present, especially in the lower part of eye. The pupil typically is horizontally oval, but it may be rhomboidal or even vertical in a few species (*Nyctibatrachus*).

Some aquatic ranids (*Euphlyctis*, *Lankanectes*, and *Occidozyga*) show pedomorphic retention of the larval lateral-line system in adults. Tadpoles of ranids have a sinistral spiracle. Most have a generalized anteroventral, ventral, or almost terminal oral apparatus, with complete keratinized structures (upper and lower jaw sheaths and labial teeth) and marginal and submarginal oral papillae. Tadpoles of a few groups lack some of these structures. Occidozyginae lack papillae, tooth rows and the upper jaw sheath; the lower jaw is recessed and semilunar in shape. Nyctibatrachinae lack tooth rows on both lips; Micrixalinae lack them on the lower lip and *Hildebrandtia* on the upper lip. All other groups have rows on both lips. The common number of tooth rows is two on the upper lip and three on the lower lip, but many variations exist. Tooth rows are usually simple, except in the genus *Hoplobatrachus*, in which they are double.

Some groups have very specialized tadpoles. Tadpoles with very elongated bodies and tails with shallow tail fins (*Indirana* and *Nannophrys*) are semiterrestrial and may use the tail or, at later stages, the hind limbs to move over long distances on the ground. Numerous tadpoles that live in streams (*Chaparana*, *Clinotarsus*, *Conraua*, *Nasirana*, *Odorrana* group, *Paa*, *Petropedetes*, and *Pseudorana*) have muscular tails with low fins and a large oral disc with numerous papillae and many tooth rows, up to 14 on the anterior lip and 12 on the posterior lip. Some tadpoles (*Amolops* group and *Pseudoamolops*) have a large sucker that includes the oral disc and extends onto the anterior part of the belly. Tadpoles of some members of the *Amolops* and *Hylarana* groups have dermal glands on the body and sometimes on the tail. Certain ranid tadpoles have a dark tail tip (e.g., *Fejervarya*) or a large, brightly colored ocellus at the lateral base of the tail (*Nasirana*) that may attract predators, thus diverting attack from the head.

Distribution

The family is distributed throughout the Holarctic, except at higher latitudes and elevations, all of Africa except most of the Sahara, the Oriental region and western Pacific islands to Fiji, northern Australia, Central America, and the northern part of South America. It is absent (except as introduced) from Madagascar, New Zealand, New Caledonia and the Pacific islands east of Fiji, most of Australia and South America, the West Indies, many oceanic islands, and the Arctic and Antarctic. Because of the human consumption of frog legs, ranids were introduced by humans into Madagascar (*Hoplobatrachus*) and a number of continental regions and oceanic islands (mostly subgenera *Aquarana* and *Pelophylax*).

Habitat

The popular image of *Rana* as a green frog bathing in the sun on a water lily leaf is misleading. Although most European and North American ranids live close to ponds or lakes or move to these lentic habitats for breeding, this does not apply to the whole family. Many tropical ranids live or breed in slowly flowing or swiftly running water, which often is virtually the only aquatic habitat available. These frogs are most diverse in the tropical and subtropical parts of the Oriental region and in sub-Saharan Africa. Most ranids live in forests or along streams, where, because of the longer survival of natural vegetation and rocky shelters than in open areas, the frogs can survive long after deforestation of the surrounding environment. Some ranids occur in savannas, grasslands, or even high-elevation habitats; however, at high elevations they tend to be aquatic, mostly because of the risk of desiccation by wind. Other aquatic ranids, either in lentic (still-water) or in lotic (running-water) habitats, also inhabit tropical and temperate regions, but most of them are primarily terrestrial, staying close to water much of the time or going to water to breed.

Several taxa that undergo direct development do not need free water for breeding and can spend most of their lives away from water; this has allowed them to conquer several rather dry oceanic islands, such as the Solomons. Although many ranids climb on rocks, bushes, or low branches of trees, none are truly arboreal, as are many members of the related Rhacophoridae, which climb and live mostly in trees. Species of *Ingerana* inhabit rocky cliffs adjacent to cascades, where they live in a permanent mist that provides the moisture necessary for the direct development of their large eggs that presumably are deposited in rock crevices.

Behavior

Most ranids are nocturnal, especially along streams and ponds; thus, they avoid desiccation from sunlight and diurnal predators. Some frogs living close to ponds and lakes, especially in temperate regions, tend to be active by day; they bask in sunlight and periodically enter water. Species living at high elevations also tend to be diurnal, because at night the temperatures may drop too low for them to maintain activity. In tropical forests, ranids feed and breed at night, especially after rains. In savannas and semiarid areas, most species spend the dry season estivating in burrows underground, but a few species concentrate in the few remaining aquatic habitats and may remain active all year round. In humid forests ranids are active most of the year, though breeding is restricted to the rainy periods. In temperate climates, these frogs hibernate either underground or at the bottom of water bodies deep enough to allow the maintenance of a layer of free water below the frozen surface. A few species that occur at high latitudes have antifreeze glycerol-like substances in their tissues.

Territorial behavior is common in ranids. Males of most species call from a permanent site, and they react to the intrusion of another male. The characteristics of the call may change by becoming more aggressive; if the intruder does not leave, physical fighting may ensue (jumping on or biting each other). Some species have hard structures (fanglike odontoids or keratinized spines on the prepollex, fingers, and chest) that may be involved in agonistic behavior. Snakes, birds, and mammals feed on ranids. Large frogs may even feed on smaller ones, including their young, and in natural habitats the young often live in different areas or are not active at the same time as adults, presumably to limit this kind of predation.

Many species are cryptically colored and remain motionless to avoid predation. In several taxa (e.g., *Fejervarya* and *Phrynobatrachus*) in which many individuals exist in close proximity around ponds or in grass, color polymorphism (e.g., with or without colored spots, lines, or bands on the dorsum) may create a visual search image for some predators, like birds, so that predation is concentrated for some time on the most common morph. This results in a gradual change in the relative frequencies of morphs over time.

A few large ranids, such as *Pyxicephalus*, may attack their potential predators or those of their larvae and bite them, but most ranids avoid predation by escape behavior. In species that aggregate around ponds, diving into water often is accompanied by an expulsion of air from the lungs; this alarm or warning call usually prompts nearby individuals to jump into the water. In ponds with soft muddy or sandy bottoms, frogs may hide in the substrate before surfacing, but in streams with nude rocky bottoms devoid of shelters, frogs commonly let the current carry them downstream, where they swim to the bank and remain motionless, protected by their coloration. A few pond-dwelling ranids (*Euphlyctis cyanophlyctis* and *Rana erythraea*) can skitter on the surface of water; they may even start from one bank, cross the pond, and jump onto the opposite bank.

Terrestrial frogs (e.g., *Ptychadena* and *Rana sensu stricto*) may leap quickly and repeatedly on the ground over several meters without stopping and disappear from sight within a few seconds. Ranid tadpoles usually are swift swimmers and escape predators by dispersing in many directions, sometimes even jumping above the water surface. In a few genera (*Aubria* and *Pyxicephalus*), however, tadpoles from the same clutch tend to remain tightly grouped in ball-like schools, which presumably reduces predation.

Feeding ecology and diet

Most ranids are sit-and-wait predators and feed on a wide variety of prey, primarily invertebrates, but the kind of prey depends mostly on the habitats of the frogs. Some terrestrial species tend to move around, both by day and night, likely increasing their chances of finding prey. Large species tend to eat larger prey, including small vertebrates (birds, mammals, reptiles, and other frogs, even their own young). Tadpoles usually feed by rasping food from the substrate with their keratinized mouthparts. Some (*Hoplobatrachus*) are carnivorous and may feed on heterospecific or conspecific tadpoles.

Reproductive biology

Most ranids have seasonal breeding activity. In temperate climates, breeding usually occurs once a year; in Europe and North America, depending on the species, breeding takes place in early spring, immediately after the melting of ice and snow, or in late spring, when waters are much warmer. In the tropics breeding may happen several times a year, usually at the beginning or end of rainy seasons. In species that breed in standing water (permanent ponds, lakes, or paddy fields) or in temporary waters (marshes or rain pools), calling males aggregate at the breeding site. Choruses may be very loud and audible from considerable distances. Different males often tend to synchronize their calls; if calling is stopped by some perturbation, it is often the same male that reinitiates calling and is followed quickly by the others. Females move to the breeding chorus only when they are ready to ovulate. They may be intercepted on their way to the chorus by peripheral (so-called satellite or parasite) silent males that benefit from the others' calls. In species that breed in lotic aquatic habitats, such as streams or torrents, calling males usually are scattered along the stream, and their calls, which are not synchronized, commonly are sequences of pure notes separated by long intervals.

Amplexus usually is axillary, but in *Phrynoglossus* it is inguinal. Egg sizes and numbers vary widely and correlate negatively for frogs of the same size. Egg diameter varies from 0.04 in (1 mm) in many small species to 0.2 in (5 mm) in *Ceratobatrachus guentheri* and eggs number from a few to about 20,000 in *Rana catesbeiana*. Fertilization is external, except perhaps in some possibly ovoviviparous *Limnonectes* and in a few poorly known members of the Paini group (*Annandia* and *Ombrana*), in which males have a ventrally directed vent surrounded by spines and females have a dorsally directed vent, thus suggesting the possibility of internal fertilization.

Eggs that are deposited in open waters have a pigmented animal pole (brown or black), which probably contributes to their heating and to protection from ultraviolet radiation. Eggs that are hidden under shelters are unpigmented. In some taxa, all ripe eggs are emitted as a single clutch, which is fertilized synchronously by the male, but in other taxa the ovarian complement is partitioned into several clutches deposited at different times in different places. Egg clutches of different females may be grouped together or isolated, often attached to vegetation or as a surface film on water. In temperate regions or at high altitudes, egg clutches are laid preferentially in shallow waters, which are much warmer than the deep parts of the ponds. The eggs of torrent-breeding frogs may be stuck by their jelly under stones or big rocks in a swift-running, richly oxygenated part of the stream. In a few species of the genus *Limnonectes*, eggs are deposited under decaying vegetation on the forest floor, where the male stays by them for a few days before carrying them to a pool or a stream, a form of behavior resembling larval transport in Dendrobatidae.

A few species build nests for their eggs. In *Babina* and some species of *Nidirana*, eggs deposited in water inside nests in paddy fields or marshes hatch as tadpoles. In *Anhydrophryne*, *Arthroleptella*, and *Taylorana*, males dig holes in mud under dead leaves or rocks, where the eggs are deposited and undergo direct development. Other direct-developing ranids deposit eggs in rock crevices (*Discodeles* and probably *Ingerana*) or on vegetation above ground (*Phrynodon*). In the latter genus the female remains near the clutch until the froglets hatch; she sometimes urinates on the eggs, behavior that not only moistens the eggs but also may protect them against fungi and parasites.

Most ranids have aquatic eggs that hatch as free-swimming tadpoles, but direct development has evolved independently in several lineages. Direct development is known in the African Cacosterninae (*Anhydrophryne* and *Arthroleptella*) and Petropedetinae (*Phrynodon*) and in the Asian Dicroglossinae Ceratobatrachini (all five genera) and Limnonectini (*Taylorana*). It also is suspected in the Asian *Batrachylodes*. At least one species of *Limnonectes* in Sulawesi might undergo direct development within the female's genital tract. In some species of *Platymantis* and *Discodeles opisthodon*, embryos have several folds on the sides of the belly, which probably serve as respiratory devices, and a hard conical tubercle is present at the extremity of snout, which allows the froglet to pierce the egg capsule at hatching.

Conservation status

According to the IUCN, three species are Extinct: *Arthroleptides dutoiti*, *Rana fisheri*, and *R. tlaloci*. In addition, seven species are Critically Endangered; six are Endangered; 14 are Vulnerable; four are Lower Risk/Near Threatened; and 12 are Data Deficient. Like most anurans, many ranids are threatened with population declines and extinction. Even in pristine national parks in North America, populations of *Rana* have declined drastically or have become extinct, possibly because of acid rains, increased ultraviolet irradiation, or spreading of pathogens or parasites. Introduction of fishes (especially salmonids) into mountain lakes and even some frogs (such as the aquatic pipid *Xenopus*) or large ranids (for example, *Rana catesbeiana* or *Hoplobatrachus tigerinus*) into fragile ecosystems have had deleterious effects on local populations of many ranids. Capture by humans for frog leg consumption has drastically reduced populations of medium-size to large species in some parts of Europe, northern Africa, and Asia.

Significance to humans

Frog legs have long been considered a delicacy in France, where their consumption used to be a regional and seasonal tradition, linked to the breeding period of brown frogs (*Rana temporaria*) and green frogs (*Pelophylax*) in other regions. Subsequent to deep-freezing technology, this consumption has become more widespread, especially in Europe and North America. Because frog "farming" is not profitable, this increased consumption is weighted more and more on natural populations of frogs, especially in southern and southeastern Asia, but several countries now limit this commerce, which is restricted by the Washington Convention for a few ranids.

Local consumption by some ethnic groups of whole frogs (not just the legs), often in soups, is a tradition in several tropical countries of Asia and Africa. In southern Africa, adults and larvae of *Pyxicephalus* are considered a great delicacy. Some species (e.g., members of the genus *Paa* in the Himalayas or adult males of *Elachyglossa* in Indochina) are considered to have medicinal value. The use of ranids, especially some European and North American *Rana*, has contributed greatly to the growth of descriptive and experimental embryology and teratology, and thus to the understanding of the development of vertebrates, and to the perfection of our techniques of intervention in this field.

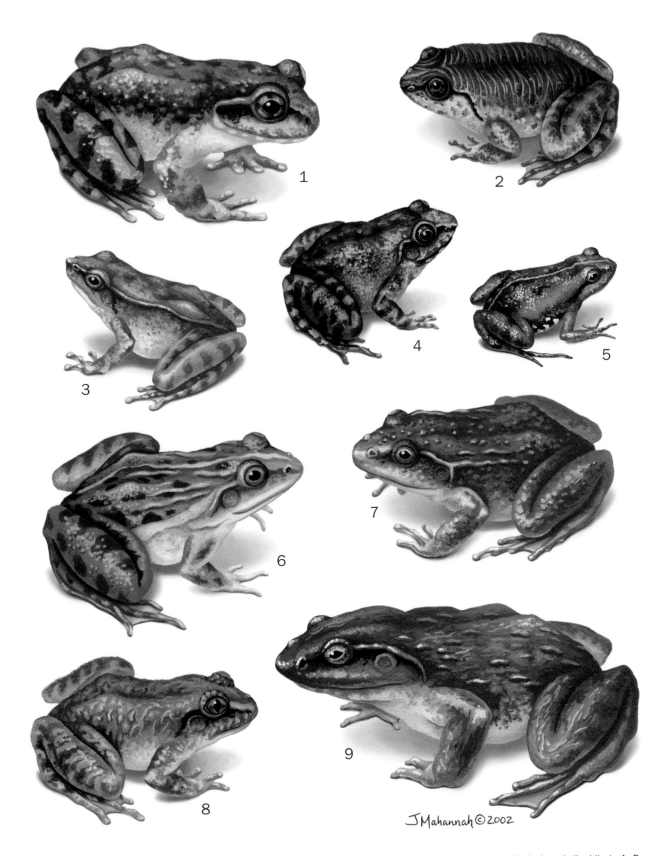

1. Spiny-armed frog (*Paa liebigii*); 2. Corrugated water frog (*Lankanectes corrugatus*); 3. Nilgiri tropical frog (*Micrixalus phyllophilus*); 4. Penang Taylor's frog (*Taylorana hascheana*); 5. Micro frog (*Microbatrachella capensis*); 6. Indian tiger frog (*Hoplobatrachus tigerinus*); 7. Faro webbed frog (*Discodeles opisthodon*); 8. Malabar night frog (*Nyctibatrachus major*); 9. Goliath frog (*Conraua goliath*). (Illustration by Jacqueline Mahannah)

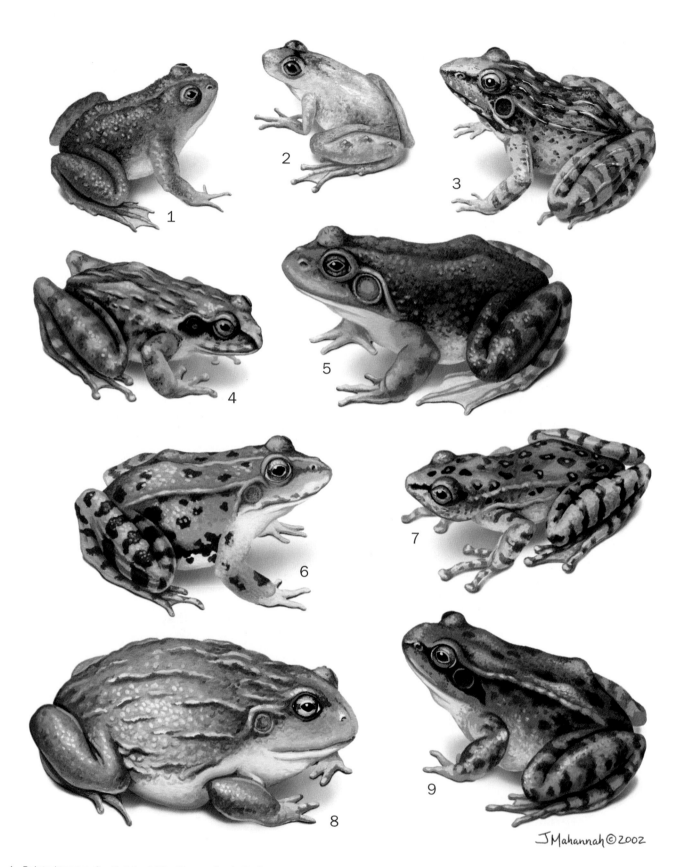

1. Pointed-tongue floating frog (*Occidozyga lima*); 2. Sanderson's hook frog (*Phrynodon sandersoni*); 3. Sharp-nosed grass frog (*Ptychadena oxyrhynchus*); 4. Beddome's Indian frog (*Indirana beddomii*); 5. Bullfrog (*Rana catesbeiana*); 6. Roesel's green frog (*Rana esculenta*); 7. Beautiful torrent frog (*Amolops formosus*); 8. African bullfrog (*Pyxicephalus adspersus*); 9. Brown frog (*Rana temporaria*). (Illustration by Jacqueline Mahannah)

Species accounts

Micro frog
Microbatrachella capensis

SUBFAMILY
Cacosterninae

TAXONOMY
Phrynobatrachus capensis Boulenger, 1910, Cape Flats, Cape Province, Republic of South Africa.

OTHER COMMON NAMES
None known.

PHYSICAL CHARACTERISTICS
This is one of the smallest anuran species in the world, with a snout-vent length in the adult of 0.4–0.7 in (10–18 mm). The dorsal coloration varies from pale to dark green, gray, fawn, russet, or black, with a dark line from the eye to the armpit, often with a narrow pale or green vertebral stripe, and sometimes with broad lateral stripes or speckles. The dorsum is slightly warty. The ventral surface is smooth, with black and white mottling. Webbing is present but leaves free two or three phalanges of the fourth toe. The male vocal sac extends over half the ventral surface and is blown out almost to the size of the body during call.

DISTRIBUTION
This species exists in the coastal lowlands of southwestern Cape Province, South Africa.

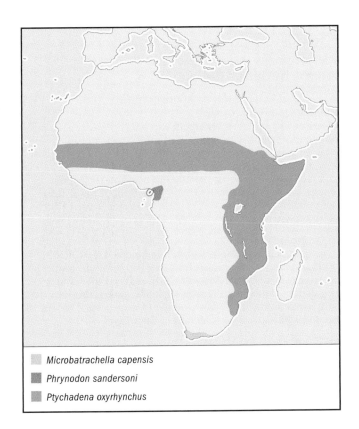

Microbatrachella capensis
Phrynodon sandersoni
Ptychadena oxyrhynchus

HABITAT
The frogs live around temporary acidic pools and ponds and in decaying roots.

BEHAVIOR
Little is known.

FEEDING ECOLOGY AND DIET
Little is known.

REPRODUCTIVE BIOLOGY
Males call from half-submerged sites in the marginal vegetation. Calls consist of a series of five to six scratches ("tschik, tschik, tschik"). The very tiny pigmented eggs are deposited in June and July in clusters of about 20, attached to vegetation below the water surface in shallow pools. Benthonic tadpoles have a rather low tail fin and 3/3 tooth rows. They reach 1 in (25 mm) in length, with a tail length of 0.7 in (18 mm). Metamorphosis takes place in December.

CONSERVATION STATUS
The species is listed as Endangered according to the IUCN. It is threatened by habitat destruction and pollution over its restricted, and apparently decreasing, range, which covers only a small area of coastal lowlands.

SIGNIFICANCE TO HUMANS
None known. ◆

Faro webbed frog
Discodeles opisthodon

SUBFAMILY
Dicroglossinae, tribe Ceratobatrachini

TAXONOMY
Rana opisthodon Boulenger, 1884, Treasury and Faro Islands, Solomon Islands.

OTHER COMMON NAMES
None known.

PHYSICAL CHARACTERISTICS
This is a large species, with a snout-vent length reaching 3 in (80 mm) in males and 5 in (125 mm) in females. The skin is smooth or warty, with olive or dark brown coloration on top. The hind limbs are short, the feet are incompletely webbed, and the digital tips are dilated into small discs. The tongue shows a median process. Males have internal vocal sacs.

DISTRIBUTION
This frog lives on the Solomon Islands.

HABITAT
Little is known.

BEHAVIOR
Little is known.

FEEDING ECOLOGY AND DIET
Little is known.

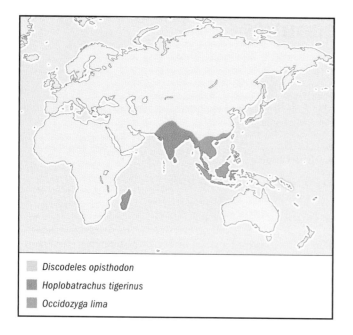

Discodeles opisthodon

Hoplobatrachus tigerinus

Occidozyga lima

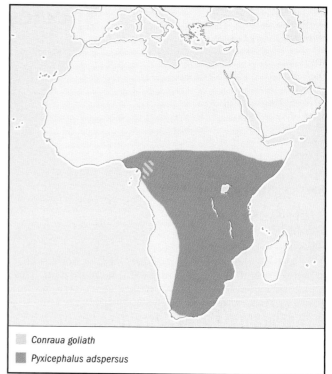

Conraua goliath

Pyxicephalus adspersus

REPRODUCTIVE BIOLOGY
Eggs, which are very large (at least 0.2 in, or 5 mm, in diameter), are deposited in moist crevices of rocks close to water. Development takes place within these transparent gelatinous balls. The embryos lack tail and gills, but on each side of the abdomen there are several regular transverse folds with a respiratory function. The tip of the snout of these tiny frogs bears a small conical protuberance, projecting slightly through the delicate envelope of the egg and used to perforate this envelope.

CONSERVATION STATUS
Not listed by the IUCN.

SIGNIFICANCE TO HUMANS
None known. ◆

Goliath frog
Conraua goliath

SUBFAMILY
Dicroglossinae, tribe Conrauini

TAXONOMY
Rana goliath Boulenger, 1906, Efulen, Cameroon.

OTHER COMMON NAMES
German: Goliatfrosch.

PHYSICAL CHARACTERISTICS
At 12.6 in (320 mm) in snout-vent length and 7 lb (3.25 kg), this is the largest species of frog still living on our planet. It is dark gray dorsally, with some spots and faintly visible dark bars on the limbs and lips; the ventral coloration is light. The skin on the dorsum and limbs is finely granular. The hind limbs are long, the hand shows slight webbing at the base of the fingers (especially between the first and second), and the foot has complete webbing, without incurvation between the toe tips, which are dilated.

DISTRIBUTION
The distribution of the goliath frog is from southern Cameroon to equatorial Guinea.

HABITAT
These frogs live in rapids and cascades of rivers in equatorial forest.

BEHAVIOR
Little is known.

FEEDING ECOLOGY AND DIET
Little is known.

REPRODUCTIVE BIOLOGY
Males of this and other species of the genus *Conraua* are devoid of vocal sacs but have developed a unique mode of calling; they emit long and powerful whistles with the mouth slightly open. Egg masses containing several hundred pigmented eggs, 0.14 in (3.5 mm) in diameter, are attached to plants in rocky pools among the rapids. Tadpoles have 7–8/5–8 tooth rows, numerous papillae, and a low tail fin. They can reach a size of 1.9 in (47 mm). Larval development takes between 85 and 95 days.

CONSERVATION STATUS
This species is considered Vulnerable by the IUCN. Recent overharvesting of the species for food, the pet trade, and habitat alteration by humans seem to have reduced the number of populations drastically, and the remainders seem threatened with extinction within a short period of time despite official legal and administrative protection of the species.

SIGNIFICANCE TO HUMANS
These frogs traditionally are hunted as food by the local people, who often approach them by boat on the river and fire at them with guns before they can jump into the water. The species also has been collected live to serve as pets for terrarium keepers in North America. ◆

Indian tiger frog
Hoplobatrachus tigerinus

SUBFAMILY
Dicroglossinae, tribe Dicroglossini

TAXONOMY
Rana tigerina Daudin, 1802, Bengal, India.

OTHER COMMON NAMES
English: Indian bullfrog, tiger Peters frog; German: Tiger-frosch, Asiatischer Ochsenfrosch.

PHYSICAL CHARACTERISTICS
This is a large frog species, with a snout-vent length up to 4.3 in (110 mm) in males and 6.3 in (160 mm) in females. The frogs have greenish coloration, longitudinal skin folds on the dorsum, and strong hind limbs with large webbing. Males show nuptial pads on the first finger and vocal sacs on both ventral sides of the throat, forming bluish longitudinal folds.

DISTRIBUTION
This frog occurs in Bangladesh, Bhutan, India, Myanmar, Nepal, and Pakistan; it was introduced into Madagascar.

HABITAT
The species lives around ponds and in paddy fields.

BEHAVIOR
Little is known.

FEEDING ECOLOGY AND DIET
Little is known.

REPRODUCTIVE BIOLOGY
At the beginning of the monsoon, breeding males, which have bright yellow coloration, gather around standing waters, where they emit loud calls that attract females. The pigmented eggs are small and numerous. Tadpoles show very strong jaw sheaths and have 3–4/4–5 double tooth rows; they are carnivorous.

CONSERVATION STATUS
This species is not listed by the IUCN. However, overexploitation of this species for frog leg consumption has resulted in steep declines in populations, especially in northern India, which has resulted in a striking increase in pest populations in paddy fields. Recent legal protection of this species has limited this decline, though it has not suppressed it entirely.

SIGNIFICANCE TO HUMANS
None known. ◆

Penang Taylor's frog
Taylorana hascheana

SUBFAMILY
Dicroglossinae, tribe Limnonectini

TAXONOMY
Polypedates hascheanus Stoliczka, 1870, Penang Island, Malaysia.

OTHER COMMON NAMES
None known.

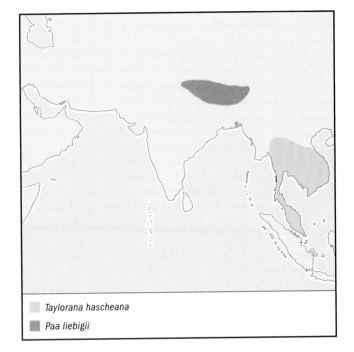

Taylorana hascheana

Paa liebigii

PHYSICAL CHARACTERISTICS
This is a small species, reaching 1.5 in (37 mm) in snout-vent length. It is yellow, orange, or brown dorsally, with dark brown spots and a mid-dorsal chevron, sometimes with a yellow vertebral streak. The extremities of the digits are dilated into small discs with dorsoterminal grooves. The feet are poorly webbed. The male is devoid of nuptial pads and vocal sacs but has a pair of fanglike odontoids on the lower jaw.

DISTRIBUTION
This species is found in Cambodia, Laos, Malaysia, Thailand, and Vietnam.

HABITAT
The frog inhabits the forest floor, often close to small rivers.

BEHAVIOR
Little is known.

FEEDING ECOLOGY AND DIET
Little is known.

REPRODUCTIVE BIOLOGY
Males dig holes in the mud under dead leaves, from where they emit isolated short notes ("kra") of 250–400 msec, separated by silences of 30 sec to several minutes. Females meet them in these "nests," where they deposit five to 13 large whitish eggs (0.12 in, or 3 mm, in diameter). The complete development takes place within the egg, from which tiny froglets emerge about one month after egg laying.

CONSERVATION STATUS
Not listed by IUCN.

SIGNIFICANCE TO HUMANS
None known. ◆

Spiny-armed frog
Paa liebigii

SUBFAMILY
Dicroglossinae, tribe Paini

TAXONOMY
Rana liebigii Günther, 1860, Nepal.

OTHER COMMON NAMES
None known.

PHYSICAL CHARACTERISTICS
This is a large species, with a snout-vent length up to 4.6 in
(117 mm). The frog is brown, yellow, reddish, or blackish in
color, with long hind limbs and well-developed, but incomplete
webbing. The iris is bright gold, with a horizontal and a vertical dark line forming a cross in the eye. Breeding adult males
have very large forelimbs and large black spines on the prepollex, the first three fingers, the arm, the forearm, and both
sides of the breast.

DISTRIBUTION
This species exists in Bhutan, western China (Xizang), northern India, and Nepal.

HABITAT
These frogs live along torrents from 5,000 ft (1,520 m) to
11,500 ft (3,510 m) in forested and nonforested areas.

BEHAVIOR
Little is known.

FEEDING ECOLOGY AND DIET
Little is known.

REPRODUCTIVE BIOLOGY
Breeding males call at night from below rocks or from the
banks of fast-running torrents. Their call consists of a long
(2.3–4.6 sec) series of 15–27 pure notes separated by long silences (10.1–37.2 sec), which are easier for females to locate
than continuous noisy calls would be. Amplexus and egg deposition take place below rocks in oxygenated parts of the torrent. The eggs are large (0.2 in, or 5 mm, in diameter) and
only slightly colored at the animal pole; inside, a sticky jelly
maintains them attached to rocks in the current. The tadpoles
have strong tail muscles, low tail fins, and 3–6/3 tooth rows.

CONSERVATION STATUS
Not listed by IUCN.

SIGNIFICANCE TO HUMANS
In central Nepal, women eat females of this species to relieve
abdominal pain. ◆

Corrugated water frog
Lankanectes corrugatus

SUBFAMILY
Lankanectinae

TAXONOMY
Rana corrugata Peters, 1863, Sri Lanka.

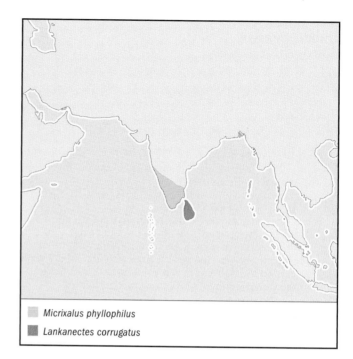

Micrixalus phyllophilus

Lankanectes corrugatus

OTHER COMMON NAMES
None known.

PHYSICAL CHARACTERISTICS
This is a stout, medium-size species, with a snout-vent length
up to 2.6 in (65 mm). The hind limbs are short and thick, with
broad but not complete webbing and slightly dilated tips. The
dorsal coloration is brown or brownish orange with dark spots.
The dorsal parts of the head and body are covered with a network of ridges, and the larval lateral-line system persists in
adults. Adult males have odontoid fangs on the lower jaw and
internal vocal sacs.

DISTRIBUTION
This species is found in Sri Lanka.

HABITAT
These frogs live along shaded, slow-flowing streams and
marshes in forested areas at elevations of 200–5,000 ft
(60–1,525 m).

BEHAVIOR
Little is known.

FEEDING ECOLOGY AND DIET
Little is known.

REPRODUCTIVE BIOLOGY
Males of this species emit dull advertisement calls ("urrm")
that can be heard from several meters in the forest habitat.
Tadpoles have 2/3 tooth rows and are 1 in (26 mm) long when
the hind limbs are fully developed.

CONSERVATION STATUS
Not listed by IUCN.

SIGNIFICANCE TO HUMANS
None known. ◆

Nilgiri tropical frog
Micrixalus phyllophilus

SUBFAMILY
Micrixalinae

TAXONOMY
Limnodytes phyllophila Jerdon, 1853, Nilgiris, southern India.

OTHER COMMON NAMES
None known.

PHYSICAL CHARACTERISTICS
This small species (1.25 in or 3.175 cm long) is brownish, with smooth dorsal skin and narrow dorsolateral folds. The hind part of the abdomen and the lower side of the legs are rose colored. Vomerine teeth are absent; the tongue bears a median process. The hind limbs are of medium length, the toes are nearly entirely webbed, and the digital tips bear small discs. The males have internal vocal sacs and nuptial pads.

DISTRIBUTION
This species ranges across southern India.

HABITAT
The frogs inhabit evergreen hill forests.

BEHAVIOR
Little is known.

FEEDING ECOLOGY AND DIET
Little is known.

REPRODUCTIVE BIOLOGY
The body of the tadpole is elongated and depressed, with a long and slender tail and low fins. The subterminal mouth bears many papillae and stout jaw sheaths, but there is only a single row of poorly developed teeth on the upper jaw.

CONSERVATION STATUS
Not listed by IUCN.

SIGNIFICANCE TO HUMANS
None known. ◆

Malabar night frog
Nyctibatrachus major

SUBFAMILY
Nyctibatrachinae

TAXONOMY
Nyctibatrachus major Boulenger, 1882, Malabar and Wynaad, southern India.

OTHER COMMON NAMES
None known.

PHYSICAL CHARACTERISTICS
This medium-size, stout species has a snout-vent length up to 2.1 in (53.6 mm). The dorsal coloration varies from light tan to dark brown, with indistinct markings. The pupil is vertical. The limbs are short and the feet nearly entirely webbed; the digital tips have discs bearing dorsoterminal folds. Adult males have well-developed femoral glands, internal vocal sacs, and nuptial pads.

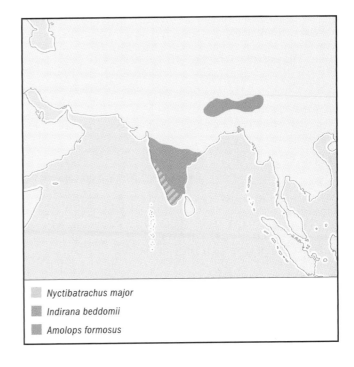

□ *Nyctibatrachus major*
■ *Indirana beddomii*
■ *Amolops formosus*

DISTRIBUTION
The species ranges across southern India.

HABITAT
This frog lives in and beside rocky hill streams at medium elevations, 360–3,020 ft (110–920 m).

BEHAVIOR
Little is known.

FEEDING ECOLOGY AND DIET
Little is known.

REPRODUCTIVE BIOLOGY
Eggs have pigmented animal poles. Tadpoles are devoid of tooth rows but have jaw sheaths and numerous papillae.

CONSERVATION STATUS
Not listed by IUCN.

SIGNIFICANCE TO HUMANS
None known. ◆

Pointed-tongue floating frog
Occidozyga lima

SUBFAMILY
Occidozyginae

TAXONOMY
Rana lima Gravenhorst, 1829, Java, Indonesia.

OTHER COMMON NAMES
English: Java frog.

PHYSICAL CHARACTERISTICS
This small frog has a maximum size of 1.5 in (39 mm). The skin is very rough, and there is persistence of the lateral-line system in the adult. The dorsum is dark olive with dark spots

and sometimes a mid-dorsal line; the rear parts of the thighs show two longitudinal dark lines enclosing a longitudinal white line. The tongue is pointed behind, and there are no vomerine teeth. The extremities of the digits are pointed, and webbing of the feet is complete. Males have nuptial pads and internal vocal sacs.

DISTRIBUTION
The frog occurs in southern China, Indochina, Indonesia, and Malaysia.

HABITAT
These frogs live in ponds, marshes, and paddy fields, where they seldom leave water.

BEHAVIOR
Little is known.

FEEDING ECOLOGY AND DIET
Little is known.

REPRODUCTIVE BIOLOGY
Males emit calls composed of two short notes. Amplexus is axillar. The eggs are small and pigmented, and the tadpole is elongated (up to 1.3 in, or 33 mm), with a pointed snout and tail tip and a high crest on the anterior tail fin. The tadpole's mouth is small, without papillae, tooth rows, or upper jaw sheath but with a horseshoe-shaped lower jaw sheath.

CONSERVATION STATUS
Not listed by IUCN.

SIGNIFICANCE TO HUMANS
None known. ◆

Sanderson's hook frog
Phrynodon sandersoni

SUBFAMILY
Petropedetinae

TAXONOMY
Phrynodon sandersoni Parker, 1935, Cameroon.

OTHER COMMON NAMES
None known.

PHYSICAL CHARACTERISTICS
This is a small species, with a snout-vent length up to 0.9 in (22 mm) in males and 1 in (26 mm) in females. The frogs have trapezoidal enlarged digital tips and femoral glands. The dorsal coloration varies widely, from translucent yellow to brownish; the lower parts are lemon yellow. Males are devoid of nuptial pads but have internal vocal sacs and odontoid fangs on the lower jaw.

DISTRIBUTION
The frogs exist in Cameroon, Fernando Póo, and West Africa.

HABITAT
This species inhabits hilly equatorial forest.

BEHAVIOR
Little is known.

FEEDING ECOLOGY AND DIET
Little is known.

REPRODUCTIVE BIOLOGY
Males are territorial and will bite intruder males in their territory. Eggs are laid on the leaves of small trees, bushes, or herbaceous plants, up to 6.6 ft (2 m) above a very humid substrate. Each clutch counts 12–17 eggs that are each 0.09 in (2.3 mm) in diameter. The female remains in the vicinity of the eggs during daylight hours and climbs on them every evening to spend the night over them. After 12 days of development, the tadpoles, which are curved narrowly within the egg's capsule, are "ejected" from the eggs. They are devoid of jaw sheaths and tooth rows, and they do not feed until metamorphosis, living on their vitelline reserves. Small froglets with fully resorbed tails develop about six weeks after egg laying.

CONSERVATION STATUS
Not listed by IUCN.

SIGNIFICANCE TO HUMANS
None known. ◆

Sharp-nosed grass frog
Ptychadena oxyrhynchus

SUBFAMILY
Ptychadeninae

TAXONOMY
Rana oxyrhynchus Smith, 1849, Natal, South Africa.

OTHER COMMON NAMES
None known.

PHYSICAL CHARACTERISTICS
This is a medium-size species, with a snout-vent length up to 2.3 in (58 mm) in males and 2.7 in (68 mm) in females. The snout is long, and the dorsum is brownish with dark spots on strongly elevated skin ridges. The hind limbs are very long, with large webbing. Males have nuptial pads and vocal sacs that protrude through lateral slits.

DISTRIBUTION
These frogs exist in most of sub-Saharan Africa.

HABITAT
The species inhabits forests and nearby savannas.

BEHAVIOR
Little is known.

FEEDING ECOLOGY AND DIET
Little is known.

REPRODUCTIVE BIOLOGY
Males emit intense, high-pitched thrills of about 0.4 sec, which are repeated every second. Tadpoles have 2/2 tooth rows.

CONSERVATION STATUS
Not listed by IUCN.

SIGNIFICANCE TO HUMANS
None known. ◆

African bullfrog
Pyxicephalus adspersus

SUBFAMILY
Pyxicephalinae

TAXONOMY
Pyxicephalus adspersus Tschudi, 1838, Cape of Good Hope, South Africa.

OTHER COMMON NAMES
English: Giant pixie; German: Gruener Grabfrosch.

PHYSICAL CHARACTERISTICS
This is a large, toadlike species; males have a snout-vent length up to 9 in (230 mm) and a weight up to 2.4 lb (1.075 kg). Males are larger than females. Adults are olive green and juveniles are bright green, with longitudinal skin folds on the dorsum, short legs, and a shovel-shaped inner metatarsal tubercle. Odontoid fangs are present on the lower jaw.

DISTRIBUTION
The species' range is southern Africa.

HABITAT
This frog inhabits open grass or bush country.

BEHAVIOR
These frogs estivate underground in a cocoon made of layers of molted skin. They emerge after heavy rains to breed.

FEEDING ECOLOGY AND DIET
These frogs are omnivorous. Because of their large size and aggressive habits, they can feed on vertebrates (mammals, birds, snakes, lizards, and frogs, including their own young or even other adults).

REPRODUCTIVE BIOLOGY
Males gather in daylight in shallow temporary pools, where they emit their loud "whoop, whoop," which recalls the lowing of cattle. Males usually fight among themselves, frequently wounding each other with their odontoids. Dominant males may fertilize the eggs of several females in their territory. Females lay 3,000–4,000 pigmented eggs that are each 0.08 in (2 mm) in diameter. Tadpoles with 4–5/3 tooth rows may reach a size of 2.8 in (71 mm). They swim together in schools of up to 3,000, attended by the father. The father can attack and bite potential predators (including lions or humans) or dig channels 49 ft (15 m) long or more, allowing tadpoles that have become isolated in peripheral puddles to return to the main pond. Metamorphosis usually takes place very quickly (as little as 18 days after egg laying). Froglets may eat each other.

CONSERVATION STATUS
Not listed by IUCN.

SIGNIFICANCE TO HUMANS
From prehistoric times to the present, adults, young, and tadpoles have been eaten by various peoples in Africa. ◆

Beautiful torrent frog
Amolops formosus

SUBFAMILY
Raninae, tribe Amolopini

TAXONOMY
Polypedates formosus Günther, 1876, Khasi Hills, Assam, India.

OTHER COMMON NAMES
English: Assam sucker frog.

PHYSICAL CHARACTERISTICS
This is a medium-size species, with a snout-vent length up to 3.3 in (85 mm). The frogs have a bright green, greenish, or olive dorsum covered with spots. The dorsal and ventral skins are smooth; dorsolateral folds are absent. The hind limbs are very long, with complete webbing. The digital tips bear large discs with ventrolateral grooves. Adult males have vocal sacs and velvety nuptial pads on the first finger.

DISTRIBUTION
The species' range is Bhutan, northern India, and Nepal.

HABITAT
The frogs live along torrents from 5,640 ft (1,720 m) to 8,700 ft (2,650 m) in forested and nonforested areas.

BEHAVIOR
Little is known.

FEEDING ECOLOGY AND DIET
Little is known.

REPRODUCTIVE BIOLOGY
Males call from the banks or rocks along or in torrents. Eggs, which are ivory white, are stuck by their jelly under rocks or stones in the rapid part of the torrent. Tadpoles are gastromyzophorous, that is, they have a large sucker that covers the anterior part of the belly and numerous tooth rows (6–7/3).

CONSERVATION STATUS
Not listed by IUCN.

SIGNIFICANCE TO HUMANS
None known. ◆

Bullfrog
Rana catesbeiana

SUBFAMILY
Raninae, tribe Ranini

TAXONOMY
Rana catesbeiana Shaw, 1802, North America.

OTHER COMMON NAMES
French: Grenouille taureau (France), Ouaouaron (Quebec); German: Nordamerikanischer Ochsenfrosch; Spanish: Rana toro americana, rana mugidora.

PHYSICAL CHARACTERISTICS
This member of the subgenus *Aquarana* is the largest North American frog, reaching 8 in (203 mm) and more than 3.3 lb (1.5 kg). It is greenish, olive, or brownish, sometimes with darker spots on the back. The tympanum is large, especially in males, and there are no dorsolateral folds. The hind limbs are long and the feet fully webbed. Males have nuptial pads, single internal vocal sacs, and yellowish throats.

DISTRIBUTION
The species inhabits eastern North America from Mexico to southern Canada. It was introduced into western North America, Central and South America, the West Indies, Japan, China, Thailand, several European countries, and several oceanic islands.

Rana catesbeiana

HABITAT
This semiaquatic frog can be found in many habitats, though it prefers larger bodies of water than most other frogs.

BEHAVIOR
Bullfrogs prefer warmer weather, digging into the mud to hibernate during cold winter weather. Adult males are aggressive and defend their shoreline territories by wrestling with other male bullfrogs.

FEEDING ECOLOGY AND DIET
Rather than actively hunting, bullfrogs wait for their prey to come to them. They eat others of their own species, frogs and tadpoles, snakes, insects, worms, and crustaceans.

REPRODUCTIVE BIOLOGY
After hibernation, males gather to emit their low, guttural calls composed of long notes. They are territorial and aggressive. Eggs, which are 0.05–0.07 in (1.2–1.7 mm) in diameter and pigmented at the animal pole, are laid in groups of 3,000–20,000. Tadpoles have 2–3/3 tooth rows and attain lengths up to 6.7 in (170 mm) before metamorphosis, which may occur after two to four years in northern latitudes (Quebec, Canada).

CONSERVATION STATUS
Not listed by IUCN.

SIGNIFICANCE TO HUMANS
This frog is consumed by humans and is used for dissection in colleges and universities. It has been introduced into a variety of regions all around the world, often with success; many of these introduced populations have had dramatic negative impacts, through competition and direct predation, on the local fauna. Because of its high level of fertility, eradication of the species once it is established in a new habitat is difficult, if not impossible. ◆

Roesel's green frog
Rana esculenta

SUBFAMILY
Raninae, tribe Ranini

TAXONOMY
Rana esculenta Linnaeus, 1758, Nürnberg, Germany.

OTHER COMMON NAMES
English: Edible frog; French: Grenouille verte; German: Teichfrosch Wasserfrosch.

PHYSICAL CHARACTERISTICS
This is the common green frog that appears in many textbooks as well as in comics and children books. It is about 2.4–3.5 in (60–90 mm) in snout-vent length. Typically, this frog is green or greenish, though sometimes other colors (brownish, grayish) are seen; there are a varying numbers of spots on the back. The frog has rather long hind limbs, and its feet are almost fully webbed. In several characters, this form is intermediate between the two species from which it originated by hybridization: *Rana lessonae* (which is smaller, with shorter hind limbs and a short, shovel-shaped internal metatarsal tubercle) and *Rana ridibunda* (which is larger, with longer hind limbs and a long and flat internal metatarsal tubercle). Males have nuptial pads and white external vocal sacs that protrude during calling through slits on the sides of the mandible close to the mouth commissure.

DISTRIBUTION
The species is distributed throughout Europe.

HABITAT
This frog lives in open habitats around medium-size or large ponds and lakes, and less often close to small ponds or along rivers.

Rana temporaria

Rana esculenta

BEHAVIOR

This frog is semiaquatic and seldom goes far from water except on rainy nights, when it may colonize new habitats. It is active both during the day and at night. It has complex social structures, behaviors, and vocal repertoires, especially since it shares its habitat with at least one of its parental species.

FEEDING ECOLOGY AND DIET

Little is known.

REPRODUCTIVE BIOLOGY

This form is not properly a species but a "stabilized hybrid," or klepton (which may be indicated by inserting "kl." before its "specific" name), that shows a modified meiosis known as hybridogenesis. As a result, the frogs produce pure gametes containing the chromosomes from only one of their original parental species, that is, either *Rana lessonae* or *R. ridibunda;* they breed with the opposite species, and, thus, individuals identical to first-generation hybrids are produced again at each generation. This frog breeds in late spring or early summer (April–June). Males gather in breeding leks, where they emit loud calls and where they are joined by females that are ready to lay eggs. Females lay 2,000–6,000 pigmented eggs that are 0.04–0.06 in (1–1.5 mm) in diameter. Tadpoles have high tail fins and 2/3 tooth rows, and they are swift swimmers. They usually reach a size of 1.6–1.77 in (40–45 mm) and metamorphose in late summer; occasionally, they hibernate and reach giant sizes of 3.5–4.7 in (90–120 mm).

CONSERVATION STATUS

This species is not listed by the IUCN. However, because of frog consumption, many populations of this frog have been drastically reduced.

SIGNIFICANCE TO HUMANS

The legs of this frog traditionally have been eaten in Europe, especially in France. Apart from this economic function, this frog is particularly significant to humans as basic research material. Despite its having been used for many years in innumerable experimental works in various fields of biology (among them, physiology, embryology, teratology), it was only in the 1960s that the extraordinary nature of this "species" was suspected and later established. Several kleptons exist among European green frogs (subgenus or genus *Pelophylax*), but in all cases one of the two parental species is *Rana ridibunda.* This phenomenon is still largely misunderstood, and research on this frog complex remains promising for the understanding of basic aspects of cell physiology and vertebrate sexuality. Other unresolved research topics related to these frogs include the massive anomalies affecting the limbs of high percentages of frogs in some populations, which have been studied for more than half a century but remain a mystery.

Brown frog
Rana temporaria

SUBFAMILY

Raninae, tribe Ranini

TAXONOMY

Rana temporaria Linnaeus, 1758, Sweden.

OTHER COMMON NAMES

English: European common frog, grass frog; French: Grenouille rousse; German: Grasfrosch; Spanish: Rana roja, rana bermeja.

PHYSICAL CHARACTERISTICS

This is the most common European species of the group of brown frogs, the distribution of which covers most of Europe from sea level in the north to above 6,562 ft (2,000 m) in the south. Over this vast area, the species shows considerable variety in most characters, and several subspecies have been recognized. It is 2.4–3.7 in (60–95 mm) in snout-vent length and displays a vast array of colorations, including brown, reddish, orange, yellow, olive, gray, and blackish; none is green. The dorsum is more or less spotted, the legs are barred, and the eye coloration varies considerably, with a basic golden iris, which may be more or less charged in melanophores. The hind limbs are short but may be longer in some southern populations or regions. The webbing is usually large but is less developed in Iberian populations. Males have nuptial pads and internal vocal sacs, and their throats are bluish during the breeding period.

DISTRIBUTION

The species is distributed throughout Europe.

HABITAT

This frog occurs in forest habitats and grasslands. At high elevations and latitudes, it lives in meadows, marshes, and peat bogs.

BEHAVIOR

This frog spends most of its life on the forest floor or in the grass, but it moves to ponds for breeding. In mountain habitats it may remain around ponds or lakes for most of the year.

FEEDING ECOLOGY AND DIET

Little is known.

REPRODUCTIVE BIOLOGY

This species breeds as soon as snow and ice melt, at widely different periods according to elevation and latitude. Males gather for calling, and egg masses often are grouped by the dozens or hundreds in shallow parts of the ponds. Each female lays 1,000–4,000 eggs that are each 0.08–0.12 in (2–3 mm) in diameter. Tadpoles, which have 3–4/4 tooth rows, may reach a length of 1.77 in (45 mm) before metamorphosis.

CONSERVATION STATUS

This species is not listed by the IUCN. However, in several countries, and especially in mountain areas, commercial exploitation of these frogs for human consumption has had drastic negative impacts on the populations.

SIGNIFICANCE TO HUMANS

This species is eaten by Europeans. ◆

Beddome's Indian frog
Indirana beddomii

SUBFAMILY

Ranixalinae

TAXONOMY

Polypedates beddomii Günther, 1876, Anamallays, Malabar, Sevagherry and Travancore, India.

OTHER COMMON NAMES

None known.

PHYSICAL CHARACTERISTICS

This medium-size species has a snout-vent length up to 2 in (49.5 mm) in males and 2.4 in (60.1 mm) in females. The dorsal skin is covered with short longitudinal glandular folds. The coloration varies; it can be yellowish, pinkish, or brownish, with irregular speckling. The hind limbs are long, the webbing is incomplete, and the tips of the digits are dilated into discs. Adult males have large tympana, vocal sacs, nuptial pads, and femoral glands.

DISTRIBUTION
The species is distributed throughout southern India.

HABITAT
The frogs inhabit the forest floor or rocky soil in evergreen forest at 330–2,950 ft (100–900 m).

BEHAVIOR
Little is known.

FEEDING ECOLOGY AND DIET
Little is known.

REPRODUCTIVE BIOLOGY
Pigmented eggs presumably are deposited outside water under shelters, such as stones, rotten vegetation, or the bark of dead trees. Tadpoles are peculiar, with an elongated body form, extremely large eyes, and a slender and pointed tail. The hind limbs develop early, and they have 4–5/4 tooth rows. From the beginning they can use their tails, and later their hind limbs, to skitter on the rocks or ground, which allows them to go from one humid terrestrial shelter or shallow pool to another.

CONSERVATION STATUS
Not listed by IUCN.

SIGNIFICANCE TO HUMANS
None known. ◆

Resources

Books
Duellman, William E., and Linda Trueb. *Biology of Amphibians.* New York: McGraw-Hill, 1986.

Passmore, N. I., and V. C. Carruthers. *South African Frogs: A Complete Guide.* Revised edition. Johannesburg, South Africa: Witwatersrand University Press, 1995.

Periodicals
Blommers-Schlösser, R. M. A. "Systematic Relationships of the Mantellinae Laurent, 1946 (Anura, Ranoidea)." *Ethology, Ecology, and Evolution* 5 (1993): 199–218.

Bossuyt, F., and M. C. Milinkovitch. "Convergent Adaptive Radiations in Madagascan and Asian Ranid Frogs Reveal Covariation between Larval and Adult Traits." *Proceedings of the National Academy of Sciences of the United States of America* 97 (2000): 6585–6590.

Clarke, B. T. "Comparative Osteology and Evolutionary Relationships in the African Raninae (Anura Ranidae)." *Monitore Zoologico Italiano (new series), supplement* 15 (1981): 285–331.

Dubois, A. "Liste des Genres et Sous-genres Nominaux de Ranoidea (Amphibiens, Anoures) du Monde, avec Identification de Leurs Espèces-types: Conséquences Nomenclaturales." *Monitore Zoologico Italiano (new series), supplement* 15 (1981): 225–284.

———. "Notes sur la Classification des Ranidae (Amphibiens, Anoures)." *Bulletin Mensuele de la Société Linnéenne de Lyon* 61, no. 10 (1992): 305–352.

Emerson, S. B., R. F. Inger, and D. Iskandar. "Molecular Systematics and Biogeography of the Fanged Frogs of Southeast Asia." *Molecular Phylogenetics and Evolution* 16 (2000): 131–142.

Kosuch, J., M. Vences, A. Dubois, A. Ohler, and W. Böhme. "Out of Asia: Mitochondrial DNA Evidence for an Oriental Origin of Tiger Frogs, Genus *Hoplobatrachus.*" *Molecular Phylogenetics and Evolution* 21 (2001): 398–407.

Marmayou, J., A. Dubois, A. Ohler, E. Pasquet, and A. Tillier. "Phylogenetic Relationships in the Ranidae: Independent Origin of Direct Development in the Genera *Philautus* and *Taylorana.*" *Comtes Rendus de l'Académie de Sciences Paris* 323 (2000): 287–297.

Ohler, A., and A. Dubois. "Démonstration de l'Origine Indépendante des Ventouses Digitales dans Deux Lignées Phylogénétiques de Ranidae (Amphibiens, Anoures)." *Comtes Rendus de l'Académie de Sciences Paris* 309 (1989): 419–422.

Vences, Miguel, and F. Glaw. "When Molecules Claim for Taxonomic Changes: New Proposals on the Classifcation of Old World Treefrogs." *Spixiana* 24, no. 1 (2001): 85–92.

Vences, Miguel, Stefan Wanke, Gaetano Odierna, Joachim Kosuch, and Michael Veith. "Molecular and Karyological Data on the South Asian Ranid Genera *Indirana, Nyctibatrachus* and *Nannophrys* (Anura: Ranidae)." *Hamadryad* 25, no. 2 (2000): 75–82.

Alain Dubois, Docteur d'Etat

Squeakers and cricket frogs

(*Arthroleptidae*)

Class Amphibia

Order Anura

Family Arthroleptidae

Thumbnail description
Mostly small, inconspicuous brown frogs

Size
The frogs are generally less than 1 in (25 mm) long, although some, like the East African species *Arthroleptis tanneri,* may exceed 2.4 in (60 mm)

Number of genera, species
8 genera; 77 species

Habitat
Forest

Conservation status
Not threatened

Distribution
Sub-Saharan Africa

Evolution and systematics

No fossils are known from this terrestrial family. There is an ongoing debate concerning the relationships of this group. The consensus appears to be that it is not a subsection of the Ranidae, and should retain its status as a discrete family. Two subfamilies are recognized—the Arthroleptinae and the Astylosterninae—although an alternative classification regards each as a distinct family. The characteristics of the Arthroleptinae include a typical hourglass pattern on the back, and the presence of an elongated third finger in males. The characteristics of the Astylosterninae include bent fingers with projecting bony tips.

Physical characteristics

These are smooth-skinned terrestrial frogs. A longitudinal middorsal fine skin ridge is characteristic of the Arthroleptinae. Webbing is absent between the toes. Some species have enlarged disks on the fingers and toes. The frogs are mostly less than 1 in (25 mm) long, although some, like the East African species *A. tanneri,* may exceed 2.4 in (60 mm). The limbs and body are gracile in most species, although some of the burrowing species are robust and have robust limbs and flattened tubercles on the heel. A characteristic arthroleptine pattern is a dark hourglass or series of diamond-shaped markings along the dorsal midline. The background color varies greatly within a species, and can range from red to olive. Adult males in the Arthroleptinae have extremely long third fingers. In some species the finger may reach 40% of the body length. The astylosternines are mostly large frogs associated with fast-flowing streams in forests. The subfamily is distinguished on small differences in anatomy; most have curved sharp terminal phalanges that protrude through the skin of the finger tip.

Distribution

The family is found throughout tropical Africa from sea level to 9,800 ft (3,000 m) in forest or wooded savanna. The ranges are decreasing as the African rainforest is being destroyed.

Habitat

The frogs are known from the moist tropics, where they are found in leaf litter. The arthroleptines are inhabitants of

Hairy frog (*Trichobatrachus robustus*) lives in sub-Saharan Africa. During the breeding season, males move to mountain streams and grow hairlike projections that aid in aquatic respiration. (Photo by R. Wayne Van Devender. Reproduced by permission.)

natural forests, but will live in any dense vegetation. Many astylosternines are associated with rapidly flowing torrents on forested slopes. There are no free-swimming larvae in Arthroleptinae, whereas large, well-muscled astylosternine larvae develop in fast-flowing streams.

Behavior

The adults are active throughout the year, with peaks of feeding and breeding after rain. They emerge only after dark in more open habitats, but can be found active in the shaded forest during the day. In areas where there is a distinct dry season they estivate. Males engage in combat with other males during the breeding season, in an effort to hold a breeding territory.

Feeding ecology and diet

The leaf litter frogs eat minute insects and other arthropods like small spiders, as well as other frogs. The larger, more robust species will eat anything that moves, providing it can be forced into the mouth. The terrestrial frogs move through the leaf litter taking small moving arthropods. The river dwellers feed along the edge of the water.

Reproductive biology

Arthroleptines are terrestrial breeders with direct development. Large, yolky eggs are laid in a hollow nest on the ground and develop into small adults without a free-swimming tadpole stage. Astylosternines deposit eggs in quiet backwaters of streams; they develop into torrent-adapted tadpoles. There are peaks of calling after rain, and most egg clutches are laid during the start to middle of the rainy season. In moist forests near rivers, breeding takes place over an extended period. Male arthroleptines call from concealed sites in leaf litter, although some species like the common squeaker sometimes call in the open from ground level. Astylosternine males call from the shallow edges of rivers. The eggs of arthroleptines are laid in small clutches under dead leaves. In these moist surroundings they hatch rapidly into juveniles, passing through a tadpole stage in the egg. There is no direct parental care, although the males of some species attract more than one female into the breeding territory, effectively placing the eggs from previous females within his care. Little is known of astylosternine breeding, but the hairy frog, *Trichobatrachus robustus*, remains underwater near the eggs, apparently to protect them from predators.

Conservation status

This is an endemic African family. Squeakers are common, and it is not unusual to see two or three along every step of a forest path or along the bank of a river. As the African forests are being logged, the available habitat is contracting, and the populations of all the forest amphibians are becoming smaller.

Significance to humans

The small arthroleptines are not of direct importance to humans; they are not eaten and they are not toxic. The larger astylosternines, such as the hairy frog, are a prized food of local people.

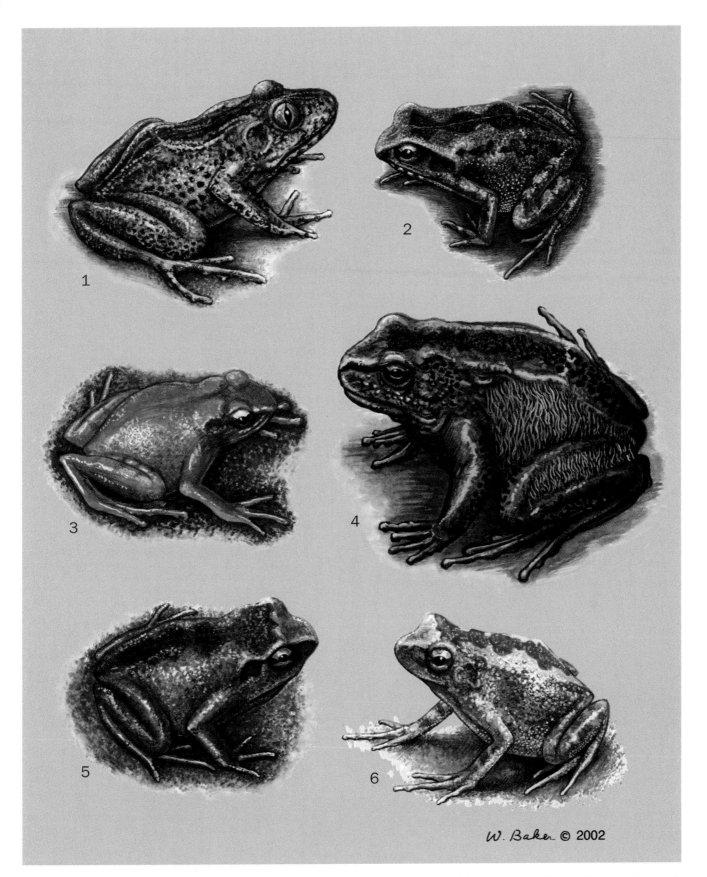

1. Crowned forest frog (*Astylosternus diadematus*); 2. Common squeaker (*Arthroleptis stenodactylus*); 3. Tanner's litter frog (*Arthroleptis tanneri*); 4. Hairy frog (*Trichobatrachus robustus*); 5. Bush squeaker (*Arthroleptis wahlbergii*); 6. Ugandan squeaker (*Schoutedenella poecilonotus*). (Illustration by Wendy Baker)

Species accounts

Common squeaker
Arthroleptis stenodactylus

SUBFAMILY
Arthroleptinae

TAXONOMY
Arthroleptis stenodactylus Pfeffer, 1893, central and southern Africa.

OTHER COMMON NAMES
English: Shovel-footed squeaker, dune squeaker, savanna squeaking frog, Kihengo screeching frog.

PHYSICAL CHARACTERISTICS
This is a robust species, with relatively short legs. The inner metatarsal tubercle is large, spadelike, and as long as, or longer than, the first toe. The pattern on the back consists of a pair of dark sacral spots, with various combinations of a three-lobed dorsal band. In some animals a pale vertebral line is present. A dark line runs from the tip of the snout to the shoulder.

DISTRIBUTION
This species is widespread, known from southern and eastern Democratic Republic of the Congo to Kenya and southward to northern South Africa, Zimbabwe, and Mozambique.

HABITAT
It is often associated with leaf litter. It can be found at altitudes from 130 to 6,600 ft (40–2,000 m). This frog is very common, and is able to live in gardens and natural vegetation.

BEHAVIOR
This species is active during the day in the wet season. The frogs move around searching for food.

FEEDING ECOLOGY AND DIET
This frog appears to eat a wide range of insect and other arthropod prey, as well as earthworms, snails, and even other frogs.

REPRODUCTIVE BIOLOGY
The male calls from concealed sites in leaf litter and under vegetation, during the day and night after rain. Eggs are deposited in hollows or burrows in damp earth, often under bushes or around the roots of trees, or under loose leaf mold. Eggs are 0.1 in (2.5 mm) in diameter, creamy white, and deposited in clutches of 33–80.

CONSERVATION STATUS
This species is widespread and common, able to live around human habitation, and not specifically threatened.

SIGNIFICANCE TO HUMANS
The common squeaker may live around human habitation, but has no direct significance as food or in any other way. ◆

Tanner's litter frog
Arthroleptis tanneri

SUBFAMILY
Arthroleptinae

TAXONOMY
Arthroleptis tanneri Grandison, 1893, west Usambara Mountains, Tanzania.

OTHER COMMON NAMES
English: Tanner's squeaker.

PHYSICAL CHARACTERISTICS
This is the largest arthroleptid—females exceed 2.4 in (60 mm). They are robust, with no expanded disks on the fingers or toes, nor webbing between the toes. The skin of the back is smooth. The back is brown with indistinct darker chevron-shaped markings. A dark band runs from the nostril through the eye to the upper arm. The limbs are crossbanded.

DISTRIBUTION
This species is confined to highland forest in the west Usambara Mountains of Tanzania.

HABITAT
This species prefers forest floor habitats and may be found along streams in the forest.

BEHAVIOR
The frogs sit and wait along streams or in leaf litter for insect prey during the day.

FEEDING ECOLOGY AND DIET
Tanner's litter frog eats forest-floor arthropods, including small spiders.

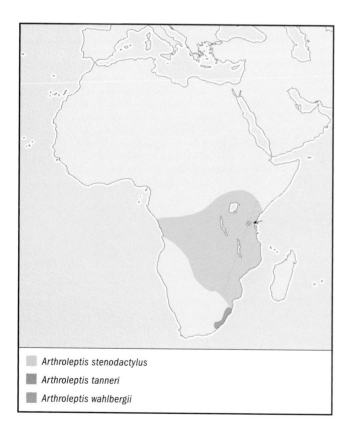

▢ *Arthroleptis stenodactylus*

▮ *Arthroleptis tanneri*

▦ *Arthroleptis wahlbergii*

REPRODUCTIVE BIOLOGY

Males call from the ground, well camouflaged in the leaf litter. The eggs are laid in clutches of about 30 eggs in hollow nests under the cover of dead leaves. The young emerge directly without a free-swimming tadpole stage.

CONSERVATION STATUS

Not listed by the IUCN, though this species is restricted to a small forest patch at Mazumbai, as the rest of the west Usambara Mountains have been cleared of natural forest.

SIGNIFICANCE TO HUMANS

None known. ◆

Bush squeaker
Arthroleptis wahlbergii

SUBFAMILY

Arthroleptinae

TAXONOMY

Arthroleptis wahlbergii Smith, 1849, eastern South Africa.

OTHER COMMON NAMES

English: Wahlberg's screeching frog.

PHYSICAL CHARACTERISTICS

Females are larger than males and attain lengths of 1 in (25 mm). The inner metatarsal tubercle is small, rounded, and less than half the size of the inner toe. The tips of the fingers and toes do not possess disks, although they may be swollen. The color pattern of the back is variable. Tan and darker brown background colors are typical. An hourglass pattern is common, and a pale vertebral stripe is found in some specimens.

DISTRIBUTION

This species is endemic to the tropical east coast of South Africa, and in suitable adjacent habitats inland.

HABITAT

It is found under leaf litter at the base of dense bushes. This species occurs in forest or thick bush, and is common under lush hedges and shrubs in gardens.

BEHAVIOR

This small frog is very secretive, rarely coming into the open, and then only after the start of the rains. Diligent searching for days for the same calling male is often fruitless.

FEEDING ECOLOGY AND DIET

This frog is known to eat a wide range of insect prey, such as crickets, cockroaches, beetles, and grasshoppers, as well as other arthropod prey like spiders and centipedes. They also eat earthworms, snails, and even other frogs.

REPRODUCTIVE BIOLOGY

The call is a long, high-pitched "wheep" or "wheepee." The eggs are pale and about 0.1 in (2.5 mm) within a capsule of 0.2 in (5 mm). Clutches of 11–80 eggs are known. Eggs are laid 0.8–1.2 in (20–30 mm) below the surface of the leaf litter, usually beneath bushes or other dense vegetation. The tadpole stage is passed in the egg. Eggs have been found in shallow burrows with an adult in attendance.

CONSERVATION STATUS

Not threatened.

SIGNIFICANCE TO HUMANS

This species is able to successfully coexist with humans even in large cities like Durban. ◆

Ugandan squeaker
Schoutedenella poecilonotus

SUBFAMILY

Arthroleptinae

TAXONOMY

Schoutedenella poecilonotus Peters, 1863, West Africa.

OTHER COMMON NAMES

English: West African screeching frog.

PHYSICAL CHARACTERISTICS

This is a small frog with a blunt snout. The head is broad and the body is squat. Females can be as large as 1.1 in (28 mm). The skin is quite smooth with small warts, although some individuals have a granular skin. There is no webbing between the toes. The color of the back varies from reddish to light tan with a dark pattern.

DISTRIBUTION

The Ugandan squeaker is found throughout the forest belt from West Africa to Uganda. There is some confusion with other species.

HABITAT

This frog is found in the forest, and also in peripheral savanna where there is lush vegetation.

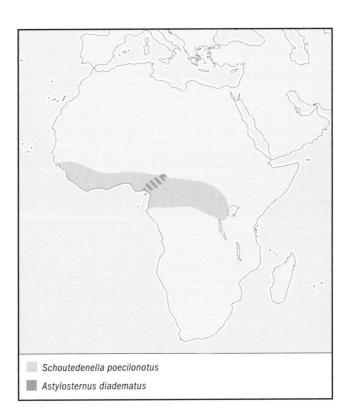

☐ *Schoutedenella poecilonotus*

◼ *Astylosternus diadematus*

BEHAVIOR
The frogs move slowly along the ground taking small prey that move nearby.

FEEDING ECOLOGY AND DIET
This squeaker feeds on small leaf-litter arthropods.

REPRODUCTIVE BIOLOGY
Males call from beneath dead leaves on the ground. Females lay clutches of 10–25 large, yolky eggs of 0.1 in (3 mm) in diameter. Each female may lay two or more clutches. The frogs only survive one breeding season, and have a recorded longevity of around six months.

CONSERVATION STATUS
This species is not threatened, although the general concerns of the loss of forest habitat apply.

SIGNIFICANCE TO HUMANS
None known. ◆

Crowned forest frog
Astylosternus diadematus

SUBFAMILY
Astylosterninae

TAXONOMY
Astylosternus diadematus Werner, 1898, Cameroon.

OTHER COMMON NAMES
None known.

PHYSICAL CHARACTERISTICS
The female is much smaller than the male. The largest frogs are 2.7 in (70 mm) in length. There is a distinct marking on

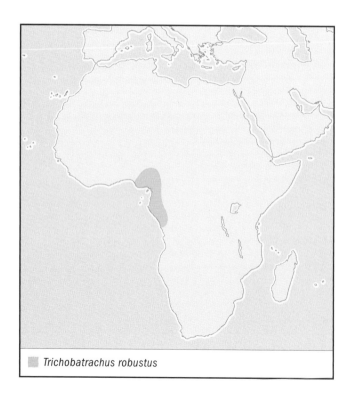

Trichobatrachus robustus

the head, and the underside is characteristically white or yellow with many dark spots.

DISTRIBUTION
This species is known from southwestern Cameroon and extreme eastern Nigeria at high elevations.

HABITAT
This frog is known from high savanna and dense mountain forest.

BEHAVIOR
The frogs are active during the day, with peaks of feeding activity after dark.

FEEDING ECOLOGY AND DIET
These frogs eat a range of small arthropods found on the forest floor.

REPRODUCTIVE BIOLOGY
Eggs are laid in quiet backwaters of streams. The tadpoles move into faster water as they grow. Although the tadpoles are found in fast-flowing streams, they do not have large sucker mouths.

CONSERVATION STATUS
Not threatened.

SIGNIFICANCE TO HUMANS
None known. ◆

Hairy frog
Trichobatrachus robustus

SUBFAMILY
Astylosterninae

TAXONOMY
Trichobatrachus robustus Boulenger, 1900, West Africa.

OTHER COMMON NAMES
French: Grenouille poilve; German: Haarfrosch.

PHYSICAL CHARACTERISTICS
The frog is stocky, up to 5.2 in (130 mm) in males, although the females only attain 3.6 in (90 mm), with darker markings on a brown background. The throat is yellow. During the breeding season, the sides of the thighs and body of the male develop small hairlike outgrowths. These increase the surface area for the uptake of oxygen. This fringe gives the frog its common name. The tadpole has an oral disk and a large suckerlike disk on the abdomen.

DISTRIBUTION
This frog is known from eastern Nigeria to Equatorial Guinea.

HABITAT
Hairy frogs are found in dense forest along streams.

BEHAVIOR
The frogs are terrestrial during most of the year, and feed along the forest floor. When the rains set in and the breeding season starts, the females remain in the forest to feed while the males move into the streams. Once the females are ready to breed they join the males in the water.

FEEDING ECOLOGY AND DIET
This species feeds along the edges of streams and on the forest floor. They eat a range of insects and other arthropods.

REPRODUCTIVE BIOLOGY
Eggs are laid in fast-flowing rivers. The male attends the egg clutches underwater, presumably to protect them from predators. The fringe of hairlike papillae enables him to remain underwater for days without needing to come to the surface for air.

CONSERVATION STATUS
Not threatened. This species is widely distributed and not in need of any conservation action.

SIGNIFICANCE TO HUMANS
Large hairy frogs are collected and eaten by local people, although not in significant numbers. ◆

Resources

Books

Channing, Alan. *Amphibians of Central and Southern Africa.* Ithaca, NY: Comstock Publishing Associates, 2001.

Passmore, Neville, and Vincent Carruthers. *South African Frogs: A Complete Guide.* Revised edition. Halfway House, South Africa: Southern Book Publishers and Johannesburg: Witwatersrand University Press, 1995.

Rödel, Mark-Oliver. *Herpetofauna of West Africa.* Vol. 1, *Amphibians of the West African Savanna.* Frankfurt: Chimaira, 2000.

Alan Channing, PhD

Shovel-nosed frogs

(Hemisotidae)

Class Amphibia

Order Anura

Family Hemisotidae

Thumbnail description
Small frogs with powerful forelimbs and a hard, sharp snout for burrowing

Size
1–3 in (25–80 mm)

Number of genera, species
1 genus; 8 species

Habitat
Savanna

Conservation status
Not threatened

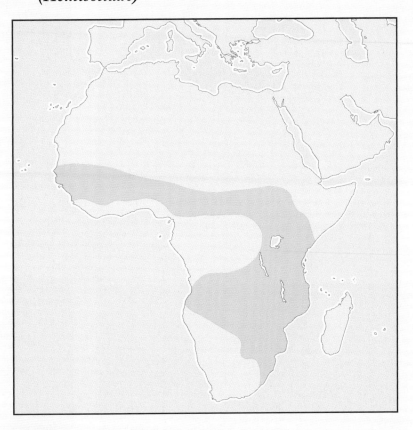

Distribution
Sub-Saharan Africa

Evolution and systematics

No fossils of this family are known. There is some evidence suggesting that this family is related closely to the rain frogs in the genus *Breviceps*, family Microhylidae. Another point of view is that these similarities follow from a common burrowing way of life and may not reflect a true relationship. No subfamilies are recognized.

Physical characteristics

These heavily built frogs have particularly robust skeletons associated with their burrowing habits. The species have a globular body, with short, muscular limbs. The well-muscled limbs end in short fingers and toes. The snout is sharp and has a hardened tip for digging, and a groove runs transversely behind the eyes. The frogs are smooth-skinned, with very small eyes. A large, flattened tubercle on the inner heel assists them in pushing headfirst into the soil. Adults are as small as 1 in (25 mm) and range in size to the largest, the spotted snout-burrower, at 3 in (80 mm). The back and sides are generally brown or purple with yellow spots or blotches.

Distribution

These frogs are found in the tropical savanna of sub-Saharan Africa, from Ethiopia, in western Africa, to South Africa and from sea level to 5,900 ft (1,800 m).

Habitat

Shovel-nosed frogs are native to open and wooded savanna where soils are sandy. The larvae are found in deep temporary pools with muddy substrates, and they occur together with tadpoles of many other species, such as *Xenopus* and *Kassina*.

Behavior

The frogs are active during the wet season, emerging from burrows after dark to feed. They are found in habitats that become very arid before the rains start. In the dry season they burrow deep into banks and the mud of hollows, where they estivate. Adults emerge after rain to feed on the surface, although they may tunnel like moles and catch underground prey, such as earthworms.

Shovel-nosed frog (*Hemisus guttatus*) is built to bury itself, head first, into the soil. Shown here in Natal, South Africa. (Photo by Animals Animals ©Austin J. Stevens. Reproduced by permission.)

First in a series of five photographs showing a common snout-burrower (*Hemisus sudanensis*) burrowing head-first into soil. (Photo by Alan Channing. Reproduced by permission.)

Snout-burrower begins by forcing its head into the soil, pushing with its strong legs. (Photo by Alan Channing. Reproduced by permission.)

Head and forelegs work themselves into the soil. (Photo by Alan Channing. Reproduced by permission.)

Head and forebody are submerged. (Photo by Alan Channing. Reproduced by permission.)

Snout-burrower is almost completely submerged into the soil. (Photo by Alan Channing. Reproduced by permission.)

Feeding ecology and diet

Shovel-nosed frogs eat nocturnal termites. In captivity they readily eat earthworms. They can be found after rain, feeding on the surface. They hunt earthworms by digging tunnels just below the surface. The hardened, sharp snout enables these frogs to move rapidly through loose soil.

Reproductive biology

Breeding is initiated by the first rains of the season. The male calls from a concealed site under vegetation at the edge of pools, usually on wet mud. The calls are prolonged buzzes. The male clasps the female and is dragged into the burrow by the larger female, who digs. The male then fertilizes the eggs in the nest. Females mate with only one male. Females remain with the developing eggs, which are laid in a burrow or under a log or stone. About 150–200 eggs are laid in a compact mass, each egg 0.08–0.10 in (2–2.5 mm) in diame-ter within a capsule 0.12–0.16 in (3–4 mm) in size. Clutch sizes may be as small as 30–35. At the top of the clutch are numerous empty egg capsules, which help protect the clutch. The nest is situated a little back from the water. Continuing rains cause the ponds to fill, and the water rises to the level of the tadpoles and liberates them.

Conservation status

Most species are widespread, and all are common. In areas where lowlands are drained and converted to housing schemes, much of the frogs' habitat is lost. This is especially true of species that are found in prime tourist areas along the east coast of Mozambique and South Africa.

Significance to humans

None known.

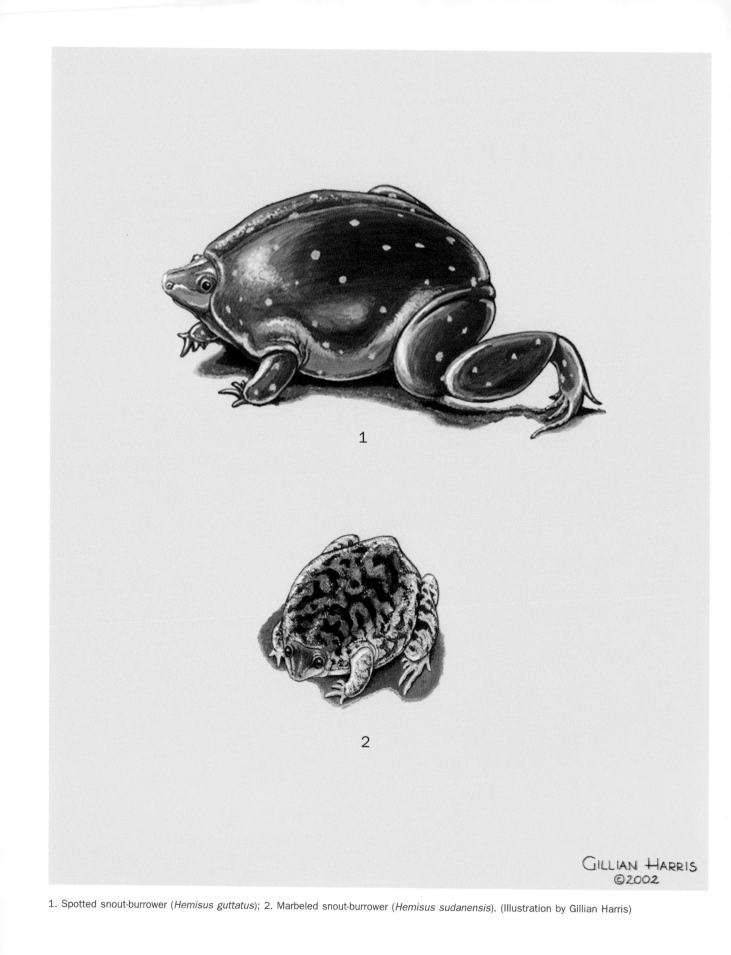

1. Spotted snout-burrower (*Hemisus guttatus*); 2. Marbeled snout-burrower (*Hemisus sudanensis*). (Illustration by Gillian Harris)

Species accounts

Marbled snout-burrower
Hemisus sudanensis

TAXONOMY
Hemisus sudanensis Steindachner, 1863, sub-Saharan Africa.

OTHER COMMON NAMES
English: Marbled shovel-nosed frog, mottled shovel-nosed frog, pig-nosed frog, mottled burrowing frog.

PHYSICAL CHARACTERISTICS
Large females reach 2.2 in (55 mm). The eyes are small, the forearms are massive, and the toes are slightly webbed. Coloration varies, with dark gray or brown marbling or spots on a paler brown background. A light vertebral line is often present.

DISTRIBUTION
Found in most of sub-Saharan Africa, excluding rainforests, from Senegal to Eritrea, western Ethiopia, and Somalia and south into southern Kenya and the northern and northeastern parts of South Africa.

HABITAT
Open savanna.

BEHAVIOR
The frogs feed on the surface or hunt prey underground by digging tunnels.

FEEDING ECOLOGY AND DIET
These frogs eat a range of small insects and feast on winged termites when they emerge. They also readily eat earthworms.

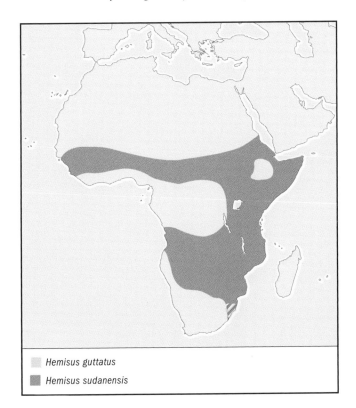

Hemisus guttatus
Hemisus sudanensis

REPRODUCTIVE BIOLOGY
Females are attracted to calling males. The male clasps the female, and she digs headfirst into the soft mud near a temporary pool. The eggs are laid and fertilized in an underground burrow. The female may remain near the eggs, which develop into tadpoles in the nest. Rain causes the pool to fill, and the tadpoles swim out of the nest as it floods. In extreme cases the tadpoles swarm onto the back of the female, who carries them to water.

CONSERVATION STATUS
Not threatened.

SIGNIFICANCE TO HUMANS
None known. ◆

Spotted snout-burrower
Hemisus guttatus

TAXONOMY
Hemisus guttatus Rapp, 1842, northeastern South Africa.

OTHER COMMON NAMES
English: Spotted shovel-nosed frog, spotted burrowing frog, eastern sharp-snouted frog.

PHYSICAL CHARACTERISTICS
The female may reach 3 in (80 mm); this is the largest species of snout-burrower. The toes are not webbed, and the back pattern is quite distinct, with a number of yellow dots on a dark purple or brown background. The head is pointed and small, with very small eyes. The snout tip is hard and used for burrowing. The arms are muscular, and the fingers are thick and strong.

DISTRIBUTION
Recorded from the KwaZulu Natal lowlands between Hluhluwe and Durban through the interior of South Africa.

HABITAT
Areas of flat, sandy soil that flood during the rains.

BEHAVIOR
Active after dark, when they feed and breed.

FEEDING ECOLOGY AND DIET
Eats burrowing prey, such as earthworms, also takes insects that are active on the surface at night.

REPRODUCTIVE BIOLOGY
The advertisement call is a long, high-pitched buzz. Eggs are laid in chambers that are 5.9 in (15 cm) below the surface. Each clutch consists of some 200 eggs. Each egg is 0.10 in (2.5 mm) in diameter within a 0.16-in (4-mm) jelly capsule. The eggs are protected by a few top layers of empty jelly capsules.

CONSERVATION STATUS
The species is not directly threatened, although parts of the coastal habitat are threatened by development.

SIGNIFICANCE TO HUMANS
None known. ◆

Resources

Books

Channing, A. *Amphibians of Central and Southern Africa.* Ithaca, NY: Cornell University Press, 2001.

Periodicals

Kaminsky, S. K., K. E. Linsenmair, and T. U. Grafe. "Reproductive Timing, Nest Construction and Tadpole Guidance in the African Pig-nosed Frog, *Hemisus marmoratus.*" *Journal of Herpetology* 33 (1999): 118–123.

Alan Channing, PhD

African treefrogs

(Hyperoliidae)

Class Amphibia

Order Anura

Family Hyperoliidae

Thumbnail description
Most species are typical treefrogs with webbing and digital discs, and live in trees or on reeds; a few are toadlike and live on and in the ground

Size
From 0.5 in (12 mm) in body length for the smallest adult male (*Hyperolius minutissimus*) to 4.3 in (110 mm) for the largest female (*Leptopelis palmatus*)

Number of genera, species
19 genera; 240 species

Habitat
Forest, woodland, and savanna

Conservation status
Vulnerable: 3 species

Distribution
Sub-Saharan Africa, Madagascar, and Seychelles

Evolution and systematics

Hyperoliidae was formerly regarded as part of the family Rhacophoridae, the Asian treefrogs, which are very similar in morphology and ecology. Based on small morphological differences, such as the shape of the metasternum, it was postulated that most of the African and some of the Madagassan members of the Rhacophoridae deserved their own family. Further studies have shown that Rhacophoridae and Hyperoliidae are not closely related, but have developed independently from the true frogs, the Ranidae.

African treefrogs are separated into four subfamilies—Hyperoliinae, Kassininae, Leptopelinae, and Tachycneminae—but the affinity of several of the genera to subfamily is disputed.

Hyperoliinae

Hyperoliinae, the largest subfamily, is distributed throughout the range of the family, except on the Seychelles. Most members are small, and most males possess vocal sacs and associated gular glands.

There are 12 genera. *Hyperolius* is the largest genus, with at least 85 species, but many subspecies are recognized, and new species and subspecies continue to be found. All have a horizontal pupil, a character separating them from the similar genus *Afrixalus*. *Hyperolius* has been called by museum zoologists "the most difficult of all frog genera" because they are so similar in morphology, but it is quite easy to separate the species by their calls and their habitat preference and color pattern, features not apparent in museum specimens. They are small, 0.5–1.6 in (1.2–4 cm). Most *Hyperolius* fall in two phases, the nature of which is not well understood. The newly metamorphosed froglets of both sexes, and some—in most species the majority—of adult calling males have the "juvenile" phase, a subdued yellow to brownish color with darker stripes or a darker hourglass pattern on the back. All adult females found at the breeding localities, as well as some of the males, have the "female" phase, which normally is very colorful and shows the characteristic color pattern of the species.

A number of small genera are similar to *Hyperolius* and probably closely related. *Nesionixalus*, with two species from the Atlantic islands (Bioko, Saõ Tomé) may not really be distinct from *Hyperolius*. *Acanthixalus* (two species), with a diamond-shaped pupil, is found in the forests of Cameroon and eastern Ivory Coast. *Alexteroon* (three species, horizontal pupil), *Arlequinus* (one species, diamond-shaped pupil), and

Several diverse color patterns of one species of frog, *Hyperolius viridi-flavus*. 1. *H. v. variabilis*; 2. *H. v. reesi*; 3. *H. v. taeniatus*. (Illustration by Emily Damstra)

The greater leaf-folding frog, *Afrixalus fornasinii*, lays its eggs on a reed or leaf, 2–3 ft (60–90 cm) above the water. It folds together and glues the leaf margins of about 2 in (5 cm) of the leaf to protect its eggs. When they hatch, the tadpoles fall to the water. (Illustration by Emily Damstra)

Chlorolius (one species, horizontal pupil) are *Hyperolius*-like frogs from Cameroon. *Chrysobatrachus* (one species, horizontal pupils) and *Callixalus* (one species, vertical pupils) are endemic to the highlands of central Africa, whereas *Cryptothylax* (one or two species, diamond-shaped pupils) is found in the western part of Central Africa.

The only genus of Hyperoliidae on Madagascar, *Heterixalus* (11 species; vertical diamond-shaped pupils), is very like *Hyperolius* in body shape and color pattern.

The genus *Afrixalus*, distributed throughout sub-Saharan Africa, consists of very small to medium-sized frogs 0.6–1.6 in

(1.5–4.1 cm). These frogs have vertical, diamond-shaped pupils. Almost all species have a pattern in dark brown and light gold, and the pattern is normally diagnostic for the species. Males and females are the same size, an unusual feature among treefrogs. *Kassinula* (one species) is superficially very similar to a tiny *Kassina*, but is probably more related to the Hyperoliinae. Its voice is quite different from that of *Kassina*.

Kassininae

Kassininae, with four or five genera, occurs nearly throughout tropical Africa. *Kassina* (12 species) are quite large frogs 1–2.5 in (2.5–6.4 cm). Most are terrestrial. They tend to run rather than leap, and are sometimes called running frogs. Their hind legs are not much longer than their forelegs. Their characteristic voice is a very brief whistle or popping sound with a fast rising frequency.

Semnodactylus from Southern Africa (one species), have a voice quite different from *Kassina*, and are terrestrial frogs superficially similar to *Kassina*. *Tornierella*, from Ethiopia (two species), and *Phlyctimantis*, (four species), from the forests from Tanzania to Sierra Leone, are very similar to *Kassina* in biology, appearance, tadpoles, and voice, and the two are probably closely related, although *Phlyctimantis* is arboreal. *Opisthothylax* (one species, vertical pupil), is very much like *Afrixalus*, and may not belong to the Kassininae.

Leptopelinae

Leptopelinae consists of the African genus *Leptopelis*. A large genus, with at least 45 species, these frogs are medium to large 1–4.3 in (2.5–11 cm). Some species live on or under ground; at the other extreme are species that live in the treetops. Other species live in bushes in the savannas, others in open forest, and many inhabit the dense evergreen forest. Most terrestrial species have a warty skin, and lack webbing and digital discs. The morphology follows the biology, in that species in more dense forest are smoother skinned, and have more webbing and larger digital discs. Most *Leptopelis* occur in two phases, a green juvenile phase, which in some species is retained by many adult specimens, and a much more subdued brownish adult phase.

Tachycneminae

Tachycneminae consists solely of *Tachycnemis*. This one species is the only treefrog on the Seychelles.

Physical characteristics

Most hyperoliids are typical treefrogs, with well-developed webbing and digital discs, but a few terrestrial species lack discs and webbing and are more toadlike. The digital discs are offset by an intercalary element between the distal and penultimate phalanges, and the pectoral girdle is firmisternal, a condition in which two elements in the breastbones are jutted together, bracing the frog against the jar of landing after jumps. Males have a well-developed vocal sac (pouch). Except in *Leptopelis*, the pouch has an area of thickened skin, as well as a gular flap or gular gland, the shape of which may differ among species. African treefrogs are similar in appearance, ecology, and anatomy to two other large families of treefrogs, Rhacophoridae, with one genus (*Chiromantis*) in Africa and two genera (*Boophis* and *Aglyptodactylus*) on Madagascar; and Hylidae, with one species in Africa north of the Sahara where the Hyperoliidae does not occur. The *Hyperolius nasutus* group is similar in appearance to Centroleniidae in tropical America, but Centroleniidae and Hylidae differ from Hyperoliidae by having arciferal pectoral girdles. Hyperoliidae range in size from 0.5 in (1.3 cm) in body length for the smallest adult male (*Hyperolius minutissimus*) to 4.3 in (11 cm) for the largest female (*Leptopelis palmatus*).

Hyperolius viridiflavus nitidulus calling in Nigeria. Note the large gular pouch with gular flap. (Photo by A. Schiøtz. Reproduced by permission.)

Distribution

Hyperoliidae occurs throughout sub-Saharan Africa, except in the central and western parts of South Africa and the dry parts of Namibia. *Heterixalus* is endemic to Madagascar, and *Tachycnemis* to the Seychelles.

Habitat

Like most frogs, hyperoliids congregate at breeding sites in the beginning of the rainy season. Breeding sites are selected on criteria based on the surrounding vegetation, so that the different species can be grouped into distinct faunas, or guilds, associated with vegetation. Three major faunas are recognized: savanna, high forest, and farmbush (or bushland) fauna.

The savanna fauna is found over a wide spectrum of landscapes, from open, treeless grassland to dense bush with many shrubs and trees. Savannas have great fluctuation in humid-

African treefrog (*Leptopelis brevirostris*) calls in Cameroon. (Photo by J.-L. Amiet. Reproduced by permission.)

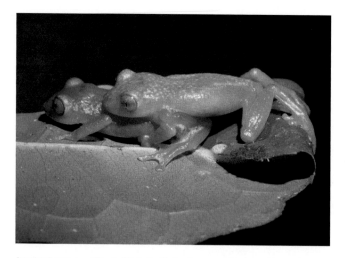

A pair of gray-eyed frogs (*Opisthothylax immaculatus*) are forest dwellers in Africa. (Photo by J.-L. Amiet. Reproduced by permission.)

ity and temperature, with very low humidity and high temperature in the daytime throughout most of the year, even during dry spells in the rainy season, whereas the forest has a more stable, cooler, and more humid microclimate.

The high forest fauna is found in the moist evergreen forest or rainforest in the southern parts of West Africa, Cameroon, and the Congo basin, and as a few isolated forests outside this area, most notably the Eastern Arc Forests in Tanzania.

The farmbush, or bushland, fauna is distinct from the savanna fauna and the high forest fauna, although it occurs in the same areas. In the savanna belt, this fauna is found in the gallery forests and in the rather dry, semideciduous forests in the coastal areas of eastern Africa. In the high forest belt, this fauna is found in clearings with farmland, or where abandoned farmland is in the process of returning to forest. This vegetation is widespread in the forest belt of Africa, so the farmbush fauna is much better known than the high forest fauna, which today is confined to isolated pockets of forest.

In addition to this separation into three faunas, a few species are confined to higher altitudes in mountains, but the

A Seychelles treefrog (*Tachycnemis seychellensis*) rests on a palm leaf, Praslin Island, Seychelles. (Photo by Lawson Wood/Corbis. Reproduced by permission.)

montane taxa can also be separated into grassland or savanna species and forest forms.

Behavior

Hyperoliids are nocturnal and emerge around dusk, either to seek food or breeding. Some species in savannas are fossorial (adapted to digging), but others spend the hot, dry daytime immovable on leaves. It is not known whether the fossorial species spend most of the dry season dormant underground, as is known for some other frogs, or emerge to hunt in the early morning when the humidity is high. In the dry season in savannas, *Leptopelis* may estivate underground; they have been dug up completely covered by a cocoon of dry shed skin. In the dry season, the *Hyperolius viridiflavus* group has an almost waterproof skin, thanks to a layer of mucus, and younger individuals can survive a water loss of up to half their body weight. Some waste products are stored in the skin rather than being excreted, and this also conserves water.

Feeding ecology and diet

Most treefrogs will eat any small animal of a suitable size, but they mostly feed on insects. The two species in the Ethiopian genus *Tornierella* and the Cameroonese *Leptopelis brevirostris* feed on snails. The East African *Afrixalus fornasinii* have be observed eating the eggs of *Hyperolius* and *Chiromantis*. *A. fornasinii* will stick its head into the foam nest of *Chiromantis* and eat some of the eggs. This behavior of feeding on immobile objects is unusual, because frogs normally react on movements of their prey, and in fact are believed to be unable to observe things that do not move.

Tadpoles of African treefrogs are probably omnivorous, eating all suitable material, primarily algae and bacteria on stones and plants, but also decomposing plants and animals.

Reproductive biology

The Hyperoliidae gather near small, temporary waterholes in the beginning of the rainy season, sometimes even before the waterholes have been formed. The males start calling and thereby attract the females. The temporary waterholes contain fewer predators than permanent waters and make it less likely for the tadpoles to be eaten by fish (although some fish in Africa also live in temporary waters). Normally, many species of frogs gather at the same ponds. The females are attracted to the voices of males of their own species, as are other males. Although almost all tadpoles live in water, there is a general tendency to keep eggs and the very young tadpoles out of reach of the many dangers in water. Most Hyperoliidae thus place their eggs out of water, glued to leaves above a pool. The tadpoles drop into the water when they start wriggling with their tails.

The common name for the genus *Afrixalus*, leaf-folding frogs, refers to their way of depositing their eggs. The male and female will place a small number of eggs on a leaf above water and fold this leaf around the eggs. The egg-jelly is sticky enough to hold the leaf together until the eggs hatch, when the small tadpoles wriggle down into the water. The tadpoles have a characteristic sharklike appearance and are agile plant eaters. *Acanthixalus* breeds in small water-filled holes in forest trees. *Kassina* places the eggs in water, and the tadpoles have a high fin. *Opisthothylax* is the only member of the family that makes a foam nest. *Leptopelis* bury their large, yolk-filled eggs in the soil, sometimes 33 ft (10 m) or more from the nearest waterhole. The tadpoles stay in the egg until the yolk is used and they have become strong enough to wriggle, eel-like, down to the water. One forest species, *L. brevirostris*, has probably foregone the free-living tadpole stage; the tadpoles metamorphose before leaving the egg.

Alexteroon has parental care; the female guards the eggs and helps the tadpoles break free of the jelly. In the South African *Afrixalus delicatus*, females can mate with several males on the same night (or several days apart), ensuring a more genetically diverse offspring. Another species, *A. brachycnemis*, has a voice consisting of a zip and a trill. The zip serves to keep the other males at a distance, the trill serves to attract the female. The other species in the genus have a similar division of the call, probably with a similar function.

Satellite males have been observed in some *Afrixalus*. These males sit quietly some distance away from a calling male and intercept and mate with an approaching female.

Conservation status

Hyperoliids are strictly bound to their preferred habitat, and although hard data on population sizes and population trends are lacking, it is safe to assume that populations are declining as their preferred habitat is reduced. Thus species living in threatened habitats are themselves threatened. This is especially true for the rich, unique fauna in the isolated Eastern Arc Forests in Tanzania, where 35 endemic species of amphibians occur, 10 of them hyperoliids. The small, dwindling forests in Ethiopia are also threatened, and so is the habitat for a number of species in South Africa with very restricted distributions. Three species are listed by the IUCN as Vulnerable: the South African *Hyperolius pickersgilli* and *Leptopelis xenodactylus;* and *Tachycnemis seychellensis* from the Seychelles.

Significance to humans

In the wet season treefrogs gather in swamps and lakes in great numbers, where they eat huge numbers of insects, especially mosquitoes. Because mosquitoes transfer one of the primary plagues of Africa, malaria, one must assume that treefrogs play an important role for humans, although studies of their importance in this respect are lacking. Apart from that, direct significance to humans is small. None of the species are eaten by humans.

1. Female Seychelles treefrog (*Tachycnemis seychellensis*); 2. Betsileo reed frog (*Heterixalus betsileo*); 3. Big-eared forest treefrog (*Leptopelis macrotis*); 4. Painted reed frog (*Hyperolius viridiflavus*, subspecies *H. v. viridiflavus*); 5. Toad-like treefrog (*Leptopelis bufonides*); 6. African wart frog (*Acanthixalus spinosus*); 7. Sharp-nosed reed frog (*Hyperolius nasutus*); 8. Greater leaf-folding frog (*Afrixalus fornasinii*); 9. Bubbling kassina (*Kassina senegalensis*). (Illustration by Emily Damstra)

Species accounts

African wart frog
Acanthixalus spinosus

SUBFAMILY
Hyperoliinae

TAXONOMY
Hyperolius spinosus Buchholz and Peters, 1875, Cameroon. No subspecies are recognized.

OTHER COMMON NAMES
None known.

PHYSICAL CHARACTERISTICS
Both sexes attain lengths up to 1.4 in (3.6 cm). The dorsum is very warty, grayish to brown, with transverse darker bands.

DISTRIBUTION
The species occurs in the northern part of the Cameroon-Congo rainforest.

HABITAT
This frog inhabits dense rainforest.

BEHAVIOR
Acanthixalus spinosus seems to spend life in small holes filled with water in tree trunks and branches. Adults spend the days submerged with their nostrils just above water and may emerge to forage at night. If attacked, the frog closes its eyes, keeps its limbs close to the body, and sticks out its orange tongue.

FEEDING ECOLOGY AND DIET
Nothing is known, although the diet most likely consists of arthropods of a suitable size.

REPRODUCTIVE BIOLOGY
This frog apparently is mute. Eight to 10 eggs are placed in a sticky jelly just above water in a small water body in a tree. The tadpoles fall into the water, where they grow very slowly (for a tropical frog), probably because of scarcity of food. Up to three months are required before metamorphosis.

CONSERVATION STATUS
Nothing is known about the conservation status of this frog, but the forests in its range are degrading rapidly.

SIGNIFICANCE TO HUMANS
None known. ◆

Greater leaf-folding frog
Afrixalus fornasinii

SUBFAMILY
Hyperoliinae

TAXONOMY
Euchnemis fornasini Bianconi, 1849, Mozambique. No subspecies are recognized.

OTHER COMMON NAMES
English: Banana frog.

PHYSICAL CHARACTERISTICS
Afrixalus fornasinii is the largest member of the genus. Both sexes have a body length of up to 1.6 in (4.1 cm). The ground color is dark brown with a pair of silverish, broad stripes, leaving a dark mid-dorsal band. In the northern half of this frog's distribution, up to half the specimens lack the dark mid-dorsal stripe, so that the back is entirely silverish.

DISTRIBUTION
This species is found in eastern Africa, from the coast of Kenya to the east coast of South Africa, and inland to eastern Zambia and Zimbabwe.

HABITAT
These frogs are typical members of the savanna community of the eastern lowlands, from the coast of Kenya to the northeastern coast of South Africa. The species is associated with rather large ponds containing reeds.

BEHAVIOR
Little is known aside from the feeding and reproductive biology.

FEEDING ECOLOGY AND DIET
Other than small insects of suitable size, *A. fornasinii* eat the newly laid eggs of *Hyperolius* and of *Chiromantis xerampelina*, a treefrog that lays its eggs in a foam nest above water.

REPRODUCTIVE BIOLOGY
The call is a creaking sound followed by a series of unmelodic clicks. It has been compared with the stuttering of a small ma-

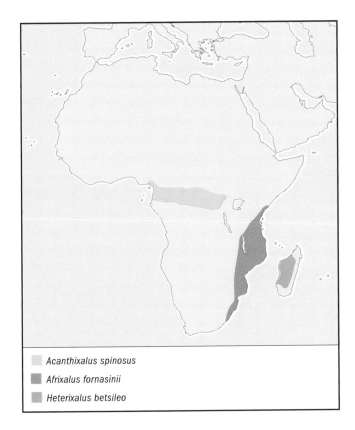

Acanthixalus spinosus

Afrixalus fornasinii

Heterixalus betsileo

chine gun. The eggs are placed on vegetation above water, and leaves are glued around the egg mass. The tadpoles later wriggle down to the water.

CONSERVATION STATUS
These frogs are very common over a large area.

SIGNIFICANCE TO HUMANS
None known. ◆

Betsileo reed frog
Heterixalus betsileo

SUBFAMILY
Hyperoliinae

TAXONOMY
Eucnemis betsileo Grandidier, 1872, Betsileo, Madagascar. No subspecies are recognized.

OTHER COMMON NAMES
None known.

PHYSICAL CHARACTERISTICS
This is a small treefrog; males are 0.75–1.1 in (1.9–2.8 cm), while females are 0.75–1.14 in (1.9–2.9 cm). The webbing is extensive, and the discs on fingers and toes are well developed. The frogs are green to yellow with yellow or white dorsolateral lines.

DISTRIBUTION
This species is found in the central plateau in Madagascar, at heights above 2,625 ft (800 m), and at lower altitudes in the western part.

HABITAT
This frog is common on the savanna and in cleared parts of the forests.

BEHAVIOR
Little is known aside from the reproductive biology.

FEEDING ECOLOGY AND DIET
Not known.

REPRODUCTIVE BIOLOGY
Breeding starts early in the wet season; breeding sites are open stagnant waters where males call in large choruses. The eggs are deposited in vegetation just above water.

CONSERVATION STATUS
The species is common over a large area.

SIGNIFICANCE TO HUMANS
None known. ◆

Sharp-nosed reed frog
Hyperolius nasutus

SUBFAMILY
Hyperoliinae

TAXONOMY
Hyperolius nasutus Günther, 1864, Duque de Braganca, Angola. Several subspecies have been described, but currently none are recognized.

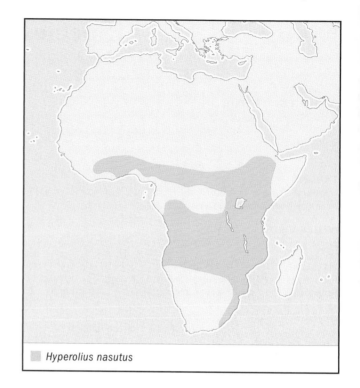

■ *Hyperolius nasutus*

OTHER COMMON NAMES
English: Long reed frog.

PHYSICAL CHARACTERISTICS
H. nasutus is a small frog, very slender and sharp-nosed; both sexes are 0.75–0.94 in (1.9–2.4 cm). The color is a transparent green, with light dorsolateral lines in males. This line sometimes also appears in females.

DISTRIBUTION
H. nasutus occurs in the savanna of tropical Africa, with the exception of the western part of West Africa, but there is some suspicion that more than one species is involved. A similar species, *H. benguellensis*, with a different voice, has been distinguished in southern Africa.

HABITAT
This frog inhabits rather dense, humid savanna.

BEHAVIOR
This delicate-looking little frog can survive harsh conditions in the dry season, probably by hiding in stems of grass and emerging only when humidity is high.

FEEDING ECOLOGY AND DIET
This frog probably feeds on arthropods of a suitable size.

REPRODUCTIVE BIOLOGY
The eggs are placed in water, a rare feature in *Hyperolius*, in batches of about 200. There are indications that males are born early in the rainy season metamorphose and grow so quickly that they can reproduce later in the same season.

CONSERVATION STATUS
Not threatened.

SIGNIFICANCE TO HUMANS
None known. ◆

Painted reed frog
Hyperolius viridiflavus

SUBFAMILY
Hyperoliinae

TAXONOMY
Eucnemis viridiflavus Duméril and Bibron, 1841, Abyssinia; *Hyperolius marmoratus* Rapp, 1842, Natal; *H. marginatus* Peters, 1854, Macanga, Mozambique; *Rappia tuberculata* Mocquard, 1897, Lambarene, Gabon. About 40 subspecies are recognized.

OTHER COMMON NAMES
English: Reed frog, sedge frog.

PHYSICAL CHARACTERISTICS
H. viridiflavus is a characteristic and abundant reed frog on the savanna, and its call—a chorus sounding like small bells or xylophones—is much more tonal than other *Hyperolius* calls. All members have a blunt snout and much webbing. Males have a very large gular sac; females have a transversal gular fold, a feature which is otherwise rare in the genus. In contrast to this morphological uniformity, the color pattern varies wildly. As a result of this variation, the group is usually subdivided into subspecies. However, the number of subspecies and the boundaries between them are not settled, and it can even be disputed whether the classical subspecies concept is appropriate here. More than 100 names have given to subspecies in this group, and more than 40 are commonly used.

These forms can be regarded as subspecies of one species, *H. viridiflavus*, but some researchers prefer to split them up into a small number of species belonging to a "superspecies." This is partly because there are a few cases of two "subspecies" occurring together, which indicates that they cannot interbreed and are thus not the same species.

The group is often split into three species: *H. viridiflavus*, distributed throughout West Africa and the northern part of Eastern Africa to southern Tanzania; *H. marginatus*, (sometimes called *H. parallelus*) found from southern Tanzania and northern Mozambique, and across Africa to Angola and the southern Congo; and *H. marmoratus*, found from the east coast of South Africa to southern Mozambique and Zimbabwe. However, the question of species relationship is far from settled. In addition to the savanna-living members of this group, *Hyperolius tuberculatus* exist in the forest in central Africa and at a single locality in West Africa. It is usually regarded as a member of the *H. viridiflavus* superspecies.

DISTRIBUTION
This frog is found throughout the savannas of sub-Saharan Africa.

HABITAT
Most forms are strictly confined to the savanna, but one group (*H. tuberculatus*) occurs in clearings in the forest belt.

BEHAVIOR
The savanna-living members of the *H. viridiflavus* group can sit exposed in the glaring sun, even in the dry season. Their skin is almost waterproof thanks to a thin layer of dried mucus, and the young are able to tolerate a water loss of up to one-half their body weight. Some waste products can be stored in the skin as a pigment, so that the skin becomes chalky white in the dry season.

FEEDING ECOLOGY AND DIET
This frog most likely feeds on all suitable arthropods.

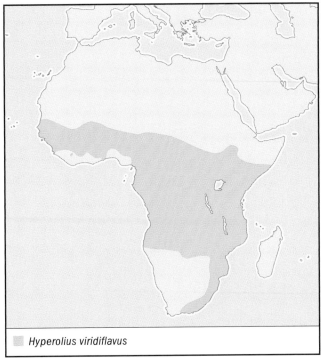

Hyperolius viridiflavus

REPRODUCTIVE BIOLOGY
Observations in captivity show that members of this group have a great capacity for producing repeated clutches of eggs with intervals of a few weeks, but whether that is also the case in nature is not known. The newly metamorphosed frogs are very large compared to the adult and to other *Hyperolius* juveniles, and are themselves able to reproduce the following rainy season, perhaps even sometimes late in the same season. At least one observer has noted the ability of this species to change sex from female to male while still maintaining the ability to produce eggs, but this remarkable observation has not been made by the many people keeping this species in terraria.

CONSERVATION STATUS
This species is widespread and common, but some subspecies are very localized.

SIGNIFICANCE TO HUMANS
The Masai in East Africa believe that cattle will die if they eat *H. viridiflavus*. It may be that the very bright colors of some subspecies are a warning coloration, and their often exposed resting places during the day may enhance the warning effect.

Bubbling kassina
Kassina senegalensis

SUBFAMILY
Kassininae

TAXONOMY
Cystignathus senegalensis Duméril and Bibron, 1841, Galam, Senegal. Several subspecies have been described, but presently it is regarded as monotypic.

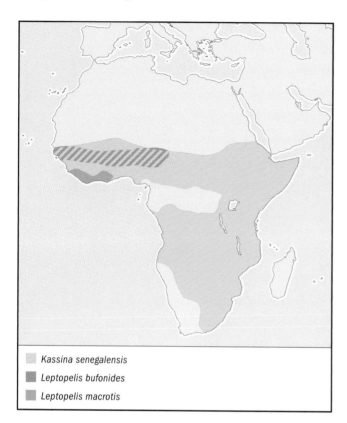

Kassina senegalensis
Leptopelis bufonides
Leptopelis macrotis

OTHER COMMON NAMES
English: Running frog.

PHYSICAL CHARACTERISTICS
These frogs are of medium size, with both sexes growing to about 1–1.9 in (2.5–4.9 cm); different populations differ much in size. The hind legs are not much longer than the forelegs, so the frogs will crawl or run rather than jump. The frogs are striped in gray and black, or spotted in part of southern Africa. There are differences in sizes and patterns throughout the vast range. However, the significance of this difference is not well understood, so *K. senegalensis* is regarded as monotypic.

DISTRIBUTION
This frog occurs throughout the savannas of Africa.

HABITAT
K. senegalensis lives on the ground in the savanna.

BEHAVIOR
The most typical night sound in the African savanna in the rainy season is the popping, melodious whistle of *K. senegalensis*. Hearing the frog is easy, but finding it is very difficult. The frog sits quietly on the ground, and its gray and black stripes and spots makes it very hard to find among the grass.

FEEDING ECOLOGY AND DIET
This frog's diet consists most likely of arthropods of a suitable size.

REPRODUCTIVE BIOLOGY
The male calls from the ground, often at the edge of shallow waterholes. The eggs are placed in water, and adhered to the vegetation. The tadpoles have a very high fin and swim gracefully in midwater.

CONSERVATION STATUS
Not threatened.

SIGNIFICANCE TO HUMANS
This species pleases humans with its melodious voice. ◆

Toad-like treefrog
Leptopelis bufonides

SUBFAMILY
Leptopelinae

TAXONOMY
Leptopelis bufonides Schiøtz, 1967, Bolgatanga, Ghana. No subspecies are recognized.

OTHER COMMON NAMES
None known.

PHYSICAL CHARACTERISTICS
A small *Leptopelis*; males are 1.1–1.3 in (2.9–3.3 cm), while females are 1.4–1.6 in (3.6–4.1 cm). The skin is warty, and the fingers and toes are without web and digital discs.

DISTRIBUTION
L. bufonides is known only from a few localities in the northern, dry part of the West African savanna, but is probably widespread in those places.

HABITAT
This frog inhabits open, dry savanna.

BEHAVIOR
L. bufonides lives on the ground and is unable to climb. It spends most of its time—perhaps the entire dry season—underground in burrows, where the humidity is not too low.

FEEDING ECOLOGY AND DIET
Not known.

REPRODUCTIVE BIOLOGY
Not known.

CONSERVATION STATUS
Not threatened.

SIGNIFICANCE TO HUMANS
None known. ◆

Big-eared forest treefrog
Leptopelis macrotis

SUBFAMILY
Leptopelinae

TAXONOMY
Leptopelis macrotis Schiøtz, 1967, Gola Forest Reserve, Sierra Leone. No subspecies are recognized.

OTHER COMMON NAMES
None known.

PHYSICAL CHARACTERISTICS
L. macrotis is a large Leptopelis; males are 1.6–1.8 in (4.1–4.6 cm), while females are up to 3.3 in (8.4 cm). This frog is

smooth-skinned, and has fully webbed feet and large digital discs.

DISTRIBUTION
L. macrotis is known from the forest of West Africa, from Ghana westward to Sierra Leone. It is probably widespread in West Africa, but very few people have looked for it, so the species is known only from few specimens and few localities. Very closely related and similar species occur in Cameroon (*L. rufus* and *L. millsoni*) and on Ihlo do Principe, an island off the Cameroon coast (*L. palmatus*).

HABITAT
This frog inhabits dense rainforest, where it lives high up in trees.

BEHAVIOR
Not known.

FEEDING ECOLOGY AND DIET
This frog feeds most likely on arthropods of a suitable size.

REPRODUCTIVE BIOLOGY
L. macrotis calls from high-up branches of trees near small watercourses. The frogs most likely emerge to the ground only to bury their large, yolk-filled eggs in the moist soil not far from water.

CONSERVATION STATUS
Although not listed by the IUCN, this species is threatened to the extent that its habitat, dense forest, is disappearing.

SIGNIFICANCE TO HUMANS
None known. ◆

Seychelles treefrog
Tachycnemis seychellensis

SUBFAMILY
Tachycneminae

TAXONOMY
Eucnemis seychellensis Duméril and Bibron, 1841, Seychelles. No subspecies are currently recognized, but the four discrete populations might deserve subspecific recognition.

OTHER COMMON NAMES
None known.

PHYSICAL CHARACTERISTICS
This is a large treefrog; males are 1.3–2 in (3.3–5.1 cm), while females are 1.8–3 in (4.6–7.6 cm). The pupils are vertical. There are differences in size, coloration, and other characters between the four island populations. On Mahé Island and Praslin, males are brown and females are green; both sexes are green on Silhouette and La Digue.

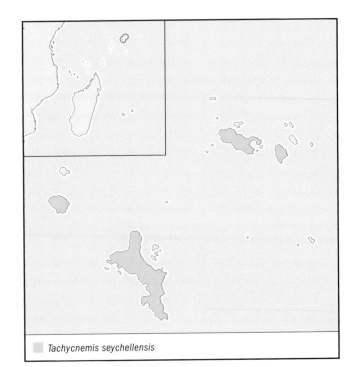

Tachycnemis seychellensis

DISTRIBUTION
T. seychellensis is the only treefrog on the isolated Seychelles Islands in the Indian Ocean. It occurs on the four largest of these granitic islands, Mahé, Silhouette, La Digue, and Praslin.

HABITAT
A forest species, this frog occurs along forest watercourses in the breeding season.

BEHAVIOR
Nothing is known aside from the reproductive biology.

FEEDING ECOLOGY AND DIET
Nothing is known.

REPRODUCTIVE BIOLOGY
T. seychellensis forms breeding aggregations; depositing 100–500 eggs on the ground or on stems of plants near streams or ponds, or in places to be flooded. The tadpoles are eel-shaped with a long, strong tail. They are similar to tadpoles of *Leptopelis* in morphology and dentition, and like them probably able to migrate to water over damp soil.

CONSERVATION STATUS
This species is listed as Vulnerable by the IUCN.

SIGNIFICANCE TO HUMANS
None known. ◆

Resources

Books

Carruthers, V. C. *Frogs and Frogging in Southern Africa.* Cape Town: Struik, 2001.

Channing, A. *Amphibians of Central and Southern Africa.* Ithaca, NY: Cornell University Press, 2001.

Passmore, N. I., and V. C. Carruthers. *South African Frogs.* Johannesburg: Southern Book & Witwatersrand U. P., 1995.

Rödel, M. O. *Herpetofauna of West Africa. I. Amphibians of the West African Savanna.* Frankfurt: Chimaira, 2000.

Schiøtz, A. *The Treefrogs of Eastern Africa.* Copenhagen: Steenstrupia, 1975.

———. *Treefrogs of Africa.* Frankfurt: Chimaira, 1999.

Stewart, M. *Amphibians of Malawi.* Albany: State University of New York Press, 1967.

Periodicals

Blommers-Schloesser, R. M. A. "Observations on the Malagasy Frog Genus *Heterixalus.*" *Beaufortia* 32 (1982): 1–11.

Drewes, R. C. "A Phylogenetic Analysis of the Hyperoliidae." *Occasional Papers of the California Academy of Science* 139 (1984): 1–70.

Laurent, R. F. "Le genre *Afrixalus* en Afrique centrale." *Annales du Musée Royal de l'Afrique Centrale* 235 (1982): 1–58.

———. "Les genres *Cryptothylax, Phlyctimantis* et *Kassina* au Zaire." *Annales du Musée Royal de l'Afrique Centrale* 213 (1976): 1–67.

———. "Le genre *Leptopelis* au Zaire." *Annales du Musée Royal de l'Afrique Centrale* 212 (1972): 1–62.

Liem, S. S. "The Morphology, Systematics and Evolution of the Old World Treefrogs." *Fieldiana Zoology* 57 (1970): 1–145.

Nussbaum, R. A., and Sheng Hai Wu. "Distribution, Variation and Systematics of the Seychelles Tree Frog, *Tachycnemis seychellensis.*" *Journal of Zoology, London* 236 (1995): 1–14.

Poynton, J. C., and D. G. Broadley. "Amphibia Zambesiaca 3, Rhacorphoridae and Hyperoliidae." *Natal Museum Annals* 28 (1987): 161–229.

Schiøtz, A. "The Treefrogs of West Africa." *Spolia Zoologica Musei Hauniensis* 25 (1967): 1–346.

———. "The Superspecies *Hyperolius viridiflavus.*" *Videnskabelige Meddelelser Dansk Naturhistorisk Forening* 134 (1971): 21–76.

Arne Schiøtz, DSc

Asian treefrogs

(*Rhacophoridae*)

Class Amphibia

Order Anura

Family Rhacophoridae

Thumbnail description
Small to relatively large treefrogs with the two halves of the pectoral girdle fused midventrally and expanded disks on the fingers and toes

Size
0.6–4.9 in (15–120 mm) in snout-vent length

Number of genera, species
13 genera; 341 species

Habitat
Both primary and disturbed forests, agricultural fields, ponds, streams, and savanna

Conservation status
Endangered: 1 species; Vulnerable: 2 species; Lower Risk/Near Threatened: 1 species; Data Deficient: 2 species

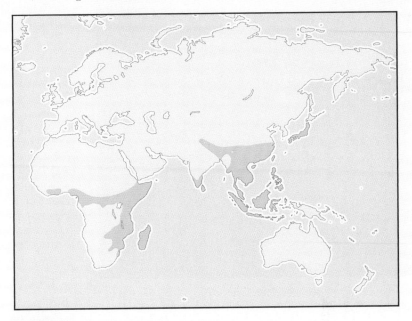

Distribution
Southeast Asia from eastern India, Sri Lanka, and Nepal, to Japan, Borneo, Celebes, and the Philippines; also in sub-Saharan Africa and Madagascar

Evolution and systematics

Asian treefrogs are most closely related to the true frogs (Ranidae) as evidenced by abutting epicoracoids in the pectoral girdle (firmisterny), the metasternum forming a bony style, and the presence of the cutaneous pectoris muscle. However, rhacophorids differ from ranids by having an intercalary element between the penultimate and terminal phalanges in the digits.

No fossils are known for the family. Relationships among the frogs currently assigned to Rhacophoridae are problematic. Results of morphological and molecular analyses are controversial. Some workers place rhacophorids as a subfamily of Ranidae; others recognize as many as three separate families, one of which also contains species usually placed in Ranidae. Herein, the 341 recognized species of rhacophorids are placed in 12 genera in three subfamilies, though 13 genera may be recognized.

Buergerinae

Flange on the third metacarpal bone and femoral glands absent; nuptial excrescences present; eggs deposited in water; free-living tadpoles. One genus (*Buergeria*) with four species in eastern Asia.

Mantellinae

Flange on third metacarpal bone absent; femoral glands present; nuptial excrescences absent; eggs deposited on ground or on vegetation; free-living tadpoles. Two genera (*Mantella* and *Mantidactylus*) in Madagascar.

Rhacophorinae

Flange on third metacarpal absent or present; femoral glands absent; nuptial excrescences present; eggs deposited in water, in arboreal cavities, or foam nests; free-living tadpoles or direct development. Nine genera: *Chiromantis* in sub-Saharan Africa; *Aglyptodactylus* and *Boophis* in Madagascar; *Chirixalus*, *Nyctixalus*, *Philautus*, *Polypedates*, *Rhacophorus*, and *Theloderma* in southeastern Asia.

Physical characteristics

Asian treefrogs have enlarged disks on the ends of the fingers and toes that aid in climbing on vertical surfaces and clinging to branches and leaves, and a head that is usually distinct from the body. They have varying degrees of webbing on the toes, and some have extensive webbing on the fingers. Some species of *Rhacophorus* with extensive webbing also have flaps of skin or fringes along the outside of the limbs and a flattened body. These characters increase their surface area, enabling them to glide or parachute from their higher perches to other trees or the ground when they jump. They have therefore been named "flying frogs." One of the most famous is Wallace's flying frog (*Rhacophorus nigropalmatus*) of Borneo and Southeast Asia, which can glide up to 24 ft (7.3 m) if it is dropped from a height of 17.7 ft (5.4 m).

Asian treefrogs usually have large eyes with horizontal pupils. The dorsal coloration varies from green, gray, and brown to white and black. Many have spots or irregular blotches on their backs. Some species have flash colors on

A jade treefrog (*Rhacophorus dulitensis*) gliding. (Photo by Stephen Dalton/Photo Researchers, Inc. Reproduced by permission.)

their sides and inside their thighs, and sometimes on the webbing between the fingers and toes. Several species have fringes on the forearm from the elbows to the outside of the fourth finger, and projections at the knees and vent. The skin may also vary from smooth to the very bumpy skin of the genus *Theloderma*. In some species the skin is co-ossified to the skull.

The species of the genus *Mantella* do not look at all like treefrogs. They are usually not arboreal but instead spend their lives on the forest floor. As a result, most of these diminutive frogs lack one of the most conspicuous characters of this family, the expanded digital disks. Mantellas have toxic skin secretions, and bright dorsal colors similar to the poison frogs (Dendrobatidae) in South America. These colors vary from bright yellow, orange, or red on the dorsum as in the golden mantella (*Mantella aurantiaca*), to yellow, orange, or red on the limbs with jet black on the dorsum as in *Mantella cowanii*. Femoral glands are present in males of *Mantella* and of the highly variable genus *Mantidactylus*, which are more like treefrogs with expanded disks on the tips of their digits. Because of the presence of femoral glands, the absence of nuptial pads on the males, and the non-amplexing mating behavior, both of these genera are considered closely related. Frogs of the Malagasy genus *Aglyptodactylus* also lack enlarged disks on the tips of the fingers and toes, which indicates that these frogs are terrestrial or possibly semifossorial.

Distribution

Most Asian treefrogs occur in south, southeast, and east Asia from eastern India, Sri Lanka, and Nepal throughout Myanmar (Burma), Thailand, Laos, Kampuchea (Cambodia), and Vietnam, southeast along the Malay Peninsula onto the islands of Sumatra, Java, Borneo, and Sulawasi, and through-

out the Philippines. They also occur in China and Japan, extending into temperate forests north of 40° latitude on the island of Honshu, Japan. *Chiromantis*, with only three species, is in sub-Saharan Africa, and four genera (*Aglyptodactylus*, *Boophis*, *Mantella*, and *Mantidactylus*) are endemic to Madagascar.

Habitat

Most Asian treefrogs occur in forests and some (*Nyctixalus*, *Philautus*, and *Theloderma*) usually are not near water. Many species of *Chirixalus*, *Polypedates*, and *Rhacophorus* also inhabit flooded rice fields and grasses or low shrubs between agricultural lands and forests. Many species of the genera *Boophis* and *Buergeria* breed in streams, others in ponds, ditches, or other sources of stagnant water. *Polypedates leucomystax* is abundant in and around human habitation, such as on buildings and in gardens, and even within cities throughout Southeast Asia. Species of *Chiromantis* inhabit dry areas of the African savanna; they can be found resting on tree limbs exposed to direct sunlight.

Behavior

Most of what is known about the behavior of rhacophorid frogs is related to mating and reproduction, since it is during the mating season that these frogs are more noticeable as males gather and call at breeding sites. In this respect, most rhacophorid frogs are active at night, when the males set up territories around a pond or stream and advertise to females. However, males can even call during the day but usually from hidden retreats. In contrast to this, males of many species of *Mantella* are usually active during the day, calling and fighting for territories in the open, calling from hidden positions under leaf litter, or foraging for ants, termites, and fruit flies.

Feeding ecology and diet

Adults probably feed primarily on insects, spiders, and other arthropods depending on relative size. Tadpoles of most species graze on algae on the rocks and debris. Tadpoles of *Philautus carinensis* and *Chirixalus eiffingeri* have been reported to feed on eggs of other frogs.

Reproductive biology

Most males call at night, or from a hidden retreat during the day. However, males of many species of *Mantella* are active and call during the day. Most species of *Chirixalus*, *Chiromantis*, *Polypedates*, and *Rhacophorus* deposit eggs in a foam mass on vegetation over ponds or swamps, and the tadpoles fall into the water below or are washed out of the foam by the next rain. In contrast to the foam-nesters, species of the genera *Nyctixalus* and *Theloderma* lay a small number of eggs on the inner walls of water-filled tree holes. The eggs hatch and the tadpoles drop into and develop within the water in the tree hole. Alternatively, the many relatively small frogs of the genus *Philautus* and some frogs of the genus *Mantidactylus* lay a small number of eggs on the ground. The embryos develop directly into froglets. All species of the genera *Aglyptodactylus*, *Boophis*, and *Buergeria* lay eggs in stagnant or moving water.

Most rhacophorids have prolonged breeding seasons, but *Aglyptodactylus* are "explosive breeders" during a few days in temporary ponds. *Mantella* and *Mantidactylus* do not engage in amplexus. Instead, there appears to be an abbreviated contact at which time the male induces the female to lay eggs by hormonal stimulation from femoral glands. For example, in *Mantidactylus depressiceps*, the male and female position themselves on vertical leaves so that the male is over the female and his thighs are touching her shoulder and back. He rubs his thighs against her and she almost immediately begins to deposit eggs on the upper surface of the leaf.

In most species there is no parental care of eggs and tadpoles. However, in some *Mantidactylus*, males sit on egg masses, apparently guarding them against possible desiccation or predation. In some foam-nesting frogs, females return to the foam nest to add more foam or urinate on the nest probably to prevent desiccation. Female *Chirixalus eiffingeri* return to tree holes to feed unfertilized eggs to their own tadpoles.

Conservation status

Due to a drastic reduction in populations and habitat, the IUCN lists *Philautus schmackeri* as Endangered and *Nyctixalus spinosus* and *Mantella aurantiaca* as Vulnerable; *P. alticola* and *P. poecilus* are listed as Data Deficient. In addition, two species of *Philautus* are listed in CITES as Near Threatened and Endangered, and *M. aurantiaca* and *N. spinosus* are listed as Vulnerable. Several species are endemic to small regions or islands. For example, the genera *Aglyptodactylus*, *Boophis*, *Mantella*, and *Mantidactylus* are all endemic to the island of Madagascar and are probably highly impacted by deforestation, as are many other animals on that island.

Significance to humans

Most rhacophorid frogs receive little attention from the people of the regions where they live. Most of the frogs are too small to eat, but the legs of some larger *Mantidactylus* appear in food markets. Because of their striking colors, frogs of the genus *Mantella* have been captured and sold in the pet trade in the same fashion as poison frogs of South America.

1. Forest bright-eyed frog (*Boophis erythrodactylus*); 2. Buerger's frog (*Buergeria buergeri*); 3. Kinugasa flying frog (*Rhacophorus arboreus*); 4. Luzon bubble-nest frog (*Philautus surdus*); 5. Eiffinger's Asian treefrog (*Chirixalus eiffingeri*); 6. Painted Indonesian treefrog (*Nyctixalus pictus*); 7. Gray treefrog (*Chiromantis xerampelina*); 8. Betsileo golden frog (*Mantella betsileo*); 9. Free Madagascar frog (*Mantidactylus liber*). (Illustration by Brian Cressman)

Species accounts

Buerger's frog
Buergeria buergeri

SUBFAMILY
Buergerinae

TAXONOMY
Hyla bürgeri Temminck and Schlegel, 1838, Japan.

OTHER COMMON NAMES
English: Kajika frog; Japanese: Kajika-Gaeru.

PHYSICAL CHARACTERISTICS
A medium-sized treefrog; males are 1.5–1.7 in (37–44 mm), and females are 1.9–2.7 in (49–69 mm) in snout-vent length. The body is slender and dorsoventrally depressed. This frog has a ground color of ash gray to brown with an irregular darker pattern on the back that blends in with the rocks on which it sits. The skin on the back has scattered irregular granules with blunt tips. The legs have a banding pattern but the abdomen is cream to white. The tips of the fingers and toes are expanded into large truncated disks.

DISTRIBUTION
This species is endemic to the mountainous regions of the islands of Honshu, Kyushu, and Shikoku, Japan.

HABITAT
Usually this species breeds in mid-sized streams with numerous boulders. Outside of the breeding season it has been seen along forest roads and in trees, and overwintering on river banks under stones and among sand.

BEHAVIOR
Male territoriality observed during the breeding season.

FEEDING ECOLOGY AND DIET
Known to feed on rather small insects and spiders.

REPRODUCTIVE BIOLOGY
This is a prolonged breeder, with males territorial on rocks within riffles of streams and calling both day and night from April into August. Females enter the stream from upland areas throughout the breeding season. Females are quickly amplexed by males when they enter a breeding site. Amplectant pairs may travel up to 130 ft (40 m) to a spawning site. Spawning occurs under rocks in the stream. The egg masses contain 200–600 eggs. The tadpoles with ventrally directed mouths feed on algae among the pebbles and rocks.

CONSERVATION STATUS
This species is relatively common throughout its range and is not considered threatened.

SIGNIFICANCE TO HUMANS
The call of this species is a high trill of 10 or more clear notes that is sometimes mistaken for that of a bird. Many hot-spring resorts in Japan take advantage of this beautiful call as a tourist attraction during the spring and summer. ◆

Betsileo golden frog
Mantella betsileo

SUBFAMILY
Mantellinae

TAXONOMY
Dendrobates betsileo Grandidier, 1872, Pays des Betsileos, Madagascar.

OTHER COMMON NAMES
English: Betsileo poison frog.

PHYSICAL CHARACTERISTICS
This is a small frog of 0.8–1.1 in (20–28 mm) in snout-vent length. The tips of the fingers and toes are not expanded into disks. It has a yellow to orange middorsal color that abruptly turns black dorsolaterally. There is a pale line along the upper lip. The venter is black with irregular blue spotting. The chin is blue. The legs are gray or brown with black bands. The upper half of the iris is golden.

DISTRIBUTION
This species is found from sea level to about 1,640 ft (500 m) in northeastern, western, and southern regions of Madagascar.

HABITAT
Lowland coastal areas, usually outside of forest.

BEHAVIOR
Active during the day; males call from exposed areas and fight with other males.

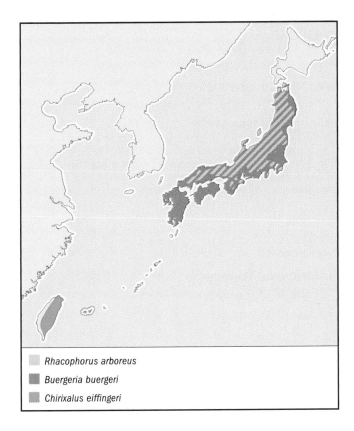

■ *Rhacophorus arboreus*
■ *Buergeria buergeri*
■ *Chirixalus eiffingeri*

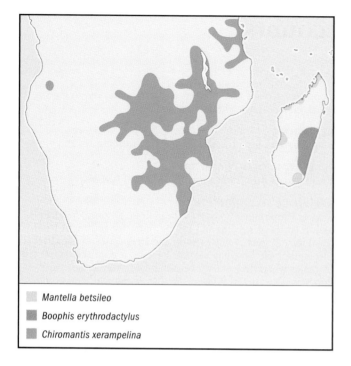

Mantella betsileo
Boophis erythrodactylus
Chiromantis xerampelina

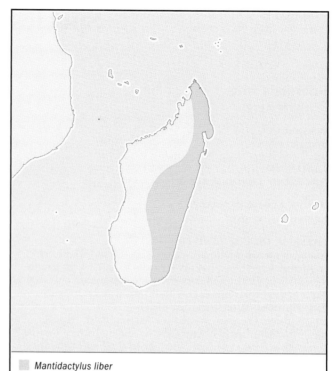

Mantidactylus liber

FEEDING ECOLOGY AND DIET
The diet consists of ants, fruit flies, and small beetles. These frogs actively search for insects during the day.

REPRODUCTIVE BIOLOGY
The call of the male consists of two short clicks with a short interval between them. Females lay clutches of eggs near streams where the tadpoles wash into pools.

CONSERVATION STATUS
Although this species is not threatened, it is exploited in the pet trade.

SIGNIFICANCE TO HUMANS
This frog may be desired as a pet, as are other mantella frogs. ◆

Free Madagascar frog
Mantidactylus liber

SUBFAMILY
Mantellinae

TAXONOMY
Rhacophorus liber Peracca, 1893, Andrangoloaka, Madagascar.

OTHER COMMON NAMES
None known.

PHYSICAL CHARACTERISTICS
This species is a relatively small treefrog with a snout-vent length of 0.8–1.1 in (21–29 mm) in males and 1.06–1.1 in (27–28 mm) in females. The fingers and toes have expanded disks. There is no webbing between the fingers and only moderate webbing between the toes. The males have either diffuse or large distinct femoral glands. They also possess a large white subgular vocal sac. The smooth dorsum varies between red, gray, or green, with a dark bar between the eyes, a pale median band, and white or yellow spots on the side under the hind legs. The legs and ventrum may be black.

DISTRIBUTION
Eastern and central Madagascar from sea level to 3,900 ft (1,200 m).

HABITAT
This species is usually found in or around water holding plants within or out of primary forests.

BEHAVIOR
Males call from vegetation near swamps, pools, and slow-moving water during the rainy season.

FEEDING ECOLOGY AND DIET
Probably feeds on small insects, spiders, and other arthropods.

REPRODUCTIVE BIOLOGY
The male's call sounds like two pebbles hitting each other. The female is attracted to a male by his call. She may nudge him from behind, and he places the ventral side of his thighs on her head and shoulders and pulsates laterally. She then begins depositing between 30 and 90 eggs on a leaf overhanging water into which the tadpoles drop after five to seven days of development.

CONSERVATION STATUS
Not threatened.

SIGNIFICANCE TO HUMANS
None known. ◆

Forest bright-eyed frog
Boophis erythrodactylus

SUBFAMILY
Rhacophorinae

TAXONOMY

Hyperolius erythrodactylus Guibé, 1953, Forêt de Mahajeby, près de Morafenobe, Ouest de Madagascar.

OTHER COMMON NAMES

None known.

PHYSICAL CHARACTERISTICS

This is a small, bright green treefrog (snout-vent length of 0.8–1.3 in [20–32 mm]), with many red spots, each surrounded by a yellow ring. The upper eyelids are yellow with brown spots, and there is a yellow line from the snout to the eye. The venter is transparent, the bones are green, and the disks of the fingers and toes are red.

DISTRIBUTION

Eastern Madagascar.

HABITAT

Usually found on leaves of trees and shrubs near rapids of large streams.

BEHAVIOR

Not known.

FEEDING ECOLOGY AND DIET

Probably feeds on insects and other small arthropods.

REPRODUCTIVE BIOLOGY

Breeds in streams, where the tadpoles live in the rapids.

CONSERVATION STATUS

Not threatened, but a limited range in southeastern Madagascar.

SIGNIFICANCE TO HUMANS

None known. ◆

Eiffinger's Asian treefrog

Chirixalus eiffingeri

SUBFAMILY

Rhacophorinae

TAXONOMY

Rana eiffingeri Boettger, 1895, Liukiu [Ryukyu] Islands, either and probably from Okinawa, the middle group, or from Ohoshima, the northern group, Japan.

OTHER COMMON NAMES

English: Big-thumbed treefrog.

PHYSICAL CHARACTERISTICS

This is a relatively small but stout treefrog, with the males 1.2–1.4 in (31–35 mm) and the females 1.4–1.6 in (36–40 mm) in snout-vent length. The rough skin on the back has scattered small round tubercles and short ridges, is pale brown to dark brown with some black-brown spotting, sometimes with a dark triangle between the eyes, an X-shaped mark on the back, and cross-bands on the legs. The tips of the fingers and toes have expanded round disks.

DISTRIBUTION

Yaeyama Island Group, Japan; Taiwan.

HABITAT

This species inhabits mountain forests not necessarily near water and occurs in groves of bamboo.

BEHAVIOR

Males usually call near tree holes or from within cut bamboo that have filled with water.

FEEDING ECOLOGY AND DIET

Presumably feeds on small insects.

REPRODUCTIVE BIOLOGY

Males call near tree holes or within cut water-filled bamboo. This species breeds throughout the year in tree holes or cut water-filled bamboo 1.6–4.9 ft (50–150 cm) above the ground. Usually the bottom of the holes is covered with rotting leaves. The eggs are laid separately or in a small mass of about 20–70 eggs above water on the inner walls of the hole. Males often stay in the breeding hole, probably to moisten and, thus, keep the eggs from desiccating. Females spawn continually over several months and periodically return to the breeding site and lay unfertilized eggs directly into the water as food for the tadpoles.

CONSERVATION STATUS

This species is not considered threatened or endangered, but it has limited distribution in Taiwan and two small islands outside of Taiwan.

SIGNIFICANCE TO HUMANS

The practice of cutting bamboo may actually create a breeding habitat for these frogs in areas where tree holes are uncommon. ◆

Gray treefrog

Chiromantis xerampelina

SUBFAMILY

Rhacophorinae

TAXONOMY

Chiromantis xerampelina Peters, 1854, Tette and Sena, Mozambique.

OTHER COMMON NAMES

English: Foam nest frog, southern foam nest treefrog, great African gray treefrog, African gray treefrog; German: Ruderfrosch.

PHYSICAL CHARACTERISTICS

This species is a relatively large tree frog with a snout-vent length of 2.8 in (72 mm) in males and 3.3 in (85 mm) in females. It is robust with long limbs. The fingers and toes have expanded disks, and the two outer fingers are opposable to the inner fingers, thereby enabling better grasp of the limbs on which it perches. The dorsum is shades of gray and brown with variable darker markings on a roughly textured skin, effectively concealing the frog against different backgrounds, especially the bark of trees. However, the color can change to almost white as temperature rises.

DISTRIBUTION

Savannas of coastal Kenya and northeastern Namibia south to Natal, Republic of South Africa.

HABITAT

This species usually occurs in warm regions at low elevations and is common in dry savanna.

BEHAVIOR

This frog and one of the two other species of this genus are unique to rhacophorids in their ability to conserve water so that they can live in the dry African savanna. Individuals orient themselves when resting on tree limbs so that their bodies do not receive full exposure to the sunlight. They tuck the arms and legs under their body, thereby decreasing the amount of surface area exposed to the air and thus reducing evaporative water loss. Also, the frogs turn almost white during the hottest times of the day to reduce heat absorption. While estivating during the dry season, this frog secretes a fluid that turns into a waterproof cocoon. In addition, physiologically this species has been shown to be more tolerant of higher temperatures, and its skin is resistant to water loss of up to 35 times that of other frogs. However, if the temperature becomes too high, this frog will produce drops of water on the skin to cool it down by evaporative cooling. Also, instead of producing urine, as in most frogs, or ammonia, as in highly aquatic frogs, this species, like reptiles, produces the semi-solid uric acid to conserve water.

FEEDING ECOLOGY AND DIET
Probably feeds on insects and other arthropods.

REPRODUCTIVE BIOLOGY
This species deposits fertilized eggs as a foam nest. The female may begin building the foam nest, stop, climb down from her perch, and rehydrate in the pool below. At this point the male usually releases her. She then climbs back up and continues building the foam nest. She may repeat this two to four times. The male that originally was on her back may not be the one that is on her back when she releases 500–1,200 eggs into the foam nest. The female may return to the nest the following night and add more foam but not more eggs to the nest to keep it from dehydrating. Communal nests result from up to 20 females and twice as many males building nests close enough together so that they coalesce. Sperm competition is thought to occur in this species (as in many foam-nesting treefrogs with multiple males spawning with one female). This is further confirmed by the presence of large testes in the males that allows them to produce and shed sperm multiple times throughout the breeding season. After four to six days the developing larvae drop from the nest into the water below.

CONSERVATION STATUS
Not threatened.

SIGNIFICANCE TO HUMANS
None known. ◆

☐ *Nyctixalus pictus*

■ *Philautus surdus*

from the edge of the snout, along the edge of the upper eyelid, and continuing partway down the side of the back. Some individuals are red or orange. The upper half of the iris is white, and the lower half is brown. The webbing on the hand is absent or only basal, and the webbing on the foot is moderate.

DISTRIBUTION
Borneo, Malaya, Sumatra, and Palawan Island, Philippines.

HABITAT
This species is found in both lowland and montane forests from near sea level to 5,400 ft (1,650 m). Adults have been found on leaves of shrubs and small trees one to three meters above the ground but probably are also higher in trees.

BEHAVIOR
Not known.

FEEDING ECOLOGY AND DIET
Presumably feeds on small invertebrates.

REPRODUCTIVE BIOLOGY
This species deposits approximately 10 eggs in a gelatinous mass on the inner walls of water-filled tree holes. The hatchling tadpoles drop into the water and feed on the detritus within the tree holes.

CONSERVATION STATUS
Not threatened.

SIGNIFICANCE TO HUMANS
None known. ◆

Painted Indonesian treefrog
Nyctixalus pictus

SUBFAMILY
Rhacophorinae

TAXONOMY
Ixalus pictus Peters, 1871, Sarawak, Malaysia (Borneo).

OTHER COMMON NAMES
English: Cinnamon treefrog, Peter's treefrog.

PHYSICAL CHARACTERISTICS
This small- to medium-sized treefrog, males 1.12–1.5 in (30–37 mm) and females 1.46–1.54 in (37–39 mm) in snout-vent length, has a relatively long pointed snout and slender limbs. The skin on the back is rough with numerous spiny tubercles, and the skin on the head is co-ossified to the skull. The dorsum is cinnamon to chocolate brown with small white spots scattered throughout, but which also form a broken line

Luzon bubble-nest frog
Philautus surdus

SUBFAMILY
Rhacophorinae

TAXONOMY
Polypedates surdus Peters, 1863, Luzon, Philippines.

OTHER COMMON NAMES
English: Common forest treefrog.

PHYSICAL CHARACTERISTICS
This small treefrog is 0.9–1.1 in (22–28 mm) in snout-vent length. The tips of the fingers and toes are expanded into disks. There is a pair of tubercles within dark spots at the shoulder and tubercles on the upper eyelids. Webbing is absent between the fingers but moderate between the toes.

DISTRIBUTION
Bohol, Mindanao, and Polillo Islands, Philippines.

HABITAT
Primary forests from 1,640 to 6,560 ft (500–2,000 m) in elevation, not necessarily near water.

BEHAVIOR
Not known.

FEEDING ECOLOGY AND DIET
Probably feeds on small insects and spiders.

REPRODUCTIVE BIOLOGY
Breeding occurs throughout the year. This species lays five to 19 large, unpigmented eggs in the leaf axils of ferns. The embryos develop directly into froglets. The tail is absorbed just before hatching.

CONSERVATION STATUS
Not threatened.

SIGNIFICANCE TO HUMANS
None known. ◆

Kinugasa flying frog
Rhacophorus arboreus

SUBFAMILY
Rhacophorinae

TAXONOMY
Polypedates arboreus Okada and Kawano, 1924, Kinugasa, Kyoto, Honshu, Japan.

OTHER COMMON NAMES
English: Forest green treefrog; Japanese: Mori-ao-gaeru.

PHYSICAL CHARACTERISTICS
This relatively large treefrog has a large head. Males are 1.7–2.4 in (42–60 mm), and the females are 2.3–3.2 in (59–82 mm) snout-vent length. The dorsum is bright green with or without black or brown spots with black edges. The abdomen is white or cream with pale brown spots. The iris is orange to brownish red. The backs of the thighs are white with black mottling reticulation. The webbing on the hand is well developed, but the webbing on the toes is moderate. The tips of the toes and fingers are expanded into large, truncated disks. The skin on the back is rough with tubercles on the upper eyelid, elbow, and shanks.

DISTRIBUTION
The species is endemic to Honshu, Japan, and the small island of Sado off northeastern Honshu; it occurs from sea level to over 6,560 ft (2,000 m) and is most common in mountainous regions.

HABITAT
Outside of the breeding season it is found perched in trees or under leaf litter, usually in forested areas but also in urban gardens. During the breeding season it is often seen in trees, grass, and on the ground near ponds and rice fields. During the winter it hibernates under moss or shallow soil.

BEHAVIOR
This frog is a prolonged breeder, beginning in April and continuing through July. Males set up territories around a breeding site, usually a pond or rice field from where they call.

FEEDING ECOLOGY AND DIET
This frog feeds on insects.

REPRODUCTIVE BIOLOGY
The males' call is a series of two to six clicks usually followed by a lower series of clucking sounds. This species deposits eggs in a foam nest on vegetation or the ground over standing water. A female exudes an albumen-based fluid from her cloaca, which she beats with her hind feet into an elliptical foam mass approximately 3.5 x 4.7 in (88 x 120 mm). The amplexing male may also participate by beating the foam with his hind feet. After the foam nest is completed, she deposits 300–800 eggs into the nest, and the male sheds his sperm over the eggs as they leave her cloaca. In some cases the female is surrounded by several males in addition to the male on her back, and all will participate in beating the fluid into a foam mass. These males then shed sperm into the foam mass along with the amplexing male. The foam mass hardens on the outside, protecting the developing embryos from desiccation and predation. After the tadpoles have hatched, the bottom of the foam nest softens by weathering or possibly enzymes released from the hatched eggs, and the tadpoles fall from the foam nest into the standing water below where they develop to metamorphosis.

CONSERVATION STATUS
This frog has some protection because of its rarity in a few prefectures in Japan (i.e., Nagano Prefecture).

SIGNIFICANCE TO HUMANS
This species is a predominant part of rural life in Japan, as the chorus of frogs signifies the long summer nights. There are a few ponds where literally hundreds of adult frogs are seen breeding day and night and are therefore set aside as tourist attractions. ◆

Resources

Books

Alcala, A. C., and W. C. Brown. *Philippine Amphibians: An Illustrated Field Guide.* Makati City, Philippines: Bookmark, Inc., 1998.

Channing, A. *Amphibians of Central and Southern Africa.* Ithaca, NY: Comstock Publishing Associates, 2001.

Glaw, F., and M. Vences. *A Fieldguide to the Amphibians and Reptiles of Madagascar.* Frankfurt, Germany: Edition Chimaira, 1999.

Inger, R. F., and R. B. Stuebing. *A Field Guide to the Frogs of Borneo.* Kota, Indonesia: Natural History Publications, 1997.

Maeda, N., and M. Matsui. *Frogs and Toads of Japan.* Tokyo, Japan: Bun-Ichi Sogo Shuppan Co., 1990.

Passmore, N. I., and V. C. Carruthers. *South African Frogs: A Complete Guide.* Johannesburg, South Africa: Witwatersrand University Press, 1995.

Schiøtz, A. *Treefrogs of Africa.* Frankfurt, Germany: Edition Chimaira, 1999.

Zug, G. R., L. J. Vitt, and J. P. Caldwell. *Herpetology: An Introductory Biology of Amphibians and Reptiles.* San Diego: Academic Press, 2001.

Periodicals

Bossuyt, F., and M. C. Milinkovitch. "Convergent Adaptive Radiations in Madagascan and Asian Ranid Frogs Reveal Covariation Between Larval and Adult Traits." *Proceedings of the National Academy of Science* 97 (2000): 6585–6590.

Brown, W. C., and A. C. Alcala. "Philippine Frogs of the Family Rhacophoridae." *Proceedings of the California Academy of Sciences 48* (1994): 185–220.

Channing, A. "A Re-evaluation of the Phylogeny of Old World Treefrogs." *South African Journal of Zoology* 24 (1989): 116–131.

Emerson, S. B., C. Richards, R. C. Drewes, and K. M. Kjer. "On the Relationships Among Ranoid Frogs: A Review of the Evidence." *Herpetologica* 56 (2000): 209–230.

Glaw, F., M. Vences, and W. Böhme. "Systematic Revision of the Genus *Aglyptodactylus Boulenger,* 1919 (Amphibia: Ranidae), and Analysis of Its Phylogenetic Relationships to Other Madagascan Ranid Genera (*Tomopterna, Boophis, Mantidactylus,* and *Mantella*)." *Journal of Zoology, Systematics, and Evolutionary Research* 36 (1998): 17–37.

Kaul, R., and V. H. Shoemaker. "Control of Thermoregulatory Evaporation in the Waterproof Treefrog *Chiromantis xerampelina.*" *Journal of Comparative Physiology and Biochemistry* 158 (1989): 643–649.

Richards, C. M., and W. S. Moore. "A Molecular Phylogenetic Study of the Old World Treefrog Family Rhacophoridae." *Herpetological Journal* 8 (1998): 41–46.

Richards, C. M., R. A. Nussbaum, and C. J. Raxworthy. "Phylogenetic Relationships Within the Madagascan Boophids and Mantellids As Elucidated by Mitochondrial Ribosomal Genes." *African Journal of Herpetology* 49 (2000): 23–32.

Wassersug, R. J., K. J. Frogner, and R. F. Inger. "Adaptations for Life in Tree Holes by Rhacophorid Tadpoles from Thailand." *Journal of Herpetology* 15 (1981): 41–52.

Jeffery Wilkinson, PhD

Narrow-mouthed frogs

(Microhylidae)

Class Amphibia

Order Anura

Family Microhylidae

Thumbnail description
Tiny to medium-size frogs with a broad spectrum of morphologic features

Size
0.5–4 in (11.5–100 mm)

Number of genera, species
67 genera; 362 species

Habitat
Forest, woodlands, scrub, savanna, grassland, semidesert

Conservation status
Endangered: 2 species; Vulnerable: 6 species; Data Deficient: 3 species

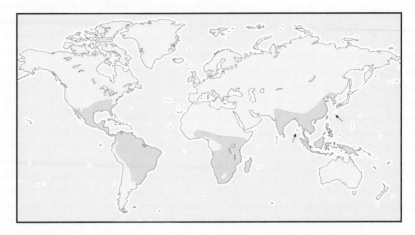

Distribution
The eastern to southwestern United States southward through much of South America; sub-Saharan Africa, Madagascar, most of India, Sri Lanka, and Southeast Asia to extreme northern Australia

Evolution and systematics

It scarcely has been questioned that the Microhylidae family constitutes a natural group most closely related to others among the ranoid frog families, such as the Ranidae, Rhacophoridae, and Hyperoliidae. Characterizing the family proves elusive, however. Definitions always include the unique character of serrated, transverse folds of skin on the palate, but some species lack these features or have them in reduced numbers and sizes. Emphasis always is placed on a distinctive suite of larval characters, though almost 30% of the genera and nearly half the species of microhylids undergo direct embryonic development, skipping a free-living larval stage. Where there are free-swimming, feeding larvae, they typically lack the cornified denticles ("teeth") and beak seen in larvae of other families and have a median, ventral spiracle (the opening through which water taken in through the mouth is discharged) rather than one on the left side. Here again there are exceptions.

The fossil record of the Microhylidae is meager, but if the assignment of fossils from the Miocene of Florida (about 24 million years ago) to the present-day genus *Gastrophryne* is accurate, it speaks to a moderately long history. The presence of Microhylidae in South America, Africa, Madagascar, India, and Australia strongly suggests a Gondwanan origin. If this is correct, the abundance of microhylids in Southeast Asia, reaching as far as Korea, remains to be explained, as does their restriction to northern Australia. One scenario has primitive microhylids riding northward on the drifting Indian subcontinent, thereby gaining access to Asia and, eventually, New Guinea and Australia. Alternatively, they could have accompanied Australia on its northward journey, spreading to New Guinea and the East Indies as those more recent terrains emerged. Until a better understanding of relationships among

the microhylid families is gained, the zoogeography of the Microhylidae will remain enigmatic.

A monograph of the family Microhylidae by H. W. Parker, published in 1934 and still the only family-wide treatment, was conservative in outlook; subsequent research has increased greatly the recognized number of genera and species. In 1954 Antenor Leitão de Carvalho published the first modern treatment of the American genera of microhylids. A phylogenetic arrangement of the New World microhyline genera presented by Zweifel in 1986 was refined by Donnelly, de Sá, and Guyer in 1990 and again by Wild in 1995. The Asterophryinae received similar attention from Zweifel in 1972 and Burton in 1986.

Relationships among the microhylid subfamilies and their genera are not well worked out. The presence of maxillary and vomerine teeth together with the retention of clavicles and procoracoid cartilages in the genera *Dyscophus* (Dyscophinae) and *Platypelis* (Cophylinae) mark these Malagasy frogs as primitive with respect to other microhylids. No firm line of relationship has been drawn, however, to other subfamilies. Relationships among the African and Asiatic microhylines have not been investigated.

Nine subfamilies are recognized; the Scaphiophrynidae of Madagascar formerly was placed as a subfamily but is now recognized as a separate family.

Asterophryinae

The vertebral column is diplasiocelous: all vertebrae are procelous (concave anteriorly, convex posteriorly) except the eighth, which is opisthocelous (convex anteriorly, convex posteriorly). There are no teeth, and the tongue is largely adherent behind. The maxillary bones tend to meet in front of the premaxillary bones or overlap them, and the pectoral girdle lacks clavicles and procoracoid cartilages. The latter are paired elements of the pectoral girdle present in the majority of frogs. All species undergo direct embryonic development. The distribution is confined to New Guinea and the Moluccas. There are eight genera with 54 species.

Brevicipitinae

The vertebral column is diplasiocelous, and there are clavicles and procoracoid cartilages. This subfamily has no teeth. The vomerine bones, paired elements in the roof of the mouth, are large, with a median expansion. Tadpoles are not free swimming. These frogs occur wholly in Africa, from Ethiopia to the Cape of Good Hope. There are five genera with 19 species.

Cophylinae

The vertebral column is procelous; there are procoracoids, but there may or may not be clavicles. This subfamily has maxillary or vomerine teeth or both. Four of the genera lack the transverse palatal folds found in almost all other microhylids. The distribution is restricted to Madagascar. There are seven genera with 35 species.

Dyscophinae

The vertebral column is diplasiocelous; clavicles and procoracoids typically are present, but they are absent in one species. The anterior median element of the pectoral girdle, the omosternum, is bony (which is unique within the Microhylidae) or cartilaginous. This subfamily has maxillary and vomerine teeth. The distribution is broadly disjunct, with one genus (*Dyscophus*, three species) in Madagascar and one genus (*Caluella*, six species) in Madagascar and Asia. Their supposed evolutionary relationship is based largely on shared primitive characters that are not phylogenetically informative, and it deserves to be reassessed.

Genyophryninae

The vertebral column is procelous; maxillary bones overlap the premaxillae only slightly. The pectoral girdle may lack clavicles and procoracoids, may have both, or may have only procoracoids. Teeth usually are lacking, but vestiges may be present on the maxilla; the tongue is at least one-fourth free behind. All species undergo direct embryonic development. The distribution is concentrated in New Guinea, with fringe species in the southern Philippines, Sulawesi, the Lesser Sunda Islands, New Britain, and northern Australia. There are 11 genera and 118 species, more species than in any other subfamily.

Melanobatrachinae

The vertebral column is procelous; there are no clavicles, and the procoracoids may be present or vestigial. This subfamily lacks a tympanum (eardrum) and teeth. Tadpoles are not free swimming. The distribution is discontinuous; the genera are dispersed between eastern Africa (Tanzania: two genera, with three species) and India (one monotypic genus).

Microhylinae

The vertebral column is diplasiocelous (rarely procelous); the pectoral girdle may lack clavicles and procoracoids, may

have both, or may have only procoracoids. There are no teeth. Vomers typically are present in separate anterior and posterior locations, but they may be absent posteriorly. Most species typically have free-swimming microhylid larva lacking cornified denticles and beak. The distribution is widespread; this is the only microhylid family in the New World apart from the Otophryninae, and it also ranges from eastern Pakistan, Sri Lanka, and India to China, Korea, the Ryukyu Islands, and into the Southeast Asian archipelago. There are 29 genera, with 115 species (second only to the Genyophryninae).

Otophryninae

The vertebral column is diplasiocelous, and there are clavicles and procoracoids. The larval spiracle is on the left side rather than median, and extends posteriorly as a tube as the tadpole grows. The larvae have dagger-like, keratinized teeth. This subfamily is distributed in northern South America from southeastern Colombia to the extreme northeast of Brazil. There is one genus, with three species. The recognition of the Otophryninae as a taxon distinct from the Microhylinae is questioned by some researchers.

Phrynomerinae

The vertebral column is diplasiocelous; procoracoids and clavicles are absent. A pad of cartilage (the intercalary cartilage) separates the distal two phalanges of each finger and toe. Tadpoles are the typical, free-swimming microhylid sort. The subfamily is wholly African in distribution, found from western Africa to Somalia and South Africa. There is one genus, with five species. The Phrynomerinae once held family status, but only the phalangeal character distinguishes *Phrynomantis* from other microhylids.

Physical characteristics

Narrow-mouthed frogs have no external physical characteristics that enable a person with a living frog in hand, but no other pertinent information, to identify it as a microhylid. Most genera of microhylids have two or three serrated folds across the palate (a feature unique to this family) and lack teeth. The pectoral girdle always has coracoid bones that meet on the midline (the so-called firmisternal condition) but often shows reduction or absence of other ventral elements. Clavicles may be as small as tiny slivers of bone or may be lacking, and the procoracoid cartilage may be absent. The omosternum (which is not always present) is, with one exception, cartilaginous rather than bony. Species with exceptions to some of these characters are presumably primitive forms in Madagascar that have teeth, lack the palatal folds, and show minimal reduction of elements of the pectoral girdle.

There are few large microhylids. Only about 8% of the species have a body length as great as 2.4 in (60 mm), whereas about 83% attain a length of less than 2 in (50 mm). A few are less than 0.5 in (13 mm) when they reach adulthood. Microhylids have adapted to a wide variety of habitats and consequently show diverse body forms. One of the most common is a teardrop shape with chunky body and narrow head ending in a pointed snout. This habitus is responsible for the name narrow-mouthed frogs, but the appellation is inappropriate for many microhylid genera. Some microhylids are true treefrogs with large eyes and expanded terminal disks on the fingers and toes that facilitate climbing. Terrestrial species include squat, small-eyed, short-legged frogs that spend most of their time in burrows or hidden in leaf litter and other, more svelte and agile forms that live in and on the forest floor amid leaf litter.

Microhylids typically are drab-colored frogs with shades of brown and dull yellow dominating dorsally; the undersides may be brighter. Species of rubber frogs (genus *Phrynomantis*) in southern Africa are exceptions; they have bright-red dorsal markings on a black or brown background. This may be a warning coloration, as at least one species produces toxic skin secretions.

Distribution

Most of the geographic range of the Microhylidae lies between the Tropics of Cancer and Capricorn, with a few species exceeding these limits in North America, South America, southern Africa, and Southeast Asia. In addition to these continental distributions and their presence in the East Indies, Philippines, and northern Australia, microhylids are an important component of the frog faunas of the large tropical islands of Madagascar and New Guinea.

Habitat

Microhylids occupy diverse habitats ranging in elevation from sea level to about 13,000 ft (4,000 m). A majority of species dwell in tropical rainforests, occupying a variety of ecological niches. Some live in burrows in the soil, from which they emerge to feed or find mates; others seem to stay largely within the shelter of the forest floor amid leaf litter. Climbing species may ascend low shrubbery at night, but more arboreal varieties may spend their entire lives high in the trees. A small number of species (genera *Oreophryne* and *Oxydactyla*) inhabit alpine grassland at elevations of more than 12,000 ft (3,700 m) in New Guinea. Temperate regions apparently lack arboreal microhylids, but frogs nevertheless range through terrestrial habitats from moist lowland forests to nearly desert conditions. Rain frogs (genus *Breviceps*) in southern Africa even burrow in sand dunes. Although aquatic habitats serve as breeding sites for many microhylid species and there are a few riparian species, no microhylid is known to be primarily aquatic.

Behavior

Given their wide range of habitats and diversity of body forms, microhylids would be expected to show a variety of adaptive behaviors, and indeed they do. Some terrestrial species have flattened, projecting "spades" on the heels that facilitate burrowing backward into the soil with a shuffling motion. Others lack these structures and burrow headfirst. Defensive activity is usually "leap and hide"—that is, dive underwater or burrow into surface litter—which is characteristic of frogs in general. A broad-headed frog of New Guinea,

The Malaysian painted frog (*Kaloula pulchra*) is active at night. The frog has no cervical vertebrate, and its lack of a neck makes the frog appear fat, giving it the common name of "chubby frog." (Photo by Animals Animals ©Zig Leszczynski. Reproduced by permission.)

Asterophrys turpicola, takes the offensive when disturbed. It faces a potential predator with its body inflated and its mouth gaping, displaying a bright blue tongue; then it bites and holds on. If attacked, the Madagascar tomato frog (*Dyscophus antongilii*) can produce slime so sticky that a small predator trying to eat it could find its eyelids and lips stuck together. In its adhesive properties, the slime is nearly five times stronger than rubber cement! The Great Plains narrow-mouthed toad (*Gastrophryne olivacea*) in the United States and a South American species, the dotted humming frog (*Chiasmocleis ventrimaculatus*), sometimes live with and are not attacked by tarantulas in their burrows. Quite likely this affords a degree of protection from various would-be predators. There may be a trade-off, too, if the frogs eat insects that might attack the spider's eggs.

Seasonal activity varies with climate. Where low temperature is not a factor, as in the tropics, rainfall is probably the chief determinant of the timing of the frogs' activity. Even in tropical rainforests there may be seasonal variation, and frogs may actively move about or call only during heavy rains. In tropical regions with a pronounced dry season, microhylids remain in burrows or other concealment and emerge only in the event of heavy rains. Few microhylid species live where freezing temperatures regularly occur, and those frogs depend on both warmth and rainfall to allow them to be active. Microhylids generally are nocturnal, but species at high elevations (and sometimes elsewhere) call in the daytime.

Feeding ecology and diet

Ants are a common prey of microhylid frogs, especially pointy-nosed species with a narrow gape. Just as in other features of their biological makeup, microhylids show diversity in diet. Even species that feed largely on ants do not ignore other tiny invertebrates. Larger species with a broad gape can eat bigger prey in addition to the usual insects; examples include lizards, frogs, and earthworms. Little is known of feeding behavior. Frogs found moving about on the forest floor may be foraging for food, but sit-and-wait behavior may be

A Sambava tomato frog (*Dyscophus guineti*) tucks its head down in a defensive posture. (Photo by Suzanne L. & Joseph T. Collins/Photo Researchers, Inc. Reproduced by permission.)

more common. Free-swimming larvae of microhylids (as opposed to those that develop in confined situations and do not feed) all feed by filtering microscopic organisms from the water. Some with a funnel-shaped mouth specialize in taking food from the water's surface.

Reproductive biology

The common (and presumably primitive) mode of reproductive behavior in frogs includes courtship, in which male vocalization plays an important part, and amplexus accompanied by external fertilization of the eggs as they are extruded into water. The adults then go their separate ways. The tadpoles grow and metamorphose into frogs and then, according to the habit of the species, remain in the water or disperse to other habitats. Except for the fact that adults do not remain in the water, many microhylids engage in this sort of reproductive behavior. As is the case with other aspects of their biological makeup, microhylid breeding is not stereotyped.

Mating in most kinds of frogs requires amplexus, in which the male, situated dorsally, grasps the female just behind the front legs (axillary position) or just anterior to the hind legs (inguinal position), putting himself in position to fertilize the spawn as it is extruded (there are variations on these positions). Surprisingly little is recorded about mating in microhylids, but axillary amplexus is known in species that deposit eggs in water. In many, perhaps most, species of frogs, males that are ready to breed possess patches of cornified areas, sometimes literally spines, on one or more of the fingers. These areas enable the frog to maintain a grip on a slippery female while mating.

Such structures are absent in most microhylids, but they are found in at least two genera, *Dyscophus* and *Hoplophryne*. In the species of another genus, *Anodonthyla*, a single spine projects from the first finger. *Gastrophryne* and some other genera have another solution: ventral glands in the male secrete a substance that glues the pair together. The African

genus *Breviceps* does this to bizarre effect. These are short-legged, rotund frogs, the female being much the larger of the two. Imagine a golf ball glued to a tennis ball. Very likely such adhesive behavior is more widespread in the Microhylidae than is recognized. Nothing at all is known about how the numerous direct-developing genyophrynine and asterophryine microhylids achieve insemination.

Breeding sites are diverse. Species with free-swimming but feeding larvae breed in quiet waters. According to the species, however, they may be permanent waters or sites that are dry but reliably fill in rainy seasons or fill only during sporadic heavy rains. Species with free-swimming but nonfeeding larvae breed in narrowly confined waters; examples include the leaf axils of bromeliads, tree holes, and pools in the crevices of logs. Species with direct embryonic development are divorced from free water and nest in burrows, leaf litter, and other sheltered terrestrial sites or in moisture-holding plants high in forest trees.

The timing of breeding is controlled primarily by the availability of adequately wet conditions and secondarily by temperature. In wet, largely tropical settings without distinct seasons, some individuals may be in a condition to breed year-round. Even here, however, there may be periodic variations in rainfall, which influence breeding activity. In drier tropical situations, breeding may be confined to discrete rainy seasons, and especially in temperate climate areas frogs may breed only at a warm, wet time of the year.

Vocalizing is one of the defining aspects of frogs, and microhylids are no exception. Probably all male microhylid frogs produce what are now called advertisement calls (formerly mating calls); even species that lack the secondary apparatus of a vocal sac with openings into the mouth cavity are known to call. Microhylid calls vary broadly, from pure-toned, high-pitched peeps to low, harsh notes, given either singly or in series. The characteristics of a call often define a species, but no correlation with higher systematic categories has been shown. Little research has been done on behavioral aspects of vocalization in microhylids. Certainly, the call serves to identify the caller to a female as an appropriate mate. In species of other families, a male's call has been shown to include aspects that may bear upon his suitability as a mate. Calling also functions in territoriality. This is to be expected, especially among forest-floor microhylids that do not migrate to water to breed.

Eliminating feeding is the first step beyond free-swimming larvae that feed themselves. In this mode, the eggs are heavily yolked, much larger than typical aquatic eggs, and are deposited in a small body of water, such as a tree hole or a leaf axil. The larva hatches (sometimes in an advanced stage of development) and subsists on the stored yolk until it metamorphoses. This mode of development evidently has evolved independently in different parts of the world, for example, in Madagascar (genera *Anodonthyla*, *Platypelis*, and *Plethodontohyla*), Southeast Asia (genus *Kalophrynus*), and South America (genus *Syncope*).

Direct development, where growth through metamorphosis occurs within the egg capsule, eliminates the necessity for depositing eggs in water but still requires a moist situation.

Approximately one-half of all species of Microhylidae have this reproductive mode. All species of the subfamilies Asterophryinae and Genyophryninae evidently engage in this mode of development, as do some genera in other parts of the world, for example, the African rain frog of the genus *Breviceps*.

Direct development carries with it advantages as well as constraints. The need for a large amount of yolk greatly restricts the number of eggs a female frog can produce at one time. Thus, a high survival rate of eggs and hatchlings is essential if the species is to persist. Guarding of egg clutches by male frogs probably protects them from predaceous and parasitic insects and may minimize dehydration as well. Guarding behavior is known in eight genera of Asterophryinae and Genyophryninae and probably is universal in these families. Frogs of two genyophrynine genera, *Aphantophryne* and *Liophryne*, have even been seen to transport newly hatched young on their backs. In moist tropical regions, direct development greatly increases the potential area a species can inhabit, because mountainous areas generally are poor in still-water habitats of the sort many microhylids require for breeding. Also, such free water as exists is likely to be in rapidly flowing streams, and no microhylid tadpole is known to have the peculiar adaptations needed for life in such streams. As long as there is sufficient rainfall, however, frogs that undergo direct development can spread, breeding from underground to treetop habitats.

Conservation status

As of 2002, the IUCN classifies only two microhylids as Endangered: the Camiguin narrow-mouthed frog (*Oreophryne nana*), found only on one small island in the southern Philippines, and the black microhylid (*Melanobatrachus indicus*) of southwestern India. In addition, six species are classified as Vulnerable and three as Data Deficient. The Cape rain frog (*Breviceps gibbosus*) and the desert rain frog (*Breviceps macrops*), both with small ranges in South Africa, are listed as Vulnerable, as is the Negros truncate-toed chorus frog (*Kaloula conjuncta negrosensis*) of Negros Island in the Philippines and the tomato frog (*Dyscophus antongilii*) in Madagascar.

Most microhylids are small, dull-colored creatures that attract little attention and so are unlikely to become well enough known to be recognized as needing formal protection. Where they live in national parks, such as in northern Australia, protection of both habitats and individual species may be afforded, at least nominally. Many countries and lesser jurisdictions have laws to protect wildlife, frogs included, but too often these laws serve to inhibit scientific research without doing much else. Most microhylid frogs live in tropical rainforests, and such forests around the world are being destroyed at an alarming rate. It is inevitable that many species will become extinct before their plight is even recognized, and many others will disappear without achieving scientific recognition.

Significance to humans

Microhylids frogs figure in the diets of indigenous peoples in New Guinea, South America, and probably elsewhere. Given the small size of most species, the contribution of microhylids to human nutrition must be meager and the danger from overutilization slight.

1. Horned land frog (*Sphenophryne cornuta*); 2. Wilhelm rainforest frog (*Cophixalus riparius*); 3. Boulenger's climbing frog (*Anodonthyla boulengerii*); 4. Fry's whistling frog (*Austrochaperina fryi*); 5. New Guinea bush frog (*Asterophrys turpicola*); 6. Bushveld rain frog (*Breviceps adspersus*); 7. Saffron-bellied frog (*Chaperina fusca*); 8. Boulenger's callulops frog (*Callulops robustus*). (Illustration by Brian Cressman)

1. Timbo disc frog (*Synapturanus salseri*); 2. Ornate narrow-mouthed frog (*Microhyla ornata*); 3. Eastern narrow-mouthed toad (*Gastrophryne carolinensis*); 4. Pyburn's pancake frog (*Otophryne pyburni*); 5. Banded rubber frog (*Phrynomantis bifasciatus*); 6. Malaysian painted frog (*Kaloula pulchra*); 7. Bolivian bleating frog (*Hamptophryne boliviana*). (Illustration by Brian Cressman)

Species accounts

New Guinea bush frog
Asterophrys turpicola

SUBFAMILY
Asterophryinae

TAXONOMY
Ceratophrys turpicola Schlegel, 1837, Lobo district, Triton Bay, Dutch New Guinea (Irian Jaya, Indonesia).

OTHER COMMON NAMES
None known.

PHYSICAL CHARACTERISTICS
This species is the antithesis of the narrow-mouthed frog; its head (and, consequently, its gape) is as broad as its wide body. The eyelids bear fleshy spines, and the limbs and sides of the body are warty. It is one of the larger microhylids, reaching a body length of 2.5 in (65 mm). Its coloration is drab, with brown and black shades dominating.

DISTRIBUTION
This frog lives at low to moderate elevations in New Guinea, from the western end in Irian Jaya (Indonesia) to eastern Papua New Guinea.

HABITAT
This is a species of the forest floor, where it calls from sites below the surface. In at least one region in Papua New Guinea it has adapted to disturbed environments and is found in suburban gardens.

BEHAVIOR
The New Guinea bush frog is noteworthy for its unusual defensive behavior. When annoyed, it may inflate the body and hold the mouth open, exposing the bright blue tongue. If annoyance persists, the frog may leap at and bite its disturber, holding on for minutes. Curiously, closely similar behavior is found in unrelated but morphologically similar species of the genera *Hemiphractus* (Hylidae) and *Ceratophrys* (Leptodactylidae) in tropical America.

FEEDING ECOLOGY AND DIET
This species has a wide-ranging diet that includes lizards, insects, and frogs.

REPRODUCTIVE BIOLOGY
Undoubtedly, the species has direct embryonic development, but no details of the reproductive habits are known.

CONSERVATION STATUS
Not threatened.

SIGNIFICANCE TO HUMANS
None known. ◆

Boulenger's callulops frog
Callulops robustus

SUBFAMILY
Asterophryinae

TAXONOMY
Mantophryne robusta Boulenger, 1898, Saint Aignan island, south of Fergusson island, British New Guinea (Misima Island, Louisiade Archipelago, Milne Bay Province, Papua New Guinea).

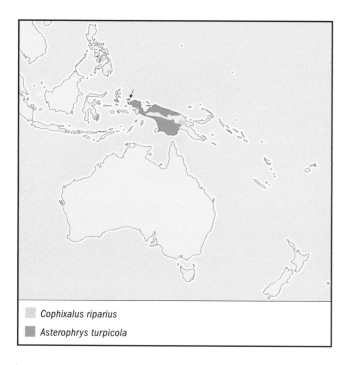

☐ *Cophixalus riparius*
■ *Asterophrys turpicola*

☐ *Callulops robustus*
■ *Sphenophryne cornuta*

OTHER COMMON NAMES
None known.

PHYSICAL CHARACTERISTICS
This is a rather toadlike species, with a relatively broad head and body, large eyes, and short legs. At a maximum body length of almost 3 in (73 mm), it is one of the larger microhylids. Many individuals are dark purplish brown dorsally with a somewhat darker facial region. Small white spots may be present laterally on the body. A variant coloring is light reddish brown all over. These different colors may represent geographic variation.

DISTRIBUTION
Boulenger's callulops frog has a wide distribution in New Guinea, from the Birds Head peninsula at the western tip of the island to islands off the eastern end. It occurs from nearly sea level to an elevation of at least 4,800 ft (1,460 m).

HABITAT
The species lives in rainforest regions, though not necessarily primary forest.

BEHAVIOR
The frogs are terrestrial, sheltering in burrows up to 3 ft (1 m) in length, which probably are appropriated by the frogs rather than constructed by them.

FEEDING ECOLOGY AND DIET
Nothing is recorded.

REPRODUCTIVE BIOLOGY
Males call from the entrances to burrows, though they sometimes leave the burrows to call. A male frog was found sitting on a clutch of 17 eggs about 0.25 in (7 mm) in diameter in a decaying tree stump. The tails of the well-developed embryos were heavily vascularized and probably served for respiration, because the embryos had no gills.

CONSERVATION STATUS
Not threatened.

SIGNIFICANCE TO HUMANS
None known. ◆

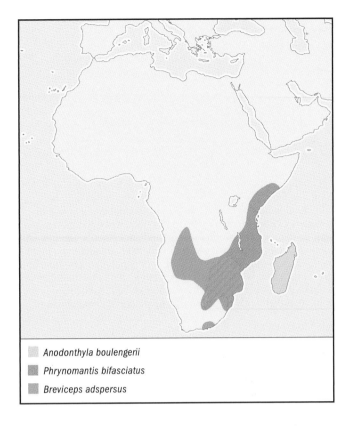

Anodonthyla boulengerii

Phrynomantis bifasciatus

Breviceps adspersus

Bushveld rain frog
Breviceps adspersus

SUBFAMILY
Brevicipitinae

TAXONOMY
Breviceps adspersus Peters, 1882, Damaraland (Namibia) and Transvaal (Republic of South Africa).

OTHER COMMON NAMES
German: Gesprenkelter Kurzkopffrosch.

PHYSICAL CHARACTERISTICS
This is a nearly globular frog with short legs and a blunt, pushed-in snout. The toes are not webbed, and there are stout "spades" (inner and outer metatarsal tubercles) on the hind feet.

DISTRIBUTION
The species ranges across southern Africa, into Namibia, Botswana, Zimbabwe, Mozambique, and northern South Africa.

HABITAT
The frog inhabits open or savanna regions where the soil is sandy.

BEHAVIOR
This is a burrowing frog that digs by using its "spades" while shuffling backward. Burrows may be as deep as 20 in (50 cm). For much of the year the frogs remain underground, living on stored fat, but they emerge to feed and call for mates in the wet summer. Such activity is mostly nocturnal but occasionally happens during the day.

FEEDING ECOLOGY AND DIET
Termites are the principal food, taken when they emerge by the thousands from their underground nests during rainy periods to mate and disperse.

REPRODUCTIVE BIOLOGY
When the rains arrive, males call from the mouths of their burrows or sometimes while walking (not hopping) about. The mating posture is with the smaller male glued (literally) to the female's back. Thus positioned, the pair digs backward into the soil and hollows out a small cavity in which about 30 eggs are laid; the female may remain with the eggs. The young develop entirely within the egg capsule, hatching as tiny frogs about 0.25 in (6 mm) in length after four to six weeks.

CONSERVATION STATUS
Not threatened. Two species of rain frogs with small ranges are considered Vulnerable, but the more widespread Bushveld rain frog apparently is more secure.

SIGNIFICANCE TO HUMANS
None known. ◆

Boulenger's climbing frog
Anodonthyla boulengerii

SUBFAMILY
Cophylinae

TAXONOMY
Anodonthyla boulengerii Müller, 1892, Madagascar.

OTHER COMMON NAMES
None known.

PHYSICAL CHARACTERISTICS
This is a small frog, with a body length of 0.85 in (22 mm).
The head is narrower than the body, and the snout is bluntly
pointed. Teeth, which are lacking in most microhylids, are present in the upper jaw. The fingers and toes (except the first in
each instance) have enlarged terminal disks; those of the fingers are much the broader. There is no webbing. The first finger of the male bears a sharp, projecting spine that may serve
to help grip the female during amplexus. The background
color is brown with varying paler or darker brown markings.

DISTRIBUTION
The species ranges across eastern Madagascar.

HABITAT
This is an arboreal species, but sometimes it is found under the
forest floor litter.

BEHAVIOR
Little is known aside from reproductive behavior.

FEEDING ECOLOGY AND DIET
Like so many other microhylids, this species feeds on ants.

REPRODUCTIVE BIOLOGY
Males are reported to call from the trunks of trees and ferns as
well as other vegetation but rarely from leaves. This is one of
many species of microhylids whose breeding habits diverge
from those more typical of frogs. The 25–30 eggs are small
(less than 0.1 in, or 2 mm) and are laid in water held in tree
cavities or leaf axils. The male remains with the eggs as they
develop and through the period of larval growth, which lasts
less than a month. The tadpoles take no food; they survive and
grow through metamorphosis on energy supplied by the yolk.

CONSERVATION STATUS
Not threatened.

SIGNIFICANCE TO HUMANS
None known. ◆

Fry's whistling frog
Austrochaperina fryi

SUBFAMILY
Genyophryninae

TAXONOMY
Austrochaperina brevipes Fry, 1915, Bloomfield River, near
Cooktown, northeastern Queensland, Australia.

OTHER COMMON NAMES
None known.

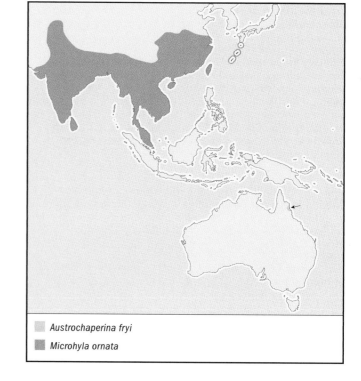

☐ *Austrochaperina fryi*
■ *Microhyla ornata*

PHYSICAL CHARACTERISTICS
This is a small frog—females attain a body length of 1.3 in (35
mm), and males are slightly smaller. The species has average
frog morphologic features: rather than the fat body, narrow
head, small eyes, and pointed snout of the classic microhylid,
the slightly rotund body is only a little wider than the head,
the eyes are relatively large, and the snout is rounded. The
toes are not webbed, and the tips of the fingers and toes are
slightly broadened. The body is brown above, with a reddish
tint, and the facial region is black.

Species of microhylid frogs often are confusingly similar in
morphologic characteristics. In this instance, *Austrochaperina
fryi* cannot be distinguished reliably by morphologic features
from another Australian species, *Austrochaperina robusta*, but the
two have distinctly different calls: *A. fryi* gives a series of brief
whistles, whereas *A. robusta* produces whistles in couplets.

DISTRIBUTION
The species occurs in northeast Queensland, Australia.

HABITAT
The habitat is the leaf litter on the floor of rainforest, where the
frogs hide by day amid the leaves or under other cover. The
range in elevation is from virtually sea level to 3,600 ft (1,100 m).

BEHAVIOR
Little is known aside from reproductive behavior.

FEEDING ECOLOGY AND DIET
Nothing is recorded, but the frogs undoubtedly eat small invertebrates.

REPRODUCTIVE BIOLOGY
At night males call from superficial cover on the forest floor,
not from deep within the leaf litter. Like other members of the
subfamily, *A. fryi* has direct embryonic development. Large
eggs, about 0.2 in (5 mm) in diameter, are deposited in shel-

tered sites on the forest floor and are attended by the male. The clutch size is seven to 12 eggs. Hatchling frogs are about 0.25 in (6 mm) in body length.

CONSERVATION STATUS
Not threatened. The presence of this species in protected areas may ensure its survival, though forest destruction may fragment the range.

SIGNIFICANCE TO HUMANS
None known. ◆

Wilhelm rainforest frog
Cophixalus riparius

SUBFAMILY
Genyophryninae

TAXONOMY
Cophixalus riparius Zweifel, 1962, at an elevation of 9,100 ft (2,774 m) beside Pengagl Creek on the eastern slope of Mount Wilhelm, Territory of New Guinea (Simbu Province, Papua New Guinea).

OTHER COMMON NAMES
None known.

PHYSICAL CHARACTERISTICS
The largest species of its genus, the Wilhelm rainforest frog reaches a body length of 2 in (50 mm). The habit is that of a treefrog: large hands, fingers and toes with broad terminal disks, and large eyes. The background color is tan, with varying darker brown markings and sometimes a purple tinge.

DISTRIBUTION
The species lives in the central ranges of Papua New Guinea at elevations of about 6,000–9,000 ft (1,900–2,800 m).

HABITAT
This species inhabits rainforest.

BEHAVIOR
The Wilhelm rainforest frog is an adaptable, scansorial (climbing) frog that is at home high in forest trees or on steep surfaces at ground level. It is found in pandanus trees and in cavities in epiphytic plants (ones that grow attached to trees) as well as in holes on steep banks, such as road cuts.

FEEDING ECOLOGY AND DIET
Nothing is recorded.

REPRODUCTIVE BIOLOGY
Like other genyophrynine microhylids, the Wilhelm rainforest frog has direct embryonic development. In one instance, a string of 27 eggs, about 0.2 in (6 mm) in diameter and joined by short cords, was found in a burrow in a soil bank with an attending frog. Male frogs call from such burrows. Eggs also are deposited in sheltered arboreal situations, where frogs can be heard calling from far out of reach.

CONSERVATION STATUS
Not threatened.

SIGNIFICANCE TO HUMANS
None known. ◆

Horned land frog
Sphenophryne cornuta

SUBFAMILY
Genyophryninae

TAXONOMY
Sphenophryne cornuta Peters and Doria, 1878, near the Wa Samson river in northern New Guinea (Irian Jaya, Indonesia).

OTHER COMMON NAMES
None known.

PHYSICAL CHARACTERISTICS
The maximum body length is about 1.6 in (41 mm); females are slightly larger than males. The body is relatively slender, with long legs and a somewhat pointed snout. Enlarged terminal disks on the fingers and toes, both of which are not webbed, give the species the aspect of a tree frog, and a pointed tubercle on each eyelid distinguishes it from other species with which it might be confused. The color varies from dark to light brown above and gray laterally, with largely gray, orange, or red undersides.

DISTRIBUTION
The species inhabits most of New Guinea at low to moderate elevations, usually below 4,100 ft (1,250 m).

HABITAT
This species inhabits rainforest.

BEHAVIOR
Males call at night from shrubs or other low vegetation but apparently not from high in trees. Females may be more terrestrial (when they are not attracted to calling males), but this is not known. Calls that differ from the presumed advertisement call have been heard in a call-response sequence between two individuals, suggesting a territorial function.

FEEDING ECOLOGY AND DIET
Food habits have not been studied, but small invertebrates must be the mainstay.

REPRODUCTIVE BIOLOGY
The presumed advertisement call is a rattling sound lasting up to three seconds. The presence of large, heavily yolked eggs in females essentially confirms that this species, like other microhylids in New Guinea, has direct embryonic development, but this remains to be verified by the discovery of eggs and an associated parent.

CONSERVATION STATUS
Not threatened. As a widely distributed species, the horned land frog is less likely than many others to be exterminated by destruction of rainforest.

SIGNIFICANCE TO HUMANS
None known. ◆

Saffron-bellied frog
Chaperina fusca

SUBFAMILY
Microhylinae

TAXONOMY
Chaperina fusca Mocquard, 1893, Sintang, Borneo (Kalimantan, Indonesia).

Kaloula pulchra

Chaperina fusca

OTHER COMMON NAMES
None known.

PHYSICAL CHARACTERISTICS
Males grow only up to 0.8 in (21 mm) and females to 0.9 in (24 mm) in body length. The body is slender to moderately stocky; the head is as wide as, or slightly narrower, than the body; and the snout is rounded. The tips of the fingers bear expanded terminal disks, and the toes have disks of similar size; the toes have scant webbing. Each elbow and heel has a conical dermal projection. For a microhylid, this is a colorful frog. All undersurfaces have large yellow spots against a network of black, and this pattern extends well up onto the sides of the body. The dorsal surfaces of the head and body are black, with a pattern of greenish yellow to silvery flecks.

DISTRIBUTION
The range includes the Malay Peninsula and Borneo northward to the southern Philippine islands of Jolo, Mindanao, and Palawan.

HABITAT
The saffron-bellied frog inhabits primarily the ground surface layer of forest. It is found both in level, lowland country and in hills up to 5,900 ft (1,800 m).

BEHAVIOR
When not breeding, these frogs hide under litter, such as fallen leaves, during the day and may climb into low vegetation at night.

FEEDING ECOLOGY AND DIET
The diet is not described, but the food undoubtedly is small invertebrates.

REPRODUCTIVE BIOLOGY
Males call in chorus by day around small, temporary pools, where the eggs are laid and the tadpoles develop.

CONSERVATION STATUS
Not threatened.

SIGNIFICANCE TO HUMANS
None known. ◆

Eastern narrow-mouthed toad
Gastrophryne carolinensis

SUBFAMILY
Microhylinae

TAXONOMY
Engystoma carolinense Holbrook, 1836, Charleston, South Carolina, and extending westward to the lower Mississippi River, United States.

OTHER COMMON NAMES
German: Carolina-Engmaulfrosch; Spanish: Ranita olivo.

PHYSICAL CHARACTERISTICS
This small frog grows up to 1.5 in (3.8 cm) in length. It has a chunky body tapering to a narrow head, with a fold of skin across the head behind the small eyes and a somewhat pointed snout. The toes lack webbing. The dorsum varies in color, generally a shade of brown more or less distinct from a paler shade laterally.

DISTRIBUTION
The eastern narrow-mouthed toad occurs in the southern and eastern United States from eastern Texas and Oklahoma to Maryland and south to the Gulf of Mexico and the tip of Florida. It has been introduced into two islands in the Bahamas. A close relative, the Great Plains narrow-mouthed toad (*Gastrophryne olivacea*), has a complementary distribution westward to southern New Mexico and southern Arizona and south on both coasts of Mexico.

Gastrophryne carolinensis

Hamptophryne boliviana

HABITAT

The species inhabits mostly coastal plain and piedmont terrain and, rarely, mountains; it ranges from seashore to forest, generally avoiding dryer conditions.

BEHAVIOR

This secretive, terrestrial species seldom is seen except in breeding choruses, unless one turns over rocks, logs, or other ground surface cover, when it is quick to escape. Activity is almost wholly nocturnal. Distasteful and possibly toxic skin secretions provide some protection from enemies, but egrets, bullfrogs (*Rana catesbeiana*), and garter snakes (species of *Thamnophis*) are known predators. The secretions help protect the frogs from attack by ants.

FEEDING ECOLOGY AND DIET

This species preys on a wide variety of small invertebrates up to about 0.25 in (6 mm) in length; ants, beetles, and termites constitute the bulk of the diet.

REPRODUCTIVE BIOLOGY

Breeding takes place during heavy rains that fill or supplement permanent or semipermanent pools. The breeding season may be from April to October in the south, but it takes place over a narrower span farther north. Males call in the water or from sheltered sites nearby. They grip females in axillary amplexus while becoming glued to them by secretions from ventral skin glands. The call, which lasts up to four seconds, is often compared to the bleat of a lamb. On average, about 500 eggs are deposited as a surface film, not necessarily in one continuous batch. The duration of larval development ranges from 20 to 67 days, with the longer periods in the northern part of the range. Newly metamorphosed frogs are less than 0.5 in (11 mm) in length.

CONSERVATION STATUS

Not threatened.

SIGNIFICANCE TO HUMANS

None known. ◆

Bolivian bleating frog

Hamptophryne boliviana

SUBFAMILY

Microhylinae

TAXONOMY

Chiasmocleis boliviana Parker, 1927, Buena Vista, Santa Cruz, Bolivia.

OTHER COMMON NAMES

None known.

PHYSICAL CHARACTERISTICS

This frog, the only species of its genus, has the classic microhylid habitus of squat body tapering to a narrow head with small eyes and a pointed snout. Females reach a body length of 1.75 in (44 mm), and males are slightly smaller. The toes have no webbing, and the tips of the fingers and toes are slightly expanded. The flanks and sides of the head are dark brown, sharply set off from the tan to reddish dorsal coloring.

DISTRIBUTION

The species ranges across South America and is distributed widely on the western and northern sides of the Amazon basin in Bolivia, Peru, Ecuador, French Guiana, Surinam, and Brazil.

HABITAT

The frog inhabits mature and secondary forests at moderately low elevations (700–1,100 ft, or 220–340 m).

BEHAVIOR

Little is known of the activities of these frogs, apart from when they are breeding. They have been found feeding at columns of ants on the forest floor at night and as high as 5 ft (1.5 m) up on the trunks and branches of trees.

FEEDING ECOLOGY AND DIET

Ants make up the bulk of the diet.

REPRODUCTIVE BIOLOGY

Following the first heavy rainfall at the onset of the rainy season, frogs congregate in great numbers for a few days at lotic waters then formed or augmented. Males begin calling in daylight as they approach the ponds and then from the water, where mating and depositing of eggs take place. Eggs are deposited in clutches about 4 in (10 cm) in diameter, with about 200 eggs in each clutch. Presumably, one female lays several clutches, as one frog was found to contain more than 2,000 mature eggs. The tadpoles hatch after about a day and a half, but the period of growth to metamorphosis has not been measured. Tadpoles are preyed upon by a variety of animals, including dragonfly larvae and freshwater crabs. The frogs leave the ponds right after breeding, but later intense rain following a dry period may initiate another short period of reproductive activity.

CONSERVATION STATUS

Not threatened.

SIGNIFICANCE TO HUMANS

None known. ◆

Malaysian painted frog

Kaloula pulchra

SUBFAMILY

Microhylinae

TAXONOMY

Kaloula pulchra Gray, 1831, China.

OTHER COMMON NAMES

German: Indischer Ochsenfrosch; Spanish: Microhilido asiático.

PHYSICAL CHARACTERISTICS

This species is rather large for a microhylid, up to 3 in (75 mm) long. The body is rotund, with a blunt, rounded snout; the legs are short and stout. The hind feet have a projecting "spade" on the sole just anterior to the heel; the toes are only slightly webbed.

DISTRIBUTION

The native range is from southern China through Indochina to Sumatra. The species' presence in Borneo and Sulawesi is thought to be due to introductions, probably inadvertent. *Kaloula taprobanica*, once considered to be a subspecies of *K. pulchra*, ranges from northeastern India to Sri Lanka, so frogs of this sort are widespread.

HABITAT

The Malaysian painted frog is unusual, in that it is abundant in human settlements but is not found in undisturbed habitats.

The frogs take shelter in burrows, rubbish heaps, and the like. This association with people may explain spotty occurrences out of the apparent natural range.

BEHAVIOR
This is a secretive, terrestrial frog that utilizes the spadelike structures (metatarsal tubercles) on its hind feet to burrow backward into the soil. It is most likely to be seen when congregating to breed.

FEEDING ECOLOGY AND DIET
Ants constitute most of the diet, along with other small insects.

REPRODUCTIVE BIOLOGY
Breeding takes place only when heavy rains fill ditches and other temporary sources of water. There the males float high in the water, giving loud, honking calls. Their rounded shape is exaggerated even more when they inflate while floating and calling. In such ephemeral habitats, larval growth typically is rapid, but it may not be fast enough to reach completion before the water dries up.

CONSERVATION STATUS
Not threatened.

SIGNIFICANCE TO HUMANS
None known. ◆

Ornate narrow-mouthed frog
Microhyla ornata

SUBFAMILY
Microhylinae

TAXONOMY
Engystoma ornatum Duméril and Bibron, 1841, Malabar coast, India.

OTHER COMMON NAMES
None known.

PHYSICAL CHARACTERISTICS
This is a small frog: males grow to about 0.8 in (23 mm) and females to 1 in (25 mm) in body length. The rather rotund body tapers to a narrow head and bluntly pointed snout. Finger and toe tips are pointed, and the toes have only scant webbing. The dorsal ground color is light olive brown and sometimes grayish or reddish. A prominent dorsal pattern includes black bands that originate behind the head and diverge as they pass posteriorly.

DISTRIBUTION
This is a widely distributed species that ranges from eastern and southern China, including Hainan, to Taiwan and the Ryukyu Islands; west to Pakistan, Nepal, India, and Sri Lanka; and south through Indochina and the Malay Peninsula.

HABITAT
Typical habitats are grassy areas near flooded rice paddies, ditches, and pools up to the lower slopes of mountains at about 3,000 ft (900 m). The association with rice paddies and similar agricultural sites may help explain this species' wide geographic distribution.

BEHAVIOR
The frogs are nocturnal, taking daytime shelter in grass or the crevices in drying soil or under surface litter. They are said to be poor swimmers but very active jumpers.

FEEDING ECOLOGY AND DIET
The diet is not described, but undoubtedly small invertebrates are taken.

REPRODUCTIVE BIOLOGY
Breeding takes place after heavy rain, when males call from the water in flooded rice fields and similar habitats. The female deposits several hundred eggs that float in a film at the surface of the water, and the tadpoles exhibit unusually rapid growth. This, no doubt, is associated with the ephemeral nature of the water in many breeding sites. Tadpoles are largely transparent, which makes them inconspicuous in shallow water.

CONSERVATION STATUS
Not threatened.

SIGNIFICANCE TO HUMANS
None known. ◆

Timbo disc frog
Synapturanus salseri

SUBFAMILY
Microhylinae

TAXONOMY
Synapturanus salseri Pyburn, 1975, Timbó, Vaupés, Colombia.

OTHER COMMON NAMES
None known.

☐ *Synapturanus salseri*
▧ *Otophryne pyburni*

PHYSICAL CHARACTERISTICS

This is a small (1.1 in, or 28 mm) frog with a stout body that tapers to a narrow head with a blunt, overhanging snout. There is a transverse fold of skin just behind the small eyes, and the tympanum is concealed completely. The toes are not webbed, and the inner metatarsal tubercle is indistinct, not enlarged into a digging spade. The dorsum is gray-brown with small spots over the body and legs varying from cream to orange.

DISTRIBUTION

The species is known only from southeastern Colombia and adjacent Ecuador and Venezuela. The range of the genus (three species) extends to eastern Brazil.

HABITAT

The species lives on the floor of rainforest.

BEHAVIOR

These are secretive frogs that live in burrows beneath the forest floor under layers of fallen leaves and root tangles; they emerge rarely, if ever, into the open.

FEEDING ECOLOGY AND DIET

These frogs are known to eat ants and spiders.

REPRODUCTIVE BIOLOGY

The call, a brief whistle, is given only during heavy rainfall, presumably from the mouth of the burrow. Reproduction is entirely terrestrial. The eggs are large and few (four to six) and are deposited in the burrow and accompanied by the male frog. The hatched tadpoles do not feed but live instead on the stored yolk until they metamorphose.

CONSERVATION STATUS

Not threatened.

SIGNIFICANCE TO HUMANS

None known. ◆

Pyburn's pancake frog
Otophryne pyburni

SUBFAMILY

Otophryninae

TAXONOMY

Otophryne pyburni Campbell and Clarke, 1998, Wacará, Vaupés, Colombia.

OTHER COMMON NAMES

None known.

PHYSICAL CHARACTERISTICS

Females grow up to 2.2 in (56 mm) in length, and males are slightly smaller. The body is rather broad, with short legs, and tapers to a sharp-pointed, projecting snout. The tympanum is distinct and slightly greater in diameter than the eye. The dorsal surfaces are reddish brown to grayish yellow, uniform or with darker markings, including a mid-dorsal stripe, and separated by a pale stripe from the darker side of the body. The general impression is that of a dead leaf.

DISTRIBUTION

The frog is found in South America. The range is from southeastern Colombia eastward through southern Venezuela and Guyana to French Guiana.

HABITAT

This species lives on the forest floor in sandy soils of the rainforest.

BEHAVIOR

Individuals probably stay underground much of the time.

FEEDING ECOLOGY AND DIET

Another species of *Otophryne* is known to feed on ants, and it is likely that Pyburn's pancake frog does also.

REPRODUCTIVE BIOLOGY

Males call during the day from sheltered sites, such as beneath leaf litter or under root tangles beside streams. The unpigmented eggs are large (0.2 in, or 5 mm) in diameter and probably are deposited in a nest cavity near a stream, as is known for another species of *Otophryne*. The tadpole has characters unique within the Microhylidae. It possesses a row of sharp, keratinized teeth in the upper and lower jaws, and the spiracle, instead of opening mid-ventrally, is on the lower-left side of the body near the base of the tail. As the tadpole grows, the spiracle extends as a tube that may reach halfway along the tail. Probably the larvae of the other two species of *Otophryne* are similar. The function of the teeth is uncertain. One suggestion is that the tadpoles are predaceous. Another, perhaps more plausible, idea is that the teeth serve to screen out sand grains as the larva filter-feeds in shallow water. Tadpoles of Pyburn's pancake frog have been found hiding under leaves in tiny, shallow streams.

CONSERVATION STATUS

Not threatened.

SIGNIFICANCE TO HUMANS

None known. ◆

Banded rubber frog
Phrynomantis bifasciatus

SUBFAMILY

Phrynomerinae

TAXONOMY

Brachymerus bifasciatus Smith, 1847, country to the east and northeast of the Cape Colony, Republic of South Africa.

OTHER COMMON NAMES

English: Red-banded frog; German: Wendehalsfrosch.

PHYSICAL CHARACTERISTICS

The banded rubber frog is of medium size, with a body length up to about 2.75 in (68 mm). This is a strikingly colored frog; the dorsum is black with red bands running from the snout over the eyelids to the rear of the body and red spotting on the limbs. The body is moderately robust, with the tips of the fingers expanded into truncate disks; the legs are short, and the toes have no webbing. The skin is smooth and shiny; its texture is responsible for the name *rubber frog*.

DISTRIBUTION

The frog is found in Africa, from Somalia and Zaire to South Africa.

HABITAT

This species inhabits open or savanna country.

BEHAVIOR

These frogs are nocturnal and spend the dry season underground in holes or termitaria. They tend to walk or run rather than hop, and they burrow backward, though they are not equipped with the large "spades" on the hind feet that are found in some other microhylids. Secretions from glands in the skin have been known to cause skin irritations in people.

FEEDING ECOLOGY AND DIET

Ants and termites are the principal food.

REPRODUCTIVE BIOLOGY

Reproduction takes place in rain pools and similar inundated sites. Males give their melodious trill from beside or in shallow water; when they mate, they hold the female in axillary amplexus. Masses of as many as 1,500 eggs are attached to submerged weeds. Development to metamorphosis at a body length of about 0.5 in (13 mm) takes about one month. The tadpole has a whiplike tail tip, which it vibrates while it hangs suspended at a steep angle in the water filtering its microscopic food. As in other midwater tadpoles (as opposed to bottom feeders), these larvae have their eyes at the sides of the head, which permits a broad range of vision both above and below.

CONSERVATION STATUS

Not threatened.

SIGNIFICANCE TO HUMANS

None known. ◆

Resources

Books

Parker, H. W. *A Monograph of the Frogs of the Family Microhylidae.* London: British Museum, 1934.

Periodicals

Blum, J. P, and J. I. Menzies. "Notes on *Xenobatrachus* and *Xenorhina* (Amphibia: Microhylidae) from New Guinea with Description of Nine New Species." *Alytes* 7, no. 4 (1988): 125–163.

Burton, T. C. "A Reassessment of the Papuan Subfamily Asterophryinae (Anura: Microhylidae)." *Records of the South Australian Museum* 19, no. 10 (1986): 405–450.

Carvalho, Antenor Leitão de. "A Preliminary Synopsis of the Genera of American Microhylid Frogs." *Occ. Pap. Mus. Zool. Univ. Michigan* 555 (1954): 1–19.

Donnelly, Maureen, Rafael O. de Sá, and C. Guyer. "Description of the Tadpoles of *Gastrophryne pictiventris* and *Nelsonophryne aterrima* (Anura: Microhylidae), with a Review of Morphological Variation in Free-swimming Microhylid Larvae." *American Museum Novitates* 2976 (1990): 1–19.

Menzies, James I. "A Study of *Albericus* (Anura: Microhylidae) of New Guinea." *Australian Journal of Zoology* 47 (1999): 327–360.

Wild, Erik Russell. "New Genus and Species of Amazonian Microhylid Frog with a Phylogenetic Analysis of New World Genera." *Copeia* 1995, no. 4 (1995): 837–849.

Zweifel, Richard G. "A New Genus and Species of Microhylid Frog from the Cerro de la Neblina Region of Venezuela and a Discussion of Relationships among New World Microhylid Genera." *American Museum Novitates* 2863 (1986): 1–24.

———. "Australian Frogs of the Family Microhylidae." *Bulletin of the American Museum of Natural History* 182, no. 3 (1985): 265–388.

———. "Partition of the Australopapuan Microhylid Frog Genus *Sphenophryne* with Descriptions of New Species." *Bulletin of the American Museum of Natural History* 253 (2000): 1–130.

———. "Results of the Archbold Expeditions. No. 97. A Revision of the Frogs of the Subfamily Asterophryinae Family Microhylidae." *Bulletin of the American Museum of Natural History* 148, no. 3 (1972): 411–546.

Richard G. Zweifel, PhD

Madagascaran toadlets

(Scaphiophrynidae)

Class Amphibia

Order Anura

Family Scaphiophrynidae

Thumbnail description
Small to medium-size toad-like frogs with or without enlarged fingertips

Size
0.8–2.4 in (20–60 mm) in snout-vent length

Number of genera, species
2 genera; 9 species

Habitat
Semidesert, dry forest, rainforest, and mountain savanna

Conservation status
Not threatened

Distribution
Madagascar

Evolution and systematics

The systematics and phylogenetic relationships of the Scaphiophrynidae are of special interest because *Scaphiophryne* seems to represent a link between two major anuran lineages, the Ranoidea and Microhyloidea, indicating that the groups are closely related. Adult *Scaphiophryne* have typically microhylid features (e.g., dilated sacral diapophyses) in addition to sharing features of the Ranidae (e.g., possession of a complete shoulder girdle). The tadpoles of *S. calcarata* likewise represent a mosaic of characters of both families and intermediate features as well. *Scaphiophryne* therefore may represent an old lineage that has conserved a step in the evolution from the ranoid to the microhylid tadpole type and can be considered a "living fossil."

Reflecting the mosaic-like character distribution of *Scaphiophryne*, their classification in the anuran system was, and is, difficult. They were considered mostly a subfamily of Microhylidae, a subfamily of Ranidae, or a separate family. As of 2001, two genera are included in the Scaphiophrynidae: *Scaphiophryne* contains six nominal species and, in addition, several newly discovered but unnamed species, whereas *Paradoxophyla* includes only a single species. Unlike *Scaphiophryne*, however, *Paradoxophyla* has an incomplete shoulder girdle and typically microhylid

tadpoles, with a median spiracle (breathing vent) and apparently without keratinized mouthparts. Further research is necessary to clarify its relationships with *Scaphiophryne* and other microhylids. No subfamilies are recognized.

Physical characteristics

Adult scaphiophrynids are 0.8–2.4 in (20–60 mm) in snout-vent length. The general body form of *Scaphiophryne* is somewhat toadlike. The legs are short, and well-developed metatarsal tubercles are present on the hind limbs. Some species (e.g., *S. marmorata* and *S. gottlebei*) have distinctly enlarged fingertips that may help them climb. The coloration varies widely, but some species are beautiful and have symmetrical markings on the back. The habitus of *Paradoxophyla* is different; at first glance, it resembles that of aquatic pipid frogs of the genus *Hymenochirus*.

Distribution

The family is endemic to Madagascar and inhabits most of the island at elevations from sea level to 6,600 ft (2,000 m),

Mocquard's rain frog (*Scaphiophryne calcarata*) lays up to 500 eggs that float on the water surface and metamorphose in just 3 or 4 weeks. (Photo from Natural History Museum, University of Kansas. Reproduced by permission.)

although records are missing from far northern and northeastern Madagascar.

Habitat

Representatives of the Scaphiophrynidae occur in all climatic regions of Madagascar; *S. calcarata*, *S. brevis*, and *S. gottlebei* inhabit the hot and arid regions of the west and south. They are found in rocky formations, deciduous dry forest, open savanna, and even dry sand dunes close to the sea. Another group of species (e.g., *S. madagascariensis*) occurs in the cold, high montane savannas of central Madagascar below and above the tree line. *Scaphiophryne marmorata* and *Paradoxophyla palmata* inhabit primarily low and mid-elevation rainforests of eastern Madagascar. Despite the different habitats of the adults, the larval habitat is similar in all species; tadpoles develop in stagnant and mostly temporary ponds and swamps.

Behavior

Scaphiophrynids are primarily nocturnal and terrestrial, spending the day buried in the ground under stones, fallen tree trunks, or other refuges. The species with expanded finger disks have some climbing abilities. The rainforest-

dwelling *S. marmorata* can be found in the leaf litter on the ground and climbing on mossy trees as well. Sometimes, it is even active during the day. Activity patterns in scaphiophrynids are highly seasonal. This is especially true for the species in arid regions and those in cold mountain habitats in which good climatic conditions for the development of tadpoles and juveniles are restricted to a short period of time. Most observations have been made at the beginning of the rainy season in December or January, when breeding takes place and activity is at its peak. After the rainy season the frogs in the arid habitats disappear for about six months and presumably estivate buried in the ground.

Feeding ecology and diet

The food seems to consist mainly of insects, but few data are available.

Reproductive biology

Scaphiophrynids are primarily explosive breeders and reproduce after heavy and prolonged rains at the beginning of the rainy season (generally in December, January, or February). Males aggregate in or at the edge of temporary ponds and often form large choruses that produce a continuous loud noise that can be heard from long distances. Before they start calling, *Scaphiophryne* greatly inflate the extremely large vocal sac and the body as well. Sometimes calling males swimming at the water's surface are unable to dive when they are disturbed, because they cannot get rid of the air quickly. Amplexus is axillary. Females lay numerous small, pigmented eggs, which generally are deposited as a film on the surface of the water. The free-swimming and mainly filter-feeding tadpoles develop quickly to froglets if the water temperature is high.

Conservation status

Several *Scaphiophryne* species (e.g., *S. gottlebei*) seem to have a quite restricted distribution, but more research is necessary to assess their conservation status reliably.

Significance to humans

Some of the beautifully colored *Scaphiophryne* species are offered regularly in the pet trade.

Species accounts

Web-foot frog
Paradoxophyla palmata

TAXONOMY
Microhyla palmata Guibé, 1974, Ambana, Chaînes Anosyennes, Madagascar.

OTHER COMMON NAMES
None known.

PHYSICAL CHARACTERISTICS
The snout-vent length is 0.8–1.0 in (20–24 mm); males are slightly smaller than females. This is a distinctive, small frog with a triangular body form, a pointed snout, small eyes, and fully webbed toes. The tympanum

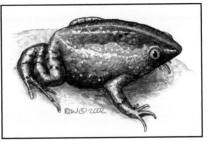

Paradoxophyla palmata

is indistinct, and the fingertips are not enlarged. The dorsum is brown, gray, or beige, with some small black spots. The venter is mostly white with distinct dark spots. Males have a dark vocal sac.

DISTRIBUTION
Rainforest in eastern Madagascar.

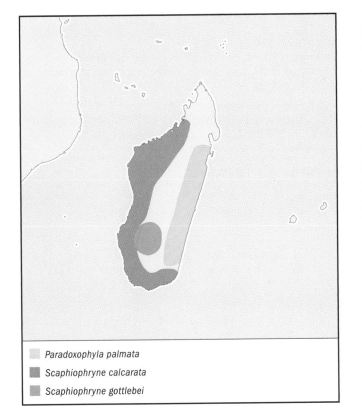

- *Paradoxophyla palmata*
- *Scaphiophryne calcarata*
- *Scaphiophryne gottlebei*

HABITAT
Known from pristine and degraded primary rainforest up to elevations of 3,300 ft (1,000 m).

BEHAVIOR
From December to February, males call after dusk at the edge of puddles or larger ponds, mainly after heavy rains.

FEEDING ECOLOGY AND DIET
Nothing is known.

REPRODUCTIVE BIOLOGY
The males' calls are rather loud trills reminiscent of crickets. Occasionally, males and couples in axillary amplexus swim on the surface of the water and dive quickly when disturbed. Females lay several hundred small pigmented eggs about 0.04 in (1 mm) in diameter surrounded by a gelatinous capsule. The gelatinous capsules and eggs emerge above the water surface. Embryonic development is rapid, and larvae hatch one day after egg deposition. The tadpoles are typical microhylid filter-feeding tadpoles and swim in open water.

CONSERVATION STATUS
Because the range of this unique species covers most of the eastern rainforest belt, including several nature reserves, it may be considered as not threatened.

SIGNIFICANCE TO HUMANS
Apparently, this frog is unknown to most indigenous people. ◆

Mocquard's rain frog
Scaphiophryne calcarata

TAXONOMY
Pseudohemisus calcaratus Mocquard, 1895, Madagascar.

OTHER COMMON NAMES
None known.

PHYSICAL CHARACTERISTICS
This is the smallest known *Scaphiophryne*; the snout-vent length is 0.8–1.1 in (20–27 mm) in males and 1.1–1.3 in (28–33 mm) in females. The dorsum is pale brown, gray, or green with or without darker sym-

Scaphiophryne calcarata

metrical markings and a pale vertebral line. The flanks are dark brown, and the venter is white; the ventral surfaces of the thighs are red to violet. The throat is black in males and marbled brown and white in females. The tips of the fingers are not enlarged. The skin on the back is smooth or slightly granular. The tympanum is indistinct, and the webbing between the toes is poorly developed. The tadpoles have a sinistral spiracle and keratinized mouthparts, although the latter are poorly developed. On the other hand, they have unperforated nares, as is typical of microhylid larvae.

DISTRIBUTION
Distributed widely in western and southern Madagascar at elevations below 1,000 ft (300 m).

HABITAT
Grassland savanna, dry forest, and other arid habitats.

BEHAVIOR
It seems nearly impossible to find this nocturnal species during the dry season. After the first heavy rains, however, large numbers of individuals walk around at night, and calling males and amplectant pairs gather in temporary, sun-exposed ponds and swamps.

FEEDING ECOLOGY AND DIET
One specimen had numerous large ants in its stomach.

REPRODUCTIVE BIOLOGY
Males aggregate in large choruses and produce loud, noisy calls. A female was observed approaching a calling male. The male became very excited and strongly enhanced the repetition rate of his vocalizations before clasping the female. Breeding activity is explosive at the beginning of the rainy season and often is finished after a few nights. Each female lays several hundred small eggs about 0.04 in (1 mm) in diameter. The tadpoles are largely transparent, mostly swim in open water, and are mainly filter feeders, but they also feed on larger particles. The larval development is rapid, in a race against the desiccation of waters. After a few weeks, metamorphosis is completed, and tiny froglets, 0.2–0.3 in (5.5–7.5 mm) in snout-vent length, emerge.

CONSERVATION STATUS
Being widely distributed and common in primary and secondary habitats, the species is not threatened.

SIGNIFICANCE TO HUMANS
At the beginning of the rainy season, *S. calcarata* occasionally may penetrate the huts of Madagascan people. ◆

Red rain frog
Scaphiophryne gottlebei

TAXONOMY
Scaphiophryne gottlebei Busse and Böhme, 1992, Isalo, Vallée des Singes, Madagascar.

OTHER COMMON NAMES
None known.

Scaphiophryne gottlebei

PHYSICAL CHARACTERISTICS
The snout-vent length reaches 1.4 in (36 mm). This is a toadlet that is colored conspicuously with contrasting colors. Four pink or red symmetrically arranged patches surrounded by black and green are present on the dorsum. The flanks and legs are white with black bands on the legs. The venter is dark grayish violet. The tips of the fingers are distinctly enlarged. The skin on the back is smooth, and the tympanum is indistinct. The webbing between the toes and the inner metatarsal tubercle is well developed.

DISTRIBUTION
Known only from a small area of the Isalo massif in southwestern Madagascar.

HABITAT
Lives in eroded sandstone formations. In the Isalo massif, humid forests persist in canyons and on the slopes, although the climate is rather arid.

BEHAVIOR
Found under stones during the day in the rainy season. It probably estivates during the dry season. The expanded terminal finger disks may indicate partial climbing habits. Disturbed specimens inflate themselves, probably as a strategy to protect against predators.

FEEDING ECOLOGY AND DIET
The diet in nature is unknown. In captivity the frog feeds on crickets and other insects.

REPRODUCTIVE BIOLOGY
Nothing is known, but probably an explosive breeder at the beginning of the rainy season. Recently metamorphosed juveniles have been found at the edge of stagnant ponds.

CONSERVATION STATUS
The species is not categorized by the IUCN and is not protected by CITES. However, because of the small known range, it may be considered potentially threatened.

SIGNIFICANCE TO HUMANS
This beautiful frog is offered regularly in the international pet trade. ◆

Resources

Books

Glaw, Frank, and Miguel Vences. *A Fieldguide to the Amphibians and Reptiles of Madagascar.* 2nd ed. Köln: Vences & Glaw Verlag, 1994.

Periodicals

Blommers-Schlösser, R. M. A. "Observations on the Larval Development of Some Malagasy Frogs, with Notes on Their Ecology and Biology (Anura: Dyscophinae, Scaphiophryninae and Cophylinae)." *Beaufortia* 24, no. 309 (1975): 7–26.

Blommers-Schlösser, R. M. A., and C. P. Blanc. "Amphibiens (Première Partie)." *Faune de Madagascar* 75, no. 1 (1991): 1–379.

Busse, K., and W. Böhme. "Two Remarkable Frog Discoveries of the Genera *Mantella* (Ranidae: Mantellinae) and *Scaphiophryne* (Microhylidae: Scaphiophryninae) from the West Coast of Madagascar." *Revue Française Aquariologie* 19, no. 1–2 (1992): 57–64.

Wassersug, R. "The *Pseudohemisus* Tadpole: A Morphological Link Between Microhylid (Orton Type 2) and Ranoid (Orton Type 4) Larvae." *Herpetologica* 40, no. 2 (1984): 138–149.

Frank Glaw, PhD

Caudata
(Salamanders and newts)

Class Amphibia

Order Caudata

Number of families 10

Number of genera, species 61 genera; 502 species

Photo: A barred tiger salamander (*Ambystoma mavortium*) eats an earthworm. (Photo by Ken Highfill/Photo Researchers, Inc. Reproduced by permission.)

Evolution and systematics

The order Caudata includes species that are generally called salamanders. Newts are terrestrial forms of some members of a single family, Salamandridae. Salamander classification is stable at the family level, but the relationship of the families is controversial. The 10 families typically are placed in three suborders. However, because the phylogenetic relationships of the species are uncertain, the suborders are seldom recognized or used by professional herpetologists. The suborder Sirenoidea includes only the Sirenidae, a small family restricted to eastern North America and generally considered the most basal lineage. The suborder Cryptobranchoidea and suborder Salamandroidea are sister taxa. Cryptobranchidae includes the small family Cryptobranchidae, of eastern Asia and eastern North America, and the large family Hynobiidae, which is restricted to Asia with the exception of one species that enters Europe in northern Russia. Salamandroidea includes most families and most species. These families are Ambystomatidae, Amphiumidae, Dicamptodontidae, Rhyacotritonidae, Plethodontidae, Proteidae, and Salamandridae. Dicamptodontidae and Ambystomatidae are restricted to North America and are thought to be sister taxa. Plethodontidae, by far the largest family, occurs in North, Central, and South America and has a few species in Mediterranean Europe. Plethodontidae has no close relatives but may be the sister taxon of the small North American family Amphiumidae. Some herpetologists consider the Rhya-

cotritonidae (a small family restricted to northwestern North America), Plethodontidae, and Amphiumidae relatively basal within the Salamandroidea. The small, gilled, permanently aquatic Proteidae occurs in North America and southern Europe. Its relationships are obscure. The Salamandridae, which is widespread in the Old World as well as in North America, may be the sister taxon to the Dicamptodontidae-Ambystomatidae.

Salamanders have evolved in fits and starts, and there has been a great deal of parallel and convergent evolution and even reversals of characters to more ancestral states. This indirect evolution has made determination of relationships uncertain, and many systematists have turned to biochemical characters. However, some features of the superorders and families are used to sort species. Sirenidae has many bizarre features, among them lack of teeth on the main jaw bones (premaxilla and dentary; the maxilla often is absent and when present is not articulated). Sirenids are permanently aquatic, have gills, and lack hind limbs; they are thought to practice external fertilization. The jaws are covered with a keratinized, beaklike structure. Cryptobranchoidea has a bone in the lower jaw, the angular, that was present in ancestral, Paleozoic forms and is lacking in all other salamanders. Cryptobranchids are the only salamanders that practice external fertilization, another ancestral trait. All Salamandroidea practice internal fertilization; the male deposits a spermatophore, which is picked

Excited rough-skinned newt males (*Taricha granulosa*) cling onto any-thing, including each other—the females are not yet in the water. (Photo by henk.wallays@skynet.be. Reproduced by permission.)

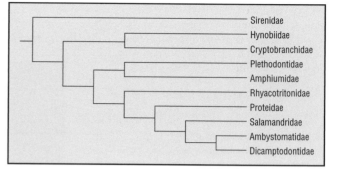

Generalized phylogenetic tree of the living families of salamanders. (Illustration by GGS. Courtesy of Gale.)

up by the vent margins of the female, and the sperm enter the reproductive tract.

Salamanders are thought more closely related to living anurans than to the gymnophionans (caecilians), mainly because of biochemical traits. The three lissamphibian groups are ancient, and there is little similarity among them. Although fossil sister taxa are known for each order, none of these fossils is of much use in assessment of relationships of the lissamphibians to Paleozoic amphibians. A number of fossil lineages are known from the Mesozoic and Cenozoic eras that fit well within the Caudata, but they do not help resolve phylogenetic relationships. A Jurassic taxon, *Karaurus*, usually is considered the sister taxon to caudates. The Caudata are monophyletic with respect to *Karaurus*, so discovery of this taxon was an important event.

Physical characteristics

Salamanders display great diversity in body form, but most of the species are small, generalized vertebrates that have a head about as wide as the body, a trunk that contains 12–20 vertebrae, four limbs that vary in length from long (overlapping when appressed to the side of the body) to very short, and a tail that usually is about the same length as the head and body combined. These generalized salamanders typically range in size from approximately 1.5 to 8 in (4–20 cm). However, some salamanders have a highly aberrant form. The permanently aquatic giant salamanders of eastern Asia may exceed 59 in (150 cm) in length and have broad, flattened bodies and heads and a strongly keeled tail. Sirens and amphiumas have very elongated bodies with many trunk

vertebrae, short tails, and diminutive limbs. Sirens lack hind limbs and have gills throughout life, whereas amphiumas have an open spiracle. Proteids are permanently aquatic, gilled forms, but they have relatively well-developed limbs that have lost some digits. The more bizarre salamanders often are known by colorful common names: hellbenders (North American cryptobranchids), mud sirens (North American sirenids), Congo eels (North American amphiumids), and olms (European proteids). Newts in the terrestrial stages are known as efts.

Herpetologists have become aware of the existence of large numbers of miniaturized salamanders, mainly within the family Plethodontidae. Most of these miniaturized species are terrestrial throughout their lives, and many are secretive. New species continue to be discovered at a rather high rate. Many of these species are less than 1.2–1.4 in (3–3.5 cm) in total length.

The most phylogenetically basal taxa have aquatic larvae with external gills and limbs that metamorphose after a growing season to produce a semiterrestrial to fully terrestrial adult. Approximately one half of the species of salamanders have abandoned the aquatic larval stage and lay eggs on land that develop directly into terrestrial juveniles. A few species are truly viviparous.

Salamanders include many species that are prototypical generalized tetrapods. The head is no broader than the trunk and usually is relatively small. The trunk is of moderate length and has 12–18 trunk vertebrae. The tail is approximately the same length as the head plus the body. The four limbs are of moderate length and just overlap when appressed to the trunk. These animals have large eyes used mainly in prey detection and capture and in predator detection. They have excellent olfactory capabilities. There is no external ear and although salamanders can hear, sound appears to play little role in their lives. Many people confuse salamanders and lizards. In the southern parts of the United States, salamanders often are called "spring lizards." Like frogs and caecilians, salamanders have moist skin that is well supplied with glands, both so-called poison or granular glands and mucous glands. Unlike that of frogs, the skin of salamanders is thick and tightly attached to underlying bone and muscle. Generalized salamanders may or may not have lungs, but in either case, most gaseous exchange takes place across the moist, vascularized skin. Salamanders that have this generalized shape can be fully aquatic, but they are more typically semiterrestrial or fully terrestrial. The terrestrial species often spend extensive periods underground, especially during nonbreeding seasons and during cold winters or dry summers.

Most salamanders fit the foregoing general description, but several are truly bizarre in appearance. The fully aquatic sirens lack hind limbs, have large, feathery external gills, and are extremely elongate. The mouth is a cornified beak that lacks teeth on the jaws, except for a tiny patch on the inside of the lower jaw that faces inward. Amphiumas, also aquatic but somewhat more terrestrial than sirens, are very elongate but generally slender. They have ridiculously small limbs with four, three, or even two toes depending on the species. Adult amphiumas lack gills but have an open spiracle, and they have formidable teeth that can produce a painful bite. Some salamanders in several families are permanently larval in morphology, but their gonads mature. Some of these animals (e.g., *Proteus* of south central Europe) are known to inhabit only underground water courses.

Unlike aquatic specialists, terrestrial salamanders can assume bizarre morphologies, such as becoming extremely slender and elongate with tails more than twice body length (e.g., the plethodontid genera *Batrachoseps*, *Lineatriton*, and *Oedipina*). Others have evolved fully webbed hands and feet, such as some species of *Bolitoglossa* and *Chiropterotriton*).

Biological attributes

Few predators specialize on salamanders, but a few snakes are known to prefer salamanders. Many salamanders are long-lived, and individuals dated by bone rings are thought to be between 20 and 30 years of age. Salamander skin is richly supplied with glands that produce bad-tasting, sticky, and sometimes dangerously poisonous secretions. Newts, especially those in the salamandrid genus *Taricha* in western North America, use a poison known as tarichatoxin, which is identical to tetrodotoxin (produced by puffer fish). This substance is one of the most potent natural toxins. One species of garter snake has evolved defenses and is capable of eating these newts, even though the poison has strong effects on the predator. Most salamanders have color patterns thought cryptic, but some salamanders have vivid colors. All these salamanders are poisonous or are Müllerian mimics of other poisonous salamanders. When bothered by a predator, a newt displays the unken reflex, depressing its trunk and raising its head and tail to expose bright ventral coloration. The newt often rocks back and forth and exudes a strong-smelling skin secretion that contains toxin.

Salamanders are generally thought to have a life cycle involving an aquatic larval and a terrestrial adult stage, but there are many variations. At least some species in all of the families except the Hynobiidae and the Rhyacotritonidae are permanently aquatic. Some species in the exceptional families are nearly completely aquatic. In contrast, true terrestriality has evolved in the Plethodontidae and to a limited extent in the Salamandridae. The most terrestrial plethodontids are found in diverse regions ranging from boreal forests to tropical rainforests. They also are found in mesic microhabitats in dry regions. The terrestrial species range

The male Vienna newt (*Triturus carnifex*) performs its nuptial dance with a female (right). Note open papiliae at his cloaca. (Photo by henk.wallays@skynet.be. Reproduced by permission.)

from those that are largely fossorial to cave dwellers and fully arboreal species.

Distribution

Salamanders are classic examples of organisms with a mainly Holarctic distribution. Only one family (Plethodontidae) has a significant presence in the tropics, and that exclusively in the New World, reaching nearly 20 degrees south. Few species reach high latitudes, and only one (*Salamandrella keyserlingii*) extends north of the Arctic Circle. The southernmost salamanders are unnamed species of the genus *Bolitoglossa* of central Bolivia. Many salamanders are montane, and some reach high elevations. The species that apparently reaches the highest elevation is the plethodontid *Pseudoeurycea gadovii*, which lives higher than 16,400 ft (5,000 m) on Pico de Orizaba in eastern Mexico. Several tropical species occur above 13,100 ft (4,000 m) in Mexico, Guatemala, and Colombia. The highest extratropical species is *Hydromantes platycephalus*, which reaches 10,500 ft (3,200 m) in California.

Because nine of the 10 families occur almost exclusively in the North Temperate Zone (a salamandrid and a few ambystomatids enter tropical Mexico; a few salamandrids barely enter tropical Asia), the greatest lineage diversity is in northern regions, especially in North America and eastern Asia. However, Plethodontidae is abundant in both temperate and tropical zones. The plethodontids of the New World tropics are numerous (207 species) and diverse, and occur from sea level to above 16,400 ft (5,000 m). In contrast, far more anurans inhabit the tropical parts of the world than the northern temperate climates. Only 90 species live in North America and 116 species in temperate Eurasia. The greatest diversity of anurans is in the neotropical region (Central America, South America, and the West Indies), which is home to approximately 2,200 species. This number is approximately three times that in tropical Asia or tropical Africa and five times that in the Australo-Papuan region.

Feeding ecology and diet

Larval salamanders and many adults that are permanently aquatic feed by using suction. The buccal cavity of the salamander is expanded by the action of the gill bar system, and the mouth is suddenly opened, drawing in water and prey. The prey are grasped with the jaws and swallowed.

Adult salamanders feed primarily on small arthropods, although ambystomatids and some salamandrids are worm specialists. Prey are located with vision, but secondarily by olfaction. Prey typically are captured with a highly mobile tongue that varies greatly in structure from one species to the next, even within the same family. The tongue has a skeleton derived from the larval gill skeleton. As many as 11 skeletal elements can be found in a salamander tongue. The tongue is propelled from the mouth by specialized protractor muscles and brought back to the mouth by extremely long retractor muscles that may arise on the back of the pelvis. The longest tongues are found in bolitoglossine plethodontids, but other families have also evolved long and fast tongues. A sit-and-wait foraging strategy typically is used. Plethodontids in particular may remain motionless until the tongue explodes out of the mouth and the prey disappears. The tongue action is too fast to be resolved by human eyes.

Reproductive biology

Many salamanders court, mate, and deposit eggs in water, but most never go into water, choosing to court, mate, and deposit eggs on land. The ancestral life cycle involves the first strategy, which includes external fertilization and a larval stage that lasts for one season. A derived state is internal fertilization, which results from the deposition of a spermatophore by the male that is picked up by the cloacal walls of the female. A more derived state is deposition of eggs on land, hatchlings being either larvae that wriggle to nearby water or are overcome by rising spring waters. A still more derived state is complete development within the egg of a hatchling that is a miniature of the adult. Some members of the family Salamandridae give birth to larvae, which may be very immature or be large and near metamorphosis. A few salamandrids retain the larvae in the oviduct for one or two years or perhaps even longer. Very large juveniles are produced that have been nourished within the female reproductive tract, first on siblings and later on secretions of the female. Larvae usually can be classified as pond or stream type. The former usually metamorphose in one season and have large, feathery gills. Stream-type larvae have a much smaller tail fin that does not extend onto the body as in pond larvae; short, sometimes inconspicuous, gills; and a depressed body with stout limbs and cornified digital tips. These larvae may persist for several seasons. All salamander larvae are carnivorous, generally eating aquatic insects but sometimes small fish and even smaller salamanders.

Resources

Books

Deban, S. M., and D. B. Wake. "Aquatic Feeding in Salamanders." In *Feeding*, edited by K. Schwenk. San Diego: Academic Press, 2000.

Duellman, William E., ed. *Patterns of Distribution of Amphibians.* Baltimore: Johns Hopkins University Press, 1999.

Duellman, William E., and Linda Trueb. *Biology of Amphibians.* Baltimore: Johns Hopkins University Press, 1994.

Petranka, J. W. *Salamanders of the United States and Canada.* Washington: Smithsonian Institution Press, 1998.

Thorn, R., and J. Raffaelli. *Les Salamandres de l'Ancien Monde.* Paris: Societe Nouvelle des Editions Boubee, 2001.

Wake, D. B., and S. M. Deban. "Terrestrial Feeding in Salamanders." In *Feeding*, edited by K. Schwenk. San Diego: Academic Press, 2000.

Zug, G. R., L. J. Vitt, and J. P. Caldwell. *Herpetology.* 2nd edition. San Diego: Academic Press, 2001.

Periodicals

Crawford, A. J., and D. B. Wake. "Phylogenetic and Evolutionary Perspectives on an Enigmatic Organ: The Balancer of Larval Caudate Amphibians." *Zoology* 101 (1998): 107–123.

David B. Wake, PhD

Sirens and dwarf sirens

(Sirenidae)

Class Amphibia
Order Caudata
Suborder None
Family Sirenidae

Thumbnail description
Small to large, elongate, aquatic salamanders
with a larval-like form and lacking hind limbs

Size
4.7–38.5 in (12–98 cm)

Number of genera, species
2 genera; 4 species

Habitat
Swamp, lake, and wetland

Conservation status
Not classified by IUCN, but local populations of
certain species may be threatened

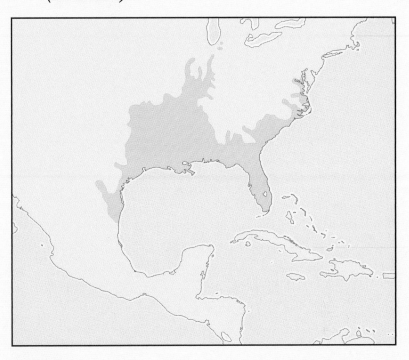

Distribution
Southeastern United States, from Texas to southwestern Michigan to Maryland

Evolution and systematics

The oldest sirenid fossils are from the upper Cretaceous
(165–65 million years ago [mya]) of Wyoming and Montana
(e.g., *Habrosaurus dilatus*) and Sudan. The genus *Siren*, with
four fossil species (Eocene [54–38 mya] of Wyoming;
Miocene [28–5 mya] of Florida and Texas; Pliocene [5–1.8
mya] of Florida), is known from the Eocene to present. *Pseudobranchus*, with two fossil species, is known from the Pliocene
to present in the southeastern United States. Vertebrae are
among the most common fossil elements collected.

In 1962 there was a proposal to place the sirenids in the
order Trachystomata, separate from the order Caudata for
the salamanders. This idea was quickly rebutted and never
gained scientific acceptance.

A consensus of where the Sirenidae should be placed within
a classification of the order Caudata has not been achieved.
The extreme larval condition of these salamanders sometimes
makes character evaluations difficult, and the lack of information on their fertilization mode adds to this dilemma. In
various recent phylogenetic analyses of salamanders, the family usually is considered to be the sister group to all other
salamanders, but information on the fertilization mode of
these animals could change that arrangement.

The genus *Siren* (sirens) with the species *S. lacertina* was
described by Linnaeus in 1767, and *S. intermedia*, now with
three subspecies, was described by Barnes in 1826.

The genus *Pseudobranchus* (dwarf sirens) with the species
P. striatus (Le Conte, 1824) was described by Gray in 1825,
and several subspecies largely based on perceived differences
in the muted, striped color pattern were subsequently named.
That situation persisted until 1993 when the species was divided into *P. axanthus*, with two subspecies each with 32 chromosomes, and *P. striatus*, with three subspecies each with 24
chromosomes. Other than the chromosome morphology and
range, the species are difficult to identify and all other information, presented mostly under the heading of *P. striatus*, ap-
plies equally well to both species.

No subfamilies are recognized.

Physical characteristics

All sirenids are paedomorphs (i.e., they attain sexual
maturity but retain a larval morphology relative to their
ancestors—no eyelids, external gills [main support structures
branched], gill slits [one to three], teeth without attachment
bases, and gill rakers present). They have a cylindrical body
with a flattened tail and lack hind limbs and all pelvic girdle
components; they have three to four fingers with cornified
tips. The maxillary bone in the upper jaw is small to absent.
There is a partial metamorphosis that is subtle and prolonged,
but they never become terrestrial. These salamanders are re-
fractory to extraneous hormone stimuli that promote metamorphosis in other salamanders and are permanently aquatic.

Greater siren (*Siren lacertina*) is the largest of the sirens, ranging from 19 to 38 in (48–97 cm). It hunts at night and spends daylight hours hidden under debris or logs. (Photo by Jack Dermid. Bruce Coleman Inc. Reproduced by permission.)

Because larviform salamanders seldom have external features representative of the sexes, it is usually impossible to sex an individual without internal examinations.

Hatchlings and small larvae have an intensely black ground color with contrasting marks in yellow, red, or silvery white; a band across the nose and one on the top of the head and stripes on the body are common. Clear, unmarked fins extending well onto the body become restricted to the tail and opaque in postlarvae.

Distribution

Sirens are distributed in the U.S. coastal plain and Mississippi River embayment from northeastern Tamaulipas, Mexico, to Maryland. Dwarf sirens occur in the coastal plain from the western Florida Panhandle to central South Carolina and throughout the Florida Peninsula.

Habitat

Sirenids occur in many types of still to slow-flowing, often swampy, sites with muddy substrates and often with floating and rooted vegetation.

Behavior

Sirens commonly find retreats in burrows in the bank during daylight hours and forage along the bottom and among vegetation at night. Dwarf sirens are most commonly caught among the fine roots of the exotic floating water hyacinth (*Eichhornia crassipes*). All sirenids swim by body and tail undulations but also move their rather small legs in walking motions when near the bottom.

Feeding ecology and diet

Sirenids have rather small mouths but will eat any animal material small enough for them to swallow. They probe about with their snouts while foraging and likely detect prey by odor. Food items are sucked into the mouth by rapidly expanding the throat and opening the mouth so that the item is carried inside with the inrush of water; the item is retained by the gill rakers and the water is expelled out the gill slits. Their feeding activities can be quite gluttonous. These animals shake food items vigorously and swallow larger organisms in a series of gulps, but items are seldom broken into pieces.

Sirenids forage along the bottom and among floating and rooted vegetation, primarily at night. Older reports of at least sirens being herbivorous surely are based on the voluminous extraneous material that sometimes is captured while sucking appropriate animal-based food items into their mouth.

Reproductive biology

The mode of sperm transfer between males and females is not known for any species, but it is generally assumed that fertilization is external. Males lack glands that make a spermatophore (a gelatinous structure containing sperm) in other salamanders, and females lack a spermotheca (reproductive sac for sperm storage). The large sperm have elongate nuclei and two axial filaments, each with an undulating membrane.

The large eggs are laid singly, sometimes attached to vegetation, or in groups.

Conservation status

Members of this family generally are not threatened, and neither the IUCN nor CITES has listings for any of the species in this family. However, some species are in decline in parts of their distribution areas, mainly because of habitat loss.

Significance to humans

Dwarf sirens have been sold as fishing bait.

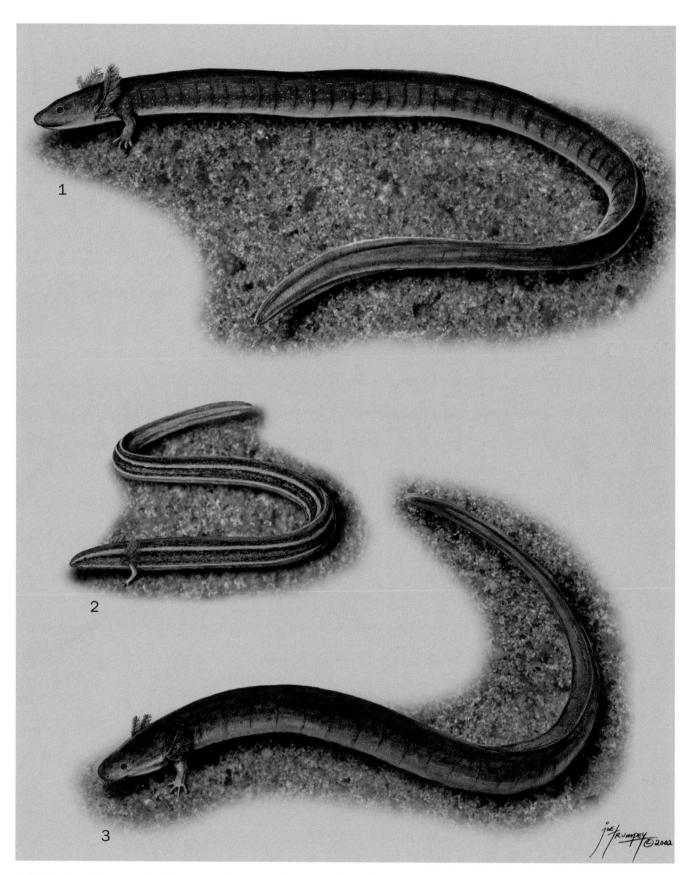

1. Greater siren (*Siren lacertina*); 2. Northern dwarf siren (*Pseudobranchus striatus*); 3. Lesser siren (*Siren intermedia*). (Illustration by Joseph E. Trumpey)

Species accounts

Northern dwarf siren
Pseudobranchus striatus

TAXONOMY
Siren striata Le Conte, 1824, Riceborough, Liberty County, Georgia, United States. Three subspecies are recognized.

OTHER COMMON NAMES
None known.

PHYSICAL CHARACTERISTICS
In addition to the familial characteristics listed above, dwarf sirens have 34–36 costal grooves (successive vertical grooves along the sides marking embryological segments), three toes, one gill slit, and a total length up to 10 in (25 cm). The coloration usually involves a series of muted stripes that extend for most of the length of the slim, cylindrical body. The head is rather pointed. Species are separated reliably only by chromosome number: 24 in *P. striatus* and 32 in *P. axanthus*.

DISTRIBUTION
The species occurs in the northern third of the Florida Peninsula and in the Gulf and Atlantic coastal plains from the western Florida Panhandle to central South Carolina.

HABITAT
These salamanders are most often caught with nets or dredges that sample emergent or floating vegetation, especially water hyacinth (*E. crassipes*), that grow in swampy wetlands, sloughs, and lakes.

BEHAVIOR
See the family behavior section.

FEEDING ECOLOGY AND DIET
Any small aquatic organism will be consumed; however, because of the small size of these salamanders, the average size of food items is smaller than for the species of *Siren*. Small worms, insect larvae, crustaceans of various sizes, and small mollusks are common food items.

REPRODUCTIVE BIOLOGY
The method of sperm transfer between male and female is not known. Eggs (lightly pigmented, about 100 per clutch, 0.1 in [2.5–2.7 mm] diameter, four jelly envelopes) with tough jelly coats and crystalline inclusions in the outer layers are laid singly and scattered and often are attached to the fibrous roots of water hyacinth and other aquatic vegetation. Larvae hatch in three to four weeks at about 0.6 in (15 mm) total length and are black with a silvery white stripe from the tip of the snout to the end of the body.

CONSERVATION STATUS
This species is threatened in at least South Carolina, and destruction of wetland habitats usually is the reason.

SIGNIFICANCE TO HUMANS
None known, other than having been used for fish bait. ◆

Lesser siren
Siren intermedia

TAXONOMY
Siren intermedia Barnes, 1826, Liberty County, Georgia, United States. Three subspecies are recognized, although the status of these names is debated (e.g., *S. i. texana*).

OTHER COMMON NAMES
German: Mittlerer Armmolch; Spanish: Sirena menor.

PHYSICAL CHARACTERISTICS
In addition to the familial characteristics listed above, this species has 31–37 costal grooves, four toes, three gill slits, and a total length up to 27 in (69 cm). The head is broadly rounded as viewed from the top. Larvae (about 0.4 in [10 mm] total length at hatching) have pronounced red bands across the tip of the snout, across the head, and on the body; postlarvae may keep a pale snout band, but the other markings disappear. The adult pattern appears to be geographically variable, but with a greenish to gray ground color with variable amounts of iridophore speckling.

DISTRIBUTION
The species inhabits the Atlantic coastal plain and Mississippi River embayment from northeastern Tamaulipas, Mexico, to southeastern Virginia.

HABITAT
Lesser sirens occur in many types of slowly flowing or still, often swampy water.

BEHAVIOR
These aquatic salamanders can be extremely numerous; they spend daylight hours burrowed into the bottom debris or the bank. Sound production is uncommon in salamanders, but a bitten individual or one routed from a hiding spot by another salamander often yelps; an individual placed in unfamiliar sur-

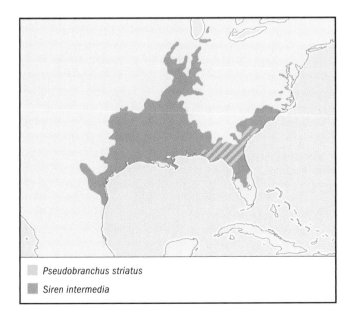

☐ *Pseudobranchus striatus*
▨ *Siren intermedia*

roundings may make several types of sounds. If oxygen concentrations become low, these salamanders come to the surface to gulp air.

If the habitat dries out, lesser sirens move into burrows in the bottom and estivate (condition of torpor analogous to hibernation but in response to hot or dry conditions). Because gills do not work effectively when not submerged, the gills reduce to small nubs under these conditions and the salamanders respire by their lungs. They produce a water-impervious cocoon by shedding their skin several times.

FEEDING ECOLOGY AND DIET
Almost any appropriately sized aquatic animal will be consumed; large individuals eat larger items than smaller salamanders, but even larger specimens eat considerable numbers of small crustaceans. Snails, worms, insect larvae, and small fishes are consumed. Individuals forage mostly along the bottom at night, and all items are sucked into the mouth and swallowed whole.

REPRODUCTIVE BIOLOGY
The method of fertilization or methods of courtship are not known, but during the presumed breeding season, most individuals large enough to be sexually mature have a number of bite marks on them that match the dimensions of the mouth. These bites are presumed to reflect either courtship or territorial disputes by males, or perhaps both. A group of up to 1,500 single, unpigmented eggs is laid in a localized spot (nest) on the bottom and is seemingly guarded by a parent. The eggs are 0.25 in [6.0–6.5 mm] in diameter with four jelly envelopes and have crystalline inclusions in the outer jelly layer. Hatching occurs in 45–75 days at about 0.4 in (10 mm) total length, and the larvae are densely black with red markings on the snout, head, and body. The clear, unmarked dorsal fin originating on the back eventually is opaque and restricted to the tail in larger individuals. Most of the red markings are lost, but the nose band may persist in muted form.

CONSERVATION STATUS
This species is threatened in Texas, probably because of the drainage of wetlands, but seems abundant in most places with the proper habitat.

SIGNIFICANCE TO HUMANS
None known, although some people fear them because they confuse them with the larger amphiuma (*Amphiuma*), which can produce a dangerous bite. ◆

Greater siren
Siren lacertina

TAXONOMY
Siren lacertina Linnaeus, 1767, Charleston, South Carolina, United States. No subspecies are recognized, but considerable variations from throughout the range are known.

OTHER COMMON NAMES
French: Sirène lacertine; German: Grosser Armmolch; Spanish: Sirena grande.

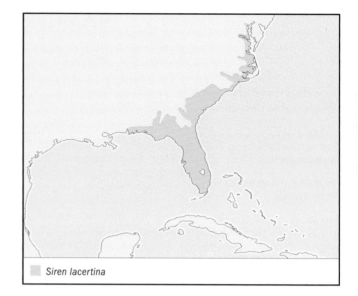

Siren lacertina

PHYSICAL CHARACTERISTICS
In addition to the familial characteristics listed above, this species has 36–40 costal grooves, four toes, three gill slits, and a total length up to 36 in (91 cm). Larvae have pronounced yellow bands across the tip of the snout and along the body; postlarvae may retain the snout band, but the other markings disappear. The adult pattern appears to be geographically variable, but with a greenish to gray ground color with variable amounts of lighter speckling.

DISTRIBUTION
The species occurs throughout the Atlantic coastal plain from western Alabama to Maryland and in all of the Florida Peninsula.

HABITAT
Usually they inhabit larger wetlands, lakes, sloughs, and ponds with soft substrates and considerable vegetation.

BEHAVIOR
Same behavior as the lesser siren. The greater sirens are also hosts to parasitic copepods.

FEEDING ECOLOGY AND DIET
Their feeding habits are similar to those of the lesser siren, although more, larger objects are common in the diet.

REPRODUCTIVE BIOLOGY
Eggs are laid singly and attached to vegetation. The larvae are similar to those of the lesser siren, but the markings are yellow and there are lateral and ventrolateral stripes on the body.

CONSERVATION STATUS
Not threatened.

SIGNIFICANCE TO HUMANS
None known.

Resources

Books

Duellman, William E., and Linda Trueb. *Biology of Amphibians.* New York: McGraw Hill, 1986.

Estes, Richard. *Gymnophiona, Caudata.* Vol. 2 of *Handbuch der Paläoherpetologie,* edited by Peter Wellnhofer. Stuttgart, Germany: Gustav Fischer Verlag, 1981.

Periodicals

Altig, Ronald. "Food of *Siren intermedia nettingi* in a Spring-fed Swamp in Southern Illinois." *American Midland Naturalist* 77, no. 1 (1967): 239–241.

Aresco, Matthew J. "*Siren lacertina* (Greater Siren). Aestivation Chamber." *Herpetological Review* 32, no. 1 (2001): 32–33.

Asquith, Adam, and Ronald Altig. "Osmoregulation of the Lesser Siren, *Siren intermedia* (Caudata: Amphibia)." *Comparative Biochemistry and Physiology* 84, no. 4 (1986): 683–685.

———. "Phototaxis and Activity Patterns of *Siren intermedia.*" *Southwestern Naturalist* 32, no. 1 (1987): 146–148.

Duke, Jeffrey T., and Gordon R. Ultsch. "Metabolic Oxygen Regulation and Conformity during Submergence in the Salamanders *Siren lacertina, Amphiuma means,* and *Amphiuma tridactylum,* and a Comparison with Other Giant Salamanders." *Oecologia* 84, no. 1 (1990): 16–23.

Gehlbach, Frederick R., Roxanne Gordon, and Judy B. Jordan. "Aestivation of the Salamander, *Siren intermedia.*" *American Midland Naturalist* 89, no. 2 (1973): 455–463.

Gehlbach, Frederick R., and Braz Walker. "Acoustic Behavior of the Aquatic Salamander *Siren intermedia.*" *BioScience* 20 (1970): 1107–1108.

Godley, J. Steven. "Observations on the Courtship, Nests, and Young of *Siren intermedia* in Southern Florida." *American Midland Naturalist* 110, no. 1 (1983): 215–219.

Goin, Coleman J. "Notes on the Eggs and Early Larvae of Three Florida Salamanders." *Natural History Miscellanea* 10 (1947): 1–4.

Moler, Paul E., and James Kezer. "Karyology and Systematics of the Salamander Genus *Pseudobranchus* (Sirenidae)." *Copeia* 1 (1993): 39–47.

Neill, Wilfred T. "Juveniles of *Siren lacertina* and *S. i. intermedia.*" *Herpetologica* 5, no. 1 (1949): 19–20.

Raymond, Larry R. "Seasonal Activity of *Siren intermedia* in Northwestern Louisiana (Amphibia: Sirenidae)." *Southwestern Naturalist* 36, no. 1 (1991): 144–147.

Reilly, Stephen M., and Ronald Altig. "Cranial Ontogeny in *Siren intermedia* (Caudata: Sirenidae): Paedomorphic, Metamorphic, and Novel Patterns of Heterochrony." *Copeia* 1 (1996): 29–41.

Reno, Harley W., Frederick R. Gehlbach, and Robert A. Turner. "Skin and Aestivational Cocoon of the Aquatic Amphibian, *Siren intermedia* Le Conte." *Copeia* 4 (1972): 625–631.

Sever, David M., Lisa C. Rania, and John D. Krenz. "Reproduction of the Salamander *Siren intermedia* Le Conte with Especial Reference to Oviducal Anatomy and Mode of Fertilization." *Journal of Morphology* 227, no. 3 (1996): 335–348.

Sullivan, Aaron M., Paul W. Frese, and Alicia Mathis. "Does the Aquatic Salamander, *Siren intermedia,* Respond to Chemical Cues from Prey?" *Journal of Herpetology* 34, no. 4 (2000): 607–611.

Ultsch, Gordon R. "Observations on the Life History of *Siren lacertina.*" *Herpetologica* 29, no. 4 (1973): 304–305.

———. "The Relationship of Dissolved Carbon Dioxide and Oxygen to Microhabitat Selection in *Pseudobranchus striatus.*" *Copeia* 2 (1971): 247–252.

Ronald Altig, PhD

Asiatic salamanders

(*Hynobiidae*)

Class Amphibia
Order Caudata
Suborder Cryptobranchoidea
Family Hynobiidae

Thumbnail description
Medium-size to small salamanders, usually dark brown to olive in color

Size
Body length 4–10 in (100–250 mm)

Number of genera, species
8 genera; 43 species

Habitat
Marshes, mountain streams, and ponds

Conservation status
Endangered: 5 species; Vulnerable: 4 species; Data Deficient: 1 species

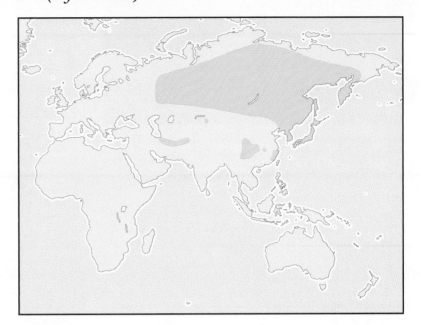

Distribution
Central and eastern Asia

Evolution and systematics

In 1995 the first fossil record of hynobiid, which resembles the extant species *Ranodon sibiricus*, was discovered from the Upper Pliocene in Kazakhstan. Later, two other fossil species were described from late Miocene and Lower Pleistocene deposits in Europe. The oldest fossil hynobiid is perhaps *Liaoxitriton zhongjiani*, which was discovered from the early Cretaceous of western Liaoning, China. In morphological features, it most closely resembles hynobiids of the genus *Batrachuperus*. It may represent an extinct branch of extant hynobiid salamanders or an early branch of ancestral hynobiids.

The current understanding about the origin of hynobiids is based on comparisons among extant salamander groups using a phylogenetic approach. Recent morphological and molecular studies have suggested that hynobiid salamanders are related most closely to the family Cryptobranchidae, which includes the hellbender (*Cryptobranchus alleganiensis*) and the Chinese giant salamander (*Andrias davidianus*). They may share a recent common ancestor and may represent one of the oldest salamander lineages. Together with the family Sirenidae, the three families usually are considered the most "primitive" salamanders.

At present, there are eight recognized genera. Most genera are well established, but the validity of *Pseudohynobius* and *Liua* is still in dispute. In addition, the genus *Batrachuperus* may not be a natural group. It may consist of two distantly related groups. About half of the described species belong to

the genus *Hynobius*, most of which occur only in Japan. The Japanese species are the best studied of the hynobiids. Several other genera may be as diverse as *Hynobius*. The small numbers of described species in these genera may be attributed more to lack of knowledge than to lack of diversity.

The species *Protohynobius puxiongensis* is morphologically very different and has been placed into its own subfamily. All other hynobiids are placed into another subfamily and sorted into two groups based on morphological characters. One group includes *Hynobius* and *Salamandrella*, and the other includes *Onychodactylus*, *Ranodon*, *Pseudohynobius*, *Batrachuperus*, and *Liua*. The former genera share several characters, such as large number of eggs and a one-year duration of larval development. The latter group has a much smaller number of eggs and two to three years of larval development. The genus *Onychodactylus* once was considered an "advanced" group, because it possesses several unique characteristics: lack of lungs, a round and slender tail, a short egg sac, and distinctive male secondary sexual characters. Data from molecular studies, however, have shown a different pattern, in which *Onychodactylus* may represent the oldest lineage of hynobiids, instead of an "advanced" group. Two subfamilies are recognized.

Subfamily Hynobiinae

This subfamily lacks an internasal bone and includes all the other hynobiids. Their distribution covers a large area of central and eastern Asia. At present there are about 42 species, which are grouped into seven genera.

Hokkaido salamander (*Hynobius retardatus*) larva. (Photo by henk .wallays@skynet.be. Reproduced by permission.)

Subfamily Protohynobiinae

This subfamily is characterized by the presence of an internasal bone. Currently, only one species (*Protohynobius puxiongensis*) is assigned to the Protohynobiinae. Its distribution is limited to western China.

Physical characteristics

Hynobiids are small to medium-size salamanders. The body length typically is between 4 and 10 in (100 and 250 mm), although some species may grow as long as 11.8 in (300 mm). Metamorphosis is complete in all species, and adults have eyelids and lack larval teeth and gill slits. Larvae have four pairs of gill slits. The coloration of most species is dull and varies from sandy brown to dark olive, although a few species, such as *Pseudohynobius flavomaculatus*, have colorful spots on their backs.

Some mountain stream dwellers have keratinized structures on their appendages. *Onychodactylus* species have claw-like structures on the fingers and toes; *Liua shihi* and a few *Batrachuperus* species have horny covers on their hands and feet. Such structures presumably aid in the grasp of the substrate by increasing friction. Other morphological variations include reduced lungs in *Ranodon*, *Liua*, and *Batrachuperus* and

Tsushima salamander (*Hynobius tsuensis*). (Photo by henk.wallays@ skynet.be. Reproduced by permission.)

a complete lack of lungs in *Onychodactylus*. Such reduction of the lung may be associated with the aquatic lifestyle. All *Batrachuperus*, *Salamandrella keyserlingii*, and some *Hynobius* have four instead of five toes. The arrangement of vomerine teeth, which are located on the roof of the mouth cavity, is an important character for identifying genera and species. Overall, the external appearance is similar and rather uniform in hynobiid salamanders.

Distribution

Hynobiids are exclusively Asian. They range from Japan, Taiwan, and the mainland of China westward to Afghanistan, Iran, and Kazakhstan in central Asia. To the north, the salamanders occur from the Kamchatka peninsula, the island of Sakhalin, Siberia, and Mongolia westward to the west of the Ural Mountains. The Siberian salamander (*Salamandrella keyserlingii*) is the only member of the family that ranges into European Russia and is the only salamander found north of the Arctic Circle. Some hynobiids, such as *Hynobius nebulosus*, are lowland species, whereas others, such as species of *Batrachuperus*, are strictly montane. The latter are particularly common at elevations of 6,500–13,000 ft (2,000–4,000 m). The highest record is for *B. tibetanus*, which occurs at an elevation of 13,940 ft (4,250 m) in western China.

Habitat

Species of the genera *Batrachuperus* and *Liua* are aquatic year-round. They occur primarily in mountain streams with cool, often fast-flowing water. During daylight hours, they frequently are found under rocks in the water. Occasionally, they are under large rocks on shore but are never far from water. Adults of other hynobiid species are terrestrial, but in the breeding season (February to June) they migrate to and congregate at breeding sites, either mountain streams with running water or ponds. Most species breed in only one of the types of water. For example, *Hynobius leechii* breeds exclusively in ponds. Nevertheless, other species, such as *Hynobius chinensis*, breed in both ponds and streams. Little is known about the activities of most of the terrestrial species out of the breeding season. They have been found under rocks or grasses or in burrows, and *Pseudohynobius flavomaculatus* and some species of *Hynobius* have been dug out of soil.

Behavior

Little is known about the behavior of this group aside from feeding and reproductive behavior.

Feeding ecology and diet

Hynobiids actively forage at night. They are carnivorous, and both larvae and adults feed on various insects and on aquatic and terrestrial invertebrates. Some species (*Hynobius retardatus* and *Batrachuperus londongensis*) practice cannibalism. *Batrachuperus mustersi* has the most bizarre diet of all hynobiids. This species is a cave dweller and often shares caves with bats; baby bats have been found in the stomachs of these salamanders. Apparently, the baby bats fall into the water, where they are eaten.

Reproductive biology

Fertilization in hynobiids is external. Males of most hynobiids release sperm into the water while the females are depositing eggs. The only exception is *Ranodon sibiricus;* the males of this species produce a spermatophore-like structure, but the eggs are fertilized externally. Eggs are deposited in two groups, representing the eggs from each oviduct, respectively; each group is contained in a gelatinous sac. The egg sacs generally are attached to rocks or vegetation in ponds, streams, or marshes. Egg sacs that fail to attach to an object are often not fertilized. The number of eggs in each sac varies within as well as across species, ranging from three in *Onychodactylus japonicus* to 105 in *Salamandrella keyserlingii.*

In most species, males actively participate in the spawning process. The female chooses an object, which she grasps firmly and to which the egg sacs adhere. After she releases part of the egg sac, she lets go of the object and floats backward. The males, waiting nearby, immediately move onto the egg sac. The males often push and kick the female with their limbs and press the egg sacs with the cloacal area to fertilize the eggs. The male's activity may help and accelerate the egg-deposition process.

Most eggs hatch in three to five weeks, although their development is temperature-dependent. Larvae of some species, such as most *Hynobius,* hatch at an early developmental stage and have balancers, which are temporary appendages and provide stability for the larvae. Larvae of most stream breeders, such as *Batrachuperus,* hatch at a more advanced stage and do not have balancers. The duration of the larval stage varies from one to three years. Water temperature plays an important role in determining duration.

The breeding season varies from late winter to early summer. Most *Hynobius* breed in late winter and early spring. The eggs develop in ice-cold water mixed with ice and snow. On the other hand, *Batrachuperus* breed in early summer. The breeding season may be as late as July for some western Chinese species. Little is known about parental care in hynobiid salamanders. It is known that males of *Hynobius nebulosus* guard and vigorously defend the egg sacs.

Conservation status

Hynobiid salamanders are highly endemic. Many are restricted to one island or mountain, and most have limited distributions. Local range reduction and fragmentation are well documented in Japan and China. For example, *Hynobius chinensis* of eastern China has disappeared from many sites. The primary threat to their survival is probably their limited distributions and small population sizes. Human habitat destruction poses another problem. Five species are listed as Endangered: *Batrachuperus mustersi, Hynobius abei, Hynobius okiensis, Hynobius takedai* and *Ranodon sibiricus.* In addition, four species are listed as Vulnerable: *Batrachuperus gorganensis, Batrachuperus persicus, Hynobius hidamontanus,* and *Hynobius dunni.* One species, *Hynobius stejnegeri,* is listed as Data Deficient.

Significance to humans

None known.

1. Japanese marbled salamander (*Hynobius naevius*); 2. Japanese clawed salamander (*Onychodactylus japonicus*); 3. Semirechensk salamander (*Ranodon sibiricus*); 4. Hokkaido salamander (*Hynobius retardatus*); 5. Tibetan stream salamander (*Batrachuperus tibetanus*); 6. Siberian salamander (*Salamandrella keyserlingii*). (Illustration by Marguette Dongvillo)

Species accounts

Tibetan stream salamander
Batrachuperus tibetanus

SUBFAMILY
Hynobiinae

TAXONOMY
Batrachuperus tibetanus Schmidt, 1925, Lintao, China.

OTHER COMMON NAMES
None known.

PHYSICAL CHARACTERISTICS
The body length is 6.7–8.3 in (170–211 mm), and the coloring is dark brown to olive. In some individuals, there are pale spots on the back. The tail is compressed laterally, which is perhaps an adaptation to an aquatic lifestyle. Like other *Batrachuperus* and differing from most other salamanders, the species has four instead of five toes.

DISTRIBUTION
This species occurs only in western China along the northeastern corner of the Tibetan plateau.

HABITAT
This salamander is aquatic throughout the year. Preferred habitats include small mountain streams and creeks at elevations of 4,900–14,000 ft (1,500–4,250 m), but mostly the salamanders live at or above 9,800 ft (3,000 m). Typical creeks are 3.3–6.6 ft (1–2 m) wide and 6–12 in (15–30 cm) deep. During the day the salamanders hide under rocks or logs in moving water. Occasionally, they are under large rocks on mountain slopes near water.

BEHAVIOR
Little is known aside from feeding and reproductive behavior.

FEEDING ECOLOGY AND DIET
The salamanders actively forage in shallow water at night. Occasionally they also forage on land. The diet includes small

crustaceans (70–90% amphipods) as well as aquatic and terrestrial insects.

REPRODUCTIVE BIOLOGY
Breeding occurs once annually. The breeding season lasts from April to June in most areas. Paired egg sacs are attached to the undersides of rocks or logs. Each egg sac typically contains 10–15 eggs. The embryos develop in moving water. The duration of larval development is two to three years.

CONSERVATION STATUS
Not threatened.

SIGNIFICANCE TO HUMANS
None known. ◆

Japanese marbled salamander
Hynobius naevius

SUBFAMILY
Hynobiinae

TAXONOMY
Salamandra naevia Temminck and Schlegel, 1838, Honshu and Shikoku islands, Japan.

OTHER COMMON NAMES
None known.

PHYSICAL CHARACTERISTICS
The body length is 2.7–5.6 in (69–142 mm). This species has relatively short and robust limbs and tail and 14 grooves be-

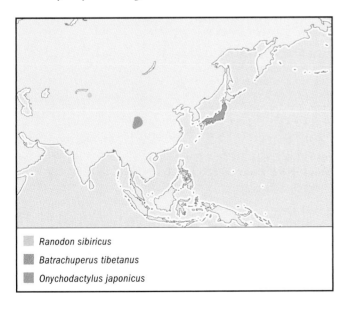

■ *Ranodon sibiricus*
■ *Batrachuperus tibetanus*
■ *Onychodactylus japonicus*

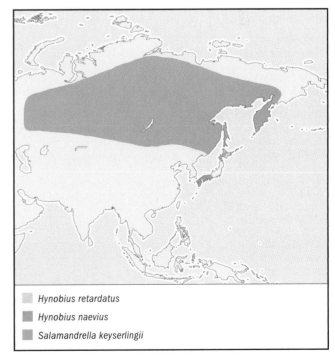

■ *Hynobius retardatus*
■ *Hynobius naevius*
■ *Salamandrella keyserlingii*

tween the forelimbs and hindlimbs on each side of the trunk. The color is dark blue with pale blue spots.

DISTRIBUTION
This species occurs on three Japanese islands: Honshu, Shikoku, and Kyushu.

HABITAT
This species is a mountain forest dweller and is particularly common at elevations of 1,600–3,200 ft (500–1,000 m). Adults are terrestrial during the nonbreeding season. They are commonly found under logs, rocks, and leaf litter near streams on mountain slopes. Eggs are deposited in mountain streams with moving water.

BEHAVIOR
Nothing is known about the behavior of this species.

FEEDING ECOLOGY AND DIET
Larvae feed primarily on larvae forms of aquatic insects. Adults feed on both aquatic and terrestrial insects and other small invertebrates.

REPRODUCTIVE BIOLOGY
The breeding season generally starts in late March and lasts until early April. Each egg sac contains 12–18 eggs. Eggs hatch within four to five weeks, and larvae complete metamorphosis in the fall.

CONSERVATION STATUS
Not listed by the IUCN.

SIGNIFICANCE TO HUMANS
None known. ◆

Hokkaido salamander
Hynobius retardatus

SUBFAMILY
Hynobiinae

TAXONOMY
Hynobius retardatus Dunn, 1923, Hokkaido Island, Japan.

OTHER COMMON NAMES
None known.

PHYSICAL CHARACTERISTICS
The body length is 4.3–7.3 in (110–185 mm). There are 11 or 12 costal grooves and a relatively long tail. The limbs and toes are long compared with other species of *Hynobius*. The dorsum is dark brown with a few indistinct spots. This species has a low diploid chromosome number of 40, compared with 56–78 in other hynobiids. Neotenic form, in which some individuals maintain the larval morphology and grow to adult size, is present in this species.

DISTRIBUTION
This species occurs on Hokkaido, the northernmost of the main Japanese islands.

HABITAT
The preferred breeding sites are slow-moving streams or ponds below elevations of 6,500 ft (2,000 m). During nonbreeding periods, the salamanders are terrestrial and are commonly found under grasses, rocks, and leaf litter on the forest floor. Differing

from other *Hynobius* species, individuals of this species frequently visit bodies of water during the nonbreeding season.

BEHAVIOR
Little is known aside from feeding and reproductive behavior.

FEEDING ECOLOGY AND DIET
Larvae feed on small aquatic invertebrates. Adults feed mainly on insects, crustaceans, some aquatic worms, and, occasionally, fish. Cannibalism is common among larvae in high-density populations. A wide head presumably is an adaptation against cannibalism; larvae with narrower heads are more vulnerable to cannibalism. The larvae have kin recognition ability; they preferentially consume nonkin and avoid siblings.

REPRODUCTIVE BIOLOGY
Breeding starts when the snow begins to melt and the water temperature is typically 37–41°F (3–5°C). In most areas, the breeding season is in April, but at higher elevations, it may be delayed until early June. Both males and females migrate to and congregate at the breeding sites. Mating and egg deposition take place at night. Paired egg sacs are attached to twigs and grass. Each egg sac usually contains 30–50 eggs, but there may be as many as 93 eggs. Larvae typically complete metamorphosis in the same year, but low water temperature sometimes delays metamorphosis until the next year or even the third year.

CONSERVATION STATUS
Not listed by the IUCN.

SIGNIFICANCE TO HUMANS
None known. ◆

Japanese clawed salamander
Onychodactylus japonicus

SUBFAMILY
Hynobiinae

TAXONOMY
Salamandra japonica Houttuyn, 1782, Honshu Island, Japan.

OTHER COMMON NAMES
English: Japanese lungless salamander.

PHYSICAL CHARACTERISTICS
The body length is 4.2–7.2 in (106–184 mm). The body is slender, and the tail is relatively long. The dorsum is brown with orange spots on the back of the head and the limbs, and there are longitudinal orange mid-dorsal stripes. The larvae have clawlike structures on the fingers and toes. Adults have the "claws" only during the breeding season. Lungs are absent.

DISTRIBUTION
This species occurs on two Japanese islands, Honshu and Shikoku.

HABITAT
Usually this species occurs at elevations of more than 3,200 ft (1,000 m). During the nonbreeding season, they are terrestrial but are very close to water. Their favorite places are beneath wet rocks or logs beside a stream. They also are found frequently under logs on the forest floor, in tree holes, and in other places with high humidity.

BEHAVIOR
Little is known aside from feeding and reproductive behavior.

FEEDING ECOLOGY AND DIET
Adults feed primarily on insects and their larvae. Other invertebrates, such as spiders, millipedes, snails, tadpoles, and larval fish, also are part of their diet.

REPRODUCTIVE BIOLOGY
Males are biennial breeders, and females probably are triennial breeders. The breeding season is in May; in some areas it extends into June. The females deposit eggs at night, generally in the headwaters of mountain streams. Each egg sac typically contains three to eight eggs; the total number of eggs produced per female is seven to 15. Larvae take three years to reach metamorphosis.

CONSERVATION STATUS
Not listed by IUCN.

SIGNIFICANCE TO HUMANS
None known. ◆

Semirechensk salamander
Ranodon sibiricus

SUBFAMILY
Hynobiinae

TAXONOMY
Ranodon sibiricus Kessler, 1866, Kazakhstan.

OTHER COMMON NAMES
English: Alatau salamander.

PHYSICAL CHARACTERISTICS
The body length is 5.9–9.8 in (150–250 mm). The tail is as long as the rest of the body and has a dorsal crest. The body is brown, with scattered black spots. The adults have small lungs.

DISTRIBUTION
This species is limited to the Ala Tau mountains and the adjacent Tien Shan mountains of eastern Kazakhstan and western China.

HABITAT
This species lives in mountain streams and marshes at elevations of 4,920–9,000 ft (1,500–2,750 m).

BEHAVIOR
Little is known aside from feeding and reproductive behavior.

FEEDING ECOLOGY AND DIET
This species forages nocturnally. Larvae begin feeding at four to eight days after hatching. Larvae mainly feed on aquatic invertebrate larvae, such as those of ostracods and trichopterans. Feeding continues during metamorphosis. The diet of newly metamorphosed individuals is exclusively aquatic invertebrates, but mature salamanders feed on both aquatic and terrestrial invertebrates.

REPRODUCTIVE BIOLOGY
The breeding season is from May to July. The duration of larval development is two to three years. *Ranodon sibiricus* is the only hynobiid salamander that produces spermatophore-like structures that bear sperm. Males attach the spermatophore to the undersides of rocks and plants. Instead of picking up the spermatophore, the females attach their egg sacs to the same objects, and the eggs are fertilized externally. In such cases, it is the male that chooses the spawning site, not the female.

CONSERVATION STATUS
The IUCN lists this species as Endangered. It is currently under protection in both Russia and China. Restricted distribution and habitat destruction are the main threats to the survival of the species. The species is doing well in captivity, and some reintroduction programs are in place.

SIGNIFICANCE TO HUMANS
None known. ◆

Siberian salamander
Salamandrella keyserlingii

SUBFAMILY
Hynobiinae

TAXONOMY
Salamandrella keyserlingii Dybowski, 1870, vicinity of Lake Baikal, Russia.

OTHER COMMON NAMES
German: Sibirischer Winkelzahnmolch.

PHYSICAL CHARACTERISTICS
The body length is 3.9–5.0 in (100–127 mm). Costal grooves, usually 13 in number, are distinctive. The tail is much shorter than the body. A mid-dorsal brown stripe extends the length of the body. This probably is a complex of species.

DISTRIBUTION
This species has the widest distribution of all hynobiid salamanders. Its range includes northern Japan, Korea, northeastern China, Mongolia, and Siberia (including Sakhalin and the Kamchatka Peninsula), and it extends westward to European Russia; it is the only species of salamander that occurs north of the Arctic Circle.

HABITAT
The Siberian salamander is a terrestrial species. During the nonbreeding season marshes are their preferred habitat; they are found in these marshes under grasses and sometimes in burrows. The females usually deposit eggs in large, shallow ponds.

BEHAVIOR
Little is known aside from feeding and reproductive behavior.

FEEDING ECOLOGY AND DIET
Siberian salamanders are nocturnal foragers, especially after rains. The diet consists mainly of insects, small snails, earthworms, and, occasionally, small fish.

REPRODUCTIVE BIOLOGY
Each egg sac contains 28–105 eggs; the total number of eggs deposited by one female is 59–189. The breeding season in most areas begins in early April. Females first grasp a twig or grass with their four limbs and then deposit the egg sacs. One end of the egg sac is attached to a twig or grass. When the egg

sacs are released halfway, the female lets go of the twig and floats backward. At this time, several males usually join in. The males and females often grab each other and form a "ball." The males do not simply fertilize the eggs; their activity often also helps the female deposit the egg sacs. Larvae complete metamorphosis in one year.

CONSERVATION STATUS
Not listed by the IUCN, but listed as near threatened in Japan by the National Strategy of Japan on Biological Diversity.

SIGNIFICANCE TO HUMANS
None known.

Resources

Books

Duellman, William E., and Linda Trueb. *Biology of Amphibians.* Baltimore: Johns Hopkins University Press, 1994.

Pough, F. Henry, Robin M. Andrews, John E. Cadle, Mart L. Crump, Alan H. Savitzky, and Kentwood D. Wells. *Herpetology.* 2nd ed. Upper Saddle River, NJ: Prentice Hall, 2001.

Weisrock, D. W., J. R. Macey, A. Larson, and T. J. Papenfuss. "Phylogenetic Relationships Among Hynobiid Salamanders: Evidence for an Old North Asian Fauna and Clock-like Evolution in the Mitochondrial Genome." In *Abstracts of the Joint Meeting of the ASIH/AE/HL/SSAR.* State College, PA: Penn State University, 1999.

Zhao, Ermi, Qixiong Hu, Yaoming Jing, and Yuhua Yang. *Studies on Chinese Salamanders.* Oxford, OH: Society for the Study of Amphibians and Reptiles, 1988.

Periodicals

Averianov, A. O., and L. A. Tjutkova. "*Ranodon cf. sibiricus* (Amphibia, Caudata) from the Upper Pliocene of Southern

Kazakhstan: The First Fossil Record of the Family Hynobiidae." *Palaeontologische Zeitschrift* 69 (1995): 257–264.

Dong, Z., and Y. Wang. "A New Urodele (*Liaoxitriton zhongjiani* gen. et. sp. nov.) from the Early Cretaceous of Western Liaoning Province, China." *Vertebrata PalAsiatica* 36 (1998): 159–172.

Hasumi, M., and F. Kanda. "Breeding Habitats of the Siberian Salamander (*Salamandrella keyserlingii*) Within a Fen in Kushiro Marsh, Japan." *Herpetological Review* 29 (1998): 150–153.

Venczel, M. "Land Salamanders of the Family Hynobiidae from the Neocene and Quaternary of Europe." *Amphibia-Reptilia* 20 (1990): 401–412.

Wakahara, M. "Kin Recognition Among Intact and Blinded, Mixed-Sibling Larvae of a Cannibalistic Salamander *Hynobius retardatus.*" *Zoological Science (Tokyo)* 14, no. 6 (1997): 893–899.

Jinzhong Fu, PhD

▲
Asiatic giant salamanders and hellbenders
(Cryptobranchidae)

Class Amphibia
Order Caudata
Suborder Cryptobranchoidea
Family Cryptobranchidae

Thumbnail description
Cryptobranchids are large aquatic salamanders with dorsoventrally flattened bodies, short limbs, and laterally compressed tails; metamorphosis is incomplete; lateral folds present on the body and limbs are highly vascularized; eyelids are absent, and the eyes are small; larvae may maintain gills from 18 months to more than three years

Size
Larvae are usually 1–1.3 in (25–33 mm) long; adult hellbenders are about 29 in (740 mm), and adult Asiatic giant salamanders can reach 5.2–5.9 ft (1.6–1.8 m) in length

Number of genera, species
2 genera; 3 species

Habitat
Typically found in cool streams and rivers with gravel and rock-covered bottoms

Conservation status
Vulnerable: 1 species; Data Deficient: 1 species

Distribution
These salamanders range across eastern North America, central China, and central Japan; fossils are known from western North America and Europe

Evolution and systematics

Cryptobranchids are believed to be derived from hynobiid-like amphibians by the retention of larval characters into adulthood. Morphologic features (fusion of the tibialis muscles, which move the lower leg, as well as the first hypobranchials with the first ceratobranchials, both found in the hyoid apparatus) and molecular studies (analysis of ribosomal RNA sequences) support a group with a single line of phylogeny (lineage) comprising the families Hynobiidae and Cryptobranchidae within the suborder Cryptobranchoidea. The fossil record of cryptobranchid salamanders begins with *Cryptobranchus saskatchewanensis* from the Upper Paleocene to the Lower Eocene, *C. matthewi* from the middle Miocene to the Miocene-Pliocene boundary, and *C. alleganiensis* from the Pleistocene and recent sites in North America. It includes *C. scheuchzeri* from the Middle Oligocene to the Upper Pliocene in Europe and *Andrias japonicus* from Pleistocene sites in Asia. *C. guiday* from the Pleistocene of Kansas probably is referable to *C. alleganiensis*.

It is impossible to compare many parameters between fossil and extant forms. For example, adult *C. alleganiensis* have a single hemoglobin (iron-containing pigment of red blood cells), and adult *Andrias davidianus* possess two different types of hemoglobin, yet nothing is known about the hemoglobin of fossil cryptobranchids. There are three extant species: the Chinese giant salamander, *Andrias davidianus* Blanchard,

1871, "Tchong Pa from the Fowho River" (Jiangyou County, Sichuan Province, China); the Japanese giant salamander, *Andrias japonicus* Temminck, 1837, mountains of Suzuka, Omi Province, en route from Tsuchiyama to Sukanoshita, Japan (lectotype, Hoogmoed, 1978); and the hellbender, *Cryptobranchus alleganiensis* Daudin, 1803, Allegheny Mountains, Virginia, United States. Two subspecies of hellbenders are recognized. No subfamilies are recognized. Cryptobranchids have a very low level of genetic variation, but hellbenders may have substantial variation of mitochondrial DNA.

Physical characteristics

Hellbenders usually maintain a single pair of gill openings, whereas the Asiatic giant salamanders lose these openings during metamorphosis. The Japanese giant salamanders are distinguished from the Chinese giant salamander by having large tubercles on the head and throat that are singular and scattered irregularly. The tubercles of the Chinese relative are smaller and typically paired. Cryptobranchids vary widely in color and pattern. The dorsum is slate, brown, greenish brown, yellowish brown, orange-red, or, rarely, albino. During the breeding season some mature hellbenders from Ozark streams change from greenish brown or yellowish brown during the day to orange at night.

1. Japanese giant salamander (*Andrias japonicus*); 2. Hellbender (*Cryptobranchus alleganiensis*); 3. Chinese giant salamander (*Andrias davidianus*). (Illustration by Brian Cressman)

Cryptobranchus alleganiensis

Distribution

The Chinese giant salamander ranges from Jiangsu to Quinghai south to Guanxi, Guangdong, and Sichuan. They are found in montane streams, usually below 4,888 ft (1,490 m), including the middle and lower tributaries of the Huang He, Yangtze, and Zhu Jiang rivers. Giant salamanders from Taiwan may be introduced. The Japanese giant salamander is found in the prefecture of Oita on Kyushu Island and from the southwestern region of Honshu Island northeast to the prefecture of Gifu, Shikoku, typically below 2,264 ft (690 m). Hellbenders occur in streams, usually below 2,641 ft (750 m), from the Susquehanna and Allegheny rivers of southern New York south to the Tennessee river drainage of northeastern Mississippi and west to the Springfield and Salem plateau streams of the Missouri and Arkansas Ozark highlands.

Habitat

Although cryptobranchids are typically cool-stream inhabitants, adults have survived damming in some lakes. Year-round water quality data from a Missouri (United States) stream with very high hellbender populations cited these figures: dissolved oxygen, 8.4–13.6 ppm; carbon dioxide, near 0–9.8 ppm; alkalinity, 122–289 ppm; and temperature, 49.6–72.5°F (9.8–22.5°C).

Behavior

All species are typically nocturnal and exhibit positive correlations with stream flow (rheotaxis) and touch (thigmotaxis). They become less photophobic during the breeding season, and their positive thigmotaxis may be related to the amount of light present. Daily activity patterns of hellbenders are similar before and during the breeding season, but individual activity may vary with gender, season, and water temperature. When hellbenders are stressed by high temperatures or low

oxygen concentrations, they rock their bodies laterally, presumably to expose the well-vascularized lateral folds to more oxygenated water. This behavior also has been observed in males that are brooding eggs. All are permanently aquatic, and although they readily leave aquatic holding tanks, especially at night and during the breeding season, they rarely are seen out of water under noncaptive conditions. Adult hellbenders use cutaneous respiration for about 95% of oxygen uptake at temperatures up to 77°F (25°C) and rarely surface in their natural habitat. In one Missouri stream hellbenders moved from still to moving water when temperatures reached 69–71°F (21–22.5°C), even though water temperature, oxygen concentrations, and pH levels were not significantly different between the sites. During the breeding season there is increased antagonistic activity, especially between mature males, which may result in serious injuries from bites.

Feeding ecology and diet

Cryptobranchids feed on a wide variety of prey, including worms, mollusks, insect larvae, crustaceans, lamprey, fish and their eggs, anurans and their tadpoles, aquatic reptiles, and small mammals. They also feed on carrion and their own shed skin and eggs, and they are cannibalistic. Usually, most of the adults' diet consists of crustaceans and fish. The symphysis (fuse) of the two mandibles is flexible, with large bundles of elastic cartilage that allow one side of the mandible to be depressed as much as 40°. Asymmetrical suction feeding is accomplished when the mandible is quickly depressed and nearby prey are sucked into the mouth.

Reproductive biology

Asiatic giant salamanders and hellbenders breed from August through January, as day length decreases or is near minimal annual day length. During much of the breeding season, the water temperature also decreases. Fertilization is exter-

Andrias japonicus

Andrias davidianus

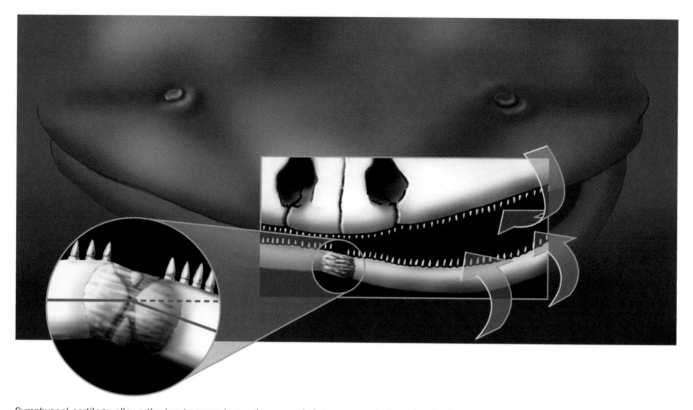

Symphyseal cartilage allows the jaw to move in a unique way during asymmetrical suction feeding. (Illustration by Brian Cressman)

nal. As the breeding season approaches, mature crypto-branchids, especially males, become more diurnally active, searching for and obtaining brooding sites. These sites are typically under rocks or logs, with the opening facing downstream. Brooding sites also may be tunnels in banks or crevices in bedrock. Typically, a dominant male establishes a position at the entrance to the nest and either allows a gravid female into the nest cavity or drives her in; he may aggressively resist the female's efforts to leave before egg laying. As two strands of eggs begin to protrude from the female's cloaca (the cavity that serves as terminal depository for excretory, reproductive, and digestive material), males are attracted, perhaps chemically, and position themselves along-

side the female. The male rocks the lower portion of his body, dispensing sperm over the egg masses. The male then guards the eggs and the nest against intruders, including other members of the same species.

Nesting sites may be limited to one male and the eggs of a single female. However, communal nesting is well documented, and some nest sites may have numerous males and females. The social structure of many conspecifics of both genders within some nesting sites is not well understood. An Asiatic giant salamander may lay 400–600 eggs, which hatch in 50–60 days. A hellbender may have 150–750 enlarged ovarian eggs and communal nests with up to 1,946 eggs. Increased numbers of eggs deposited usually correlates with larger salamander size. Asiatic giant salamanders may live more than 60 years, and hellbenders are known to live 29 years in captivity and probably more than 30 years in the wild. Sexual dimorphism is evident just before and during the breeding season. During this period the cloacal glands of adult males exhibit enlarged swelling around the margins of the vent.

Conservation status

All extant cryptobranchid species have shown drastic population declines and fragmentation. The largest individuals within populations are much smaller and presumably younger than those cited in earlier studies. The IUCN categorizes the Japanese giant salamander (*Andrias japonicus*) as Vulnerable, and lists the Chinese giant salamander (*Andrias davidianus*) as Data Deficient.

Chinese giant salamanders (*Andrias davidianus*) are the largest salamanders in the world. (Photo by Animals Animals ©Zig Leszczynski. Reproduced by permission.)

Both the Chinese and Japanese giant salamanders are listed as CITIES I species and also receive other national and regional protection in their respective countries. Hellbenders (*Cryptobranchus*) are listed as species of special concern, threatened, or rare and endangered by state regulations in the United States. During the past century much of the giant salamander and hellbender habitat has been destroyed or degraded by channeling, damming, increased siltation, and pollution. There is evidence that overcollecting is decimating populations. The Japanese giant salamander was bred at the Amsterdam Zoo at the beginning of the twentieth century. Propagation efforts at the Asa Zoo in Japan began in the 1970s. The Asa Zoo teams have produced substantial husbandry information and were successful in breeding these salamanders during the 1980s. There are several propagation programs in Japan and China that harvest eggs from streams and hatch and raise these salamanders. In the United States substantial numbers of population and reproduction studies have been completed in Missouri's Ozark streams.

Significance to humans

Extant cryptobranchids have been used as a human food source for centuries. In parts of Asia they were used in some medical and religious practices and were considered culinary delicacies until they gained protected status. In North America they were used for food, fish bait, and witchcraft (perhaps medicinal purposes). Large adult cryptobranchids are capable of delivering a severe bite to a finger or hand.

Resources

Books

Nickerson, M. A., and C. E. Mays. *The Hellbenders: North American Giant Salamanders.* Milwaukee: Milwaukee Public Museum Press, 1973.

Petranka, James W. *Salamanders of the United States and Canada.* Washington, D.C.: Smithsonian Institution Press, 1998.

Periodicals

Cundall, D., J. Lorenz-Elwood, and J. D. Grooves. "Asymmetric Suction Feeding in Primitive Salamanders." *Experientia* 43 (1987): 1229–1231.

Haker, K. "Haltung und Zucht des Chinesischen Riesensalamanders *Andrias davidianus.*" *Salamandra* 33 (1997): 69–74.

Naylor, B. G. "Cryptobranchid Salamanders from the Paleocene and Miocene of Saskatchewan." *Copeia* 1 (1981): 76–86.

Nickerson, M. A. "Maintaining Hellbenders in Captivity: The Evolution of Our Knowledge." *Proceedings of the American Association of Zoological Parks and Aquaria, Little Rock, Ark.* 1977–1978: 396–399.

Nickerson, M. A., K. L. Krysko, and R. D. Owen. "Ecological Status of the Hellbender (*Cryptobranchus alleganiensis*) and the Mudpuppy (*Necturus maculosus*) Salamanders in the Great Smoky Mountains National Park." *Journal of the North Carolina Academy of Sciences* 118, no. 1 (2002): 27–34.

Peterson, C. L., D. E. Metter, and B. T. Miller. "Demography of the Hellbender *Cryptobranchus alleganiensis* in the Ozarks." *American Midland Naturalist* 119, no. 2 (1988): 492–496.

Other

"About Andrias." (11 July 2002) <http://www3.ocn.ne.jp/herpsgh/aboutandrias.html>

Organizations

Asa Zoological Garden. Asa–cho, Asakita–ku, Hiroshima, 731-33 Japan. Phone: (082) 838-1111. Web site: <http://www.tourism.city.hiroshima.jp/english/level7/h040200008.html>

Max A. Nickerson, PhD

Pacific giant salamanders
(Dicamptodontidae)

Class Amphibia
Order Caudata
Suborder Salamandroidea
Family Dicamptodontidae

Thumbnail description
Large salamanders that have a multiyear aquatic larval stage and usually metamorphose into terrestrial adults with very large heads, short tails, stout bodies, and strong limbs

Size
6.7–13.8 in (17–35 cm)

Number of genera, species
1 genus; 4 species

Habitat
Wooded areas with clear, permanent streams for larvae

Conservation status
No species listed by the IUCN

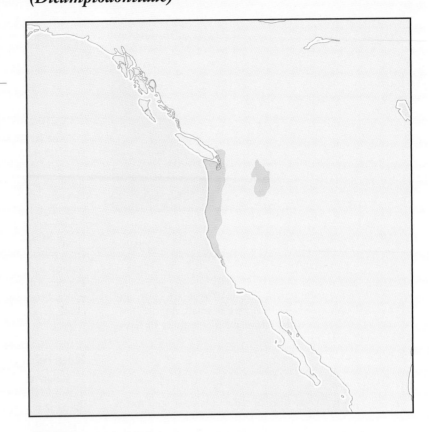

Distribution
Northwestern North America

Evolution and systematics

These large salamanders occupy a somewhat enigmatic position phylogenetically. They long were thought to be allied with mole salamanders (Ambystomatidae) and torrent salamanders (Rhyacotritonidae) and were classified in a single family, Ambystomatidae, but there are no uniquely derived traits shared by these three groups. Despite the fact that each contains only a single genus, at present three families are typically recognized. Some molecular evidence suggests that dicamptodontids and ambystomatids are sister taxa. Some recent classifications place the dicamptodontids and rhyacotritontids in a single family, Dicamptodontidae, but there is no morphological or molecular evidence in support of a sister-group relationship of these two distinct lineages. Dicamptodontids differ from most ambystomatids in having a larval stage that lasts for more than one year, but there are species in both families that never metamorphose and breed as permanently gilled forms. Dicamptodontids also differ from ambystomatids in having relatively much larger and more heavily ossified skulls, with more skull bones; in the anatomy of the tongue skeleton; and in having trunk vertebrae that are not pierced by the spinal nerves. Members of both families have species that reach the largest size of any metamorphosed, terrestrial salamander (one plethodontid is as nearly as

large), but dicamptodontids are more massive. Dicamptodontids differ from rhyacotritontids in being much larger and in having less well developed skulls and limbs.

The single genus, *Dicamptodon*, includes four species (*D. aterrimus, D. copei, D. ensatus,* and *D. tenebrosus*). No subfamilies are recognized.

Physical characteristics

These large, robust salamanders have a massive head, well developed eyes, and large, well developed limbs. Fore and hind limbs fail to overlap when adpressed to the trunk in *D. copei* as well as in some individuals of the other species, but overlap slightly by as many as four costal folds in the other species. The tail is relatively short, always much shorter than head plus body length, and it is laterally compressed with a distinct keel. Metamorphosed adults are dark in coloration, often attractively mottled or marked with different shades of gray. Larvae are somewhat flattened dorsoventrally and are darkly pigmented. The relatively short, robust tail is somewhat flattened at the base but then becomes laterally compressed with a modest fin. The fin terminates at the base of

A coastal giant salamander (*Dicamptodon tenebrosus*) eats a large banana slug (*Ariolimax columbianus*) in northern California, USA. (Photo by Karl H. Switak/Photo Researchers, Inc. Reproduced by permission.)

the tail, well behind the pelvis. Gills of larvae are short and relatively inconspicuous in larvae living in small, rapidly flowing streams, but can become large and filamentous in larvae living in lakes and larger streams.

Distribution

The family is narrowly distributed and is restricted to the Pacific Northwest region of North America. An isolated group of populations of *D. ensatus* occurs on the central and southern San Francisco peninsula in California. The species are more or less continuously distributed in the coastal ranges from north of San Francisco to the northern end of the Olympic Peninsula in Washington, and from the southern Cascade Mountains of Oregon into the coastal mountains of southwestern British Columbia (but not on Vancouver and neighboring islands). One species, *D. aterrimus*, is disjunctly distributed and occurs in the mountains of northern Idaho and northwestern Montana, west of the Continental Divide.

Habitat

Dicamptodontids are restricted to wooded areas that have clear, permanent streams in which their larvae live. Typically they live in coniferous woodlands that are in relatively steep terrain. They do especially well in areas dominated by Douglas fir and coast redwood. Larvae do best in small, troutless streams, but larger larvae may live in rivers (such as the Willamette) and small lakes. Occasionally they are caught on hook and line by fishermen.

Behavior

Adults are frequently found walking by day in dark, moist forests, but more typically they are nocturnal. Large individuals can be aggressive and engage in head-butting and tail-lashing. They are capable of inflicting a dangerous bite

because they have strong jaws and small but numerous and well developed teeth. Nothing is known concerning courtship and mating, but fertilization is internal by means of a spermatophore, probably deposited in a terrestrial site, and eggs are deposited singly but in large groups of 80 or more under large rocks and logs in headwater streams. Adults are capable of vocalizing. They produce a sharp "bark," but the function of this behavior is unknown.

Feeding ecology and diet

The larvae are opportunistic feeders of benthic larvae of insects (those found at the bottom of the stream or river), but they also take other stream-dwelling organisms. Because the larvae grow to large size, they feed on larger prey as well, including larvae of ambystomatid salamanders and small fish. They are considered to be the most abundant vertebrate predators in headwater streams throughout their range. Adults display a wide range of sizes. As small adults they eat a wide array of terrestrial invertebrates, which they catch with their protrusible tongue. As they grow larger they prey on vertebrates, such as slender salamanders, lizards, shrews, mice, and even snakes, which they seize with their strong jaws. They forage widely and climb vegetation as high as 6.5 ft (2 m) on tree trunks.

Reproductive biology

Relatively little is known concerning the reproduction of dicamptodontids. Courtship has not been observed, but most likely it is aquatic because *D. copei* typically reproduces as a gilled form. Eggs are hard to find. The few discoveries have been under large rocks and logs in or at the edge of streams. Eggs develop slowly, and hatching does not occur for many weeks. Newly hatched larvae have large amounts of abdominal yolk and probably do not feed for several weeks. The larval stage is at least two years and may be as long as four years occasionally. Metamorphosis is rare in *D. copei*. Reproductive larvae have been reported in the other species, but courting and egg laying has not been observed, and the gravid larvae may metamorphose before reproducing.

Conservation status

Pacific giant salamanders are dependent on forests and clear, unpolluted streams. Areas in which they occur have been and are now undergoing great habitat modification as a result of forestry practices, road building and other construction, and urbanization. Salamanders are most abundant in old growth forests, but they survive following logging and even increase in density so long as streams remain relatively unsilted. Eventually their numbers decline, and as the forest regrows they appear not to recover, and their populations in second growth forests are much lower than in primary forests. These salamanders may spend most of their lives as stream-dwelling larvae, so stream quality is a major factor in their long-term survival. Different species are protected to some degree in various preserves, including national and state parks, and none of the species is at risk of extinction at present.

Significance to humans

Pacific giant salamanders are rarely seen by humans, but rare encounters are memorable because the animals are impressively large and do not attempt to escape unless molested. Adults can be encountered during periods of light rain on the floors of dense coniferous forests.

Species accounts

Coastal giant salamander
Dicamptodon tenebrosus

TAXONOMY
Dicamptodon tenebrosus Baird and Girard, 1852, Oregon, United States.

OTHER COMMON NAMES
English: Pacific giant salamander; German: Pazifisher Riesen-Guerzahnmolch.

PHYSICAL CHARACTERISTICS

Adults of this species may be the largest terrestrial salamanders, with head plus body lengths of more than 7.5 in (19 cm) and total length of at least 13.5 in (34 cm). Projected mean adult sizes based on statistical analyses by Nussbaum (1976) are about 8 in (20 cm) total length. The

Dicamptodon tenebrosus

largest recorded specimens, however, are larvae found in large rivers; these exceed 8 in (20 cm) in head plus body length and are nearly 14 in (36 cm) in overall length. While tiger salamanders (*Ambystoma tigrinum*) also reach this approximate size, they are differently proportioned, having longer tails and smaller heads, so giant salamanders are perceived as being larger and are certainly more massive. Pacific giant salamanders are geographically variable in proportions, with populations from California having longer legs. The species is highly variable in coloration of metamorphosed individuals. Coloration is variable, but the ground color is dark in larvae and in adults, and as metamorphosis approaches, light coloration appears over the dark base and produces a marbling effect of light (silvery to dull golden in color) on dark. The marbling varies from fine to coarse, and in extreme cases it is so coarse that the underlying ground color is obscured.

DISTRIBUTION
This species occurs in and near streams from southwestern British Columbia, Canada (south of the Fraser River), southward generally west of the crest of the Cascade Mountains to Mendocino and Sonoma counties in northern California, where it is abruptly replaced by the closely related and morphologically nearly identical California giant salamander, *D. ensatus*. In some streams hybridized populations occur, but there is apparently no gene flow between the two species. Some populations occur in isolated habitats in north central Oregon, east of the Cascade crest.

HABITAT
Coniferous woodlands.

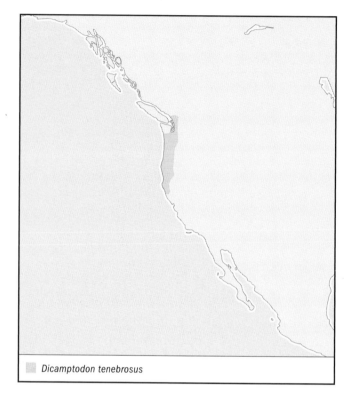

Dicamptodon tenebrosus

BEHAVIOR
Adult salamanders are generally nocturnal and secretive, but they can be encountered on rainy days in densely forested regions walking through leaf litter. When approached they may issue a distinctive "bark," for they are one of the very few salamanders with a voice. They also can be found on rainy nights attempting to cross roads in areas near breeding sites.

FEEDING ECOLOGY AND DIET
These salamanders are voracious eaters and readily take frogs and small mammals, but they also eat worms and arthropods.

REPRODUCTIVE BIOLOGY
Very little is known concerning the breeding habits of Pacific giant salamanders. They have internal fertilization, so males are presumed to produce a spermatophore, but it has not been observed. Females typically lay large numbers of large, yolky eggs under large rocks that are at least partially submerged in streams. Eggs take several months to hatch, and they appear to be guarded by the female during this time.

CONSERVATION STATUS
Not threatened. The greatest danger to this species is destruction of forests and siltation of streams. Larvae require at least two full years in clear streams.

SIGNIFICANCE TO HUMANS
The species is rarely encountered and is little known to humans. ◆

Resources

Books

Petranka, James W. *Salamanders of the United States and Canada.* Washington, DC: Smithsonian Institution Press, 1998.

Periodicals

Nussbaum, Ronald A. "Geographic Variation and Systematics of Salamanders of the Genus *Dicamptodon* Strauch

(Ambystomatidae)." *Miscellaneous Publications of the Museum of Zoology of the University of Michigan* 149 (1976): 1–94.

Parker, M. S. "Feeding Ecology of Stream-dwelling Pacific Giant Salamander Larvae *Dicamptodon tenebrosus.*" *Copeia* 1994 (1994): 705–718.

David B. Wake, PhD

Mole salamanders

(Ambystomatidae)

Class Amphibia
Order Caudata
Suborder Salamandroidea
Family Ambystomatidae

Thumbnail description
These are medium to large, stocky salamanders, generally with both aquatic larval and terrestrial metamorphosed stages; often boldly patterned as adults, with well-developed costal grooves (successive vertical grooves on the sides of the body)

Size
3.5–13.8 in (90–350 mm) in total length

Number of genera, species
1 genus; 33 species

Habitat
Ambystomatids inhabit woodlands and grasslands, including semi-arid pine and juniper woodland with vernal pools, ponds, or occasionally streams for breeding; they are absent from arid deserts within their range

Conservation status
Critically Endangered: 1 species; Vulnerable: 3 species

Distribution
North America from southern Canada to the mountains bordering the Mexican Plateau in central Mexico

Evolution and systematics

Ambystomatids have had a complex taxonomic history, with up to three subfamilies and six genera recognized. The current understanding of the family removes both *Dicamptodon* and *Rhyacotriton* from the ambystomatids and subsumes the old generic names *Rhyacosiredon*, *Bathysiredon*, and *Siredon* into the single genus *Ambystoma*. All species that occur in Mexico (about half of the total in the genus) are close relatives of the tiger salamander (*Ambystoma tigrinum*) and generally are placed in the tiger salamander complex based on molecular and morphologic evidence. Amazingly, these Mexican species originally were placed in four genera, reflecting their diverse ecological and morphologic characteristics. Molecular and fossil evidence suggest that the members of the tiger salamander complex are a recently derived adaptive radiation of salamanders. However, the other members of the family are an old, deeply differentiated set of species that have a fossil record extending from the Lower Oligocene (30 million years ago) through the Pleistocene. No subfamilies are recognized.

Physical characteristics

Ambystomatids are small to large heavy-bodied salamanders with broad heads, small protuberant eyes, well-developed costal grooves, and a long, laterally flattened tail. Many species are brightly colored as metamorphosed adults, with yellow, orange, or silver spots, bars, and frosted patterns on a black background. In some species large poison glands are found on the head and along the body, and transformed adults of all species are distasteful to predators. Lungs are present in all metamorphosed animals. The tiger salamander, at 13.8 in (350 mm) in total length, is one of the largest terrestrial salamanders in the world.

All ambystomatids also have an aquatic larval stage, characterized by filamentous external gills, a large tail-body fin, and small eyes that lack moveable eyelids. In many species of the tiger salamander complex in Mexico and the United States, metamorphosis from the larval stage never occurs, and adults breed and remain in the larval phase throughout their lives. This fascinating life history pattern goes by many names, but we use

Northwestern salamanders (*Ambystoma gracile*) in amplexus in Oregon, USA. (Photo by Dr. Paul A. Zahl/Photo Researchers, Inc. Reproduced by permission.)

the term *paedomorphosis*. Some Mexican species are obligatorily paedomorphic (individuals never go through a full metamorphosis in nature), whereas others from Mexico and the United States are facultative paedomorphs, with some individuals metamorphosing and others failing to do so. The most famous case of an obligate paedomorph is the Mexican axolotl, *Ambystoma mexicanum*. Recent genetic work suggests that metamorphosis is controlled by one or a few genes and that paedomorphosis has evolved several times within the family. Facultative paedomorphosis also is seen in *A. gracile* and *A. talpoideum*.

Distribution

Ambystomatids are distributed from southern Canada south through the Mexican Plateau to just south of Mexico City. Within that range, they occur in most habitats, except the arid deserts of the Great Basin, the southwestern United States, and the central deserts of the Mexican Plateau.

Habitat

Ambystomatid salamanders occur in two primary kinds of habitat. The tiger salamander complex (18 species currently recognized) represent grassland species that shun woodlands but can survive in relatively dry grassland habitat. Paedomorphic species also are found in large, permanent lakes in central Mexico, as long as they are free of predatory fish. The remaining members of the family (15 species) are woodland species found primarily in the eastern and central United States and Canada (13 species), with two representatives restricted to western North America (*A. gracile* and *A. macrodactylum*). Most species spend the vast majority of the year in underground rodent burrows and emerge only on rainy nights to feed or migrate to breeding sites, where they may remain for several weeks.

Behavior

Two key types of behavior of ambystomatid salamanders are their antipredator responses and their migratory movements. Many metamorphosed ambystomatids assume characteristic defensive postures and actions when confronted by predators, including the head-down stance of *A. talpoideum* and *A. gracile* and tail lashing (seen in most transforming species). Both behaviors present a potential predator with parts of the body that are heavily laden with poison-secreting glands.

Ambystomatid salamanders are also famous for their migration to breeding ponds. In some species, literally hundreds of animals may migrate on a single rainy night to a breeding site, presenting a spectacular display of salamanders crossing the landscape. In other species, migrations are more protracted and may take many weeks to complete. Generally, males migrate before females and remain in the pond longer.

Feeding ecology and diet

Like all salamanders, ambystomatids are strictly predators both as larvae and as adults. They consume a wide variety of invertebrate and vertebrate prey, ranging from insects, earthworms, and crustaceans to frog tadpoles and even baby rodents.

Reproductive biology

Most ambystomatids breed in the winter or spring, although montane forms breed in the summer. Terrestrial adults move into vernal pools, ponds, or, more rarely, streams to breed; two species mate and lay eggs on land. Courtship is relatively simple, with males often competing for the opportunity to mate with females. Sperm transfer is via a spermatophore, a packet of sperm on a cone-shaped protienaceous base that is deposited on the substrate and picked up by a female. One male may deposit more than 30 spermatophores during a single night. Inseminated females lay their eggs either singly, attached to the pond bottom or vegetation, or in large clusters. Larvae spend several months to several years in the water before metamorphosing and assuming a terrestrial lifestyle.

Conservation status

The IUCN lists one species (*Ambystoma lermaense*) as Critically Endangered and three species (*A. californiense*, *A. cingulatum*, and *A. mexicanum*) as Vulnerable. In addition, many other Mexican species are suspected of being Endangered. Primary threats are terrestrial and aquatic habitat loss, introduced predatory fish that consume the larvae, and, possibly, a chitrid fungal disease. Paedomorphic species are particularly susceptible, because they often occur in a single lake, where introduced fish, pollution, or draining can endanger an entire species.

Significance to humans

Although they are consumed widely in Mexico, ambystomatids are generally of little direct significance to humans. Like many amphibians, they are considered an important indicator of overall environmental quality.

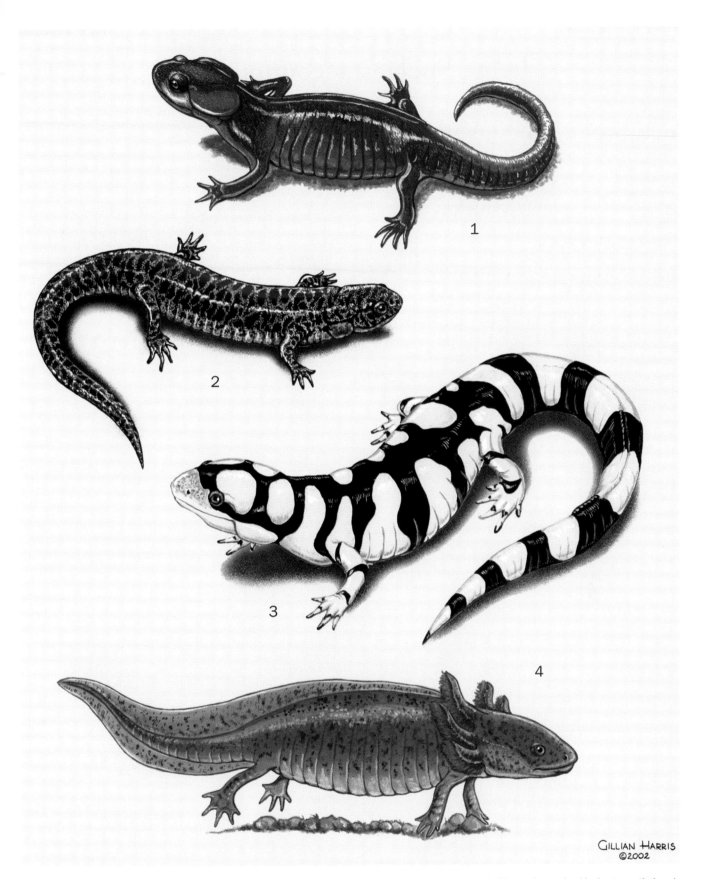

1. Northwestern salamander (*Ambystoma gracile*); 2. Flatwoods salamander (*Ambystoma cingulatum*); 3. Tiger salamander (*Ambystoma tigrinum*); 4. Mexican axolotl (*Ambystoma mexicanum*). (Illustration by Gillian Harris)

Species accounts

Flatwoods salamander
Ambystoma cingulatum

TAXONOMY
Ambystoma cingulatum Cope, 1868, Grahamville, South Carolina, United States. Some authors recognize two subspecies, and others recognize none.

OTHER COMMON NAMES
English: Reticulated salamander.

PHYSICAL CHARACTERISTICS
This small species grows to only 5.3 in (135 mm) in length. The head is relatively narrow. The flatwoods salamander is dark gray to black, with grayish or silvery lines or flecks that form a reticulate or frosted pattern on the back.

DISTRIBUTION
This salamander is restricted to the southeastern United States, where it ranges across northern Florida and southern Alabama east through southern Georgia and the extreme southern part of South Carolina.

HABITAT
The flatwoods salamander inhabits seasonally wet pine flatwoods with vernal pools. Originally it was associated with a unique community of longleaf pine/wire grass, but much of this habitat is now replaced by slash pine plantations.

BEHAVIOR
This species spends most of the year in underground crayfish burrows or tunnels left by dead roots. The larval period lasts about four months, from about January to April.

FEEDING ECOLOGY AND DIET
Larvae of this species feed on small zooplankton and other invertebrates. After metamorphosis, adults spend most of their lives in underground burrows, where they feed on earthworms and small invertebrates.

REPRODUCTIVE BIOLOGY
This is one of two species of *Ambystoma* (the other is *A. opacum*) that court and lay eggs on land. Females lay eggs in dry pond bottoms; the embryos develop and hatch when the ponds fill after heavy rains. This strategy is thought to give young larvae a head start over potential predators that might exclude them from ponds if they hatched later in the season.

CONSERVATION STATUS
The flatwoods salamander is listed as Vulnerable by the IUCN and as Endangered under the U.S. Endangered Species Act. It is extirpated from Alabama, and its remaining stronghold is in Florida.

SIGNIFICANCE TO HUMANS
None known. ◆

Northwestern salamander
Ambystoma gracile

TAXONOMY
Siredon gracilis Baird, 1859, Cascade Mountains, near latitude 44° north, Oregon, United States. Two subspecies generally are recognized, although scientific evidence for this is weak.

OTHER COMMON NAMES
None known.

PHYSICAL CHARACTERISTICS
These fairly large salamanders grow to 8.7 in (220 mm) in length. They are uniformly brown or black and breed both as metamorphs and paedomorphs. Metamorphosed adults have extensive poison glands in the parotoid region and on the base of the tail; when disturbed, they often secrete a white, sticky, toxic secretion.

DISTRIBUTION
The northwestern salamander occurs in wet fir and redwood forests of northwestern North America from Sonoma County, California, United States, to British Columbia, Canada.

HABITAT
Paedomorphs of this species are most common in permanent lakes at higher elevation, whereas metamorphs tend to occur in conifer forests at lower elevations. Unlike many other ambystomatids, paedomorphs can coexist with predatory fish by shifting their activity patterns and becoming nocturnal.

BEHAVIOR
Individuals of this species spend most of their adult life in underground burrows, although they may be found on the surface during rains. When these salamanders are disturbed, they as-

◻ *Ambystoma gracile*
◼ *Ambystoma cingulatum*

sume a rigid posture with the tail partially raised and secrete a white toxic liquid from the parotoid region of the head and from the upper ridge of the tail.

FEEDING ECOLOGY AND DIET

Larval northwestern salamanders feed on zooplankton, a wide variety of aquatic invertebrates, and frog tadpoles; larger individuals take larger prey. Terrestrial adults presumably feed on earthworms and other invertebrates.

REPRODUCTIVE BIOLOGY

Reproductive maturity is reached in two to several years, depending on elevation. Populations vary in terms of metamorphosis/paedomorphosis; it is not known whether the two types interbreed.

CONSERVATION STATUS

No obvious declines have been documented, although there may be evidence that populations are reduced in logged or secondary-growth forests.

SIGNIFICANCE TO HUMANS

None known. ◆

Ambystoma tigrinum

Ambystoma mexicanum

Mexican axolotl

Ambystoma mexicanum

TAXONOMY

Gyrinus mexicanus Shaw, 1789, Mexico. *Ambystoma mexicanum* is a member of the tiger salamander complex. It was known for many years as *Siredon mexicanum*.

OTHER COMMON NAMES

German: Axolotl; Spanish: Ajolote.

PHYSICAL CHARACTERISTICS

This is a large paedomorphic species that occasionally metamorphoses in captivity but apparently is an obligate paedomorph in the wild. Adults are dark brown, often with faint black reticulations. Captive strains are available in a variety of colors, including white, gold, black, and albino.

DISTRIBUTION

This species is known only from Lake Xochimilco (the "Floating Gardens") and associated canals and springs immediately southeast of Mexico City, Mexico. This area has been highly modified by human activities for centuries, but the salamanders remain in moderate numbers.

HABITAT

Because it is an obligate paedomorph, this species occurs only in permanent aquatic habitats. In the vicinity of Lake Xochimilco, it is found frequently near vegetation.

BEHAVIOR

This species has not been well studied in the wild. These salamanders apparently feed and grow throughout the year and are able to coexist with introduced carp and other fish.

FEEDING ECOLOGY AND DIET

Like many other members of the tiger salamander complex in Mexico, the axolotl was probably the top aquatic predator in its habitat before fish were introduced. They feed on insects, snails, worms, tadpoles, and small fish.

REPRODUCTIVE BIOLOGY

In the wild, most Mexican members of the tiger salamander complex, including the Mexican axolotl, breed between November and February. In captivity they breed during most months of the year, except June, July, and August.

CONSERVATION STATUS

Listed as Vulnerable by the IUCN, the axolotl is protected under CITES from international trade and is protected in Mexico. This species is the best-studied species of salamander in the world and has been a "model system" in developmental biology for well over 100 years. Although it is endangered in the wild, the axolotl is commonly reared in captivity for the scientific and pet trades. Salamanders with albino, gold, and wild-type color patterns are frequently sold in the pet trade.

SIGNIFICANCE TO HUMANS

The axolotl was an important species to the Aztec cultures that were centered in the valley of Mexico. All of the paedomorphic Mexican species of ambystomatids continue to be exploited locally for food and medicine in central Mexico, and the axolotl is an important study system for developmental biology throughout the world. ◆

Tiger salamander

Ambystoma tigrinum

TAXONOMY

Salamandra tigrina Green, 1825, Near Moore's town (Moorestown), New Jersey, United States. Up to six subspecies are recognized in the United States. Formerly, the California tiger salamander, *Ambystoma californiense*, was considered a subspecies of *A. tigrinum*. Considerable debate remains over whether the species should be considered one or several species.

OTHER COMMON NAMES
English: Mud puppy, water dog; French: Ambystome tigré
German: Tigerquerzahnmolch; Spanish: Salamandra tigre.

PHYSICAL CHARACTERISTICS
This is a large, robust species that grows to 13.8 in (350 mm) in length. Adults vary widely in color pattern, from black with bright yellow spots and bars or indistinct yellow flecks and reticulations to pure brown or black. In the central United States and Rocky Mountains, tiger salamanders may breed as paedomorphs or metamorphs; elsewhere (including *A. californiense*) they always metamorphose.

DISTRIBUTION
The tiger salamander is the most widely distributed salamander in North America, ranging from southern Canada south, roughly to the border of Mexico and the United States. It is absent from the Appalachian Mountains, the northeastern United States, parts of the southern United States, and the Great Basin, including the Mojave Desert. In California it is replaced by the California tiger salamander, *A. californiense*.

HABITAT
Primarily a grassland-associated species, the tiger salamander is found in prairie and open, dry woodland habitats. It ranges from sea level to an elevation of more than 11,000 ft (3350 m).

BEHAVIOR
Adults of this species spend virtually all of their lives in underground rodent burrows. They emerge and migrate to breeding ponds during spring rains and sometimes can be found on the surface at night during heavy rains.

FEEDING ECOLOGY AND DIET
A fearsome predator, the tiger salamander is a feeding generalist. As larvae, they eat prey ranging from tiny zooplankton to tadpoles and even each other. In captivity they consume prey almost as large as they are. On land they eat all kinds of invertebrates and small vertebrate prey.

REPRODUCTIVE BIOLOGY
Like many other members of the tiger salamander complex and the northwestern salamander group, many tiger salamander populations vary in terms of metamorphosis, with both metamorphs and paedomorphs coexisting in permanent or semipermanent water bodies.

CONSERVATION STATUS
In many areas tiger salamanders are abundant and under no obvious threats, whereas in other regions they are Endangered. The Sonoran tiger salamander, *A. t. stebbinsi* is listed as Endangered under the U.S. Endangered Species Act, and the related California tiger salamander (*A. californiense*) is listed as Vulnerable by the IUCN. Across its range the tiger salamander cannot coexist with predatory fish, and the introduction of bass, catfish, and other species poses a threat to these salamanders.

SIGNIFICANCE TO HUMANS
Larval tiger salamanders are often the top predator in the vernal pools and ponds where they live, and they are therefore an important part of many aquatic ecosystems. In many parts of the United States, tiger salamander larvae are commercially valuable as fish bait. Particularly in the southwestern United States, these "water dogs" are sold in large numbers. ◆

Resources

Books

Bishop, Sherman C. *Handbook of Salamanders: The Salamanders of the United States, of Canada, and of Lower California.* Ithaca, NY: Comstock, 1994.

Duellman, William E., and Linda Trueb. *Biology of Amphibians.* New York: McGraw-Hill Book Co., 1986.

Frost, Darrel R. *Amphibian Species of the World: A Taxonomic and Geographical Reference.* Lawrence, KS: Allen Press and Association of Systematics Collections, 1985.

Petranka, James W. *Salamanders of the United States and Canada.* Washington, DC: Smithsonian Institution Press, 1998.

Shaffer, H. Bradley. "*Ambystoma gracile.*" In *Status and Conservation of U.S. Amphibians,* edited by M. J. Lannoo. Vol. 2, *Species Accounts.* Berkeley, CA: University of California Press, 2001.

Periodicals

Brodie Jr., E. D., and L. S. Gibson. "Defensive Behavior and Skin Glands of the Northwestern Salamander, *Ambystoma gracile.*" *Herpetologica* 25 (1969): 187–194.

Collins, J. P., J. B. Mitton, and B. A. Pierce. "*Ambystoma tigrinum*: A Multispecies Conglomerate?" *Copeia* no. 4 (1980): 938–941.

Eagleson, G. W. "A Comparison of the Life Histories and Growth Patterns of Populations of the Salamander *Ambystoma gracile* (Baird) from Permanent Low-Altitude and Montane Lakes." *Canadian Journal of Zoology* 54 (1976): 2098–2111.

Means, D. Bruce, John G. Palis, and Mary Baggett. "Effects of Slash Pine Silviculture on a Florida Population of Flatwoods Salamander." *Conservation Biology* 10, no. 2 (1996): 426–437.

Shaffer, H. Bradley. "Natural History, Ecology, and Evolution of the Mexican 'Axolotls.'" *Axolotl Newsletter* 18 (1989): 5–11.

———. "Systematics of Model Organisms: The Laboratory Axolotl, *Ambystoma mexicanum.*" *Systematic Biology* 42, no. 4 (1993): 508–522.

———, and M. L. McKnight. "The Polytypic Species Revisited: Genetic Differentiation and Molecular Phylogenetics of the Tiger Salamander, *Ambystoma tigrinum* (Amphibia: Caudata), Complex." *Evolution* 50, no. 1 (1996): 417–433.

Taylor, J. "Orientation and Flight Behavior of a Neotenic Salamander (*Ambystoma gracile*) in Oregon." *American Midland Naturalist* 109 (1983): 40–49.

Titus, T. A. "Genetic Variation in Two Subspecies of *Ambystoma gracile* Baird (Caudata: Ambystomatidae)." *Journal of Herpetology* 24, no. 1 (1990): 107–111.

H. Bradley Shaffer, PhD

Newts and European salamanders

(Salamandridae)

Class Amphibia

Order Caudata

Suborder Salamandroidea

Family Salamandridae

Thumbnail description
Long, slender body with long tail and well-developed limbs

Size
3–14 in (7–35 cm)

Number of genera, species
15 genera; 59 species

Habitat
Damp places close to ponds and streams, where breeding takes place

Conservation status
Critically Endangered: 1 species; Vulnerable: 5 species; Lower Risk/Conservation Dependent: 1 species; Data Deficient: 4 species

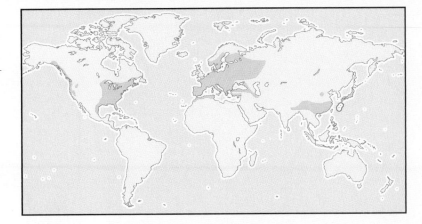

Distribution
Discontinuous across the Northern Hemisphere

Evolution and systematics

Salamandrids originated in the late Cretaceous or early Paleocene in Europe and later dispersed to Asia and North America. The oldest fossil salamandrids come from the Cenozoic in Europe.

The family is sometimes divided into three groups: (1) the *Salamandra* group (*Chioglossa*, *Mertensiella*, *Salamandra*, and *Salamandrina*), (2) the *Triturus* group (*Cynops*, *Euproctus*, *Neurergus*, *Notophthalmus*, *Pachytriton*, *Paramesotriton*, *Taricha*, and *Triturus*), and (3) *Pleurodeles* and *Tylototriton*.

Physical characteristics

Variable in size and appearance, most salamandrids have a long, slender, flexible body and a long tail. The limbs are well developed. Many salamandrids develop dorsal body and tail fins when they enter water. There are four toes on the forelimbs and four or five on the hind limbs. The skin usually is rough, except in the aquatic phase, in which the skin becomes smooth, thin, and slimy, serving as a route by which oxygen is taken up from water. In the aquatic phase, the skin is shed frequently. Newts often are seen eating the discarded skin. Many species have well-developed skin glands, which often are large and prominent on the head (parotid glands). There are no costal grooves on the body. All salamandrids have toxic or distasteful skin secretions. Many of these animals are brightly colored and have distinctive defensive postures. The eyelids are moveable. Lungs are present in juveniles and adults; larvae have feathery external gills.

There is no simple way to differentiate newts and salamanders. All members of the Salamandridae are salamanders, but species that spend a prolonged period each year living in water and becoming temporarily adapted to life in water are called newts. Newts include the European *Triturus* species, *Notophthalmus* and *Taricha* in North America, and *Cynops* in eastern Asia.

Some populations of some species of salamandrids are paedomorphic, meaning they become sexually mature adults while retaining a number of larval features, such as external gills and a large, finned tail. Adults do not become terrestrial but remain in water throughout life. Paedomorphosis occurs in some European (*Triturus*) species and in the three North American (*Notophthalmus*) species. Why some populations of these species are paedomorphic is not known.

This newt embryo is about to hatch. (Photo by Nuridsany et Pérennou/Photo Researchers, Inc. Reproduced by permission.)

Warty newt courtship—the male wafts pheromones with its tail. (Illustration by Wendy Baker)

A female newt lays single eggs and immediately wraps a waterweed leaf around each egg to hide it from predators such as fish. (Illustration by Gillian Harris)

Rough-skinned newt (*Taricha granulosa*) during breeding season on Olympic Peninsula, Washington, USA. (Photo by Lee Rentz. Bruce Coleman Inc. Reproduced by permission.)

Distribution

Fragmented in Northern Hemisphere, including western and eastern North America, Europe, Asia, north Africa, and Japan.

Habitat

Salamandrids are found in a variety of habitats, including woodland, grassland, and heath. In the terrestrial phase, salamandrids need damp conditions and are generally confined to dense vegetation or crevices under rocks and logs, where conditions remain moist at the drier times of year. Because the larvae are aquatic, all salamandrids need water for reproduction. Many breed in ponds; some breed in larger lakes and others in mountain streams. The larvae are vulnerable to predation, and

The enlarged gills of a smooth newt (*Triturus vulgaris vulgaris*) larva. (Photo by Nigel Cattlin/Holt Studios International/Photo Researchers, Inc. Reproduced by permission.)

many salamandrids thrive best in ponds that dry up during the summer, because these ponds cannot support populations of fish, dragonfly larvae, and other aquatic predators.

Behavior

Little is known about the behavior of salamandrids during the greater part of the time they live on land, because they are rarely seen. At least some species, notably the eastern newt (*Notophthalmus viridescens*) have highly developed powers of orientation that enable them to return to breeding ponds each spring. This involves the ability to detect at least one aspect of the environment that provides directional information, including smell, the position of the sun, the pattern of light polarization in the sky, and the direction of the magnetic field of the earth.

The most striking and best-studied aspect of salamandrid behavior is mating. Salamandrids achieve internal fertilization with spermatophores. During mating, the male deposits a spermatophore close to the female and then places, pushes, or entices her over it, so that the sperm is taken up into her cloaca. The female stores the sperm in special storage organs called spermathecae. The female thus controls when and where she lays the eggs.

There is much diversity among species in the behavior that accompanies sperm transfer. Many male salamandrids restrain the female before and during sperm transfer. This behavior involves grasping the female, a behavior called amplexus. European newts (*Triturus*) do not exhibit amplexus. Unable to

The defensive position of a spectacled salamander (*Salamandrina terdigitata*) is used to alarm predators and expose bright warning coloration. (Illustration by Michelle Meneghini)

Male smooth newt (*Triturus vulgaris*) in breeding dress. (Photo by Adrian Davies. Bruce Coleman Inc. Reproduced by permission.)

constrain the female or to control her movements, the male European newt must attract the female with the intensity and complexity of his displays. Physical differences in appearance between the sexes (sexual dimorphism) are much more marked in *Triturus* species than in any other tailed amphibians.

Indirect sperm transfer by means of a spermatophore has two interesting consequences. First, it is unreliable; in some species, many spermatophores are missed by females. Second, rival males can interfere. For example, in several species rival males mimic female behavior, eliciting spermatophores that are not found by females. Much of the diversity and complexity in salamandrid sexual behavior can be interpreted as adaptations that increase the reliability of sperm transfer or that counteract sexual interference (sexual defense). For ex-

ample, *Taricha* males defend females by picking them up and carrying them away if a rival male approaches.

Chemical communication is important in salamandrid mating. Males have glands that produce courtship pheromones. In some species the pheromones are carried on the head; in others they open into the cloaca. Male pheromones alter the hormonal state of the female, making her receptive to males.

Feeding ecology and diet

All salamandrids feed on small invertebrate prey, including insects, earthworms, slugs, and snails. In the aquatic phase, newts feed on aquatic insects and are voracious predators of frog tadpoles. Feeding under water requires changes in the shape of the eye for seeing prey and of the mouth for sucking prey into the mouth. In the aquatic phase, newts develop lateral line organs in the skin. These organs enable the newt to direct tiny water currents and thus locate moving prey, even in the dark or in muddy water. Larval salamandrids eat small invertebrates, such as water fleas.

Reproductive biology

Most salamandrids are terrestrial as adults but migrate to water to breed. In terms of life history, salamandrids are a diverse family. The species vary greatly in the proportion of life spent in water and on land. Newts that lay eggs singly (e.g., *Triturus* and *Notophthalmus*) have long breeding seasons because it takes many weeks for a female to lay all her eggs. In contrast, female *Taricha* lay eggs in clusters and spend little time in the water.

Four European salamandrids are viviparous. That is, the eggs are retained in the female's body, where they develop into large larvae or, in some instances, miniature adults. Viviparous salamandrids have small clutches, so only a small proportion of eggs complete development. In the Caucasian salamander (*Mertensiella luschani*) only two, fully developed young are born after a gestation period of three or four years. The fire salamander (*Salamandra salamandra*), alpine sala-mander (*S. atra*), and Lanza's alpine salamander (*S. lanzai*) also are viviparous.

Conservation status

Most salamandrids are threatened by loss of habitat as the result of deforestation, urbanization, and intensive agriculture. Some species can coexist with humans where agriculture takes a traditional form, involving the creation of hedgerows and of ponds for livestock. Modern agricultural methods, however, are disastrous for amphibians. Ponds are filled in, hedges are torn up, and pesticides, herbicides, and chemical fertilizers kill amphibians.

The IUCN 2002 Red List includes 11 species. One species, *Euproctus platycephalus*, is categorized as Critically Endangered. Five are listed as Vulnerable; one as Lower Risk/Conservation Dependent; and four as Data Deficient.

Significance to humans

Because they taste bad or are toxic, salamandrids are not eaten by humans. Several species are popular as pets, in which context they are well known for the ability to escape from all but the most secure aquarium or terrarium.

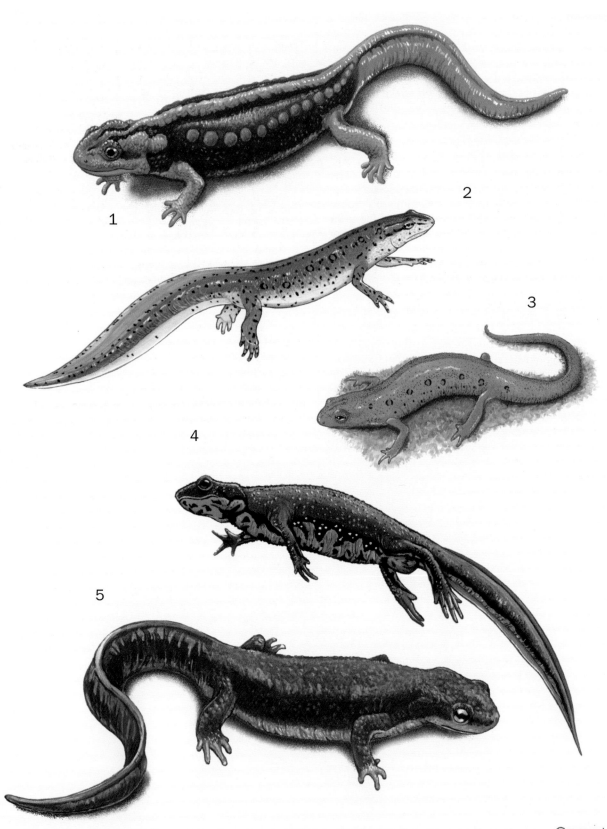

1. Mandarin salamander (*Tylototriton verrucosus*); 2. Eastern newt (*Notophthalmus viridescens*); 3. Red eft (juvenile) form of eastern newt; 4. Japanese fire-bellied newt (*Cynops pyrrhogaster*); 5. California newt (*Taricha torosa*). (Illustration by Gillian Harris)

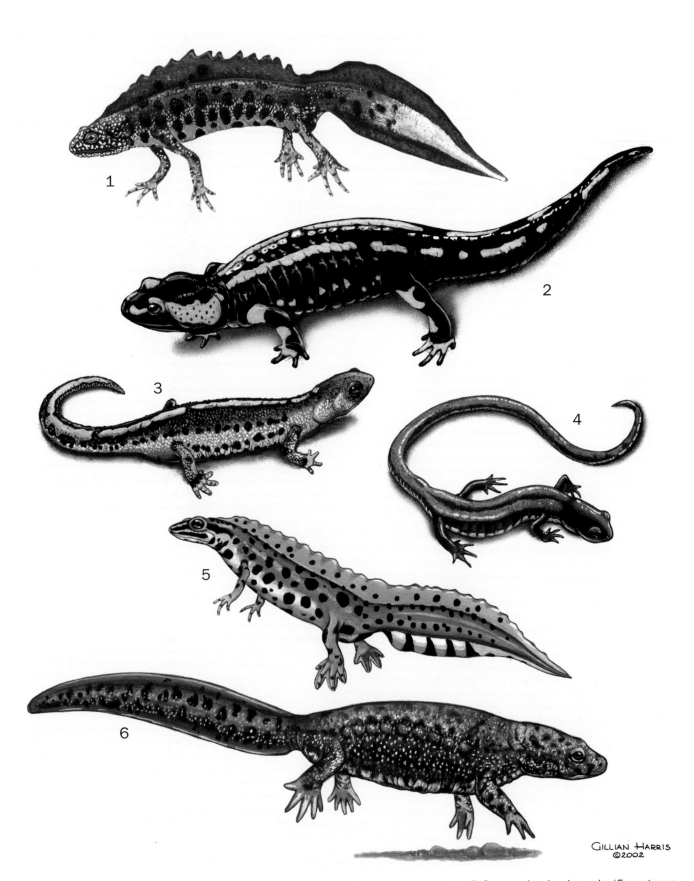

1. Great crested newt (*Triturus cristatus*); 2. European fire salamander (*Salamandra salamandra*); 3. Pyrenean brook salamander (*Euproctus asper*); 4. Golden-striped salamander (*Chioglossa lusitanica*); 5. Smooth (common) newt (*Triturus vulgaris*); 6. Spanish sharp-ribbed newt (*Pleurodeles waltl*). (Illustration by Gillian Harris)

Species accounts

Golden-striped salamander
Chioglossa lusitanica

TAXONOMY
Chioglossa lusitanica Bocage, 1864, Coimbra, Portugal.

OTHER COMMON NAMES
English: Gold-striped salamander; French: Chioglosse; German: Goldstreifen salamander; Spanish: Salamandra rabilarga.

PHYSICAL CHARACTERISTICS
The golden-striped salamander can grow to 6 in (16 cm) in length. It has a long, slender body and tail. The tail constitutes 67% of the total length. Because of its shape and rapid movements, this salamander resembles a lizard. It is dark brown and has two golden-brown stripes on the back that merge to form one stripe on the tail. On some salamanders, the stripes are broken into lines of spots. The golden-striped salamander has a long, narrow head, large eyes and a long, sticky tongue for catching prey.

DISTRIBUTION
Northern Portugal and northwest Spain.

HABITAT
The golden-striped salamander inhabits wet, mountainous areas.

BEHAVIOR
Nocturnal in its habits, the golden-striped salamander is active only when it is damp and is thus confined to areas of heavy rainfall. It hibernates underground or in caves during the winter and estivates (is dormant) during dry summer periods. If attacked, the golden-striped salamander can run quickly. If

caught, it often drops its tail. The tail regrows but never reaches the previous length. This salamander produces a milky, toxic skin secretion when attacked.

FEEDING ECOLOGY AND DIET
The golden-striped salamander uses a long, protrusible tongue to feed on flies and other insects.

REPRODUCTIVE BIOLOGY
Terrestrial for most of its life, the golden-striped salamander breeds in water, laying clumps of as many as 20 eggs in summer or autumn under rocks in springs and streams. Males develop swellings on the upper parts of the forelimbs during the breeding season. The larvae remain in water over winter.

CONSERVATION STATUS
This rare species is listed as Vulnerable. It is threatened by habitat loss, land drainage, replacement of natural forest by plantations, and agricultural pollution.

SIGNIFICANCE TO HUMANS
None known. ◆

Japanese fire-bellied newt
Cynops pyrrhogaster

TAXONOMY
Molge pyrrhogaster Boie, 1826, Nagasaki, Japan.

OTHER COMMON NAMES
German: Japanischer Feuerbauchmolch; Spanish: Tritón vientre de fuego.

PHYSICAL CHARACTERISTICS
The Japanese fire-bellied newt reaches a total length of up to 5 in (12 cm). It has a long tail with a large fin that enables the

Chioglossa lusitanica
Triturus cristatus
Euproctus asper

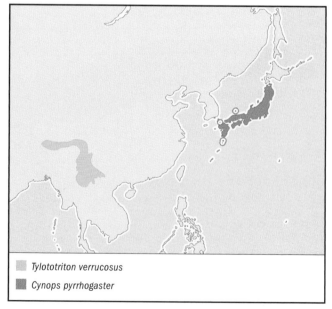

Tylototriton verrucosus
Cynops pyrrhogaster

salamander to swim powerfully. The tail of the male is tipped by a thin filament. Black above, the salamander has a bright red, spotted belly that acts as warning coloration. When attacked, the salamander produces toxic skin secretions, especially from large glands on the head.

DISTRIBUTION
Honshu, Shikoku, and Kyushu, Japan.

HABITAT
The highly aquatic Japanese fire-bellied newt inhabits ponds and pools, often reaching high population density.

BEHAVIOR
Not known.

FEEDING ECOLOGY AND DIET
The Japanese fire-bellied newt feeds on wide variety of small invertebrates.

REPRODUCTIVE BIOLOGY
Mating takes place in water and does not involve amplexus. The male stands in front of the female and may restrain her with one hind foot. In this position the male beats the tip of his tail, producing a water current that carries odor from glands in his swollen cloaca to the female's snout. Eggs are laid in water and attach to submerged vegetation.

CONSERVATION STATUS
Not listed by the IUCN. One of the seven species in *Cynops*, *C. wolterstorffi*, once was found in China, but it has likely become extinct as the result of destruction and degradation of the aquatic habitat, particularly through chemical pollution.

SIGNIFICANCE TO HUMANS
These brightly colored newts may be found in the pet trade. ◆

Pyrenean brook salamander
Euproctus asper

TAXONOMY
Triton glacialis Phillipe, 1847, Lac Bleu, Pyrenees.

OTHER COMMON NAMES
French: Euprocte de Pyrénées; German: Pyrenäen Gebirgsmolch; Spanish: Tritón pirenaico.

PHYSICAL CHARACTERISTICS
Total length is up to 5.5 in (14 cm). A slender animal with a long tail, the Pyrenean brook salamander is dark gray, brown, or black on the back, often with pale yellow markings that may form a continuous or broken stripe along the midline. The belly is yellow or orange. The tail is flattened laterally, enabling the salamander to swim well in flowing water. There is little difference in appearance between the sexes.

DISTRIBUTION
France and Spain in Pyrenees Mountains.

HABITAT
The Pyrenean brook salamander is one of three species in its genus adapted for life in fast-flowing mountain streams. It inhabits streams and mountain lakes at an altitude higher than 1,970 ft (600 m) and free of ice for more than four months each year.

BEHAVIOR
Almost nothing is known about the behavior of the Pyrenean brook salamander.

FEEDING ECOLOGY AND DIET
The Pyrenean brook salamander feeds on insects and other invertebrates.

REPRODUCTIVE BIOLOGY
During mating, members of this genus exhibit a unique form of amplexus in which the male restrains the female by wrapping his prehensile tail around her. This posture may be maintained for many hours while spermatophore transfer is achieved. Females lay 20–40 eggs under rocks. These hatch into streamlined larvae that have small external gills. The larvae undergo metamorphosis and leave the water when they are 2.0–2.4 in (50–60 mm) long.

CONSERVATION STATUS
Not threatened. The Pyrenean brook salamander is locally common, but its range has contracted. Pollution of streams presents a risk. This salamander is protected by the Convention on the Conservation of European Wildlife and Natural Habitats (Berne Convention).

SIGNIFICANCE TO HUMANS
None known. ◆

Eastern newt
Notophthalmus viridescens

TAXONOMY
Triturus viridescens Rafinesque, 1820, Lake Champlain, New York, United States.

OTHER COMMON NAMES
English: Red-spotted newt; French: Tritón vert.

Notophthalmus viridescens

Taricha torosa

PHYSICAL CHARACTERISTICS
Green above and yellow on the belly, the body is decorated with a number of bright red spots, each ringed in black. The tail is one-half the total length of the newt, which is up to 4.8 in (12 cm). The tail of aquatic adults bears a large fin, which is larger on males than on females.

DISTRIBUTION
Eastern North America. The black-spotted newt (*Notophthalmus meridionalis*) occurs only in coastal areas of Texas and Mexico. The striped newt (*N. perstriatus*) is confined to southern Georgia and northern Florida.

HABITAT
The eastern newt lives and breeds in all kinds of permanent and semipermanent water.

BEHAVIOR
In its juvenile stage, which lasts one to four years, the eastern newt acquires a vivid red coloration, is highly toxic, and is called a *red eft*. This newt has the most complex and variable life history of any amphibian. It typically goes through four stages: egg, aquatic larva, eft, and terrestrial adult, returning annually to water to breed. There is much variation in this basic pattern, however, from one part of the range to another. Some populations have no eft stage. In other populations, some adults enter water at maturity but then do not leave. In other populations, some adults are never terrestrial but are paedomorphic. Adult eastern newts are only mildly toxic in comparison with red efts. When attacked, adults exhibit the unken reflex whereby they twist themselves into a circle to expose a bright yellow belly. Paedomorphosis is widespread in this species, in which the red and black markings take the form of stripes.

FEEDING ECOLOGY AND DIET
The eastern newt feeds on a variety of small invertebrates and on frog tadpoles.

REPRODUCTIVE BIOLOGY
Mating behavior is complex and variable. In the breeding season, males develop large and powerful hind limbs that have horny patches on the inner surfaces. Males use the limbs to capture females in a remarkably rapid movement. The male holds the female in amplexus for a long time, stimulating her by rubbing large glands on his head over her snout. When the female is responsive, the male dismounts and deposits a spermatophore. Sometimes a quicker form of mating is used in which the male quickly "tests" the female by waving his tail in front of her. If the female responds, the male proceeds immediately to spermatophore deposition. The female lays eggs singly, attaching them to water weeds. It is thought that clutch size is 200–300 eggs.

CONSERVATION STATUS
Not threatened. The eastern newt is by far the most widespread of the three species in this genus, although it has declined over much of its range as a result of habitat loss and pollution.

SIGNIFICANCE TO HUMANS
None known. ◆

Spanish sharp-ribbed newt
Pleurodeles waltl

TAXONOMY
Pleurodeles waltl Michahelles, 1830, Cádiz, Spain.

☐ *Pleurodeles waltl*
■ *Triturus vulgaris*

OTHER COMMON NAMES
French: Pleurodèle de Waltl; German: Spanische Rippenmolch; Spanish: Gallipato.

PHYSICAL CHARACTERISTICS
Total length is up to 12 in (30 cm). The Spanish sharp-ribbed newt is one of the largest salamandrids, growing very large in some localities. Greenish gray with darker blotches, this newt has a row of pale spots along each side of the body. These spots mark the ends of the ribs.

DISTRIBUTION
Southern Spain and Portugal, coastal areas of Morocco.

HABITAT
The Spanish sharp-ribbed newt lives in ditches, ponds, and lakes that contain water plants. If the pond habitat dries up in summer, this newt is able to survive buried in mud.

BEHAVIOR
The ribs of the Spanish sharp-ribbed newt are sharp-tipped, providing a unique form of defense. When attacked, the newt twists its body, and the sharp-tipped ribs protrude through the skin to impale the attacker. The Japanese spiny newt (*Echinotriton andersoni*) uses a similar form of defense.

FEEDING ECOLOGY AND DIET
The Spanish sharp-ribbed newt is a voracious predator of pond-living invertebrates.

REPRODUCTIVE BIOLOGY
In the spring, rough patches develop on the forelimbs of males that enable the newt to grasp a female during mating. Eggs are laid in water in spring and summer.

CONSERVATION STATUS
Not threatened. The range has contracted owing to loss of some ponds and to pollution and degradation of others.

SIGNIFICANCE TO HUMANS
None known. ◆

European fire salamander

Salamandra salamandra

TAXONOMY

Lacerta salamandra Linnaeus, 1758, Nuremberg, Germany.

OTHER COMMON NAMES

French: Salamandre terrestre; German: Feuersalamander; Spanish: Salamandra pintada.

PHYSICAL CHARACTERISTICS

With a total length up to 11 in (28 cm), the European fire salamander is a robustly built animal with a relatively short tail. The species shows considerable variation in color and skin pattern. Individuals may be black with yellow markings or yellow with black or occasionally red or orange markings. The markings may be spots or stripes. The limbs are short and stout with broad toes, and the tail is cylindrical and shorter than the body. Females are slightly larger than males.

DISTRIBUTION

Europe.

HABITAT

The European fire salamander inhabits deciduous and, occasionally, coniferous forests at 656–3,280 ft (200–1,000 m).

BEHAVIOR

Once it has completed its larval stage, the European fire salamander lives entirely on land. Individuals live in burrows and are territorial, defending the ground around the burrow against intrusion by neighbors. Striking color patterns act as warning coloration. Two rows of poison glands run along the body, and a cluster of poison glands is present on each side of the head behind the eyes. When attacked, fire salamanders can squirt toxin from these glands over a considerable distance.

FEEDING ECOLOGY AND DIET

Fire salamanders are active at night. They emerge from the burrow when conditions are damp to forage for worms, insects, insect larvae, and slugs.

Salamandra salamandra

REPRODUCTIVE BIOLOGY

During mating, which takes place on land, the male grasps the female from below. He stimulates the female with glands on his head and when the female is receptive deposits a spermatophore. The male then flips his tail to one side so that the female falls onto it. The eggs develop inside the female and are eventually released into ponds or streams as larvae in clutches of 12–50 eggs. In a few, high-altitude populations, the larvae are retained in the female throughout development and are released as miniature adults. During development in the oviduct, larvae may be cannibalistic, eating smaller siblings. As a result, only a few individuals in each clutch of eggs complete development.

CONSERVATION STATUS

Not threatened. The range has contracted in some areas as a result of deforestation. The population is declining in northern Spain for unknown reasons.

SIGNIFICANCE TO HUMANS

None known. ◆

California newt

Taricha torosa

TAXONOMY

Triton torosa Rathke, 1833, San Francisco Bay, California, United States. Two subspecies are recognized.

OTHER COMMON NAMES

German: Kalifornischer Gelbbauchmolch; Spanish: Tritón de California.

PHYSICAL CHARACTERISTICS

Heavily-built with a total length up to 8 in (20 cm), the California newt is light to dark brown on the back and flanks and has a bright yellow, orange, or, in the case of the red-bellied newt (*T. rivularis*), red belly. In the terrestrial phase, the skin is dry and warty, but in the breeding season, aquatic males develop smooth, slimy skin, a pale body color, a generally plump appearance, a large tail fin, and a swollen cloaca.

DISTRIBUTION

California coast, Sierra Nevada. The two populations that inhabit these regions are recognized as subspecies, *Taricha torosa torosa* and *T. torosa sierrae*.

HABITAT

For much of their lives, California newts live underground in the burrows of ground squirrels and other animals. They emerge into the open only on rainy nights. In the spring, the newts migrate to ponds and lakes, where they can build up large populations. In northern parts of the range, California newts inhabit mesic forests. In the south they live in drier areas, including oak woodland and grassland.

BEHAVIOR

All three *Taricha* species are extremely poisonous. The skin secretes the powerful nerve poison tetrodotoxin, which is lethal to humans. When attacked, California newts exhibit the unken reflex, raising the head and tail up and over the body, extending the limbs, and closing the eyes to reveal the bright belly.

FEEDING ECOLOGY AND DIET

The California newt feeds on a variety of small invertebrates and on eggs and larvae of its own species.

REPRODUCTIVE BIOLOGY

Before mating, the male California newt grasps the female from above, holding her firmly beneath him with both pairs of limbs. The pair can remain in this posture (amplexus), for many hours. The male rubs glands on his chin over the female's head and body until she adopts a receptive posture, at which point the male releases the female and deposits a spermatophore. As soon as the female has picked up the spermatophore, the male grasps her again. This post-mating amplexus, which is unique to this genus, guards the female from mating attempts by other males. The female lays eggs in gelatinous clusters of seven to 30 attached to water plants. The larvae are yellowish brown with bushy external gills and large tail fins.

CONSERVATION STATUS

Not threatened. The California newt is at risk in some parts of the range, however. In southern California, this newt is a species of special concern. Threats include loss of breeding habitat, reduction in hatching success due to increases in ultraviolet-B radiation, and predation on eggs and larvae by introduced crayfish and mosquitofish. In some places, road kill is an important cause of adult mortality. Important stretches of road sometimes are closed to allow newts to migrate safely to ponds and streams.

SIGNIFICANCE TO HUMANS

None known. ◆

Great crested newt
Triturus cristatus

TAXONOMY

Triton cristatus Laurenti, 1768, Nuremberg, Germany.

OTHER COMMON NAMES

French: Triton á crête; German: Kammolch; Spanish: Tritón crestado.

PHYSICAL CHARACTERISTICS

The crested newt (total length up to 6 in [16 cm]) gets its name from the large, deeply notched crest that runs along the back of of the breeding male. The male also has a deep tail decorated with a conspicuous white stripe. As does its close relative, the green and black marbled newt (*Triturus marmoratus*), the crested newt has a remarkable abnormality of chromosomes. As a result of the abnormality, 50% of all young die as early embryos. This may be one reason crested newts have declined more rapidly than have European newts not handicapped in this way.

DISTRIBUTION

Europe.

HABITAT

The crested newt needs dense cover when terrestrial and large, deep ponds for breeding.

BEHAVIOR

Crested newts live as long as 16 years. They spend much of their lives on land, and little is known of their habits. Crested newts are markedly distasteful. When the newt is handled, glands in the skin produce a bitter-smelling milky secretion that humans and potential predators, such as water birds and hedgehogs, find highly aversive. The bright orange and black pattern on the belly of crested newts appears to be warning coloration; predators associate the color with the distastefulness and do not attack the newt.

FEEDING ECOLOGY AND DIET

Crested newts feed on variety of small invertebrates, on frog tadpoles, and on the larvae of other newts.

REPRODUCTIVE BIOLOGY

Adults migrate to ponds early in the spring. In Sweden they have been observed moving over snow and entering ponds that are still partially covered in ice. Females start the breeding season full of large, yolk-filled eggs, but it takes males several weeks to fully develop the deep tail and crest. Males that emerge from winter hibernation with larger fat reserves develop larger crests, and it is likely they are more attractive to females than are males with small crests.

While in breeding ponds, crested newts are secretive by day and mate at dusk. The male takes up a position in front of the female and displays to her with rhythmic beats of his tail. This wafts a pheromone, secreted by a large gland in the cloaca, toward the female's snout. The display also presents visual stimuli, particularly the white tail-stripe, which is conspicuous in dim light. If the female responds to the displays by moving toward him, the male turns and deposits a spermatophore on the floor of the pond. The female moves over it and picks it up with her open cloaca.

Two or three days after mating, the female begins to lay eggs, a process that takes many weeks. Crested newts produce 70–600 eggs, usually 150–200, laid individually carefully wrapped in the leaf of a water plant. After two to three weeks, the eggs hatch into tiny larvae, which, once they have used up the reserves of yolk, start to feed on tiny aquatic animals, such as water fleas. Larval development takes two to three months, and the young emerge from the pond as miniature adults in late summer and autumn. Female newts mate several times during the breeding season, interrupting egg-laying to replenish the supply of sperm.

CONSERVATION STATUS

This species is listed as Lower Risk/Conservation Dependent. Like the other five species of large-bodied newts widely distributed across Europe, the population of crested newts has declined over much of the range as a result of changes to the habitat. This newt is the victim of changes in land use and agricultural practices. At the southwestern edge of its distribution, however, the crested newt is slowly expanding its range. In central France, the crested newt overlaps with the marbled newt, and hybrids between the two species are quite common. In some parts of France, the crested newt seems to be coping better than the marbled newt with new patterns of land use and is expanding into ponds previously used only by marbled newts, which are declining as a result.

SIGNIFICANCE TO HUMANS

None known. ◆

Smooth newt
Triturus vulgaris

TAXONOMY

Lacerta vulgaris Linnaeus, 1758, Sweden.

OTHER COMMON NAMES

English: Common newt; French: Triton ponctué; German: Teichmolch; Spanish: Tritón vulgar.

PHYSICAL CHARACTERISTICS

The smooth newt is small and slender (length up to 4 in [11 cm]). The tail constitutes approximately one-half the total length. In the terrestrial phase, this newt is brown or dark gray.

DISTRIBUTION

Europe.

HABITAT

The habitat is variable, including woodland, grassland, parkland, hedgerows, gardens, heath, and moorland. The smooth newt breeds in small ponds.

BEHAVIOR

The skin secretions of smooth newts are distasteful rather than toxic and provide little defense against predation. The newts are eaten by birds and other animals.

FEEDING ECOLOGY AND DIET

The smooth newt feeds on a wide variety of small invertebrates and on frog tadpoles.

REPRODUCTIVE BIOLOGY

Smooth newts return to ponds to breed in early spring and remain aquatic for several months. This species, like other *Triturus* species, exhibits marked sexual dimorphism during the breeding season. The male develops a high dorsal crest that runs along the back and tail. This crest has a jagged edge and, like the rest of the body, is marked with large, dark spots. Parallel stripes of red and blue decorate the lower edge of the male's tail, just behind the greatly swollen and dark cloaca. The toes on the hind limbs of the male develop flaps of skin. These flaps help the male swim fast in pursuit of females.

Females lay several hundred eggs during the breeding season. Each egg is laid individually, carefully wrapped in a folded leaf. The eggs hatch into tiny carnivorous larvae, which grow over the summer months to leave the water in late summer at a length of approximately 0.8 in (2 cm). The offspring spend the next two or three years on land before they return to breed as mature adults.

CONSERVATION STATUS

Not threatened. Although they have lost many breeding ponds throughout Europe as the result of modern methods of agriculture, smooth newts remain common in many areas. They have a remarkable ability to colonize any new pond soon after it forms.

SIGNIFICANCE TO HUMANS

None known. ◆

Mandarin salamander
Tylototriton verrucosus

TAXONOMY

Tylototriton verrucosus Anderson, 1871, western Yunnan, China.

OTHER COMMON NAMES

English: Crocodile newt; German: Burma-Krokodilmolch.

PHYSICAL CHARACTERISTICS

The mandarin salamander is robustly built with a total length up to 7 in (18 cm). It has a large head with prominent glandular ridges. The long, laterally compressed tail in the aquatic phase bears a well-developed fin. The salamander is black or dark brown and is covered with two rows of large brown, orange, or red tubercles. This striking coloration is aposematic (conspicuous and serving to warn). Mandarin salamanders produce a distasteful skin secretion. The skin has a granular texture.

DISTRIBUTION

China, India, Nepal, Thailand, and Vietnam.

HABITAT

The mandarin salamander lives in hills and mountains. The natural habitat is damp woodland and forest, but the salamander also inhabits a variety of habitats that are the result of human activity, such as rice fields, tea gardens, and meadows.

BEHAVIOR

Little is known.

FEEDING ECOLOGY AND DIET

The mandarin salamander feeds on a variety of small invertebrates.

REPRODUCTIVE BIOLOGY

Terrestrial for most of its life, the mandarin salamander migrates to ponds and other water bodies in March or April when the monsoon rains begin. Mating occurs in water, the male clasping the female before spermatophore transfer. The female lays 30–60 eggs in water. There are reports that the female guards her eggs. Sexual maturity is achieved at three to five years of age.

CONSERVATION STATUS

Not threatened. The mandarin salamander, however, has declined in abundance, primarily as the result of loss and change of its natural habitat. Collection for the pet trade has had a negative effect.

SIGNIFICANCE TO HUMANS

This species appears frequently in the international pet trade. ◆

Resources

Books

Griffiths, R. A. *Newts and Salamanders of Europe.* London: T. & A. D. Poyser, 1996.

Petranka, J. W. *Salamanders of the United States and Canada.* Washington, DC: Smithsonian Institution Press, 1998.

Tim R. Halliday, PhD

Olms and mudpuppies

(Proteidae)

Class Amphibia
Order Caudata
Suborder Salamandroidea
Family Proteidae

Thumbnail description
Permanently aquatic, medium-size to large salamanders with a somewhat elongated "squared-off" snout, small limbs, reduced numbers of toes, and large, bushy, red gills; most olms are pale (rarely dark) and nearly eyeless, whereas mudpuppies are dark with large spots and small eyes

Size
12–16 in (30.5–40.6 cm)

Number of genera, species
2 genera; 6 species

Habitat
Freshwater

Conservation status
Vulnerable: 1 species

Distribution
Southeastern Europe and eastern North America

Evolution and systematics

Fossil proteids, including species of two extinct genera, are known from the Miocene and Pleistocene of Europe, the Upper Paleocene and Pleistocene of North America, and the Miocene of Asia. Fossils of *Necturus* date to the Paleocene of Canada and the Pleistocene of Florida. Fossils of *Proteus* date from the Pleistocene of Germany.

As of the year 2002, one species of *Proteus* and five species of *Necturus* were recognized. These are enigmatic salamanders that have long been shrouded in mystery in both popular and scientific circles. The European blind cave salamander or olm (*Proteus anguinus*) has been known for centuries as the "human fish" (because of its pale skin color), haunting the subterranean waters of the Dinaric karst in Slovenia and adjacent areas. Olms were first mentioned in the scientific literature more than 300 years ago. North American mudpuppies (species of the genus *Necturus*) likewise have been known for centuries as "waterdogs"; They have been familiar to scientists since at least 1799 and are one of the most extensively used vertebrates for courses in comparative anatomy and physiology. And yet the phylogenetic relationships of proteids to other salamanders and systematics within the family remain problematic.

The relationship between *Necturus* and *Proteus* has been questioned and has long been the subject of controversy. Part of the problem with this group is that it represents ancient lineages of permanently aquatic, neotenic salamanders that retain their larval morphologic characters into adulthood. Neoteny is a well-known phenomenon in which adults maintain larval morphologic features and breed in that state. Some scientists believe that the morphologic similarities between *Proteus* and *Necturus* are the result of convergent (or parallel) evolution toward neoteny in permanent aquatic habitats.

Studies of similarities in DNA sequences, proteins, and chromosomes in proteid salamanders are ongoing and have not yet produced conclusive results. In addition to general morphologic similarities, *Necturus* and *Proteus* share the same diploid chromosome number (38), which is not found in any other salamanders. All species of *Necturus* have distinctive X and Y sex chromosomes that are unlike those of other salamanders, but it is not known whether olms have this feature. Working out the precise phylogenetic relationships of these strange salamanders depends on additional research.

Taxonomy and evolutionary relationships within both genera are in a state of flux. In 1986 a population of dark-colored *Proteus* was discovered in Slovenia. Salamanders from this population have enough distinct features that some scientists consider them to be a subspecies (*Proteus anguinus parkelj*), but others suspect that they may constitute a different species altogether. Within the genus *Necturus*, comparative biochemical (protein) and karyological (chromosome) analyses indicate that individual species fall into three main lineages: *N. lewisi*, *N. punctatus*, and a group of three closely related species consisting of *N. alabamensis*, *N. beyeri*, and *N. maculosus*. The relationships among these three species and especially the taxonomic status of *N. alabamensis* are uncertain. It is clear that both genera need to be analyzed further. No subfamilies are recognized.

Physical characteristics

All proteid salamanders are permanently aquatic and have larval morphologic features, including three pairs of large, bushy, red gills; a relatively short, laterally compressed tail with a tail fin; and reduced eyes, even as adults. These are moderately large to large salamanders that can reach more than 16 in (40.6 cm) in total length (from the tip of the snout to the tip of the tail). *Proteus anguinus* is the most striking-looking species. It is long and skinny with pale, pinkish white skin, and it has a flat, narrow head with tiny degenerate eyes and small limbs with only three digits on the forelimbs and two digits on the hind limbs. These features are thought to reflect adaptation to aquatic, subterranean habitats. Thus, *Proteus* often is described as "troglomorphic." Reportedly, *Proteus* has some pigment and turns darker when exposed to light. Dark-colored individuals of *Proteus*, which are thought to be a separate subspecies (*P. a. parkelj*), are less strongly troglomorphic and have larger eyes. *Proteus* achieves a total length of approximately 12 in (30.5 cm).

Species of *Necturus* are somewhat larger than *Proteus*, reaching nearly 19 in (48.3 cm) in total length. They are fully pigmented. They are also more robust, with a wider head and a thicker body and limbs. Most *Necturus* species are dark rusty brown or grayish brown; they have large, irregular spots of black or blue-black on the dorsum (back) of the animal and a paler venter (belly), with a dark stripe passing through the eye. *Necturus* is colored cryptically against the dark bottom of lakes, rivers, and streams, but the exact color pattern depends on the species. The morphologic features of *Necturus* are at least somewhat reminiscent of *Proteus*, including small, larval eyes; a short tail; reduced limbs with four digits on both pairs; and a peculiar-looking "squared off" snout, reflecting the absence of maxillary (upper jaw) bones.

Distribution

The distribution of proteids is disjunct: *Necturus* is found only in the New World, whereas *Proteus* is exclusively Old World. All species of *Necturus*, with the exception of *N. maculosus*, are distributed along the coastal plain of the southeastern United States, from southeastern Virginia to eastern Texas. *Necturus maculosus* is by far the most widely distributed species, with a fanlike range extending from an apex in Louisiana and broadening northward to southeastern Manitoba in the west and southestern Quebec, Canada, in the east, essentially encompassing the entire Mississippi River drainage system. The combined ranges of these species result in a more or less continuous distribution of *Necturus* over most of eastern North America, interrupted in the east by the Appalachian Mountains, which form a wedge separating the two coastal species, *N. lewisi* and *N. punctatus*, from inland populations of *N. maculosus*. The two most enigmatic species, *N. alabamensis* and *N. beyeri*, are distributed near the southern limits of the ranges of *N. maculosus* to the west and *N. punctatus* to the east.

Proteus anguinus is known from approximately 250 localities, from the limestone cave systems along the Adriatic seaboard from western Slovenia and northeastern Italy in the north to Montenegro in the south. The majority of localities

are in western Slovenia. The putative subspecies, *Proteus a. parkelj*, is from the Bela Krajina region of western Slovenia.

Habitat

Proteus inhabits underground streams and lakes in limestone caves in eternal darkness, where the water is cold year-round (usually about 46.5°F, or 8°C). These salamanders are thought to congregate in their main habitat in deep crevices and fissures that are largely inaccessible to exploration by humans. Most sightings and captures of olms appear to have been in marginal habitats, where the salamanders either were flushed out by heavy rains or were hunting for food. Mudpuppies inhabit a wide variety of permanently aquatic habitats, including muddy canals; ditches; large, rocky, fast-flowing streams; reservoirs; and large, cool lakes. They are most active at night and may be found during the day by lifting or disturbing rocks and other cover. *Necturus* may be found at all seasons of the year and are even active beneath the ice in mid-winter.

Behavior

Proteid salamanders have been neglected in terms of detailed behavioral studies. Mudpuppies often are seen crawling slowly over the bottom of streams or lakes, but they can swim rapidly when frightened. In captivity, mudpuppies are secretive and prefer to hide beneath any available objects, including each other. They appear to be repelled by light. In poorly aerated water, mudpuppies constantly fan their gills, which may become large, bushy, and bright red. Under such conditions, the salamanders often rise to the surface to take gulps of air. In well-oxygenated water the gills tend to be held motionless against the sides of the neck and eventually shrink in size. There is some evidence that mudpuppies are capable of homing behavior. Even less information is available for olms. Olms are gregarious, at least when they are not breeding, and tend to congregate in deep fissures. They apparently use chemosensory cues to mark and find their "home shelters."

Feeding ecology and diet

Mudpuppies are generalist predators; their natural food includes fish, fish eggs, crayfish, worms, small mollusks, and aquatic insects, in short almost anything that moves and will fit into their mouths. They are especially fond of sculpins and sometimes can be found gorged with these fish. Most information about the ecology and other aspects of the biology of *Proteus* is based on observations of animals raised in captivity. Little is known about the feeding ecology of these salamanders in the wild, except that they seem to feed on amphipods, insect larvae, and other small invertebrates. Apparently, *Proteus* uses chemosensory cues to locate prey in total darkness.

Reproductive biology

Relatively limited information is available on the reproductive biology of proteid salamanders. The animals tend to

A mudpuppy (*Necturus maculosus maculosus*) has different gills, depending on its environment. Mudpuppies living in the fast moving waters of rivers and streams develop small, compressed gills. In warm or slow moving rivers and lakes they have big, bushy gills. (Photo by Jack Dermid. Bruce Coleman Inc. Reproduced by permission.)

be gregarious, at least when they are not breeding. The breeding season for *Necturus* is in the fall or winter, depending on the species and location. Sexually active males have swollen, "inflamed" cloacas (urogenital openings), with a pair of enlarged papillae (finger-like appendages) that project posteriorly. Breeding in *Proteus* appears to be aseasonal, reflecting the stability of their subterranean habitat. Olms seem to be much more territorial during breeding than mudpuppies. All species for which information is known show some kind of courtship ritual in which the males and females stimulate each others' cloacas. Courtship culminates in the male depositing a packet of sperm (spermatophore), which the female picks up with her cloaca. The female may store sperm in special structures inside her cloaca called spermathecae for six months or more. The eggs usually are attached beneath some object, such as a rock or a log, and are guarded by the female. The eggs are large (0.2 in, or 5–6 mm), full of yolk, and unpigmented; they may number 100 or more per clutch.

The incubation period lasts two to six months, depending on the species and the temperature. *Proteus* reportedly is capable of some degree of viviparity (live-bearing), giving birth to a pair of well-developed young. The larvae, which in mudpuppies may have strikingly different pigmentation from the adults, develop gradually into the adult form, with no distinct metamorphosis. The age of sexual maturation is not known for mudpuppies, but for olms it is reported to be seven years. Little is

known about longevity in proteid salamanders; reports range from nine years to nearly 60 years, depending on the species.

Conservation status

Proteus is listed by the IUCN as Vulnerable, mostly because of the restricted range and apparently small sizes of most known populations. Olms have been protected in Slovenia since 1949. The main threats to olms are economic development, industrial pollution, and overcollecting. Although it is almost impossible to assess populations of olms with certainty, given the inaccessibility of most of the natural habitat, population declines have been reported in Italy and Slovenia. None of the species of *Necturus* are listed by the IUCN as of 2002, and most populations seem to be doing well. Mudpuppies, however, are vulnerable to factors that are known to affect amphibians adversely, including chemical pollution, habitat alteration, and overcollecting.

Significance to humans

Proteids have long been used for scientific studies, and *Necturus* is the most commonly used amphibian in comparative anatomy courses. Proteids are also important in the pet trade and have been used to promote ecotourism, particularly in Slovenia. Because of their thin skin and dependence on clean, well-oxygenated water, proteids may be good indicators of water quality.

1. Dwarf waterdog (*Necturus punctatus*); 2. Olm (*Proteus anguinus*); 3. Mudpuppy (*Necturus maculosus*); 4. Neuse River waterdog (*Necturus lewisi*). (Illustration by Joseph E. Trumpey)

Species accounts

Olm
Proteus anguinus

TAXONOMY

Proteus anguinus Laurenti, 1768, Magdalene Cave, near Adelsberg Cave, Slovenia. Two subspecies are tentatively recognized.

OTHER COMMON NAMES

English: Blind cave salamander; German: Grottenolm.

PHYSICAL CHARACTERISTICS

The body is slender, elongated, and pinkish white, with a long, angular, "squared-off" snout; degenerate eyes; a short, laterally compressed tail; and three pink external gills on each side of the head. The limbs are thin, with a reduced number of digits: three digits on the forelimb and two digits on the hind limb. The average size of adults is approximately 10 in (25.4 cm).

DISTRIBUTION

The species occurs in Italy, Slovenia, Croatia, Bosnia, and Herzegovina, with a captive population in France.

HABITAT

Olms inhabit subterranean lakes and rivers in limestone caves of the Dinaric Alps, from Slovenia and Italy in the north to Montenegro in the south.

BEHAVIOR

These salamanders are gregarious except during the breeding season, when males are territorial. Generally, they are secretive and rarely seen, except in marginal habitats either when feeding or because of flooding.

FEEDING ECOLOGY AND DIET

Olms feed at night, using chemosensory cues to find small arthropods and other invertebrates.

REPRODUCTIVE BIOLOGY

Breeding is aseasonal. Fertilization is internal via spermatophores after courtship; eggs are large and yellowish and are laid under rocks and other cover and guarded by the female. The incubation period is up to six months; larvae develop directly into adults without metamorphosis.

CONSERVATION STATUS

The species is listed by the IUCN as Vulnerable and in Appendix II of the Convention of European Wildlife and Natural Habitats of 1979. It is not considered critically endangered or endangered but faces a high risk of extinction in the wild in the not too distant future.

SIGNIFICANCE TO HUMANS

Olms are an ecotourist attraction; they are popular in the pet trade and are used in scientific research. ◆

Neuse River waterdog
Necturus lewisi

TAXONOMY

Necturus maculosus lewisi Brimley, 1924, Neuse River, near Raleigh, North Carolina, United States.

Proteus anguinus

Necturus lewisi
Necturus maculosus
Necturus punctatus

OTHER COMMON NAMES
English: Lewis' mudpuppy.

PHYSICAL CHARACTERISTICS
This is a medium-size mudpuppy, 6–11 in (15.2–28 cm) in total length. It has a rusty, yellowish brown dorsum with large, dark spots scattered over the back and sides. The venter is paler, with fewer and smaller blotches.

DISTRIBUTION
The range is restricted to the Neuse and Tar river systems in North Carolina, United States.

HABITAT
This salamander prefers relatively wide, fast-flowing streams with a high oxygen content and a hard substrate.

BEHAVIOR
The Neuse River waterdog is active at night and retreats into burrows in the stream bank or under large rocks during the day. Activity decreases at high stream temperatures. The skin produces noxious secretions that may defend against predation.

FEEDING ECOLOGY AND DIET
The diet consists of small invertebrates and vertebrates, including crustacea, mollusks, annelid worms, aquatic insect larvae, small fish, and other amphibians. Like other mudpuppies, Neuse River waterdogs are sit-and-wait predators that use "gape and suck" feeding mechanics.

REPRODUCTIVE BIOLOGY
The breeding season is from December through March and possibly also in the spring. Eggs are deposited on the undersurface of large rocks in fast-flowing water. Hatching occurs in July.

CONSERVATION STATUS
Not threatened.

SIGNIFICANCE TO HUMANS
Waterdogs are collected in great numbers by biological supply houses, probably with little regard to exact species or locality; they also are seen in the pet trade. ◆

Mudpuppy
Necturus maculosus

TAXONOMY
Sirena maculosa Rafinesque, 1818, Ohio River, United States (state not recorded). Two subspecies are recognized.

OTHER COMMON NAMES
English: Common mudpuppy, waterdog; French: Necture tacheté; German: Gefleckter Furchenmolch.

PHYSICAL CHARACTERISTICS
This is the largest member of the genus, reaching 8–19 in (20.3–48.3 cm) in total length. Coloration varies from deep rusty brown to gray or even black, with scattered black or bluish black spots and blotches. The spots sometimes may form two fairly regular rows along the back. As with other mudpuppies and waterdogs, a dark bar extends through the eye to the gills. The venter is paler, with or without dark spots. The margins of the tail commonly are tinged with reddish orange.

DISTRIBUTION
This species has by far the widest distribution in the genus, encompassing essentially the entire Mississippi River drainage system, from southern Manitoba and Quebec, Canada, in the north to Georgia, Alabama, Mississippi, and Louisiana, United States, in the south.

HABITAT
These salamanders inhabit a wide variety of permanently aquatic habitats, including rivers, streams, canals, and lakes.

BEHAVIOR
Mudpuppies are active all year round and have been seen moving around beneath the ice in mid-winter. Adults are mostly active at night, when they forage, and they hide under rocks and other objects or in burrows during the day.

FEEDING ECOLOGY AND DIET
Mudpuppies feed on a variety of small aquatic invertebrates and vertebrates, including crayfish and other crustaceans, mollusks, worms, insect larvae, fish, and amphibians.

REPRODUCTIVE BIOLOGY
As with all mudpuppies and waterdogs, the sex of adults can be determined by examining the morphologic characteristics of the vent, especially during the breeding season, when the male's vent is swollen. The male's vent also is equipped with two nipple-like papillae that project posteriorly. The vent of the female is a simple slit. The mating season is in the autumn or winter, possibly extending into spring, depending on the locality. In May or June the eggs are attached to the undersurfaces of large rocks, where they are attended by the female, who apparently defends them against predators. Hatching takes place in one or two months, depending on the temperature of the water. The newly hatched larvae are approximately 1 in (25.4 mm) in length and have two lateral yellow stripes on a dark ground color.

CONSERVATION STATUS
Not threatened.

SIGNIFICANCE TO HUMANS
Mudpuppies are collected in great numbers by biological supply companies for use in classrooms and laboratories around the world; they are seen often in the pet trade as well. ◆

Dwarf waterdog
Necturus punctatus

TAXONOMY
Menobranchus punctatus Gibbes, 1850, southern Santee River, South Carolina, United States.

OTHER COMMON NAMES
English: Southern waterdog, Carolina waterdog.

PHYSICAL CHARACTERISTICS
This is the smallest species of *Necturus*, reaching only about 7 or 8 in (17.8–20.3 cm) in total length. The coloration is slate-gray to dark brown or black, with no spots or only a few small, pale spots. The venter is pale and has no spots. Unlike other species of *Necturus*, the larvae are not striped.

DISTRIBUTION
This species occurs along the coastal plain from southern Virginia to central Georgia in the United States. The distribution overlaps that of *N. lewisi*.

HABITAT
The dwarf waterdog prefers small and medium-size streams, especially in deeper, slower sections with leaf beds and other debris.

BEHAVIOR
Little is known about the life history and behavior of this species. Adults have been observed to congregate in leaf beds during winter.

FEEDING ECOLOGY AND DIET
Dwarf waterdogs feed on a variety of small invertebrates and vertebrates, including crayfish, worms, arthropods, mollusks, and other amphibians. They may compete for food with *N. lewisi* where the two inhabit the same streams.

REPRODUCTIVE BIOLOGY
Little is known about the reproductive biology of this species, but it is thought that mating occurs in winter, followed by egg laying sometime between March and May.

CONSERVATION STATUS
Not threatened.

SIGNIFICANCE TO HUMANS
Dwarf waterdogs are probably not collected in large numbers for educational or scientific purposes, but they may enter the pet trade. ◆

Resources

Books

Arnold, E. N., and J. A. Burton. *Field Guide to the Reptiles and Amphibians of Britain and Europe.* 2nd ed. London: HarperCollins Publishers, 1998.

Bishop, S. C. *Handbook of Salamanders.* Reprint. Ithaca, NY: Cornell University Press, 1994.

Conant, Roger, and Joseph T. Collins. *A Field Guide to Reptiles and Amphibians, Eastern and Central North America.* New York: Houghton Mifflin, 1991.

Duellman, William E., and Linda Trueb. *Biology of Amphibians.* New York: McGraw Hill Book Company, 1986.

Noble, G. K. *The Biology of the Amphibia.* New York: Dover Publications, 1954.

Petranka, James W. *Salamanders of the United States and Canada.* Washington, D.C.: Smithsonian Institution Press, 1998.

Periodicals

Arntzen, J. W., and Boris Sket. "Morphometric Analysis of Black and White European Cave Salamanders, *Proteus anguinus.*" *Journal of Zoology* (London) 241 (1997): 699–707.

Bishop, Sherman C. "The Salamanders of New York." *New York State Museum Bulletin* 324 (1941): 1–365.

Guttman, S. I., L. A. Weight, P. A. Moler, R. E. Ashton Jr., B. W. Mansell, and J. Peavy. "An Electrophoretic Analysis of *Necturus* from the Southeastern United States." *Journal of Herpetology* 24 (1990): 163–175.

Hecht, M. K. "A Case of Parallel Evolution in Salamanders." *Proceedings of the Zoological Society Calcutta,* Mookerjee Memorial Volume (1957): 283–292.

——— "A Synopsis of the Mud Puppies of Eastern North America." *Proceedings of the Staten Island Institute of Arts and Sciences* 21, no. 1 (1958): 1–38.

Sessions, S. K., and J. E. Wiley. "Chromosomal Evolution in Salamanders of the Genus *Necturus.*" *Brimleyana* 10 (1985): 37–52.

Sket, Boris. "Distribution of *Proteus* (Amphibia: Urodela: Proteidae) and Its Possible Explanation." *Journal of Biogeography* 24 (1997): 263–280.

Other

AmphibiaWeb. (May 8, 2002) <http://elib.cs.berkeley.edu/aw>

Maddison, David R., ed. *The Tree of Life Web Project.* (May 8, 2002) <http://tolweb.org/tree/phylogeny.html>

Stanley K. Sessions, PhD

Torrent salamanders
(Rhyacotritonidae)

Class Amphibia
Order Caudata
Suborder Salamandroidea
Family Rhyacotritonidae

Thumbnail description
Small, short-tailed, greenish yellow, large-eyed salamanders found near cool water in seeps, springs, or flowing streams

Size
3–4.5 in (75–11.5 cm)

Number of genera, species
1 genus, 4 species

Habitat
The name *torrent salamander* is derived from direct translation of the scientific name (Greek *rhyakos,* "stream," and *triton,* the Greek sea god). These salamanders rarely are found in torrential streams, although they are found in gravelly habitat beside such streams. More typically they are encountered in seeps and springs, especially where clear, cool water flows or drips over crumbling rocks. These habitats almost always are in closed-canopy forests often dominated by coniferous trees, but some are in riparian areas dominated by maples and alders.

Conservation status
Not classified by the IUCN

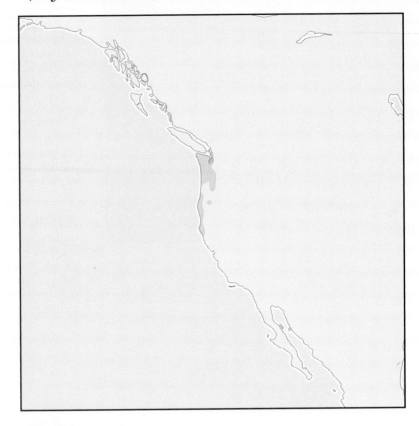

Distribution
Northwestern United States

Evolution and systematics

Torrent salamanders constitute a unique lineage of salamanders that have no close relatives. Morphological and molecular data attest to the distinctiveness of this clade, but there is no clear sister taxon. These salamanders may be remnants of an early radiation of the Salamandroidea. When first discovered early in the twentieth century, these salamanders were thought members of the family Hynobiidae (well represented in eastern Asia but not otherwise known in the New World). They then were included in the Ambystomatidae and later in the Dicamptodontidae. In several respects, they resemble plethodontids, and they may be related to that group. At present torrent salamanders are recognized as a separate family.

Physical characteristics

These are small to medium-sized (3–4.5 in; 7.5–11.5 cm) semiaquatic salamanders with relatively stocky bodies and broad heads with protuberant eyes and a short snout. Limbs are small but robust, and the tail is short and laterally compressed with a small keel. These salamanders have vestigial lungs.

Distribution

This family occurs from the Olympic Peninsula in northwestern Washington in the coast ranges to southern Mendocino County in northern California, and in the Cascade range from the vicinity of Mount Saint Helens, Washington, to Lane County, central Oregon.

Habitat

Torrent salamanders are aquatic and semiaquatic, inhabiting usually densely forested areas in small, clear, rapidly flowing streams, seeps in rocky areas, and rock crevices with thin layers of water cascading over the surface.

Behavior

The behavior of this family is not well known. These salamanders are secretive and are seldom seen.

Feeding ecology and diet

Little is known about the feeding ecology of this family. The diet probably consists of aquatic and semiaquatic insects, especially larvae, as well as other invertebrates.

Olympic torrent salamanders (*Rhyacotriton olympicus*) live in the clear, cold streams of the Olympic Mountains in Washington, USA. (Photo by Animals Animals ©Maresa Pryor. Reproduced by permission.)

Reproductive biology

Fertilization is internal. Large, unpigmented, yolky eggs are laid in cold, clear water under rocks or in crevices. Eggs develop slowly, as do the aquatic larvae, which live for three or four years. Metamorphosis occurs at close to adult size, but it is not known how long it takes metamorphs to mature. Metamorphosis is a gradual and mild transformation. Adults are characterized by morphology more typical of juveniles of other taxa, especially with respect to skulls and limbs.

Conservation status

No species are listed by the IUCN. Clearing of forests is the greatest risk to torrent salamanders, because it leads to habitat degradation. One species, *Rhyacotriton variegatus*, is protected in California.

Significance to humans

The four species of this family are probably similar to each other in ecology and other biological features. They form a unique clade that harbors unique parasites. The primary significance to humans is a contribution to understanding of the dimensions of biodiversity in the Pacific Northwest region of California.

Species accounts

Cascade torrent salamander
Rhyacotriton cascadae

TAXONOMY
Rhyacotriton cascadae Good and Wake, 1992, base of Wahkeena Falls, Multnomah County, Oregon, United States.

OTHER COMMON NAMES
English: Cascades salamander.

PHYSICAL CHARACTERISTICS
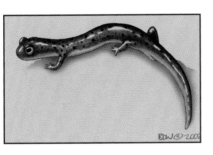

Rhyacotriton cascadae

Moderately small (3–4.5 in; 7.5–11 cm in total length) with a relatively stout body, a moderately broad head with prominent, protruding eyes and a relatively short snout, and a laterally compressed, keeled tail that is shorter than head plus body. Somewhat variable in coloration but usually rich brown above and yellowish below with greenish yellow in some specimens. The dorsal surfaces are richly marked with darker blotches and speckles. There is a sharp distinction between the brown coloration of the back and flanks and the yellow pigment of the belly. White flecking is found in the flank region above the transition to the yellow venter. The ventral surfaces are much less spotted than the dorsal ones. However, dark spots are present, as is fine gray flecking on the throat and chest. Males have swollen glands in the margins of the vent, and these produce a characteristically squared-off, conspicuous pair of structures.

DISTRIBUTION
Cascade Mountains of Washington and Oregon from near Mount Saint Helens in Washington to central Oregon. Generally found at elevations below 2,000 ft (620 m).

HABITAT
Streams, usually in heavily forested areas. These salamanders avoid large streams but may be found near them in small, rapidly flowing tributaries, where they live under moss-covered rocks and in coarse gravel, even in talus in areas that are very moist. Water often is flowing through the rocks in thin sheets. Adults venture onto land but rarely go more than a few feet (1 m) from water. They may inhabit wet rock crevices. Larvae exist in the same habitat as adults but are strictly aquatic.

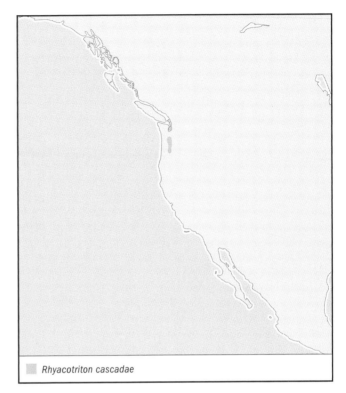

Rhyacotriton cascadae

BEHAVIOR
Not known. Extremely secretive; not seen unless actively sought through turning of rocks at the edges of the habitat.

FEEDING ECOLOGY AND DIET
Not known. Likely eat small invertebrates, especially aquatic insect larvae and mollusks.

REPRODUCTIVE BIOLOGY
Not well known. A related species lays eggs singly in small groups in cold water flowing through rocks and rock crevices. Females lay approximately eight relatively large, unpigmented, yolky eggs. Eggs probably are slow to hatch in the cold water. Larvae grow slowly, taking three or four years to metamorphose when they are relatively large (1.5–1.8 in; 37–45 mm).

CONSERVATION STATUS
Not threatened. The greatest risk for this species is clear cutting of forests, which severely affects local habitats by causing small watercourses to heat and dry.

SIGNIFICANCE TO HUMANS
None known. ◆

Resources

Books
Petranka, James W. *Salamanders of the United States and Canada.* Washington, DC: Smithsonian Institution Press, 1998.

Periodicals
Good, David A., and David B. Wake. "Geographic Variation and Speciation in the Torrent Salamanders of the Genus *Rhyacotriton* (Caudata: Rhyacotritonidae)." *University of California Publications in Zoology* 126 (1976): 1–91.

David B. Wake, PhD

Lungless salamanders

(Plethodontidae)

Class Amphibia
Order Caudata
Suborder Salamandroidea
Family Plethodontidae

Thumbnail description
Small to medium-sized salamanders with generalized body form including four limbs with four fingers and four or five toes and a medium to long tail

Size
1–13 in (2.5–25 cm)

Number of genera, species
28 genera; 346 species

Habitat
Forest, woodlands, streams, springs, and caves

Conservation status
Critically Endangered: 1 species; Endangered: 4 species; Vulnerable: 11 species; Lower Risk/Near Threatened: 7 species; Data Deficient: 2 species

Distribution
North, Middle, and South America; Central Mediterranean Europe

Evolution and systematics

The Plethodontidae was long thought to include the most derived salamanders, because of the terrestrial nature of so many of the species. However, studies of phylogenetic relationships using molecular markers have led biologists to question this assumption, and it is now generally recognized that the family is likely relatively old and derived near the base of the clade that constitutes the suborder Salamandroidea. There are few fossils earlier than the Pliocene and Pleistocene; the oldest are vertebrae from the Lower Miocene that are assigned to the living genera *Aneides* and *Plethodon*, thus showing that these close relatives were already differentiated by that time. Using estimates obtained from molecular evolutionary clocks, the family may be more than 100 million years old.

Plethodontids are thought to have arisen in what is present-day Appalachia, the ancient mountainous region of the southeastern United States. Lunglessness is thought to have evolved as an adaptation for life in flowing water. Larvae are small, and lungs would tend to act as air sacs that might make the animal float in the water column. This would dislodge them and threaten their survival. In well-aerated water such as a stream, respiration can take place readily through the skin, and so there would be little countervailing selection to retain lungs. While there are other hypotheses concerning lung loss, the flowing-water hypothesis is strengthened by similarities between plethodontids and other families: lungless salamanders in other families have larvae that live in flowing water, and usually the adults also live in or near streams.

At present two subfamilies are recognized.

Desmognathinae

This subfamily includes two genera, *Desmognathus* (with 17 species) and *Phaeognathus* (with only a single species). These salamanders have highly specialized heads and necks used for burrowing, for wedging under rocks and in stream beds, and for courtship, but they retain some ancestral traits as well. The desmognathines are restricted to eastern North America, where they extend as far west as Texas. Most of the species have an aquatic larval stage that varies in length from a few months to three years. However, at least three species have abandoned the larval stage and have direct development, with miniatures of the adult hatching from eggs laid in moist terrestrial to semiaquatic habitats. It long was assumed that larvae were an ancestral retention in the desmognathines, but

Female ensatina (*Ensatina eschscholtzii*) on her nest with eggs in Jefferson County, Washington, USA. (Photo by Suzanne L. Collins/Photo Researchers, Inc. Reproduced by permission.)

now it appears that larvae may have been re-evolved within the subfamily because the species with terrestrial development include the most basal members of the clade. Either direct development evolved independently several times, or larvae reappeared.

Plethodontinae

This subfamily is a large and heterogeneous group that is not so well characterized as the Desmognathinae. Some molecular data suggest that the Plethodontinae may not be monophyletic, but it does contain some well-defined clades, notably a large assemblage of salamanders from the American tropics, the west coastal region of North America, and some small parts of central Mediterranean Europe that are collectively placed in the tribe Bolitoglossini and termed bolitoglossines. Two other tribes are recognized: the Plethodonini, which includes fully terrestrial species that occur in North America, and the Hemidactyliini, which includes species that have an aquatic larval stage and occur mainly in eastern North America. Whereas the monophyletic status of the Bolitoglossini is well established, the other two have only weak support.

Hemidactyliines were long thought to be most similar to ancestors of the Plethodontidae in structure and general biology. Many of them live in or near streams and have larvae that live in streams, springs, or seeps. However, there are also a few hemidactyliines that have pond larvae and more that have larvae and adults that are restricted to caves. *Hemidactylium scutatum* differs in many ways from other hemidactyliines in that it is terrestrial as an adult and has an ephemeral pond-dwelling larva. Most fully metamorphosed hemidactyliines have a projectile tongue that can be fired rapidly and for a relatively great distance, but *Hemidactylium* has a much less specialized tongue. It may represent an entirely separate clade from the other hemidactyliines. The remainder of the Hemidactyliini seems to form a clade. All of them have aquatic larvae, many of which live in streams. The rest live in springs, seeps, or caves, with the exception of two species that have pond larvae: *Stereochilus marginatus* and *Eurycea quadridigitata* (which appears to be a complex of species, not yet clearly diagnosable), both of which live on the coastal plain. *Stereochilus* lacks a highly specialized tongue and may be relatively basal in the clade. Among the members of this clade are large spring salamanders (*Gyrinophilus*, with 4 species) and mud salamanders (*Pseudotriton*, with 2 species), but most of the species are relatively small. The dominant genus is *Eurycea*, a large and complicated group of 23 species, including a number of species that fail to metamorphose and spend their entire lives as larval forms that become sexually mature (these are termed perennibranchiate, referring to the retention of gills throughout life). These permanent larvae are surprisingly species rich, especially in and near the Edwards Plateau region of central Texas, where some of them have become restricted to underground waters. The cave species lack eyes and most pigment and may have bizarrely formed limbs and snouts. Cave-restricted permanently larval species have also evolved in the genera *Gyrinophilus* and *Haideotriton*. The genus *Typhlotriton* starts life as an *Eurycea*-like larva, but as it metamorphoses its eyes degenerate, its eyelids fuse, and it loses pigment. The adult is restricted to terrestrial habitats in caves. Both *Haideotriton* (1 species) and *Typhlotriton* (1 species) are closely related to *Eurycea*, which may be paraphyletic.

The Plethodontini includes species that have a relatively unspecialized tongue and are strictly terrestrial, with no larval stage and having direct development. They have long been considered to be close relatives, and some have even recommended that they be placed in the same genus. While *Plethodon* and *Aneides* are close relatives, *Ensatina* now appears to be only distantly related to the others. *Plethodon* is a very large genus (54 species in at least 3 and probably more major clades); *Aneides* is smaller (6 species) and appears on the basis of molecular evidence to be nested within the paraphyletic *Plethodon*. Exact relationships are uncertain, but eventually there may be taxonomic changes in this assemblage. Many of the species of *Plethodon* are cryptic and can only be distinguished from other members of the genus by molecular evidence. However, geographic distribution is distinctive for every species, and thus locality information aids immeasurably in their identification. The species of *Aneides* are generally well differentiated in morphology, but one pair of species (*A. ferreus*, *A. vagrans*) is virtually identical in morphology; they, too, have distinct geographic distributions. At present it appears that *Ensatina* is the sister taxon of *Plethodon* and *Aneides*; this may change as more molecular data become available.

The tribe Bolitoglossini is the only group of plethodontids that does not occur in the presumptive ancestral home, eastern North America. Instead, this tribe is widely dispersed, with many species in tropical America, a number in western North America, and a few in restricted parts of central Mediterranean Europe. All members of this group are terrestrial, with no larval stage and direct terrestrial development. There are three well supported clades recognized as supergenera:

Supergenus (SG) *Hydromantes*. This genus contains two distinct lineages, treated as genera or subgenera by different authors. SG *Hydromantes* has three species in California, and SG *Speleomantes* has five species, three on the island of Sardinia and two on the Italian mainland, also extending into extreme southeastern France.

Supergenus *Batrachoseps*. There are 20 species in this supergenus, placed in a single genus *Batrachoseps*, which in turn is divided into two subgenera that are biochemically distinct.

Supergenus *Bolitoglossa*. All tropical plethodontids are placed in this genus, which includes about 200 species. This is by far the largest salamander taxon. One genus, *Bolitoglossa*,

A spring salamander (*Gyrinophilus porphyriticus*) eats a dusky salamander (*Desmognathus fuscus*). (Photo by Gary Meszaros/Photo Researchers, Inc. Reproduced by permission.)

includes more than 80 species and has an enormous geographic range, from northeastern Mexico to central Bolivia and Brazil. Other genera contain fewer species and have smaller ranges. *Pseudoeurycea* includes about 40 species and ranges from northwestern and northeastern Mexico into southern Guatemala. *Lineatriton* includes three extremely elongate, slender species from eastern Mexico, and it is a close relative of *Pseudoeurycea*.

Also closely related to *Pseudoeurycea* are two small genera from eastern Mexico (*Parvimolge*, with one species) and southern Mexico (*Ixalotriton*, with two species). The former is a diminutive species, among the smallest of terrestrial vertebrates at about 1.5 in (3.8 cm) total length, while the latter includes one species (*I. niger*) that is an active, scansorial, leaping animal with a long whiplike tail and long legs and digits. *Chiropterotriton* includes about 14 species of diverse habitats, from cave-dwelling and arboreal to terrestrial, all from eastern and southern Mexico. *Thorius*, the minute salamanders, include 22 species from eastern and southern Mexico. All are very small and some are even smaller than *Parvimolge*, achieving sexual maturity at less than 1 in (2.5 cm) total length.

Dendrotriton (6 species), *Bradytriton* (1 species), *Cryptotriton* (6 species), *Nototriton* (12 species), and *Nyctanolis* (1 species) all occur in Middle America, with some entering southern Mexico and one, *Nototriton*, reaching central Costa Rica. Most of these are small, inconspicuous salamanders that are rarely seen; many of them occur in arboreal bromeliads and moss mats. *Nyctanolis* contrasts sharply in being relatively large, long-limbed, and spectacularly colored (a kind of harlequin pattern of red, yellow, and cream spots on a shiny black ground color). The final genus, *Oedipina* (21 species), differs dramatically from all other genera in having 18 to more than 20 rather than 14 trunk vertebrae and in being extremely slender and elongate, with tails of some species being more than twice head plus body size. These mainly fossorial animals occur from southern Mexico to Ecuador and are most numerous in Costa Rica.

Physical characteristics

These are diverse organisms that include the smallest and nearly the largest terrestrial salamanders. All are lungless and breathe through their skin. All have four limbs and a tail, but some are permanently larval and some of these are blind. The vast majority develop directly with no larval stage. Many are fully terrestrial, but a number are semiaquatic and some have become secondarily aquatic as adults. There are many fossorial as well as arboreal species.

Distribution

With the exception of six species in the middle western Mediterranean region of Europe, these are New World salamanders that occur from southern Canada throughout much of the United States and Mexico (except the north-central parts of these countries), through Central America and into southern South America (central Bolivia and Brazil). Species are most numerous in the eastern United States and Middle America.

Habitat

These are salamanders that thrive in wooded montane areas, but they occur in many other kinds of habitats. Some terrestrial species occur in desert areas that receive far less than 10 in (25 cm) of rainfall yearly, whereas others occur in rainforests, both temperate and tropical. Many species are semiarboreal to fully arboreal.

Behavior

Plethodontids are typically secretive by day and active by night. They have small home ranges, and seasonal migrations are limited to a few species that use aquatic breeding sites. Stream salamanders are more active than terrestrial species, but most species are capable of moving quickly when disturbed, and they are good at escaping capture. The more terrestrial species and especially the tropical species rely more on stealth to avoid detection and capture, and do not move as quickly as aquatic and semiaquatic species. Social behavior has been studied in only a few species, and in these groups, individuals display territoriality and aggression. All species have complex courtship and mating behavior, and courtship can take many hours.

Feeding ecology and diet

These salamanders typically feed on small arthropods but occasionally take worms, and the large species can eat members of smaller species. Prey is captured by very rapid movements of the tongue.

Reproductive biology

More generalized species lay eggs in or near shallow water, typically moving water, and eggs hatch into aquatic larvae that remain as larvae for a few months to as long as three years. Metamorphosis is somewhat more profound than in other families. A handful of species have larvae that live in standing water. More than half the species are strictly terrestrial and lay large yolky eggs that are hidden in cavities under rocks or logs, or deposited in moss mats, balls of moss hanging in vegetation, or arboreal plants including bromeliads. Hatching typically takes many weeks, and miniatures of the adult are produced.

Conservation status

Most plethodontids are secretive animals, and their conservation depends on maintenance of habitat. However, in recent years populations of many species have dramatically declined for unknown reasons, even in protected habitats. The 2002 IUCN Red List includes 25 plethodontid species.

Significance to humans

These are inconspicuous organisms that are not very often seen. However, they commonly occur in high density and typically are the most numerous vertebrates in a given region.

1. Talamancan web-footed salamander (*Bolitoglossa pesrubra*); 2. Mt. Lyell salamander (*Hydromantes platycephalus*); 3. Two-lined salamander (*Eurycea bislineata*); 4. Bell's salamander (*Pseudoeurycea bellii*); 5. Female ensatina (*Ensatina eschscholtzii*); 6. Red Hills salamander (*Phaeognathus hubrichti*); 7. Red-backed salamander (*Plethodon cinereus*); 8. Dusky salamander (*Desmognathus fuscus*). (Illustration by Gillian Harris)

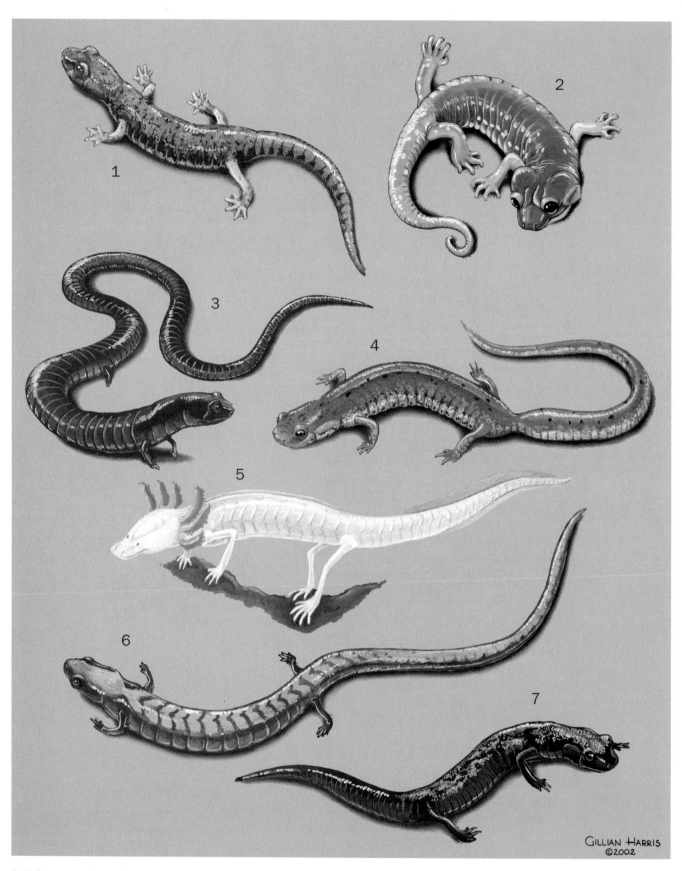

1. Italian cave salamander (*Hydromantes italicus*); 2. Arboreal salamander (*Aneides lugubris*); 3. Costa Rican worm salamander (*Oedipina uniformis*); 4. Four-toed salamander (*Hemidactylium scutatum*); 5. Texas blind salamander (*Eurycea rathbuni*); 6. Golden thorius (*Thorius aureus*); 7. Inyo Mountains salamander (*Batrachoseps campi*). (Illustration by Gillian Harris)

Species accounts

Dusky salamander
Desmognathus fuscus

SUBFAMILY
Desmognathinae

TAXONOMY
Desmognathus fuscus Green, 1818, type locality not given but thought to be near Princeton, New Jersey, United States.

OTHER COMMON NAMES
English: Northern dusky salamander; French: Salamandre sombre du nord; German: Brauner Bachsalamander.

PHYSICAL CHARACTERISTICS
This is a medium-sized salamander (to about 5.5 in or 14 cm total length) with short limbs and stocky proportions. The hind limbs are much larger than the forelimbs. The head is wedge-shaped and has prominent, protruding eyes. Jaw and neck muscles are well developed, and there is no discernable neck. The tail has a low fin and terminates in a sharp point, but often it is at least partially regenerated and then is blunt-tipped. Coloration is variable, but in general the upper surfaces are darker than the cream-colored lower surfaces, and the trunk may be mottled gray and black, striped with various shades of tan to yellow-brown to brown. A black stripe extends from the eye diagonally back to the angle of the jaw.

DISTRIBUTION
The species ranges from Quebec and New Brunswick in eastern Canada south and west as far as Indiana and South Car-

olina; a related form currently treated as a subspecies extends further south and west to peninsular Florida and Louisiana.

HABITAT
Larvae live in seeps, springs, and small streams. Adults are more terrestrial but spend most of their time in seeps or on the margins of springs and small streams, where they are found in rocky streambanks or under logs or other cover objects.

BEHAVIOR
Dusky salamanders are active animals, and when one attempts to capture them they move very rapidly and are elusive. Most of their normal surface activity occurs in the early evening, but when conditions are warm and moist they may be active throughout the night. By day they typically are found only under cover objects. They defend themselves from predators either by remaining motionless or in some instances by biting. They lack effective chemical defenses.

FEEDING ECOLOGY AND DIET
Larvae eat small invertebrates such as larvae of aquatic insects but also copepods and tiny clams. Adults eat more terrestrial prey, principally small arthropods, but they also feed on aquatic insects when in more aquatic sites. Larger individuals eat increasingly larger prey but they continue to eat small prey. Occasional cannibalism occurs, especially on larvae. Adults capture their prey by rapidly flicking their tongues and then snapping their jaws.

REPRODUCTIVE BIOLOGY
Courtship behavior is well studied and takes place on land. Following extensive behavioral interactions between the male and female, a spermatophore is deposited by the male and taken up by the female. The female stores sperm internally. Eggs are deposited in mid- to late summer in moist, hidden sites in seeps or at the edges of springs and small streams. Eggs are laid in clusters of from five to six to as many as 30 or more, and they are guarded by the female, typically until hatching, which occurs after 45 days or more. Larvae hatch with a good yolk supply and do not feed immediately. Larvae grow slowly in the fall and winter but rapidly in the spring and metamorphose in about nine months.

CONSERVATION STATUS
This is one of the most common and widely encountered salamanders in eastern North America, and it adapts well to habitat modification so long as appropriate microhabitats remain.

SIGNIFICANCE TO HUMANS
None known. ◆

Red Hills salamander
Phaeognathus hubrichti

SUBFAMILY
Desmognathinae

TAXONOMY
Phaeognathus hubrichti Highton, 1961, 3 mi (4.8 km) northwest of McKenzie on US Route 31, Butler County, Alabama, United States.

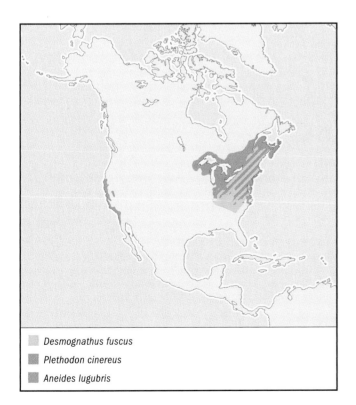

Desmognathus fuscus
Plethodon cinereus
Aneides lugubris

Eurycea rathbuni

Eurycea bislineata

Phaeognathus hubrichti

Arboreal salamander

Aneides lugubris

SUBFAMILY
Plethodontinae

TAXONOMY
Aneides lugubris Hallowell, 1849, Monterey, California, United States.

OTHER COMMON NAMES
Spanish: Salamandra arbórea.

PHYSICAL CHARACTERISTICS
Arboreal salamanders are large (to a little over 7 in or 17.8 cm total length), muscular animals with long limbs that overlap when adpressed to the trunk and a relatively long, strongly prehensile tail. Large adults have a formidable appearance, with a heavily muscularized head and body and long limbs and digits. The very long, prehensile digits have expanded, somewhat recurved tips. The large eyes bulge from the head in front of greatly enlarged jaw muscles. The head is nearly triangular in appearance, with the long snout having skin that is co-ossified to the underlying bone. The upper and lower jaws bear large, saberlike teeth with recurved tips that are capable of inflicting a serious wound. These salamanders are gray-brown to brown in coloration, with lighter ventral surfaces. Yellow spots are always present, but they may be small and scattered or large and rather densely arranged.

DISTRIBUTION
Arboreal salamanders occur mainly in California, where they are found in the coastal mountains and valleys from the northwestern part of the state continuously to the extreme northwestern part of Baja California Norte, Mexico. They are found on some off-shore islands in the Pacific Ocean. They have a disjunct distribution, with another group of populations in the foothills of the Sierra Nevada.

HABITAT
These salamanders are mainly found in oak woodland habitats, where they utilize holes in the trees for nesting sites and escape from unfavorably dry conditions. They are also found in sycamore woodlands near creeks in the southern parts of their range. They are commonly found under the bark of fallen oak logs and in rocky areas under rocks and in underground cavities.

BEHAVIOR
Arboreal salamanders are aggressive. Both sexes have enlarged jaw muscles and teeth that are used in territorial disputes and against predators. They are adept climbers but are most often found in terrestrial situations.

FEEDING ECOLOGY AND DIET
Despite their large jaws and teeth, these salamanders mainly eat arthropods, although generally an array somewhat larger than would be predicted for related less specialized species of similar size. Rarely they eat slender salamanders.

REPRODUCTIVE BIOLOGY
These fully terrestrial animals lay grape-like clusters of large, yolky eggs that are suspended from roofs of cavities, underground, in large decaying logs, or in holes in trees. Hatching takes place just before fall rains, three to four months after laying.

OTHER COMMON NAMES
None known.

PHYSICAL CHARACTERISTICS
This large, dark, elongate animal has a large head with protrusive eyes, short legs, a very long trunk, and a relatively short, round tail. They exceed 10 in (25 cm) in length and are the longest desmognathine.

DISTRIBUTION
This species is known only from a small part (Red Hills region) of southern Alabama.

HABITAT
The Red Hills salamander occurs in ravines in mature forests with closed canopies. These are fossorial salamanders that construct burrows in rich, friable soil.

BEHAVIOR
The species stays underground by day but partly emerges from its retreats at night to forage.

FEEDING ECOLOGY AND DIET
This salamander feeds mainly on small arthropods and snails.

REPRODUCTIVE BIOLOGY
This is a strictly terrestrial species and it lays large, yolky eggs.

CONSERVATION STATUS
The Red Hills salamander is classified as Endangered and is protected by federal law. It occupies a special habitat that is very limited in extent, and the greatest threat is deforestation and associated disturbances.

SIGNIFICANCE TO HUMANS
None known. ◆

CONSERVATION STATUS
Not threatened.

SIGNIFICANCE TO HUMANS
None known. ◆

Inyo Mountains salamander
Batrachoseps campi

SUBFAMILY
Plethodontinae

TAXONOMY
Batrachoseps campi Marlow Brode, and Wake, 1979, Long John Canyon, NE Lone Pine, Inyo County, California, United States.

OTHER COMMON NAMES
None known.

PHYSICAL CHARACTERISTICS
This species differs greatly from most other members of this genus in being relatively robust, long-legged, and broad headed, with a tail that is shorter than head plus body length. Maximum size is about 4.25 in (10.8 cm). The long legs fail to overlap by two to five costal folds when adpressed to the sides of the trunk. As with other members of this genus, there are only four toes. The eyes are prominent and protrude from the flattened head. Ground color is black, both above and below, but typically there are silvery or greenish gray patches on the head and the front part of the back, especially over the forelimbs. However, there is much individual and geographic variation, and in some individuals there is a general suffusion of the light pigment over most dorsal surfaces, giving the impression of a silvery gray coloration.

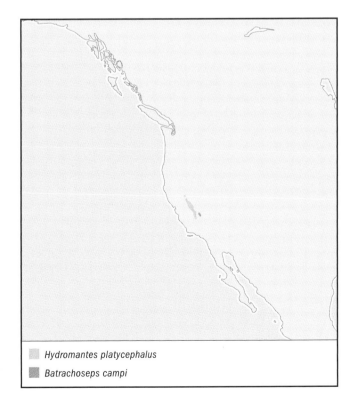

    ▨ *Hydromantes platycephalus*
    ■ *Batrachoseps campi*

DISTRIBUTION
This species has a remarkable distribution in strict desert environment throughout the Inyo Mountains in eastern California. It is found both on western facing slopes descending into the Owens Valley, and on eastern facing slopes descending into the Saline Valley, and occurs at elevations ranging from about 1,800 to 8,500 ft (550–2,590 m).

HABITAT
This species is most commonly found in moist soil near permanent streams, but it is also known from mossy limestone crevices beneath large rocks on open desert slopes. This species never enters water voluntarily but can be found in very wet soil. However, it appears not to be dependent on flowing water. It has been found in willow patches.

BEHAVIOR
Almost nothing is known concerning the biology of this remarkable species.

FEEDING ECOLOGY AND DIET
Nothing is known concerning diet, but the salamanders most likely resemble other members of the genus in eating primarily small arthropods, especially insects.

REPRODUCTIVE BIOLOGY
Large yolky eggs are produced and development is direct with no larval stage.

CONSERVATION STATUS
Classified as Endangered because of its apparent limited habitat and the fact that it is a narrow endemic, but the species is strongly differentiated genetically throughout its small range, indicating that the populations may be large and certainly that the populations are old.

SIGNIFICANCE TO HUMANS
None known. ◆

Talamancan web-footed salamander
Bolitoglossa pesrubra

SUBFAMILY
Plethodontinae

TAXONOMY
Bolitoglossa pesrubra Taylor, 1952, Cerro de la Muerte (at crossing with the Pan American Highway), Costa Rica.

Until recently this species was known as *Bolitoglossa subpalmata*, but molecular studies demonstrated that two species were inappropriately placed in the same taxon.

OTHER COMMON NAMES
None known.

PHYSICAL CHARACTERISTICS
These relatively stocky salamanders reach a size of about 4.5 in (11.4 cm) total length and have a tail about the same length as the head plus body. The tail is strongly constricted at its base. Limbs fail to overlap when adpressed to the side of the trunk and bear hands and feet that are moderately webbed, with all digits having one to three phalanges extending beyond the web and bearing subdigital pads. Color pattern is extremely variable but is basically dark brown with mottled, striped, or marbled coloration, often contrasting gray and black, or it may be uni-

Oedipina uniformis

Bolitoglossa pesrubra

formly brown. The lower surfaces are lighter, but usually the belly is relatively dark gray and the throat region is much lighter and often bears yellowish pigment. The basal parts of the limbs are usually dark red to red-orange.

DISTRIBUTION
This species is known only from the Cordillera de Talamanca in central and eastern Costa Rica, generally at elevations above 7,500 ft (2,286 m). It is the best known of the many tropical salamanders because it has been observed by generations of students in the classes organized by the Organization for Tropical Studies.

HABITAT
This species is found under the bark of logs and under surface debris in oak forests, but it also has survived in many areas where habitats have been destroyed. It can be locally abundant in rubbish heaps. It has also been found in arboreal bromeliads and in moss mats on trees and roadside banks. It was once common at very high elevations (around 10,000 ft or 3,050 m), even in completely open areas where it was found under rocks, slabs of disused concrete, and other surface objects, but in recent years it has become scarce.

BEHAVIOR
Little is known concerning the behavior of this species. It is nocturnal and forages and mates by night.

FEEDING ECOLOGY AND DIET
The vast majority of prey are small terrestrial insects that are caught with a very fast, highly projectile tongue that is fired with great accuracy.

REPRODUCTIVE BIOLOGY
Like all tropical plethodontids, this is a direct-developing species. Eggs have been found throughout the year. They are laid in small clusters of 13–38 eggs (average 22.5). The eggs are large and yolky, and take a very long time to develop, partly because of the cool temperatures typical of the montane habitat. Females guard the eggs, which hatch after many

months into tiny miniatures of the adult, well supplied with yolk.

CONSERVATION STATUS
Not threatened. Once this species was thought to be extremely tolerant of human activities and thrived even along heavily disturbed roadsides, but in recent years it has disappeared from much of its previous range. It is still found in deep forests.

SIGNIFICANCE TO HUMANS
None known. ◆

Ensatina
Ensatina eschscholtzii

SUBFAMILY
Plethodontinae

TAXONOMY
Ensatina eschscholtzii Gray, 1850, Monterey, California, United States.

This is the celebrated "ring-species" of western North America, a polytypic species comprised of seven subspecies that are distributed in a ringlike manner surrounding the inhospitable Central Valley of California. Coastal populations are generally mottled to unmarked whereas inland populations are spotted, blotched, or banded. Coastal and inland populations either hybridize or live in sympatry as distinct species, but continuous interactions around the ring suggest that one species, in various stages of species formation, is represented. Some researchers recommend breaking the complex up into 2, 4, or 11 or more species, and the taxon is under active study.

OTHER COMMON NAMES
English: Large-blotched salamander, Monterey ensatina, Oregon ensatina, painted salamander, Sierra Nevada ensatina, yellow-blotched salamander, yellow-eyed ensatina; Spanish: Salamandra ensatina.

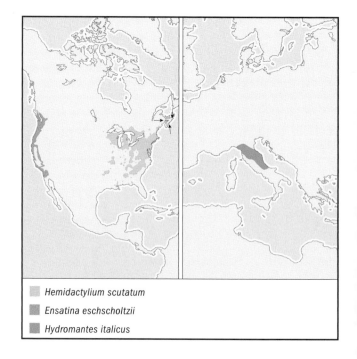

Hemidactylium scutatum

Ensatina eschscholtzii

Hydromantes italicus

PHYSICAL CHARACTERISTICS

These salamanders are relatively large and stout with long legs, a large head, and a long, well-developed tail. The tail has a marked constriction at its base and is sexually dimorphic, long and thin in males and relative short and stout in females. Males in courtship season have greatly swollen upper lips and nasolabial projections. Color varies geographically. Oregon salamanders are dull brown or yellow-brown with some highlights of lighter pigment. Painted salamanders are complexly mottled with different colors, from black to yellow and orange. Yellow-eyed salamanders have reddish brown heads and backs and bright orange lower surfaces; they have a bright yellow spot in the upper eye and often have yellow-orange eyelids. Monterey salamanders are similar in color but are less vivid and more pinkish and have black eyes. Sierra Nevada salamanders are dark brown above with numerous small to moderate sized spots or blotches of red-orange to dull red. Yellow-blotched salamanders are black with lemon yellow blotches, typically larger and less numerous than those of Sierra Nevada salamanders. Large-blotched salamanders are black with a few yellow to dull orange to flesh-colored bold blotches or bands.

DISTRIBUTION

This group occurs from Vancouver Island and the mainland of southwestern British Columbia, Canada, southward west of the Cascade-Sierra Nevada crest, to the Sierra San Pedro Mártir in northern Baja California, Mexico. It is not present in the Central Valley of California.

HABITAT

These salamanders usually occur in woodlands, but the coastal forms occur in coastal sage scrub and chaparral as well. Blotched forms usually occur in mixed conifer forests with closed canopies. The species has an enormous ecological scope and occurs in areas that receive well over 150 in (381 cm) of rainfall per year to areas in southern California and northern Mexico that receive less than 10 in (25 cm) of rain per year. The species is typically found with rotting or decaying wood but also under rocks and other cover.

BEHAVIOR

This is a nocturnal species, but the blotched forms can be found by day during periods of rain. The species is aggressive toward conspecifics and bites and eats the tail tip of adversaries. The tail is richly supplied with poison glands and exudes a milky substance that many predators find repulsive. The tail is strongly constricted at the base, and when seized by a predator it can be autotomized (separated from the body by reflex), but autotomy is rarely observed. Nevertheless, about 20% of adults have regenerated tails.

FEEDING ECOLOGY AND DIET

This species generally feeds on small terrestrial arthropods, which it consumes by modulated use of a highly projectile tongue.

REPRODUCTIVE BIOLOGY

Courtship behavior is poorly known but involves deposition of a spermatophore following an elaborate series of maneuvers by both members. Eggs are laid in grape-like clusters (generally 12–25 eggs each) in underground cavities and are guarded by the female. Eggs are especially large and yolky, and females turn onto their backs and aid oviposition with leg movements.

CONSERVATION STATUS

Not threatened.

SIGNIFICANCE TO HUMANS

None known. ◆

Two-lined salamander

Eurycea bislineata

SUBFAMILY

Plethodontinae

TAXONOMY

Eurycea bislineata Green, 1818, probably Princeton, New Jersey, United States.

OTHER COMMON NAMES

French: Salamandre á deux linges; German: Zweistreifiger Gelbsalamander.

PHYSICAL CHARACTERISTICS

Two-lined salamanders are small (5–5.5 in or 12.7–14 cm total length), slender salamanders with long, tapered, laterally compressed tails and limbs of moderate length. The eyes are prominent. Coloration ranges from greenish yellow to yellow or orange-brown. A broad band extends from behind the eye along the trunk to near the tip of the tail. This band is marked with dark brown or black speckles and spots in no apparent pattern, and it is bound on either side by a prominent brown or black stripe that extends from the eye well onto the tail, above the limb insertions. The lateral flanks are light with dark spots, and the venter is typically bright yellow with scattered dark spots.

DISTRIBUTION

This species ranges from northeastern Canada southwest through northeastern United States to Ohio, West Virginia, and Virginia. Closely related species sometimes combined with this species extend to the Gulf coastal plain, from Louisiana to Florida.

HABITAT

Larvae live mainly in small springs and seeps and in some places in ponds, where they are benthic (occurring near the bottom of the lake or stream). Metamorphosed young and adults generally stay near streams in forested areas but move out into the forest and can be found some distance from water. As adults they are often fully terrestrial for much of the year.

BEHAVIOR

Two-lined salamanders are nocturnal and forage in the forest at night. There is some evidence that they are territorial, but the species has not been intensively studied by behavioral ecologists.

FEEDING ECOLOGY AND DIET

Larvae feed on small aquatic insect larvae but also eat other small arthropods. Adults feed mainly on small arthropods but will also eat snails.

REPRODUCTIVE BIOLOGY

Courtship is thought to occur on land, and females store sperm until eggs are laid, singly, beneath rocks in small streams. Nests are formed but are not guarded. Nests can contain over 100 eggs, but clutches generally are around 50. Eggs take up to 10 weeks to hatch. Length of the larval period varies geographically but generally is two years (one year in southern parts of the range).

CONSERVATION STATUS

Not threatened.

SIGNIFICANCE TO HUMANS

None known. ◆

Texas blind salamander
Eurycea rathbuni

SUBFAMILY
Plethodontinae

TAXONOMY
Eurycea rathbuni Stejneger, 1896, subterranean waters near San Marcos, Texas, United States.

Until recently this extraordinary salamander was placed in the genus *Typhlomolge*, but molecular systematic analyses have shown that it is a highly derived member of *Eurycea*. When first discovered it was placed in the family Proteidae, but it was soon recognized as a bizarrely specialized plethodontid.

OTHER COMMON NAMES
English: San Marcos salamander, white salamander; German: Rathbunscher Brunnenmolch.

PHYSICAL CHARACTERISTICS
This blind and pigmentless salamander is strictly subterranean, where it lives in aquifers and can be observed in waters in caves and sinkholes. It is gilled and shiny white, with long, extraordinarily slender legs that strongly overlap when addressed to the trunk. It is large for a larval member of this genus, reaching a total length of about 5–5.5 in (12.7–14 cm). Eyes are absent and the large head is strongly depressed with a broad, blunt snout.

DISTRIBUTION
This species is restricted to a small area on the edge of the Edwards Plateau, near San Marcos in south-central Texas.

HABITAT
The species is known only from underground streams and pools that are in sinkholes, caves, and recesses.

BEHAVIOR
Little is known, but individuals are active and have been observed swimming in the water column of recesses and using their long limbs to grab and cling to the rocky sides of the cavity containing the water column.

FEEDING ECOLOGY AND DIET
Known food items include snails, amphipods, and cave-adapted shrimp.

REPRODUCTIVE BIOLOGY
Nothing has been reported concerning the reproductive biology of this species.

CONSERVATION STATUS
This species is listed as Vulnerable by the IUCN. One of the first species named to the federal endangered species list, this salamander is threatened by human withdrawals of water from the aquifers it occupies and by pollution from surface sources.

SIGNIFICANCE TO HUMANS
None known. ◆

Four-toed salamander
Hemidactylium scutatum

SUBFAMILY
Plethodontinae

TAXONOMY
Hemidactylium scutatum Temminck and Schlegel, 1838, Nashville, Tennessee, United States.

OTHER COMMON NAMES
French: Salamandre á quatre doigts; German: Vierzehensalamander.

PHYSICAL CHARACTERISTICS
This small (maximum total size about 4 in or 10 cm) terrestrial species is immediately identified by having only four toes on its hind feet, having a strong constriction at the base of its tail, and in having lower surfaces that are white with conspicuous black marks, usually round spots.

DISTRIBUTION
This species is found all over eastern North America from southern Canada to Oklahoma, Louisiana, and Florida.

HABITAT
This salamander is most commonly found in forested areas near bogs, swamps, and vernal pools. The species has a wide tolerance of environmental temperatures and occurs in seasonally very cold areas of southern Canada and northern Minnesota, Wisconsin, and Michigan, to very hot parts of Louisiana, Mississippi, Alabama, Florida, and Georgia. In the southern parts of its range its distribution is highly fragmented and discontinuous, and adults are rarely seen outside the spring breeding season.

BEHAVIOR
There have been extensive studies of nesting behavior. Females may lay eggs in communal nests, in isolated nests that are guarded, or abandon the eggs. They occasionally form dense aggregations as winter approaches and overwinter in subterranean sites. When attacked this salamander coils tightly and hides its head, exposing the tail, which is well supplied with glands that produce a noxious secretion. The tail can be autotomized, and it then whips rapidly back and forth.

FEEDING ECOLOGY AND DIET
Only small arthropods have been recorded as food items.

REPRODUCTIVE BIOLOGY
Raised moss mats in boggy or swampy ground are favored nesting sites. More than 1,000 eggs have been found in communal nests in such habitats. Eggs hatch after about five weeks, and larvae are small and inconspicuous, metamorphosing after a brief period of as little as two and as much as six weeks, with geographic variation.

CONSERVATION STATUS
Not threatened, although specialized breeding habits make this species vulnerable to human disturbance, such as forestry and wetland drainage and conversion.

SIGNIFICANCE TO HUMANS
None known. ◆

Italian cave salamander
Hydromantes italicus

SUBFAMILY
Plethodontinae

TAXONOMY
Hydromantes italicus Dunn, 1923, Apuan Alps and Appenines, Italy.

European members of this genus are frequently placed in the genus *Speleomantes*. European and American species are each other's closest phylogenetic relatives. All of the species involved are very similar in general morphology, which is highly specialized and distinguishes them instantly from all members of all other genera.

OTHER COMMON NAMES
German: Italienischer Höhlensalamander; Italian: Salamandra cavernicola italiana.

PHYSICAL CHARACTERISTICS
This is a stocky, short-bodied salamander with long, partly webbed hands and feet and a relatively short tail. It grows to about 4.5 in (11.4 cm) in total length. Arms and legs overlap when adpressed, and the arms are nearly the same length as the legs. The fingers and toes are relatively long and blunt-tipped. The color pattern is a rather dull gray-brown but with highlights of reddish or yellow-brown. The ventral surfaces are dark.

DISTRIBUTION
This species is endemic to the Appenine Alps of central and northern Italy, including San Marino.

HABITAT
This completely terrestrial species is associated with limestone and is most commonly found in caves. However, it also is found outside of caves, especially on wet, rocky slopes where it is active on cool, rainy nights.

BEHAVIOR
The Italian cave salamander is an agile climber, using its long limbs, large webbed hands and feet, and short tail to maneuver on the vertical walls of caves and on rock faces.

FEEDING ECOLOGY AND DIET
This salamander has a spectacular tongue that is especially long and fast and that it uses to feed on active terrestrial arthropods.

REPRODUCTIVE BIOLOGY
Small clusters of large yolky eggs are laid in crevices and underground cavities; these hatch after several months into miniatures of the adults. Egg-laying is so secretive that for many years the species was thought to give birth to living young.

CONSERVATION STATUS
Not threatened. Because the species is rarely observed, it is often thought to be rare, but in fact it is locally abundant and is very widespread. It does not appear to be in any danger but could be harmed locally by overcollection.

SIGNIFICANCE TO HUMANS
Italian cave salamanders are collected as pets. ◆

digits are extensively webbed, although all of the digits are free of the webbing for much of their length. The breadth of the body is partly the result of the generally flattened nature of the organism but is enhanced by the elongate epibranchials that lie above the shoulders lateral to the main trunk. These epibranchials extend to the middle of the trunk and are the main structural elements of the tongue, the longest of any salamander.

DISTRIBUTION
This species occurs in the Sierra Nevada of California, from the southern end of Sequoia National Park to the Sierra Buttes in the north, above about 3,000 ft (910 m) in elevation, ranging upward to about 12,000 ft (3,660 m).

HABITAT
Most species belonging to this genus are associated with limestone, but this species is a granite specialist that occurs exclusively in the high Sierra Nevada. It is very cold-tolerant and is found near melting snow, where it hides under flat pieces of granite. Much of its habitat is above tree line, and it is most frequently found in moist areas, near seeps or small streams or near snow-melt, in areas dominated by low shrubs with some scrubby trees.

BEHAVIOR
Locomotion is exceptional in these salamanders. They move along moist, slick rocks at steep angles and use their short, blunt-tipped tails as a fifth appendage, bracing themselves with it as they move. When exposed by day they roll up into a ball that often rolls downhill, away from a potential predator.

FEEDING ECOLOGY AND DIET
Very little is known concerning its diet except that it feeds using its extraordinarily long tongue and concentrates on arthropods, mainly insects.

REPRODUCTIVE BIOLOGY
Little is known, but these are direct-developing species that lay large, yolky terrestrial eggs.

CONSERVATION STATUS
Much of the range of this species is in large national parks—Yosemite, Kings Canyon, and Sequoia—as well as in wilderness areas, and the species is not threatened.

SIGNIFICANCE TO HUMANS
None known. ◆

Mt. Lyell salamander
Hydromantes platycephalus

SUBFAMILY
Plethodontinae

TAXONOMY
Hydromantes platycephalus Camp, 1916, head of Lyell Canyon, 10,800 ft (3,291 m) altitude, Yosemite National Park, California, United States.

OTHER COMMON NAMES
None known.

PHYSICAL CHARACTERISTICS
These salamanders have broad, relatively flattened heads and bodies and short tails. The limbs are relatively long and the

Costa Rican worm salamander
Oedipina uniformis

SUBFAMILY
Plethodontinae

TAXONOMY
Oedipina uniformis Keferstein, 1868, Costa Rica. Several other species are now included in this taxon, although recently two sets of lowland populations were separated on biochemical grounds and recognized as separate species, *O. pacificensis* and *O. gracilis* (the latter is the common species at La Selva Biological Station in northeastern Costa Rica).

OTHER COMMON NAMES
None known.

PHYSICAL CHARACTERISTICS
The extraordinarily slender form and tiny limbs and digits characterize this and related species. The species reaches 6–6.5 in (15.2–16.5 cm) in total length, two-thirds of which may be its tail. The head, body, and tail are of roughly equivalent width, and the tail does not taper until near its tip. Eyes are small and inconspicuous and the head is small. Typically the species is completely black with inconspicuous white speckles and occasionally light spots behind the eyes.

DISTRIBUTION
This species is known only from the region surrounding the Meseta Central of Costa Rica, generally above 4,000 ft (1,219 m) elevation.

HABITAT
This species is fossorial, found most commonly in well rotted logs and under moss mats on roadside banks, usually in forested areas.

BEHAVIOR
This salamander lives in the interstices between soil, rotting logs, leaf litter, and mats of moss and other plants covering soil and fallen logs. When uncovered the salamanders whip their tails back and forth violently in attempting to escape, and will readily autotomize any length of the tail if it is grabbed.

FEEDING ECOLOGY AND DIET
These salamanders feed on small arthropods.

REPRODUCTIVE BIOLOGY
The species lays large yolky eggs in cavities in the soil.

CONSERVATION STATUS
Not threatened. This species was once common but now is rarely seen, in part because of extensive conversion of original habitat for agriculture and urbanization.

SIGNIFICANCE TO HUMANS
None known. ◆

Red-backed salamander
Plethodon cinereus

SUBFAMILY
Plethodontinae

TAXONOMY
Plethodon cinereus Green, 1818, New Jersey, United States.

OTHER COMMON NAMES
English: Lead-backed salamander; French: Salamandre rayée; German: Rotrücken-Waldsalamander.

PHYSICAL CHARACTERISTICS
This is a small species (3.5–5 in or 9–13 cm total length) with a small head, short limbs, and an elongate, narrow trunk and tail. It occurs in two morphs, a striped or red-backed phase that features a long, even-sided dull tan to an orange or reddish stripe extending nearly the full length of the trunk and tail, and a lead-back phase in which the dull gray-black ground color of the sides extends all over the upper surfaces. Lower surfaces are light gray with numerous black speckles.

DISTRIBUTION
This species ranges from eastern Canada to western Ontario and eastern Minnesota in the North, to North Carolina and eastern Tennessee in the South, and eastern Illinois in the West.

HABITAT
This is a woodland salamander, and it is often extremely common. It is found under cover objects by day, usually wood debris but also rocks and leaf litter. It forages actively in the early evening hours and is common when conditions are moist and cool.

BEHAVIOR
Red-backed salamanders have been the subject of extensive behavioral ecological studies by Robert Jaeger and associates. They have a rich social life and are very aggressive toward conspecifics (members of their own species) as well as other species, actively protecting feeding territories and retreats. They modulate their behavior depending on familiarity, and they select mates according to diverse criteria including quality of the food eaten by potential mates. They are active foragers, and they are capable of modulating their tongue projection mechanisms depending on the size, distance, and other characteristics of the prey.

FEEDING ECOLOGY AND DIET
Food is basically small terrestrial arthropods, but occasionally they eat snails, slugs, and even small earthworms.

REPRODUCTIVE BIOLOGY
Courtship is terrestrial, and following acceptance of a spermatophore by a female, eggs are laid in secretive sites, typically underground cavities or crevices in logs or rocks. The grape-like clusters of three to 14 eggs are guarded by the female and hatch in about six weeks.

CONSERVATION STATUS
Not threatened. This is a highly adaptable species, but it does require wooded environments and does best in closed canopy forests.

SIGNIFICANCE TO HUMANS
None known. ◆

Bell's salamander
Pseudoeurycea bellii

SUBFAMILY
Plethodontinae

TAXONOMY
Pseudoeurycea bellii Gray, 1850, Mexico.

OTHER COMMON NAMES
English: Bell's false brook salamander.

PHYSICAL CHARACTERISTICS
This is a spectacular species that is the largest lungless salamander and close to the largest terrestrial salamander (it reaches nearly 14 in or 36 cm total length). It is shiny dark black with a pair of red to red-orange spots on the back of its head and paired rows of similarly colored spots along the back to the base of the tail. There is usually a chevron-shaped mark at the beginning of the paired rows. The tail is long and large and is basally contricted. The limbs are long and well developed and the overall appearance of the animal is massive.

Pseudoeurycea bellii
Thorius aureus

Golden thorius
Thorius aureus

SUBFAMILY
Plethodontinae

TAXONOMY
Thorius aureus Hanken and Wake, 1994, 0.7 mi (1.1 km) east of Cerro Pelon, Oaxáca, Mexico.

OTHER COMMON NAMES
None known.

PHYSICAL CHARACTERISTICS
This tiny salamander is one of the largest species of a group known as the minute salamanders, among the smallest known tetrapods. In comparison to other members of the large genus *Thorius*, this is a robust and colorful species that reaches a size of about 2.25 in (5.7 cm) total length. The legs are slender and short, separated by six to eight costal interspaces when adpressed to the sides of the body. Digits are joined to neighboring digits for most of their short length but are free at the tips. Nostrils of many members of this genus are very large, but in this species they are relatively small. This species has teeth on its upper jaw, but these are missing in most members of this genus. The bones of this species, as in other members of the genus, are weak and poorly articulated, and there is a large fontanelle in the back of the head, over the brain. Eyes are large and well developed. This is a colorful species, with a golden band or broad stripe on its back and tail and with relatively light ventral surfaces.

DISTRIBUTION
This species is known only from high elevations, from 7,500 to 10,000 ft (2,286–3,048 m) in the Sierra de Juarez, northern Oaxaca, Mexico, where a rich diversity of species of *Thorius* is found.

HABITAT
This species is found in high cloud forest under rocks, logs, and surface debris, and can be found under the bark of fallen logs and in moss on road banks.

BEHAVIOR
When uncovered these animals make a tight coil that hides the head. They have a constriction at the base of the tail and are capable of autotomy at this site, or at any point along the length of the tail.

FEEDING ECOLOGY AND DIET
Nothing is known.

REPRODUCTIVE BIOLOGY
Eggs have never been observed, but it is known that the species practices direct development.

CONSERVATION STATUS
Not threatened.

SIGNIFICANCE TO HUMANS
None known. ◆

DISTRIBUTION
Bell's salamander is widely distributed from northwestern and northeastern Mexico into central Mexico, usually at relatively high elevations (above 4,000 ft or 1,220 m).

HABITAT
This is a strictly terrestrial species that is found under large surface objects such as logs and rocks in relatively moist woods. It utilizes terrestrial burrows and can be found in holes in road banks.

BEHAVIOR
Almost nothing is known of the behavior of this species, except that it is nocturnal.

FEEDING ECOLOGY AND DIET
No information is available. It most likely feeds mainly on insects, which are caught with its freely projectile tongue.

REPRODUCTIVE BIOLOGY
Almost nothing is known except that it lays clutches (more than 20) of large, yolky eggs.

CONSERVATION STATUS
This species is widespread and was once common in many parts of Mexico. Not threatened, although in recent years it has become scarce.

SIGNIFICANCE TO HUMANS
None known. ◆

Resources

Books

Bruce, R., L. D. Houck, and R. G. Jaeger. *The Biology of Plethodontid Salamanders*. New York: KluwerAcademic/Plenum Publishers, 2000.

Griffiths, R. A. *Newts and Salamanders of Europe*. San Diego: Academic Press, 1996.

Petranka, James W. *Salamanders of the United States and Canada*. Washington, DC: Smithsonian Institution Press, 1998.

Periodicals

Lanza, B., V. Caputo, G. Nascetti, and L. Bullini. "Morphologic and Genetic Studies of the European Plethodontid Salamanders: Taxonomic Inferences (genus *Hydromantes*), Monografie." *Museo Regionale di Scienze Naturali, Torino* XVI (1995): 15–366.

Wake, D. B. "Adaptive Radiation of Salamanders in Middle American Cloud Forests." *Annals of the Missouri Botanical Garden* 74 (1987): 242–264.

David B. Wake, PhD

Amphiumas

(*Amphiumidae*)

Class Amphibia
Order Caudata
Suborder Salamandroidea
Family Amphiumidae

Thumbnail description
These are elongate, cylindrical, medium-sized to very large semilarval salamanders with very short limbs and one to three toes per limb, no eyelids, and a gill slit is present in the pharyngeal region; fertilization is internal

Size
Adults reach 13–46 in (33–117 cm) in length, depending on the species

Number of genera, species
1 genus; 3 species

Habitat
Streams, lakes, ditches, ponds, swamps, and marshes

Conservation status
Not threatened

Distribution
Southeastern Virginia to eastern Texas along the coastal plain and northeast to southeastern Missouri in the Mississippi Valley drainage

Evolution and systematics

Amphiumidae dates from the Upper Cretaceous and *Amphiuma* from the Upper Paleocene, but the fossil record does not give a clear picture of the evolution of the family. Concise phylogenetic placement of the family has not been possible, despite various external and internal structural features, specific muscles present or absent, chromosome numbers and appearance, and biochemical studies. Electrophoretic analysis shows that the living species *A. means* and *A. tridactylum* have considerable genetic similarity, but *A. pholeter* is genetically quite distinct from the two larger species. When larval characteristics are not considered, different cladograms show that the relationships of Amphiumidae to Dicamptodontidae and Proteidae change positions. Amphiumidae is placed in the salamander suborder Salamandroidea based on several structural features of living species, but other features that might be considered are not available from the fossil material. The fact that the family members are semilarval in construction also is an obstacle to phylogentic interpretation. Fossils of *A. means* are known from Florida, and a questionable fossil of this species is known from Texas, well out of its current range. *Amphiuma* contains the longest and most massive salaman-

ders in the United States. Local people use colloquial names, such as congo eel, lamper eel, ditch eel, lamprey, and congo snake with reference to the salamander. *Amphiuma* has legs and eyelids and lacks fins, whereas eels are legless and have fins and eyelids; the lack of scales in the congo snake preclude the validity of such a name. No subfamilies are recognized.

Physical characteristics

Amphiumas are elongate, cylindrical, eel-like salamanders with four small limbs that are usually less than 0.4 in (1 cm) long. They have one to three toes on each foot, depending on the species. The head is pointed, and the snout is somewhat depressed in two species. The tail is laterally compressed and makes up about 20–25% of the total body length. Adults have glandular skin that exudes slippery mucus. Metamorphosed individuals retain some larval features—lack of eyelids and tongue and presence of four gill arches with a single spiracular opening between the third and fourth arches. Lungs are present, but amphiumas also can breath via the pharynx and skin. When the young hatch, they retain gills for a few days. Hatchlings are a little more than 2 in (5.1 cm) in length, and metamorphosed in-

Two-toed amphiuma (*Amphiuma means*) with her eggs. She wraps her body around the eggs to protect them from predators and keep them moist. Eggs take five months to hatch. (Photo by Allan Blank. Bruce Coleman Inc. Reproduced by permission.)

dividuals may be as short as 2.3 in (5.8 cm). Adults reach 46 in (117 cm) in length. The trunk contains 57–60 costal grooves, each of which indicates a vertebra. The vertebrae are amphicelous, that is, they are concave on each end. A few anterior vertebrae bear ribs. Teeth are present on the premaxillary, maxillary, vomerine, and mandibular bones. A lateral line system is present on the body and head. The diploid chromosome number is 28. The dorsum is dark reddish brown to gray or black, and the belly may contrast with the dorsum or be almost as dark.

Distribution

These animals are found from southeastern Virginia southward along the coastal plain and throughout Florida, westward along the coastal plain; and from southwestern Alabama and all of Mississippi and Louisiana to the easternmost part of Texas and most southeastern part of Oklahoma northward to the extreme southeastern portion of Missouri. During the Cretaceous and the Upper Miocene, amphiumids were distributed widely in the United States, but since the Pleistocene, they have been restricted to their present ranges.

Habitat

These animals normally are aquatic and nocturnal. They are especially common in swamps, ditches, lakes, and sluggish streams, and one species typically is found in watery muck. They can be quite common in cities, where they occur in ditches and canals, including situations where the water is

temporary. They may hide among aquatic plants but prefer crayfish holes. In rainy weather they may crawl around on wet surfaces.

Behavior

If a ditch or pond goes dry, amphiumids hide in holes where they can estivate; they have been excavated from as deep as 3.3 ft (1 m). They can go for up to three years without food and are known to live at least 27 years. They may lie in wait for passing prey or prowl in search of prey. The skin is shed periodically and may be eaten, thus helping to sustain them. Adult males may fight during the reproductive season, and many show scars from fighting. Locomotion is via lateral undulations. The animals are sensitive to vibrations that are likely detected by the lateral line system. Animals out of water occasionally emit a whistling sound. The mud snake (*Farancia abacura*) feeds almost exclusively on amphiumas. These salamanders can be captured with dip nets, seines, minnow traps, electroshock equipment, and by hand. The skin is so slippery that cotton gloves must be used to keep hold of the animal long enough to place it in a container. Care in handling is recommended, because the bite can be painful.

Feeding ecology and diet

Most activity takes place when water temperatures are above 41°F (5°C). The salamanders are strictly carnivorous and eat worms, aquatic insects and aquatic insect larvae, frogs, salamanders, fishes, and any other small vertebrates. A favorite prey is crayfish. Amphiumas normally wait in holes for passing prey, with the front part of the body protruding. The strong teeth and powerful bite assist in subduing prey. Amphiumas are preyed upon mostly by snakes and large wading birds.

Reproductive biology

Females have smooth cloacal walls, and males have papillae lining the walls. During the breeding season the male cloaca is swollen. Males may fight during the courtship season, and, as a result, they may show scars on the body. Spermiogenesis occurs from October to May. Courtship has been observed in one species. The female makes a nest in a moist place, usually under logs, leaves, or other cover. The eggs are laid in rosary-like strings with constrictions between each egg. Fifty to 200 eggs usually constitute a clutch, but as many as 354 eggs might be produced. The female coils around and guards the eggs. Incubation of eggs may take up to six months. The eggs and their gelatinous outer layers are approximately 0.4 in (1 cm) in diameter in large species. Nests in Florida have been found in the nest mounds of alligators. Females apparently reproduce biennially and males annually.

Conservation status

No species of Amphiumas are listed by the ICUN. Although human activity has decimated much of the habitat of *Amphiuma*, it also has increased habitat by the building of aquatic sites, ponds, ditches, canals, lagoons, and lakes. Amphiumas can survive in waters with fish, and they may be major predators in some aquatic habitats. Apparently, amphiumas are not now in need of protection, except for *A. pholeter*, which is scarce and restricted to a small area.

Significance to humans

Amphiuma flesh is edible and tastes much like frogs' legs. Few people eat them, because the skin is difficult to strip from the flesh. *Amphiuma* cells, especially the red blood cells, are the largest known in vertebrates, and they have long been used in physiological studies and in the classroom. The chromosomes also are very large and useful for study. The bite of amphiumas is considered to be poisonous by some rural inhabitants.

1. One-toed amphiuma (*Amphiuma pholeter*); 2. Two-toed amphiuma (*Amphiuma means*); 3. Three-toed amphiuma (*Amphiuma tridactylum*). (Illustration by Dan Erickson)

Species accounts

Two-toed amphiuma
Amphiuma means

TAXONOMY

Amphiuma means Garden, 1821, Charleston, South Carolina, United States.

OTHER COMMON NAMES

German: Zweizehen-Aalmolch; French: Amphiume.

PHYSICAL CHARACTERISTICS

This large species reaches 46 in (117 cm) in length. The snout is somewhat depressed. The dorsum is usually dark reddish brown, black, or gray and contrasts little with the venter. A faint dark patch is present on the throat.

DISTRIBUTION

This species is distributed widely in the southeastern United States along the coastal plain from southeastern Virginia to southeastern Louisiana, including all of peninsular Florida.

HABITAT

This large aquatic salamander inhabits swamps, lakes, ditches, and sluggish streams.

BEHAVIOR

When it is active, this species swims by undulating movements. It estivates in holes.

FEEDING ECOLOGY AND DIET

Crayfish are the primary items eaten, but the species also feeds on other aquatic organisms.

REPRODUCTIVE BIOLOGY

Spermiogenesis occurs from October to May. Fifty to 200 eggs usually constitute a clutch. Courtship has not been observed. Incubation of eggs may take up to six months. Hatchlings are about 2 in (5.1 cm) in length, and metamorphosed individuals may be as short as 2.3 in (5.8 cm). Females apparently reproduce biennially and males annually.

CONSERVATION STATUS

The two-toed amphiuma is not threatened. No immediate conservation is needed for the species, but it is considered to be rare in several states and thus may warrant protection.

SIGNIFICANCE TO HUMANS

Like other members of the genus, this species is edible but rarely eaten, and it is used for classroom study. ◆

One-toed amphiuma
Amphiuma pholeter

TAXONOMY

Amphiuma pholeter Neill, 1964, 4.5 mi (7.2 km) east-northeast of Rosewood, Levy County, Florida, United States.

OTHER COMMON NAMES

None known.

PHYSICAL CHARACTERISTICS

This is the smallest species of the Amphiumidae; adults reach only 8.5–13 in (22–33 cm) in total length. Each foot has a single toe. The entire body is dark reddish brown or gray to grayish brown with no notable contrast between dorsum and venter. The head is cylindrical, and the snout is not depressed.

DISTRIBUTION

The species has a narrow range in the coastal plain of southeastern Mississippi through the Florida panhandle and the northern part of the Florida peninsula to the extreme southern portion of Georgia in the United States.

HABITAT

Amphiuma pholeter lives primarily in the liquid muck of swampy streams and the swamps of smaller alluvial streams.

BEHAVIOR

Little is known, but one might expect the species to behave similarly to the two larger species. This species is more likely to be an air breather, because of the anaerobic conditions of its mucky habitat.

FEEDING ECOLOGY AND DIET

Presumably, the diet is like that of other amphiumas, but its small size apparently limits its diet to small clams, earthworms, larval aquatic insects, and small beetles.

REPRODUCTIVE BIOLOGY

Courtship apparently takes place in winter or spring, and the eggs are probably laid in June and July, with hatching in late summer or early fall. Because of the anaerobic habitat, the young may emerge from the eggs fully metamorphosed. Hatchlings reared in the laboratory reach adult size in about two years.

Amphiuma means
Amphiuma tridactylum
Amphiuma pholeter

CONSERVATION STATUS

The limited geographic distribution and relative rarity justify a need for protection. Georgia has placed the species under protection. Sediments from runoff in the course of home and road construction tend to destroy the muck habitats. Collecting also may be a detriment to the species.

SIGNIFICANCE TO HUMANS

None known. ◆

Three-toed amphiuma

Amphiuma tridactylum

TAXONOMY

Amphiuma tridactylum Cuvier, 1827, New Orleans, Louisiana, United States.

OTHER COMMON NAMES

None known.

PHYSICAL CHARACTERISTICS

This species has three toes on each foot. The limbs of *A. tridactylum* are longer than those of *A. means* in relation to body length. In *A. tridactylum* the body is 35–37 times the length of the forelimbs, compared with 44–50 times in *A. means*, and the body is 22–25 times longer than the hind limbs compared with 31–34 times in *A. means*. Hatchlings are 1.7–2.5 in (4.3–6.4 cm) in total length, and metamorphosed individuals may be as short as 2.4 in (6.1 cm). Adults may reach 40.5 in (103 cm) in length. The dorsum is dark brown to black, and the venter is substantially lighter, thus creating a bicolor appearance. The throat has a conspicuous dark patch.

DISTRIBUTION

The species inhabits a narrow belt of eastern Texas less than 75 mi (122 km) wide and ranges from southeastern Oklahoma to southeastern Missouri and southwestern Alabama.

HABITAT

The three-toed amphiuma lives in swamps, lakes, ditches, and sluggish streams.

BEHAVIOR

Although *A. tridactylum* generally stays in a restricted area, marked individuals have been known to move as far as 1,300 ft (396 m) from the original point of capture.

FEEDING ECOLOGY AND DIET

Crayfish are the primary items eaten, but the species also feeds on other aquatic organisms.

REPRODUCTIVE BIOLOGY

The male courts the female by rubbing his snout against her. The female then rubs her nose along the male's body and coils her body under his, so that the two cloacae are joined. The male produces a spermatophore, and the female picks up the sperm with her cloaca. Examination of the female spermatheca shows that sperm is present throughout the year, but whether the sperm are fresh or carryovers from previous times is unknown. Egg development in female *Amphiuma tridactylum* occurs from November to September and spermiogenesis in the male from August to May. Females apparently have biennial egg-laying seasons. The eggs number 42–150 in salamanders from Louisiana, but examination of the ovaries indicates a potential for 106–354 eggs; the larger numbers are produced by larger females. Courtship and mating occur January to July and nesting from February to June in southeast Louisiana. Eggs are laid January to September. Hatching takes place from November to December, usually after a five-month incubation. One major study showed that most of the *A. tridactylum* eggs were laid in burrows. On one occasion a naturalist found a female with eggs retained in the body and well-developed larvae. *A. tridactylum* probably reach maturity in three to four years.

CONSERVATION STATUS

The species is in no immediate danger, except in a few states where they are rare and warrant protection. Because *A. tridactylum* and *A. means* may occupy the same habitat, they may lend themselves to an interesting potential study of resource partitioning.

SIGNIFICANCE TO HUMANS

Like other members of the genus, this species is edible but rarely eaten, and it is used for classroom study. Several new species of flatworms (Trematoda) and tapeworms (Cestoda) taken from *Amphiuma tridactylum* have been described. ◆

Resources

Books

Duellman, William E., and Linda Trueb. *Biology of Amphibians.* New York: McGraw-Hill Book Co., 1986.

Dundee, Harold A., and Douglas A. Rossman. *The Amphibians and Reptiles of Louisiana.* Baton Rouge: Louisiana State University Press, 1989.

Petranka, J. W. *Salamanders of the United States and Canada.* Washington, DC: Smithsonian Institution Press, 1998.

Periodicals

Baker, C. L. "The Natural History and Morphology of Amphiumae." *Report of the Reelfoot Lake Biological Station* 9 (1945): 55–91.

Cagle, Fred R. "Observations on a Population of the Salamander *Amphiuma tridactylum* Cuvier." *Ecology* 29, no. 4 (1948): 479–491.

Harold A. Dundee, PhD

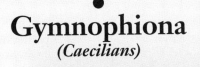

Gymnophiona
(Caecilians)

Class Amphibia

Order Gymnophiona

Number of families 5

Number of genera, species 33 genera, 165 species

Photo: Rio Cauca caecilian (*Typhlonectes natans*) eating cooked fish. Caecilians resemble earthworms but, unlike earthworms, have jaws and teeth. (Photo by henk.wallays@skynet.be. Reproduced by permission.)

Evolution and systematics

The fossil record for caecilians is primarily of a few vertebrae and other small elements; however, exquisite fossils from the Kayenta formation of Arizona include nearly complete skeletons that have small limbs, a moderately long tail, the vertebral column, and complete skulls. The fossils are late Jurassic in age, and have features that clearly identify them as gymnophiones. Vertebrae of approximately the same age have been found in the Sahara of Africa, and in Paleocene, Miocene, and Quaternary sites in South America and Central America, respectively. The Kayenta fossils clearly demonstrate a transition toward the limblessness, body elongation, and tail reduction characteristic of extant taxa. Only the Quaternary vertebra from southern Mexico has been assigned to a family, the Caeciliidae, definitively.

Five families are recognized according to current phylogenetic treatments (e.g., Hedges, et al. 1993). They include the basal Rhinatrematidae of South America, which with the family Ichthyophiidae of southeast Asia constitutes the sister group to all other caecilians, the Indian Uraeotyphlidae, the pantropical and phylogenetically poorly known Caeciliidae, and the east and west African Scolecomorphidae. However, most experts on caecilian biology also recognize a sixth family, the Typhlonectidae, because of its special features and because of the multiple paraphyly of the family Caeciliidae. Resolution of the relationships of genera now included in the Caeciliidae is likely to result in the designation of additional families.

Physical characteristics

Caecilians are limbless, elongate, usually tailless amphibians. They have a somewhat flattened head with a large, usually underslung mouth, anterior nostrils, and an extrusible organ called the tentacle between the nostril and the eye region. The tentacle is a chemosensory and tactile structure, composed largely of "remnant" elements of the eye, but with a fluid-filled channel that runs from the tip to a sensory chamber that opens into the olfactory lobe of the brain. The eyes of caecilians are small and covered by skin; they are also covered by skull bone in some species. The eyes of many species have lost some to all of their musculature, and some have modified or lost the lens, and reduced the retina and the optic nerve. However, most caecilians apparently can distinguish light and dark.

The mouths of caecilians have two rows of teeth on the upper jaw, and one or two on the lower. The tooth crowns have different shapes among species, but all are hinged, and usually recurved backward, apparently to prevent the loss of prey items. The features of the bones, and to a lesser degree, the muscles, and other organ systems provide characters for identifying species, as well as understanding the biology of caecilians. The bodies of caecilians are encircled by grooves, correlated loosely with the number of vertebrae. There is always one groove per vertebra beginning a few vertebrae behind the head; some species have two grooves per vertebra, especially on the posterior, and the most basal species have three grooves per vertebra the length of their bodies. The basal species also have short tails, which also bear grooves.

Most caecilians are a dark gray to gray-brown to deep purple color, often with a lighter head and venter. However, some are more brightly colored; for example, a species on the African island of Sao Tome, *Schistometopum thomense*, is bright yellow. A South American species, *Siphonops annulatus*, is deep blue-purple with bright white body grooves. Several of the basal species are dark gray to gray-brown to brown-black with bright yellow side stripes. Most adult caecilians are between 11.8 and 23.6 in (300 and 600 mm) long; however, some

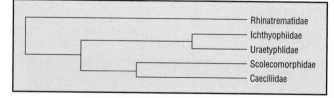
Generalized phylogenetic tree of the living families of caecilians. (Illustration by GGS. Courtesy of Gale.)

species are much smaller or larger. The smallest adult caecilians, the miniaturized *Idiocranium russeli*, are about 4.5 in (115 mm) long. The largest, *Caecilia thompsoni*, are more than 63 in (160 cm) long.

Distribution

In the New World, caecilians occur from mid-Mexico through Peru, Bolivia, Paraguay, and northern Argentina. The aquatic or semi-aquatic "typhlonectids" inhabit primarily the Orinoco and Amazonian drainages, but occur from Colombia to northern Argentina and Uruguay. Old World species are found in East and West Africa south of the Sahara and north of the Temperate Zone, the Seychelles Islands, India, Sri Lanka, Southern China, Cambodia, Laos, Vietnam, and much of Malaysia to the southern Philippines, but are not known from Madagascar, much of central Africa, and the Australian-Papuau Region.

Habitat

Caecilians typically are fossorial, living in moist organic soil, leaf litter, and (rarely) in the axils of plants just above the substrate. The "typhlonectids" are aquatic or semi-aquatic, and nose about in the substrate to find food, or scrape it from logs, rocks, and the like in the water or stream banks.

Behavoir

The skins of caecilians have mucous and serous glands; the latter secrete a substance ("poison") that is toxic to many potential predators.

Feeding ecology and diet

Caecilians are carnivores; terrestrial species prey on animals that they can reach in the substrate, such as earthworms, termites, orthopteran instars, and many other invertebrates. Larger caecilians are known to eat lizards and baby rodents. Snakes and birds are predators that feed on caecilians, and some snakes apparently specialize on them.

Reproductive biology

Members of the two basal families, the Rhinatrematidae and the Ichthyophiidae, as well as the Uraeotyphlidae, one of the two genera of the Scolecomorphidae, and several members of the large family Caeciliidae, are egg layers with free-living larvae that have small gills and tail fins. So far as is known, eggs are laid in burrows or under grass or litter on land, the mother guards the clutch, and the newly hatched larvae wriggle into nearby streams which they inhabit until they metamorphose, when they again become terrestrial. Some caeciliid species have direct development, in which the embryos of the land-laid clutch develop through metamorphosis, so that juveniles hatch and the aquatic larval period is avoided. Derived species of caeciliids and scolecomorphids, and the "Typhlonectidae," are viviparous, retaining the developing eggs in the female's oviducts, where she provides nutrients to the embryos after the yolk of the eggs has been resorbed. Pregnancies are seven to 11 months, depending on the species, and fully metamorphosed juveniles are born. The embryos of species with larvae have three pairs of gills, which are reduced at hatching, though one or more gill slits remain open. Embryos of terrestrial direct-developing and viviparous species also have triramous gills; in the "typhlonectids," the gills are fused during development and form large sac-like structures that may aid in intraoviductal transport of gases and nutrients. The fetuses of viviparous species, and apparently some with direct development or larvae, have a fetal dentition in which the tooth crowns have a distinct morphology and distribution that is quite different from that of the adults. The live-bearers apparently use the dentition to ingest the nutrient material secreted by the epithelium of the oviducts and to stimulate its production. The fetal teeth are shed at or near birth. Several studies have added new information about the developmental biology of various species, following nearly 100 years of little research on caecilian development.

Resources

Books

Duellman, William E., and Linda Trueb. *Biology of Amphibians.* Baltimore: Johns Hopkins University Press, 1994.

Himstedt, W. *Die Blindwuehlen.* Marburg: Die Neue Brehm-Buecherei Bd. 630, Westart Wissenschaften and Heidelberg: Spektrum Akademischer Verlag, 1996.

Taylor, Edward H. *The Caecilians of the World.* Lawrence, KS: University of Kansas Press, 1968.

Wake, Marvalee H. "The Osteology of Caecilians." In *Amphibian Biology.* Vol. 5, *Osteology*, edited by Harold

Heatwole and Margaret Davies. Chipping Norton, Australia: Surrey Beatty and Sons.

Periodicals

Duenker, Nicole, Marvalee H. Wake, and Wendy M. Olson. "Embryonic and Larval Development in the Caecilian Ichthyophis kohtaoensis (Amphibia, Gymnophiona): A Staging Table." *Journal of Morphology* 243 (2000): 3–34.

Hedges, S. Blair, Ronald A. Nussbaum, and Linda R. Maxson. "Caecilian Phylogeny and Biogeography Inferred from Mitochondrial DNA Sequences of the 12S rRNA and 16S

Resources

rRNA Genes (Amphibia: Gymnophiona." *Herpetology Monographs* 7 (1993): 64–76.

Jenkins, Farish A., and Dennis M. Walsh. "An Early Jurassic Caecilian with Limbs." *Nature* 365 (1993): 246–250.

Nussbaum, Ronald A., and Mark Wilkinson. "On the Classification and Phylogeny of Caecilians (Amphibia: Gymnophiona), a Critical Review." *Herpetolgy Monographs* 3 (1989): 1–42.

Savage, J. M., and M. H. Wake. "A Re-evaluation of the Status of Taxa of Central American Caecilians (Amphibia: Gymnophiona), with Comments on Their Origin and Evolution." *Copeia* 2001 (2001): 52–64.

Wake, Marvalee H., and David B. Wake. "Early Developmental Morphology of Vertebrae in Caecilians (Amphibia: Gymnophiona): Resegmentation and Phylogenesis." *Zoology—Analysis of Complex Systems* 103 (2000): 68–88.

Marvalee H. Wake, PhD

American tailed caecilians
(Rhinatrematidae)

Class Amphibia

Order Gymnophiona

Family Rhinatrematidae

Thumbnail description
Characterized by the presence of a tail and a terminal mouth; some species are more or less uniform lead gray in coloration, while others are lead gray with yellowish lateral stripes

Size
Adults range in size from 7.7 to 12.9 in (195–328 mm) in length

Number of genera, species
2 genera; 9 species

Habitat
Leaf litter and burrows in tropical rainforests

Conservation status
Not classified by the IUCN; population data unknown

Distribution
Northern South America

Evolution and systematics

Before 1968 all caecilians were placed in a single family. In 1968 E. H. Taylor described two new families, Ichthyophiidae and Typhlonectidae; the former family included caecilians now placed in the family Rhinatrematidae. Taylor's Ichthyophiidae was characterized by the presence of a tail (all other groups have no tails) and breeding habits that included egg laying and an aquatic, feeding larval stage in the life cycle. Because no information was available about the life history of the majority of species, the features of egg laying and a larval stage in the life cycle were highly speculative. Even today nothing is known about the life history of most species of Taylor's "Ichthyophiidae."

The "Ichthyophiidae," as envisioned by Taylor, included caecilians from Southeast Asia and South America. This posed an interesting biogeographic question, because "ichthyophiids" were not known to occur in the geographic regions between southwestern Asia and South America (Africa, Madagascar, Seychelles). In the 1970s R. A. Nussbaum embarked on a broad-based anatomical investigation of caecilians with the aim of gaining a better understanding of their evolutionary and biogeographical relationships. An important

result of these studies was the discovery that Taylor's "Ichthyophiidae" contained two highly distinct, relatively primitive groups, one (Ichthyophiidae) restricted to Southeast Asia and the other (Rhinatrematidae fam. nov.) restricted to northern South America. Nussbaum published these results in 1977, along with information about the rediscovery of the dual jaw-closing mechanism, which characterizes all caecilians and hence identifies the Gymnophiona.

All vertebrates, except caecilians, have a single pair of muscles used for closing the jaws. These are the paired adductor mandibulae muscles that pull up on the lower jaws via their attachment in front of the jaw joint. Caecilians have this primitive mechanism of jaw closure, but they also have a second, novel component of jaw closure that enlists (in the evolutionary sense) a new muscle to assist in closing the jaws. This new muscle, the interhyoideus posterior, pulls down on a part of the lower jaw (retroarticular process) that projects behind the jaw joint, causing the forward, toothed portion of the lower jaw to close by rotating upward. The action is much like that of a teeter-totter.

The two genera of the Rhinatrematidae have a relatively primitive dual jaw-closing mechanism, in which the retroar-

Two-lined caecilian (*Rhinatrema bivittatum*) from South America. These caecilians are small, up to about 11 in (30 cm), and have short tails. (Photo by Renaud Boistel. Reproduced by permission.)

ticular process is short and straight and the interhyoideus muscle is relatively small and unmodified. The ichthyophiid genera (sensu Nussbaum) have a longer and dorsally curved retroarticular process of the lower jaw and a well-developed interhyoideus muscle, suggesting a more highly developed and evolutionarily advanced mechanism. Other attributes of the rhinatrematids also suggest that they are the basal or most ancestral group of caecilians and that, by comparison, the Southeast Asian ichthyophiids are advanced. For example, rhinatrematids have terminal mouths in which the lower jaw is as long as the upper jaw; in ichthyophiids, the lower jaw is recessed, presumably for better burrowing efficiency. The rhinatrematids also have sensory tentacles in the presumed ancestral position adjacent to the eyes, compared with ichthyophiids, in which the tentacles are positioned somewhere between the eyes and the nostrils. No subfamilies are recognized.

Physical characteristics

Rhinatrematids are medium-sized caecilians with true, albeit short, tails. Radiographic studies show that rhinatrematids have a few vertebrae posterior to the cloacal opening, hence a true tail. Other caecilians have fleshy "terminal shields" that project behind the cloaca, but these shields contain no vertebrae and therefore are not true tails. Rhinatrematids have primary annuli, which are subdivided by secondary and tertiary annuli. All annuli are complete and orthoplicate (straight folds). The mouth is terminal, and the tentacles are adjacent to the eyes. The skulls are distinctly zygokrotaphic (with temporal openings through which the adductor mandibulae muscles bulge). The lower jaws have short, straight retroarticular processes, which attach to a weakly developed interhyoideus muscle that aids in jaw closure. The hyoid apparatus of adults has only two or three elements, which decrease in size posteriorly. The ground color is either lead gray (purplish lead gray in life) or lead gray with yellowish lateral stripes along the body.

Distribution

Northern South America, including parts of Brazil, Colombia, Ecuador, French Guiana, Guyana, Peru, Surinam, and Venezuela.

Habitat

All species are denizens of tropical forests, where they occur in moist habitats associated with leaf litter, rotten logs, and burrows in soil. Larvae occur mainly in streams but occasionally are found in sluggish waters associated with streams.

Behavior

Nothing is known of the behavior of rhinatrematids beyond the facts that they burrow in soil and leaf litter and that they sometimes twist their bodies rapidly when subduing prey organisms that have been grasped in the mouth.

Feeding ecology and diet

As with most caecilians, nothing much is known about the feeding habits of rhinatrematids. Digestive tracts of museum specimens contain large amounts of soil, suggesting that earthworms have been ingested; occasionally undigested earthworms have been noted. Remains of arthropods, mostly insects, also have been found in their guts.

Reproductive biology

There are no studies available concerning the reproductive biology of rhinatrematids. The absence of developing embryos in the oviducts of museum specimens and the presence of large eggs in the ovaries suggest that most species are egg layers. Recently, larvae of *Rhinatrema bivittatum* were discovered in French Guiana. At least one species of *Epicrionops*, *E. marmoratus*, also has indirect development with functional larvae in the life cycle. All records of caecilians that deposit eggs in terrestrial nests indicate that the females engage in maternal care, coiling around the developing eggs. Although this has not been recorded for rhinatrematids, it likely occurs, based on phylogenetic history.

Conservation status

Although not threatened according to the IUCN, caecilians (all species) are rarely observed. For the most part, it is unclear whether this is because they are rare or because they are highly secretive and difficult to find. For this reason, and because so few people have directed their efforts toward caecilian research, the conservation status of most caecilians is completely unknown.

Significance to humans

None known.

Species accounts

Marbled caecilian
Epicrionops marmoratus

TAXONOMY
Epicrionops marmoratus Taylor, 1968, Santo Domingo de los Colorados, Ecuador.

OTHER COMMON NAMES
None known.

PHYSICAL CHARACTERISTICS
Relatively large (up to 11.8 in [300 mm] in length), stocky rhinatrematid with a long tail (up to 0.9 in, or 22 mm, long). The dorsal ground color is dark lavender with scattered yellowish blotches;

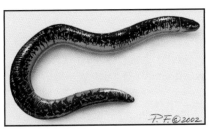
Epicrionops marmoratus

the sides and ventral surfaces are yellow with scattered dark lavender spots.

DISTRIBUTION
Occurs on the Pacific slope of Ecuador.

HABITAT
Inhabits pristine rainforest at middle elevations and also lives along streams in deforested areas.

BEHAVIOR
The behavior of this species, other than burrowing and feeding in captivity, is completely unknown. These caecilians can form their own burrows in moist soil in terraria. They discover earthworms and crickets by scent and lunge forward to grasp them in their jaws. Larger earthworms, capable of struggling when grasped, elicit a twisting response, in which the caecilian rapidly spins on its longitudinal axis. This often results in the earthworm being twisted in half; the grasped part then is swallowed. This behavior, which is characteristic of many caecilians, is reminiscent of the technique used by crocodilians to subdue and rip apart their prey.

FEEDING ECOLOGY AND DIET
Earthworms and small litter and soil arthropods have been found in the digestive tracts of marbled caecilians.

REPRODUCTIVE BIOLOGY
As with most caecilians, the reproductive biology of this species is poorly understood. Larvae of this species have been found in leaf litter and stone rubble on the bottoms of small streams. Along with the absence of embryos in the oviducts of adults, this finding suggests that this is an egg-laying species.

CONSERVATION STATUS
Not threatened.

SIGNIFICANCE TO HUMANS
None known. ◆

Rhinatrema bivittatum

Epicrionops marmoratus

Two-lined caecilian
Rhinatrema bivittatum

TAXONOMY
Caecilia bivittatum Guérin-Méneville, 1829, Guyane [Cayenne], French Guiana.

OTHER COMMON NAMES
None known.

PHYSICAL CHARACTERISTICS
These are relatively small (7.4–8.5 in [188–215 mm] in length), slender rhinatrematids with a transverse cloacal opening and a short tail (less than 0.11 in [2.8 mm] long).

Rhinatrema bivittatum

They are lead gray in coloration with paired yellowish lateral stripes.

DISTRIBUTION
These caecilians occur in French Guiana, Surinam, Guyana, and Brazil.

HABITAT
They inhabit the litter layer and soil in tropical rainforest.

BEHAVIOR
Nothing is known about their behavior.

FEEDING ECOLOGY AND DIET
The stomachs of museum specimens contain remains of earthworms and insects.

REPRODUCTIVE BIOLOGY
Their reproductive habits are largely unknown. Larvae of this species have been discovered, showing that the species has indirect development.

CONSERVATION STATUS
Not threatened. The IUCN and CITES do not classify these caecilians. Although they are rare in collections, reports suggest that the species is not uncommon in French Guiana.

SIGNIFICANCE TO HUMANS
None known. ◆

Resources

Books

Taylor, Edward Harrison. *Caecilians of the World: A Taxonomic Review.* Lawrence: University of Kansas Press, 1968.

Periodicals

Nussbaum, R. A. "The Evolution of a Unique Jaw-closing Mechanism in Caecilians (Amphibia: Gymnophiona) and Its Bearing on Caecilian Ancestry." *Journal of Zoology (London)* 199 (1983): 545–554.

———. "Rhinatrematidae: A New Family of Caecilians (Amphibia: Gymnophiona)." *Occasional Papers of the Museum of Zoology, University of Michigan* no. 683 (1977): 1–30.

Nussbaum, R. A., and M. S. Hoogmoed. "Surinam Caecilians, with Notes on *Rhinatrema bivittatum* and the Description of a New Species of *Microcaecilia* (Amphibia, Gymnophiona)." *Zoologische Mededelingen* 54, no. 14 (1979): 217–235.

Nussbaum, R. A., and M. Wilkinson. "On the Classification and Phylogeny of Caecilians (Amphibia: Gymnophiona): A Critical Review." *Herpetological Monographs* 3 (1989): 1–42.

Ronald A. Nussbaum, PhD

Asian tailed caecilians

(Ichthyophiidae)

Class Amphibia
Order Gymnophiona
Family Ichthyophiidae

Thumbnail description
Relatively primitive, medium-size to large
caecilians that have a true tail and a
subterminal (recessed) mouth; are either
unicolor (lavender gray) or unicolor with paler
lateral stripes

Size
Adults range in size from 6.7 to 21.7 in (170 to
550 mm) in total length

Number of genera, species
2 genera; 39 species

Habitat
The primary habitat is the forest floor (leaf litter
and soil) of tropical rainforests; many species
do well, however, in deforested areas under
cultivation

Conservation status
Endangered: 1 species; Vulnerable: 1 species

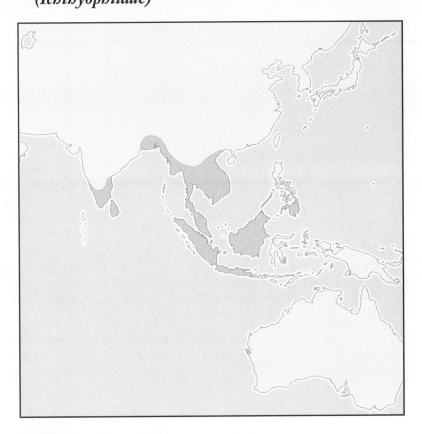

Distribution
India, Sri Lanka, and Southeast Asia

Evolution and systematics

E. H. Taylor established this family in 1968. He believed
ichthyophiids to be the most primitive of caecilians based on
the presence of a tail, numerous subdivided annuli, many dermal scales in the annular grooves, and small but distinct eyes.
He also assumed that all species had the ancestral (primitive)
life history pattern, which includes oviparity (egg laying) and
aquatic larvae that metamorphose into terrestrial adults. The
life history, however, was and still is known for very few
species. Taylor's Ichthyophiidae contained both Southeast
Asian (*Ichthyophis* and *Caudacaecilia*) and South American (*Epicrionops* and *Rhinatrema*) genera. In 1977 R. A. Nussbaum removed the South American genera to their own family
(Rhinatrematidae) and argued that the South American taxa
are relatively more primitive than the Southeast Asian
ichthyophiids based on morphologic characteristics. Subsequent molecular studies have supported this argument. No
subfamilies are recognized.

Physical characteristics

Ichthyophiids have true tails, stegokrotaphic skulls (without temporal openings), subterminal mouths, and a relatively

advanced dual jaw-closing mechanism in which the retroarticular process curves upward and the interhyoideus muscle
is well developed. Primary annuli are subdivided into complete secondary and tertiary annuli. The annuli (up to 420 in
some species) are orthoplicate (straight) posteriorly but an-
gled anteriorly on the ventral surface of the anterior portion
of the body. Numerous dermal scales are found in all the annular grooves. The tentacular openings are positioned in front
of the eyes, usually no more than halfway to the nostrils.
Species are either nearly unicolor (lavender-gray) or unicolor
with yellow-cream lateral stripes.

Distribution

These caecilians occur in India, Sri Lanka, the Philippines,
southern China, Thailand, Myanmar, Laos, Vietnam, and the
Indo-Malayan Archipelago west of Wallace's Line.

Habitat

Asian tailed caecilians are always associated with moist soil
or leaf litter or both in tropical rainforests or disturbed areas
near rainforest.

Asian tailed caecilians (*Ichthyophis kohtaoensis*) are widespread over Southeast Asia and may grow to be nearly 20 in (50 cm). (Photo by henk.wallays@skynet.be. Reproduced by permission.)

Behavior

The behavior of ichthyophiids is poorly studied. All species are burrowers. Maternal guarding of embryos is known for some species. Newly hatched larvae are attracted to light. In terraria adults leave their burrows at night and crawl on the surface. They also have been found on the surface at night during heavy rains in their natural habitats.

Feeding ecology and diet

The feeding habits of ichthyophiids are poorly known. Guts of museum specimens contain large amounts of soil, probably from ingesting earthworms. Partially digested earthworms often are seen, as are parts of insects. In captivity ichthyophiids can be maintained solely on earthworms. They also eat crickets and even strips of meat (beef), fish, and chicken.

Reproductive biology

As with all caecilians, fertilization is internal. Spermatozoa are placed inside the female's cloaca via the male's phallodeum (copulatory organ). Large white eggs strung together by gelatinous strands are deposited in hidden nests, where the female attends them until they hatch. Upon hatching, the larvae leave the nest and wriggle to a stream, where they spend an unknown amount of time feeding on small aquatic organisms until they metamorphose into subadults. After metamorphosis, they leave the streams and take up a terrestrial, burrowing lifestyle.

Conservation status

Ichthyophis glandulosus is Endangered and *I. mindanaoensis* is Vulnerable.

Significance to humans

None known.

1. Ceylon caecilian (*Ichthyophis glutinosus*); 2. Bannan caecilian (*Ichthyophis bannanicus*); 3. Pattipola caecilian (*Ichthyophis orthoplicatus*); 4. Koh Tao Island caecilian (*Ichthyophis kohtaoensis*). (Illustration by Marguette Dongvillo)

Species accounts

Bannan caecilian
Ichthyophis bannanicus

TAXONOMY
Ichthyophis bannanicus Yang, 1984, Mengla County, Xishuang-banna, Yunnan, China.

OTHER COMMON NAMES
None known.

PHYSICAL CHARACTERISTICS
This is a medium-size to large caecilian; adults grow to 7.5–15.2 in (190–386 mm) in length. This striped species has 322–388 annular folds along the body. The tentacle is positioned in front of the eye, about one-third to one-half the distance from the eye to the nostril. It is difficult to distinguish from several other striped *Ichthyophis* species from Southeast Asia.

DISTRIBUTION
This species is known only from the region of the type locality near Mengla, Yunnan, China.

HABITAT
Larvae have been collected in small streams and pools in deforested areas near rice paddies. Adults have been found under logs and in mud adjacent to pools and streams after rain.

BEHAVIOR
The behavior of this species is largely unknown. Adults burrow in moist soil in terraria. Larvae can swim, but normally they burrow and crawl in litter and silt on the bottoms of pools and streams.

FEEDING ECOLOGY AND DIET
Museum specimens (adults) have earthworms in their stomachs. They readily eat earthworms and crickets in captivity.

REPRODUCTIVE BIOLOGY
Courtship, mating, and nests are unreported. The life cycle includes a larval stage.

CONSERVATION STATUS
Not threatened.

SIGNIFICANCE TO HUMANS
None known. ◆

Ceylon caecilian
Ichthyophis glutinosus

TAXONOMY
Caecilia glutinosa Linnaeus, 1758, habitat in Indies.

OTHER COMMON NAMES
None known.

PHYSICAL CHARACTERISTICS
This is a medium-size to large caecilian that attains a length of 9.1–16.1 in (230–410 mm). It is a striped species; adults have 342–392 annuli along the body. The tentacular aperture is distinctly closer to the eye than to the nostril and close to the margin of the mouth. The species is similar to *Ichthyophis bannanicus* and a few other striped *Ichthyophis* species from Southeast Asia.

DISTRIBUTION
The species occurs in Sri Lanka.

HABITAT
Most specimens have been taken from deforested agricultural areas. They have been found in piles of rotting vegetation and

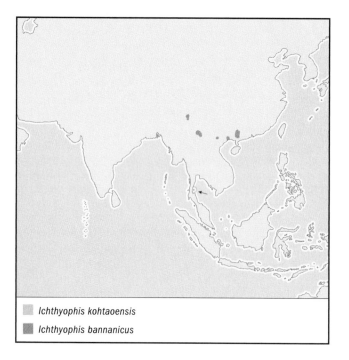

Ichthyophis kohtaoensis
Ichthyophis bannanicus

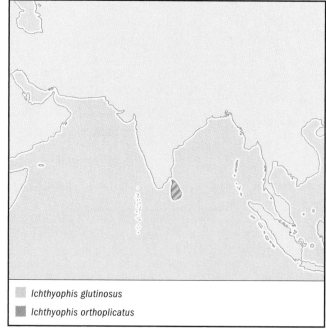

Ichthyophis glutinosus
Ichthyophis orthoplicatus

manure and in loose, wet soil. One individual was dug up from the soil of a moist meadow.

BEHAVIOR
The behavior of this species is poorly known. When prey (earthworms) are grasped on the surface, the caecilian retreats backward into the burrow while vigorously twisting its head and neck to subdue the prey. Sometimes the caecilian spins on its longitudinal axis, which may break the prey into small, more manageable pieces.

FEEDING ECOLOGY AND DIET
Subadults and adults eat mainly earthworms but also other small litter and soil invertebrates. The food of larvae is unknown, but they eat small bloodworms and earthworms in captivity.

REPRODUCTIVE BIOLOGY
Details of courtship and mating have not been reported, but the species is oviparous. Females deposit 25–38, large white eggs in jelly strings. The eggs are placed in hidden nests (cavities in soil), and the female coils around them until they hatch. Larvae range in size from about 2.8 to 4.5 in (70 to 115 mm) in total length. The larvae metamorphose (in captivity) after about 280 days.

CONSERVATION STATUS
Not threatened.

SIGNIFICANCE TO HUMANS
None known. ◆

Koh Tao Island caecilian
Ichthyophis kohtaoensis

TAXONOMY
Ichthyophis kohtaoensis Taylor, 1960, Koh Tao Island (west side), Gulf of Siam.

OTHER COMMON NAMES
German: Koh-Tao-Blindwuehle.

PHYSICAL CHARACTERISTICS
The Koh Tao Island caecilian is a medium-size to large caecilian; it attains a length of 7.6–13.8 in (192–350 mm). This striped *Ichthyophis* species has 362–366 annular folds along the body. This caecilian is similar to *I. bannanicus* and *I. glutinosus*, but it has a smaller head with a more rounded snout.

DISTRIBUTION
The species occurs on Koh Tao Island in the Gulf of Siam and mainland peninsular Thailand.

HABITAT
It inhabits the forest floor of tropical rainforest and deforested agricultural sites.

BEHAVIOR
These caecilians are burrowers. Sometimes they are seen on the surface at night.

FEEDING ECOLOGY AND DIET
The diet is largely unstudied, but this species eats earthworms and other small invertebrates.

REPRODUCTIVE BIOLOGY
Females deposit nine to 47 large white eggs in jelly strings in hidden nests, usually cavities in the soil. The female remains coiled around the eggs until they hatch after about 70–80 days. The hatchlings grow from about 2.8 to 5.9 in (70 to 150 mm) in length in 10–14 months, at which time they metamorphose.

CONSERVATION STATUS
Not threatened.

SIGNIFICANCE TO HUMANS
None known. ◆

Pattipola caecilian
Ichthyophis orthoplicatus

TAXONOMY
Ichthyophis orthoplicatus Taylor, 1965, Pattipola, Central Province, Ceylon.

OTHER COMMON NAMES
None known.

PHYSICAL CHARACTERISTICS
The Pattipola caecilian is a medium-size to large species that grows to 8.3–12.1 in (210–307 mm) in length. This unstriped caecilian is a uniform lavender-gray with 282–335 annular folds along the body. The tentacular opening is distinctly closer to the eye than to the nostril.

DISTRIBUTION
This species occurs in Sri Lanka.

HABITAT
Adults have been found in piles of rotting vegetation and manure and in soil along streams.

BEHAVIOR
Their behavior is largely unknown, although they are burrowers.

FEEDING ECOLOGY AND DIET
The stomachs of these caecilians contain soil and remains of earthworms and insects.

REPRODUCTIVE BIOLOGY
The details of their reproductive biology are unknown, but it is presumed that they are similar to the habits of other species of *Ichthyophis*.

CONSERVATION STATUS
Not threatened.

SIGNIFICANCE TO HUMANS
None known. ◆

Resources

Books

Himstedt, W. *Die Blindwühlen.* Magdeburg, Germany: Wolf Graf von Westarp, 1996.

Taylor, Edward Harrison. *Caecilians of the World: A Taxonomic Review.* Lawrence: University of Kansas Press, 1968.

Periodicals

Breckenridge, W. R., and S. Jayasinghe. "Observations on the Eggs and Larvae of *Ichthyophis glutinosus.*" *Ceylon Journal of Science (Biological Science)* 13, nos. 1 and 2 (1979): 187–202.

———, S. Nathanael, and L. Pereira. "Some Aspects of the Biology and Development of *Ichthyophis glutinosus* (Amphibia: Gymnophiona)." *Journal of Zoology (London)* 211 (1987): 437–450.

Gans, C., and R. A. Nussbaum. "On the *Ichthyophis* (Amphibia: Gymnophiona) of Sri Lanka." *Spolia Zeylanica* 35, parts I and II (1980): 137–154.

Nussbaum, R. A. "The Evolution of a Unique Jaw-closing Mechanism in Caecilians (Amphibia: Gymnophiona) and Its Bearing on Caecilian Ancestry." *Journal of Zoology (London)* 199 (1983): 545–554.

———. "Rhinatrematidae: A New Family of Caecilians (Amphibia: Gymnophiona)." *Occasional Papers of the Museum of Zoology, University of Michigan* no. 683 (1977): 1–30.

Nussbaum, R. A., and M. Wilkinson. "On the Classification and Phylogeny of Caecilians (Amphibia: Gymnophiona): A Critical Review." *Herpetological Monographs* 3 (1989): 1–42.

Ronald A. Nussbaum, PhD

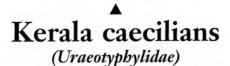

Kerala caecilians

(*Uraeotyphlidae*)

Class Amphibia
Order Gymnophiona
Family Uraeotyphylidae

Thumbnail description
Small to medium-sized caecilians with strongly
subterminal mouths; short tails; small, distinct
eyes; tentacular openings far forward, below the
nostrils; primary annuli mostly or entirely
subdivided by secondary annuli; and numerous
dermal scales; either nearly uniformly dark lead-
gray in coloration or bicolor, with undersurfaces
that are whitish to yellowish cream

Size
Subadults and adults range in total length from
5.7 to 11.9 in (145 to 303 mm)

Number of genera, species
1 genus; 5 species

Habitat
Moist soil and litter in rainforests and
deforested areas

Conservation status
Not threatened

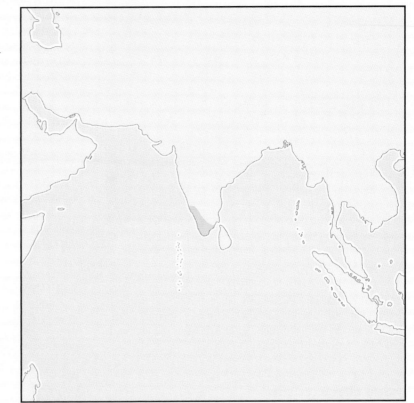

Distribution
Western Ghats, peninsular India

Evolution and systematics

Until 1968 all caecilians were placed in a single family. In
that year, E. H. Taylor established two new families, the
Ichthyophiidae and Typhlonectidae, leaving the majority of
genera and species in the original family, the Caeciliidae. Taylor retained the genus *Uraeotyphlus* and its four species in the
Caeciliidae but thought that, based on morphologic similarities, it might belong to the Ichthyophiidae. He was reluctant
to place *Uraeotyphlus* in the Ichthyophiidae, because the life
histories of all species of the genus were unknown. In 1979
R. A. Nussbaum transferred *Uraeotyphlus* to the Ichthyophiidae and placed it in its own subfamily, Uraeotyphlinae. W.
E. Duellman and L. Trueb raised Uraeotyphlinae to family
rank in 1986, and in 1996 M. Wilkinson and R. A. Nussbaum
provided morphologic evidence that Uraeotyphlidae and
Ichthyophiidae are sister groups. No subfamilies are recognized.

In morphologic features uraeotyphlids appear to be intermediate between the relatively more primitive rhinatrematid
and ichthyophiid caecilians and the more advanced caeciliids.
Relatively primitive characteristics of uraeotyphlids include
numerous skull bones, weakly fused skull bones, eyes not cov-

ered by bone, the presence of a tail, and numerous scales. Advanced features include a strongly recessed mouth, far forward and subnarial (below the nostrils) position of the
tentacular apertures, imperforate stapes (stapes bone not perforated), and lack of tertiary subdivision of the primary annuli. Based on molecular genetic data, uraeotyphlids are
phylogenetically intermediate between the more ancestral
ichthyophiids/rhinatrematids and the more derived caeciliids.

The skull of uraeotyphlids, with its projecting snout and
strongly recessed mouth, suggests that they are better burrowers than rhinatrematids and ichthyophiids are, but they
are not nearly so well adapted for burrowing as caeciliids. The
latter have fewer and more solidly fused skull bones, which
presumably increase burrowing efficiency compared with the
uraeotyphlids.

Studies of the evolutionary relationships of uraeotyphlids
and ichthyophiids based on molecular genetic data show that,
most likely, they originated and diversified in isolation on the
Indian continent as it drifted northward toward Asia during
the Cenozoic. After contacting Asia, ichthyophiids spread out
from India across Southeast Asia and the Indo-Malayan Arch-

Uraeotyphlus narayani live in the southern tip of India and grow to 12 in (30 cm). (Photo by John Measey. Reproduced by permission.)

ipelago. The alternative scenario, that India was without ichthyophiids and uraeotyphlids when it was isolated in the Indian Ocean and received them from Asia after contact with Asia, is not supported by the genetic data.

Physical characteristics

These are small to medium-sized caecilians with short, true tails; weakly stegokrotaphic skulls (solid-skull roof), with the roofing bones nearly covering the underlying adductor mandibulae (jaw-closing) muscles; stapes not perforated by the stapedial artery; and a recessed (subterminal) mouth. The tentacular apertures are far forward of the eye, below the nostrils. Most primary annuli are subdivided by secondary annuli; the most anterior few primaries may not be subdivided. There are no tertiary annuli, and annular grooves normally do not completely encircle the body. Numerous scales are present in the annular grooves.

Distribution

These caecilians inhabit the Western Ghats in Kerala State, peninsular India.

Habitat

Their habitat is rainforest and disturbed, deforested areas within the rainforest belt. They usually are found in moist soil near streams, marshes, or other bodies of water.

Behavior

Other than their burrowing locomotion, nothing is known about their behavior.

Feeding ecology and diet

The guts of adults contain soil, earthworms, and fragments of insects. Larvae also have insect remains and mineral soil in their digestive tracts.

Reproductive biology

Almost nothing is known about the reproductive habits of uraeotyphlids. It had been assumed that they are egg layers with direct development (lacking a larval stage), because of the presence of large ovarian eggs full of yolk in some specimens, the lack of oviductal embryos in museum specimens, and the lack of reported larvae. At least one species, *Uraeotyphlus oxyurus*, however, has a larval stage, which suggests that the other species also might have one. The presence of larvae is considered to be an ancestral character state, which helps us understand the evolutionary position of uraeotyphlids.

Conservation status

Not threatened.

Significance to humans

None known.

1. Kannan caecilian (*Uraeotyphlus narayani*); 2. Red caecilian (*Uraeotyphlus oxyurus*). (Illustration by John Megahan)

Species accounts

Kannan caecilian
Uraeotyphlus narayani

TAXONOMY
Uraeotyphlus narayani Seshachar, 1939, Kannan, Travancore, India.

OTHER COMMON NAMES
None known.

PHYSICAL CHARACTERISTICS
Adults range in size from 7.8 to 9.3 in (199 to 237 mm) in total length. These bicolor caecilians have a dark blue-gray dorsum and a pale, flesh-colored venter. There are 169–178 primary annuli and 77–83 secondary annuli; only the most anterior primaries are not subdivided by secondary annuli.

DISTRIBUTION
These caecilians are found in Kannan and Kottayam, Kerala State, India.

HABITAT
Details of the habitat are unknown, but the Kannan caecilian lives within the rainforest belt, presumably in moist soil and forest floor litter. Like many caecilians, this species probably occurs in agricultural areas carved out of rainforest.

BEHAVIOR
Their behavior has not been described.

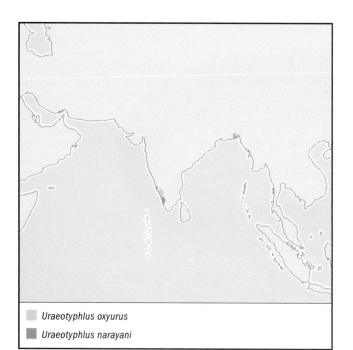

Uraeotyphlus oxyurus

Uraeotyphlus narayani

FEEDING ECOLOGY AND DIET
The diet is unknown, but presumably it consists of earthworms and small litter or soil invertebrates.

REPRODUCTIVE BIOLOGY
The reproductive habits of these caecilians are not known.

CONSERVATION STATUS
Not threatened.

SIGNIFICANCE TO HUMANS
None known. ◆

Red caecilian
Uraeotyphlus oxyurus

TAXONOMY
Caecilia oxyura Duméril and Bibron, 1841, Côte du Malabar.

OTHER COMMON NAMES
None known.

PHYSICAL CHARACTERISTICS
This relatively stout-bodied species ranges in size from 7.3 to 11.9 in (185 to 303 mm) in total length. It is nearly uniformly dark bluish gray and slightly paler below. The common name is a misnomer, because there is no red coloration. There are 98–107 primary annuli and 89–104 secondary annuli. Nearly all of the primary annuli are subdivided by secondary annuli, except occasionally the most anterior one or two primaries. Numerous scales are present in the annular folds.

DISTRIBUTION
The species ranges across Taliparamba, Wynaad, Tinnivelly, Allur near Trichur, and Anamallai Hills in Kerala, India.

HABITAT
These caecilians inhabit the rainforest belt and are found in moist soil and forest floor litter in and adjacent to forest.

BEHAVIOR
Their behavior is unknown, but presumably they are burrowers.

FEEDING ECOLOGY AND DIET
The diet is not known, but earthworm and insect remains have been found in the guts of museum specimens. Larvae also contain insect remains.

REPRODUCTIVE BIOLOGY
The reproductive habits of these caecilians are largely unknown, but the species has functional (feeding) larvae that metamorphose at about 3.5 in (90 mm) in total length.

CONSERVATION STATUS
Not threatened.

SIGNIFICANCE TO HUMANS
None known. ◆

Resources

Books

Taylor, Edward Harrison. *Caecilians of the World: A Taxonomic Review.* Lawrence: University of Kansas Press, 1968.

Periodicals

Gower, D. J., A. Kupfer, O. V. Oommen, et al. "A Molecular Phylogeny of Ichthyophiid Caecilians (Amphibia: Gymnophiona: Ichthyophiidae): Out of India or Out of Southeast Asia?" *Proceedings of the Royal Society of London B.* (in press).

Nussbaum, R. A. "The Evolution of a Unique Jaw-closing Mechanism in Caecilians (Amphibia: Gymnophiona) and Its Bearing on Caecilian Ancestry." *Journal of Zoology, London* 199 (1983): 545–554.

———. "The Taxonomic Status of the Caecilian Genus *Uraeotyphlus* Peters." *Occasional Papers of the Museum of Zoology, University of Michigan* no. 687 (1979): 1–20.

Nussbaum, R. A., and M. Wilkinson. "On the Classification and Phylogeny of Caecilians (Amphibia: Gymnophiona): A Critical Review." *Herpetological Monographs* 3 (1989): 1–42.

Wilkinson, M. "On the Life History of the Caecilian Genus *Uraeotyphlus* (Amphibia: Gymnophiona)." *Herpetological Journal* 2 (1992): 121–124.

Wilkinson, M., and R. A. Nussbaum. "On the Phylogenetic Position of the Uraeotyphlidae (Amphibia: Gymnophiona)." *Copeia* (1996): 550–562.

Ronald A. Nussbaum, PhD

Buried-eyed caecilians
(Scolecomorphidae)

Class Amphibia
Order Gymnophiona
Family Scolecomorphidae

Thumbnail description
Small to medium-sized caecilians with a recessed mouth, tentacular apertures far forward on the snout, and eyes attached to the base of tentacles, which move with the tentacles; usually dark lavender-gray above and cream to flesh-colored below

Size
Adult scolecomorphids range in size from 6.3 to 18.2 in (160–463 mm) in total length

Number of genera, species
2 genera; 6 species

Habitat
Tropical rainforests and deforested areas, generally in mountainous regions

Conservation status
Not classified by the IUCN; population data unknown

Distribution
Western (Cameroon) and eastern (Malawi and Tanzania) equatorial Africa

Evolution and systematics

In 1968 E. H. Taylor showed that caecilians are far more diverse than their rather uniform external morphologic features had suggested. At the time, caecilians were placed in a single family, the Caeciliidae. Taylor described two new families but left the majority of genera in the Caeciliidae, noting that various genera eventually might have to be removed from that family. In 1969 Taylor removed the genus *Scolecomorphus* from the Caeciliidae and placed it in a new family, the Scolecomorphidae. Species of this new family have several unusual and highly derived characteristics that set them broadly apart from all other caecilians, including a mobile eye and lack of a stapes (sound-conducting bone) in the middle ear. No subfamilies are recognized.

At the time Taylor described the Scolecomorphidae, he recognized six species in a single genus. In 1981 R. A. Nussbaum described a seventh species of *Scolecomorphus*, and in 1985 he partitioned the latter genus into two genera, *Scolecomorphus* with three species (two were lost to synonymy) and a new genus, *Crotaphatrema*, with two species. Subsequently (2000), D. P. Lawson described a third species of *Crotaphatrema*.

Scolecomorphids, along with another caecilian family (Caeciliidae), are evolutionarily derived compared to the "tailed caecilians" (Rhinatrematidae, Ichthyophiidae, and Uraeotyphlidae). Species of the former two families are advanced in several ways. They lack tails and have fewer and more solidly fused skull bones. The skin segmentation is reduced; only primary and secondary annuli (rings or folds) occur and often only primaries. There are also fewer scales and usually no larval stage in the life cycle. Scolecomorphids are perhaps the most specialized of the two advanced families. They differ from caeciliids in several uniquely derived features, including lack of a stapes in the middle ear and presence of an eye that is attached to the base of the tentacle, which can be moved outside the skull when the tentacle is protruded.

Physical characteristics

Scolecomorphids lack stapes and internal processes (bony projections) on the pseudoangular bones of the lower jaw. There are no secondary annuli; the number of primary annuli ranges from 120 to 153 and the number of vertebrae from 131

Tropical caecilians (*Scolecomorphidae*) live in tropical, sub-Saharan Africa. The eye is covered with bone and there is no tail. (Photo by Daniel Boone. Reproduced by permission.)

to 165. There are typically no dermal scales in the annular folds; rarely, a few tiny vestigial scales may be present in the posterior folds. Annuli are usually complete anteriorly but may be fused dorsally, ventrally, or both, especially along the middle and posterior parts of the body. The mouth is subterminal. The vestigial eyes are attached to and move with the tentacles and may be exposed when the tentacles are extruded. Normally, the eyes are under bone, as there are no orbits. The tentacular apertures are positioned ventrolaterally on the snout, below the nostrils and even with, or slightly anterior to, the anterior margin of the mouth. Each tentacle has an external subglobular base that is surrounded partly or entirely with a groove and a central opening through which the body of the tentacle passes when it is extruded or retracted. There is no tail; instead, there is a terminal shield without annuli. The terminus is bluntly rounded and flattened ventrally. The longitudinal vent lies in a shallow, oval depression only a few millimeters longer than the vent. The tongue lacks narial plugs. Temporal openings may be present (*Scolecomorphus*) or absent (*Crotaphatrema*). Some species have calcified spines on the phallodeum (penis), but this has not been seen in all species.

R. A. Nussbaum noted that some scolecomorphids have an interesting pattern of sexual dimorphism in which the females are the larger (longer) sex. This occurs because females have considerably more vertebrae (and primary annuli) than males; it may be advantageous, because the elongation of the body provides more space for developing fetuses. Males have larger heads than females of comparable size. R. A. Nussbaum and M. Pfrender found this to be true of several species of caecilians in different genera and families. It may be related to male combat in competition for mates or territory, as suggested by bite marks found on caecilians in captivity and in nature.

Distribution

Scolecomorphids are restricted to eastern and western equatorial Africa. *Scolecomorphus* occurs in Tanzania and

Malawi, whereas *Crotaphatrema* is restricted to Cameroon. No caecilians of any kind have been found in central equatorial Africa. This distribution pattern is anomalous, because the vast region of the upper Congo seems ideally suited for caecilians. Caecilians likely occur in this region, and scolecomorphids are among the most likely candidates to be found there.

Habitat

Like most caecilians, scolecomorphids inhabit tropical rainforests and adjoining deforested areas. They usually are found in moist areas under logs and in leaf litter on the forest floor. They also can be dug up from moist soil. Most specimens were seen in hilly or mountainous regions. In Tanzania and Malawi, they have been found in turned soil and piles of vegetation in farming regions.

Behavior

These caecilians are seen rarely, and little is known about their behavior. They are excellent burrowers, and they pump their tentacles in and out when they are moving and otherwise investigating their environment. As with all caecilians, their tentacles are thought to be chemosensory organs used for "tasting" their immediate surroundings.

Feeding ecology and diet

Little is known about the feeding habits of scolecomorphids. Soil, earthworms, and insects have been found in their digestive tracts. They readily eat earthworms and small crickets in captivity.

Reproductive biology

The three species of *Scolecomorphus* are viviparous. The young are retained in the oviducts, were they are thought to feed on "uterine milk," a nutritious substance secreted by their mother's oviducts and ingested with the aid of specialized embryonic or fetal teeth. These teeth are comblike, with multiple crowns, and they also may be used to stimulate the oviduct to secrete "milk" near the mouth of the feeding fetus. This remains to be established. The reproductive biology of the three species of *Crotaphatrema* is unknown. Because they have large, yolky, ovarian eggs, it seems likely that they are egg layers with direct development (lacking a larval stage) and female parental care.

Conservation status

Not classified by IUCN—population data are unknown.

Significance to humans

None known.

Species accounts

Kirk's caecilian
Scolecomorphus kirkii

TAXONOMY
Scolecomorphus kirkii Boulenger, 1883, East Africa, probably from the vicinity of Lake Tanganyika.

OTHER COMMON NAMES
None known.

PHYSICAL CHARACTERISTICS
This is the largest species of *Scolecomorphus;* adults attain a length of 8.5–18.2 in (215–463 mm). There are 130–152 primary annuli. The dorsal lavender-gray coloration extends ventrally past the midlateral line, encroaching on the sides of the venter; the midventral sur-

Scolecomorphus kirkii

faces are flesh- to cream-colored. The top and sides of the head are dark, like the rest of the dorsal body, but a light area is visible along the tract of the tentacle; the black retina of the eye at the base of the tentacle can be seen through the skin and skull bones.

DISTRIBUTION
This species occurs in eastern equatorial Africa in Malawi and Tanzania.

HABITAT
The species inhabits tropical rainforest and agricultural areas, generally in mountainous regions. It is found under and in surface litter and in the soil.

BEHAVIOR
The behavior of this species is not well known. They are efficient burrowers. The protrusion of the tentacles while investigating the environment has been observed and filmed. The latter studies proved that scolecomorphids can project their eyes outside their skull bones. Previously, this had been surmised from anatomical studies of museum specimens.

FEEDING ECOLOGY AND DIET
Mineral soil and remains of arthropods have been found in the guts of museum specimens.

REPRODUCTIVE BIOLOGY
Courtship and copulation are not reported. The species is viviparous.

CONSERVATION STATUS
Not threatened.

SIGNIFICANCE TO HUMANS
None known. ◆

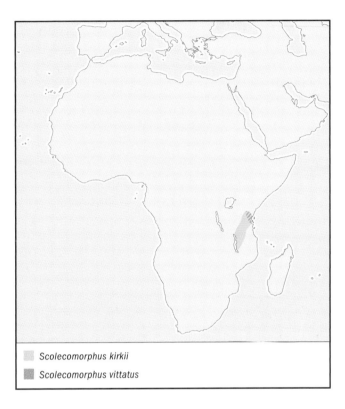

☐ *Scolecomorphus kirkii*
▨ *Scolecomorphus vittatus*

Banded caecilian
Scolecomorphus vittatus

TAXONOMY
Bdellophis vittatus Boulenger, 1895, Usambara, [Tanga Division] German East Africa (Tanzania).

OTHER COMMON NAMES
None known.

PHYSICAL CHARACTERISTICS
This is the smallest species of *Scolecomorphus;* adults grow to 5.6–14.8 in (141–376 mm) in total length. There are 120–148 primary annuli. A dorsal lavender-gray band extends ventrally only to the midlateral line and often is confined above the midlateral line; the sides and

Scolecomorphus vittatus

venter are a yellowish cream to flesh color.

DISTRIBUTION
The species occurs in eastern equatorial Africa in Tanzania (Usambara, Uluguru, and northern Pare Mountains).

HABITAT

The habitat is tropical rainforests and cleared agricultural areas in mountainous regions. They are found under and in litter on moist soil and in soil.

BEHAVIOR

The behavior of this species has not been studied, but they clearly are burrowers.

FEEDING ECOLOGY AND DIET

The diet has not been studied in detail, but soil (indicating the ingestion of earthworms) and the remains of arthropods (external skeletons) have been observed in the guts of museum specimens.

REPRODUCTIVE BIOLOGY

The species is viviparous.

CONSERVATION STATUS

Not threatened.

SIGNIFICANCE TO HUMANS

None known. ◆

Resources

Books

Taylor, Edward Harrison. *Caecilians of the World: A Taxonomic Review.* Lawrence: University of Kansas Press, 1968.

Periodicals

Lawson, D. P. "A New Caecilian from Cameroon, Africa (Amphibia: Gymnophiona: Scolecomorphidae)." *Herpetologica* 56 (2000): 77–80.

O'Reilly, J. C., R. A. Nussbaum, and D. Boone. "Vertebrate with Protrusible Eye." *Nature* 382 (1996): 33.

Nussbaum, R. A. "The Evolution of a Unique Jaw-closing Mechanism in Caecilians (Amphibia: Gymnophiona) and Its Bearing on Caecilian Ancestry." *Journal of Zoology London* 199 (1983): 545–554.

———. "*Scolecomorphus lamottei*, a New Caecilian from West Africa (Amphibia: Gymnophiona: Scolecomorphidae)." *Copeia* (1981): 265–269.

———. "Systematics of Caecilians (Amphibia: Gymnophiona) of the Family Scolecomorphidae." *Occasional Papers of the Museum of Zoology, University of Michigan,* no. 713 (1985): 1–49.

Nussbaum, R. A., and M. Wilkinson. "On the Classification and Phylogeny of Caecilians (Amphibia: Gymnophiona), a Critical Review." *Herpetological Monographs* 3 (1989): 1–42.

Taylor, E. H. "A New Family of African Gymnophiona." *University of Kansas Science Bulletin* 48, no. 10 (1969): 297–305.

Ronald A. Nussbaum, PhD

Tailless caecilians
(Caeciliidae)

Class Amphibia

Order Gymnophiona

Family Caeciliidae

Thumbnail description
Miniaturized to very large caecilians, mostly lacking a tail, with relatively few body rings

Size
4–63 in (10–160 cm)

Number of genera, species
26 genera, 107 species

Habitat
Tropical forests, grasslands, streambanks

Conservation status
Not classified by the IUCN, but several species are declining and some are thought to be extinct

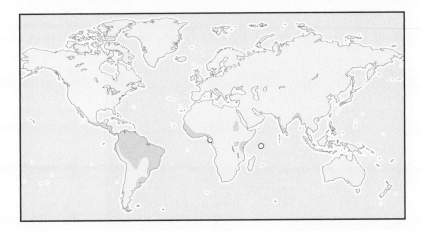

Distribution
Virtually pantropical; Central and South America, East and West Africa, the Seychelles Islands, India and Sri Lanka, Southeast Asia from the East Indies to southern Philippines, southern China through the Malay Peninsula

Evolution and systematics

The phylogenetic relationships of members of the family Caeciliidae are not well known. Various phylogenetic hypotheses have appeared in the literature, some based on morphological data and some on biochemical or molecular data. Because of either methodological problems or the sampling of very few taxa (especially with biochemical or molecular data), most recently advanced hypotheses have been rejected by experts on caecilian biology. These experts are working toward assembling more complete data sets in order to fully analyze the relationships of the members of the family. It is clear, though, that the three genera found on the Seychelles Islands form a distinct lineage. Their relationship to other caecilians, and of the African, Indian, and American caeciliids to each other, are not yet resolved.

Several classification issues exist for the family Caeciliidae. Both morphological and molecular data suggest that the family is multiply paraphyletic; that is, that some groups have members that occur not as monophyletic clades, but within other clades. A particular problem is that of the semi-aquatic to aquatic caecilians of northern, central, and western South America, long recognized as a separate family, the Typhlonectidae. The features that distinguish them are particular to that group, so they do not provide information about relationships. They share some morphological and molecular characters with certain genera placed in the family Caeciliidae. Therefore, taxonomic purists refer to the typhlonectids as a subfamily of the caeciliids, rather than a separate family. This account follows that purism, with some hesitation, because most experts on caecilians reject that conclusion. They continue to recognize the family Typhlonectidae, because they are convinced that adequate data will reveal that the current family Caeciliidae is composed of several lineages that

are likely to be designated new families. In the absence of those data, this account includes the aquatic lineage as a subfamily, the Typhlonectinae, of the Caeciliidae. The subfamily name Caeciliinae has been proposed for some of the remaining taxa; it is not appropriate to all the remaining genera, thus posing a taxonomic inequity of names. Other subfamilies have been proposed but rejected on methodological grounds. The data necessary to resolve these issues are steadily accruing, so in the near future, we should have a much better understanding of the phylogenetic relationships within the Caeciliidae and of the family with other caecilians.

Physical characteristics

Caeciliids have the features typical of most caecilians—elongate limbless bodies, very short to no true tail, the head flattened with the mouth underslung. The eyes are covered by skin, or in some species, by skin and by bone. The chemosensory/tactile tentacle lies between the eye and the nostril. Perhaps the most striking feature of caecilians is the series of rings around the body. All caeciliids have such rings, but many have a primary series that runs the length of the body, and a secondary series that can be as long as the primary series, to only a few rings posteriorly, to complete loss of the secondary series.

Distribution

This family is virtually pantropical in its distribution. Its range includes Central and South America; East and West Africa; the Seychelles Islands; India and Sri Lanka; Southeast Asia from the East Indies to southern Philippines; and southern China through the Malay Peninsula.

Gaboon caecilian (*Geotrypetes seraphini*) from Cameroon. (Photo by henk.wallays@skynet.be. Reproduced by permission.)

Habitat

Terrestrial caeciliids live in habitats characterized by loose, moist, organically rich soil and leaf litter. They are often found under stones, logs, and debris, such as piles of coffee hulls. Many species are found near streams. Some species occur in savanna areas and are found by rolling the grass layer from the soil. The semiaquatic to aquatic typhlonectives live in the banks of streams and rivers and variously venture onto nearby land or out into the bodies of slow-moving waters. They hide under hanging branches, logs, and other floating materials.

Tailless caecilian (*Gymnopis multiplicata*) in the rainforest of Costa Rica. (Photo by M.P.L. Fogden. Bruce Coleman Inc. Reproduced by permission.)

Behavior

Little is known about the behavior of terrestrial caeciliids because of their secretive, soil-dwelling nature. Members of several species are known to emerge from deeper in the soil or leaf litter to forage at dusk or dawn, often during light rain. They appear to be capable burrowers, digging head-first through moist organic soil. Species may differ in their ability to burrow efficiently in different kinds of soils. They appear to spend most of their time in their burrows, but they are capable of considerable movement as well. There also is limited information on the behavior of the semiaquatic to aquatic typhlonectines, except for some laboratory observations, because they typically live in rather slow-moving streams and rivers that have a lot of organic material in the water, thus making observation difficult in the field.

Feeding ecology and diet

Caeciliids are "sit-and-wait" predators, staying in their burrows or on the substrate surface where they seize prey items that wander near them. They are carnivores, eating earthworms, termites, a diversity of other small invertebrates, and even small lizards and rodents. They lunge at their prey, grabbing it with their strong jaws and powerful jaw muscles. They propel prey items into their mouths and progressively swallow them. Several species have been observed to retreat backward into their burrows, turning rapidly on their body axis in a corkscrew motion, so that the prey item may be sheared to a bite-sized morsel.

Reproductive biology

Caeciliids include several modes of reproduction: egg-laying with free-living larvae; egg-laying with direct development (the eggs laid on land, and development through metamorphosis occurs before hatching, so that there is no free-living larval stage); and viviparity (retention of the developing embryos in the oviducts of the mother, provision of nutrient material in the oviducts, and birth of fully metamorphosed juveniles). All caeciliids apparently have internal fertilization (as it is assumed is characteristic of all caecilians); the male inserts a cloacal intromittent organ into the vent of the female so that sperm is transported directly to her reproductive tract. Courtship and mating are not reported in the literature, save for that of typhlonectines. They have been observed in aquaria to coil around each other before the male inserts his intromittent organ. It is not known how males and females recognize each other, though pheromones have been suggested for *Typhlonectes*. Some species of caeciliids are known to provide care of either their laid eggs (*Idiocranium russeli* females curl their bodies around the clutch) or their altricial young (*Siphonops annulatus* and *Geotrypetes seraphini*). Idiocranium, *Boulengerula*, *Hypogeophis*, some *Grandisonia*, and perhaps some *Caecilia* and *Oscaecilia* have direct development, so that juveniles hatch; *Siphonops annulatus* also may have direct development, but its newly hatched young apparently feed on skin secretions of the mother, as do the newborns of *Geotrypetes seraphini*. Viviparity has probably evolved at least three times in caecilians, including all of the "typhlonectids," so far as is known, and several caeciliids (e.g., *Dermophis*, *Gymnopis*, *Geotrypetes*), as well as the scolecomorphid genus *Scolecomorphus*. Gestation periods are lengthy (e.g., seven to nine months in *Typhlonectes*, 11 months in *Dermophis*), and young are born fully metamorphosed.

Conservation status

Too little is known of the ecology and biology of nearly all species of caecilians to evaluate their conservation status. However, anecdotal information indicates that land use change is severely restricting the ranges of several species, some populations are succumbing to the chytrid fungus, and some species may be extinct; they have not been collected for some time at localities where they had been observed previously. To date, no species are listed by the IUCN.

Significance to humans

Caecilians are little-appreciated biological control carnivores; they forage on small arthropods and so forth, thus potentially helping to control such populations. Because they actively burrow, rather than following root channels or other ready-made holes in the ground, they aid in turning the soil and maintaining good soil condition. In most areas of the world, the indigenous lore about caeciliids is that they are nasty, dangerous animals; in contrast, however, humans eat them in some parts of Southeast Asia.

1. Frigate Island caecilian (*Hypogeophis rostratus*); 2. Cayenne caecilian (*Typhlonectes compressicauda*); 3. Mexican caecilian (*Dermophis mexicanus*). (Illustration by Brian Cressman)

Species accounts

Mexican caecilian
Dermophis mexicanus

TAXONOMY
Siphonops mexicanus Duméril and Bibren, 1841, Mexico. *Dermophis mexicanus* is a member of the Caeciliidae; it has been placed in a subfamily Dermophiinae by some workers, but most experts reject the use of subfamilial designations until generic relationships of caecilians are better understood.

OTHER COMMON NAMES
Spanish: Dos cabezas, solda con solda.

PHYSICAL CHARACTERISTICS
Adults are medium-length for caecilians (11.8–19.7 in [300–500 mm] total length) and fairly stout bodied. They are dark gray dorsally, with paler venters and jaw and tentacle markings. Their numerous body annuli are more darkly pigmented; this feature distinguishes *Dermophis mexicanus* from most other species in the genus.

DISTRIBUTION
Mexico, from lowlands and the mountains of Guerrero on the Pacific versant, and Veracruz on the Atlantic, to northern Panama.

HABITAT
Moist, friable soil; leaf litter.

BEHAVIOR
Little is known of the behavior of caecilians, including *D. mexicanus*. They spend most of their time in burrows in loose, moist soil; they often emerge at dusk in a light rain to forage on the surface. They make their own burrows and are effective in a diversity of soil types. The internal concertina and lateral undulation modes of locomotion of *D. mexicanus* have been analyzed.

FEEDING ECOLOGY AND DIET
These carnivores are sit-and-wait predators; their diet is composed of invertebrates and certain vertebrates that live or travel on soil or leaf litter and so on, including earthworms, termites, orthopteran instars, even small lizards and baby mice, depending on the size of the predator. Their feeding mechanics have been studied experimentally.

REPRODUCTIVE BIOLOGY
Dermophis mexicanus is a viviparous species. Nothing is known of mate attraction or courtship. They are sexually mature at two to three years of age. Like all caecilians, *Dermophis* have internal fertilization; the male inserts the extruded rear part of its cloaca into the cloaca of the female, thus transporting sperm directly to her reproductive tract. Males are spermatogenic 11 months of the year; however, females in a population are in synchrony in the developmental stages of their oviductal embryos and young, and they all give birth at about the same time, in May–June when the rainy season begins. There is no evidence of sperm storage by the females. Their pregnancies are 11 months long. They have three to 16 young, which are born at 3.9–5.9 in (10–15 cm) (the mother is only 11.8–17.7 in

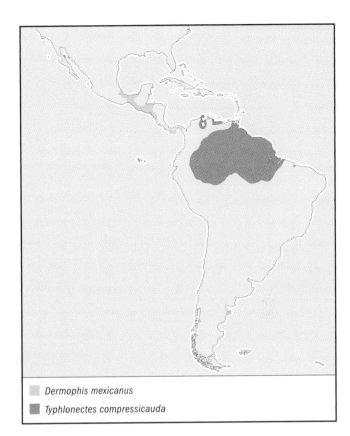

Dermophis mexicanus

Typhlonectes compressicauda

[30–45 cm]). The embryos exhaust the yolk supply of their small eggs (0.1 in [2 mm] diameter) about three months into the gestation period; the mother secretes a nutrient material from the glands of the lining of her oviducts. The developing young move around in the oviducts, ingesting the secretions. The fetuses have a special dentition that they apparently use to stimulate the secretion and to help take it into their mouths. The fetal dentition is shed at birth, and the adult teeth, which are very different in shape and distribution from those of the fetus, erupt within a few days. The fetuses have gills; they have three branches with numerous extensions. The gills and the skin are the organs of gaseous exchange in the oviducts.

CONSERVATION STATUS
The species remains locally abundant in some areas, but its habitat is being changed considerably as forests are removed for agriculture. It seems to adapt reasonably well to some kinds of farm use; for example, it has been abundant on some coffee fincas, where the coffee hulls are thrown in piles to decay, thus forming the moist organic soil in which *Dermophis* and its prey, earthworms, survive well.

SIGNIFICANCE TO HUMANS
Dermophis is valuable to humans; it can turn soil as it makes its burrows and eats insects and the like that might otherwise become super abundant. ◆

Frigate Island caecilian
Hypogeophis rostratus

TAXONOMY
Coecilia rostrata Cuvier, 1829, Mahé, Seychelles. *Hypogeophis*, including only one species, is a member of a monophyletic lineage that has radiated on the granitic islands in the Seychelles archipelago. Molecular, morphological, biochemical, and chromosome data all support the monophyly of the Seychelles genera, but do not give clues as to the relationships of the Seychelles caecilians to those of Africa, India, and southeast Asia.

OTHER COMMON NAMES
English: Sharp-nosed caecilian.

PHYSICAL CHARACTERISTICS
Hypogeophis rostratus is a relatively small (7.9–14.6 in [200–370 mm] total length as adults) caecilian, dark black–brown in coloration, slightly paler ventrally. Its body has the characteristic annuli; it does not have a tail. The head is small and rather pointed, the eyes covered by skin, the mouth underslung, and the chemosensory tentacle apparent on the side of the head. The structure and development of the skull, teeth, and vertebral column and many organs of the body are well known, through the work of the German biologist Harry Marcus during the early part of the twentieth century, and more recent work by Swiss and United States scientists.

DISTRIBUTION
Hypogeophis rostratus is a relatively abundant species, inhabiting all of the granitic islands in the Seychelles group.

HABITAT
The species occurs in a diversity of habitats, including moist leaf litter and soil, under wood and rocks, in debris piles, and occasionally in streams.

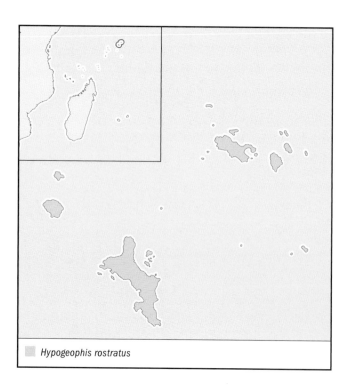

Hypogeophis rostratus

BEHAVIOR
Little is published about the behavior of the Seychelles caecilians.

FEEDING ECOLOGY AND DIET
Specific information on the natural diet of *Hypogeophis* is not available. Presumably they eat earthworms and terrestrial arthropods, as do most caecilians; there are anecdotal reports that they also consume frogs. *Hypogeophis* in captivity are strong-bite, sit-and-wait predators, suggesting that this is their mode in nature as well, similar to the majority of caecilians.

REPRODUCTIVE BIOLOGY
Males and females court in streams. *Hypogeophis rostratus* is a direct developer: females lay internally fertilized eggs in burrows on land and coil around them to guard them. They apparently can breed at any time of year. The eggs are large, approximately 0.3 in (8 mm) diameter. The first cell divisions take place atop the yolk, so that a disk of cells that will become the embryo is formed, similar to birds. The embryos develop inside their egg membranes, have gills and so forth, and then undergo metamorphosis, all before hatching as juvenile caecilians.

CONSERVATION STATUS
Hypogeophis rostratus seems to be abundant.

SIGNIFICANCE TO HUMANS
The carnivorous habits make it useful as an insect/arthropod control mechanism, and its burrowing aids in soil turning and aeration. ◆

Cayenne caecilian
Typhlonectes compressicauda

TAXONOMY
Caecilia compressicauda Duméril and Bibron, 1841, French Guiana.

Typhlonectes compressicauda is a member of a group of five genera thought to be closely related because of certain shared, derived characters, mostly associated with semi-aquatic to aquatic habits. The group is recognized as a family by most biologists expert on caecilians, but some phylogenetic analyses place them within the paraphyletic family Caeciliidae. It is difficult to distinguish *compressicauda* from *T. natans*; much of the information in the literature purportedly describing *T. compressicauda* and its biology really pertains to *T. natans*.

OTHER COMMON NAMES
None known.

PHYSICAL CHARACTERISTICS
Typhlonectes compressicauda, like most "typhlonectids," lacks secondary body annuli, and its annuli are distinct. Animals are gray to dark blue–black, and nearly uniform in color. Adults are 11.8–21.7 in (300–550 mm) total length. They have moderately long, flattened heads. They also have a slight dorsal "fin" or ridge that extends from the anterior third of the body to its end. The vent (cloacal) region is flattened, forming a disk that is paler in color. Features associated with its aquatic habit include the body fin, lateral body compression, large choanae, and development of both lungs.

DISTRIBUTION
Typhlonectes compressicauda is found throughout the Guianas and the Amazon region; the closely related and very similar *T. natans* is restricted to Colombia and northwestern Venezuela.

HABITAT

This caecilian is fully aquatic, inhabiting slow-moving, warm tropical rivers and streams.

BEHAVIOR

Individuals are known to share burrows and leave them at sunset to forage for food. They have many mucous glands all over their bodies, and the secretion is apparently toxic and distasteful to fishes. Predators include large fish, snakes, and birds. There are several reports in the literature of observation of courtship by nudging and coiling around each other, intromission, and of birth in captivity the ability to view the animals in aquaria has allowed more extensive observation of their feeding, swimming, and reproductive behavior than that of the more secretive terrestrial caecilians. Pheromones have been suggested to be involved in mate attraction.

FEEDING ECOLOGY AND DIET

The animals root around in the mud of the sides and bottoms of the waterways they inhabit. They eat arthropods of various sorts, including shrimp, insect pupae, and so on, and small fish. They have the strong-bite mechanism typical of caecilians. In captivity, they feed on pieces of earthworms and liver. They do not seem to use chemosensory cues extensively to find food items; they seem to perceive presence of prey by touch or motion.

REPRODUCTIVE BIOLOGY

Typhlonectes compressicauda is viviparous, as all typhlonectids are suspected to be. The species has a seven to nine month gestation period; a female can have six to 14 developing young, with a mean of 10. The gills of typhlonectid embryos fuse into large sacs. The gills may function in uptake of nutrients as well as gaseous exchange. The gills are shed shortly after birth. The fetal dentition, different in structure and arrangement from that of the adults, is used to ingest oviductal secretions. Fetuses are fully metamorphosed at birth; even their hemoglobin has changed from the embryonic to the adult.

CONSERVATION STATUS

This species has been collected extensively, apparently with no indication of diminution of its numbers. It has been taken by fishermen (and scientists) using nets to catch fish. No formal designation of its conservation status has been attempted, but the species apparently remains locally abundant, so far as is known.

SIGNIFICANCE TO HUMANS

The ecological significance of the species has not been determined; presumably its carnivorous habits and abundance make it an effective part of the riverine food web. Many animals have been imported and sold in aquarium stores as rubber eels or black eels; only rarely are they identified as amphibians, and then they are usually called *T. compressicauda*. In fact, the species that has been most imported is *T. natans*, so many of the reports in aquarium journals bear an incorrect species name. ◆

Resources

Books

Duellman, William E., and Linda Trueb. *Biology of Amphibians.* Baltimore: Johns Hopkins University Press, 1994.

Exbrayat, Jean-Marie. *Les Gymnophiones Ces curieux Amphibiens.* Paris: Societe Nouvelle des Editions Boubee, 2000.

Himstedt, W. *Die Blindwuehlen.* Marburg: Die Neue Brehm-Buecherei Bd. 630, Westart Wissenschaften and Heidelberg: Spektrum Akademischer Verlag, 1996.

Nussbaum, Ronald A. "Amphibians of the Seychelles." In *Biogeography and Ecology of the Seychelles Islands,* edited by David R. Stoddart. The Hague, Netherlands, and Boston: W. Junk Publishers (1984): 379–415.

Taylor, Edward H. *The Caecilians of the World.* Lawrence, KS: University of Kansas Press, 1968.

Periodicals

Brauer, August. "Beitraege zur Kenntniss der Entwicklungsgeschichte und der Anatomie der Gymnophionen. II. Die Entwicklung der auessern Form." *Zoologische Jahrbuch der Anatomie* 12 (1899): 477–408.

Marcus, Harry, O. Winsauer, and A. Huebner. "Beitrage zur Kenntnis der Gymnophionen. XVIII. Der kinetische Schadel von Hypogeophis und die Gehoerknoekelchen. " *Zeit. Anat. Entw.* 100 (1933): 149–193.

Moodie, G. E. E. "Observations on the Life History of the Caecilian *Typhlonectes compressicaudus* (Dumeril and Bibron) in the Amazon Basin." *Canadian Journal of Zoology* 56 (1978): 1005–1008.

Nussbaum, Ronald A., and Mark Wilkinson. "On the Classification and Phylogeny of Caecilians (Amphibia: Gymnophiona), a Critical Review." *Herpetology Monographs* 3 (1989): 142.

Sammouri, R., Sabine Renous, Jean-Marie Exbrayat, and Jean Lescure. "Developpement embryonnaire de *Typhlonectes compressicaudus* (Amphibia Gymnophiona)." *Annimal Science Nature Zoology Paris* 11 (1990): 135–163.

Savage, J. M., and M. H. Wake. "A Re-evaluation of the Status of Taxa of Central American Caecilians (Amphibia: Gymnophiona), with Comments on Their Origin and Evolution." *Copeia* 2001 (2001): 52–64.

Summers, Adam P., and James C. O'Reilly. "A Comparative Study of Locomotion in the Caecilians *Dermophis mexicanus* and *Typhlonectes natans* (Amphibia: Gymnophiona)." *Zoological Journal of the Linnean Society* 121 (1997): 65–76.

Wake, Marvalee H. "Reproduction, Growth and Population Structure of the Central American Caecilian *Dermophis mexicanus.*" *Herpetologica* 36 (1980): 244–256.

Wake, Marvalee H., and James Hanken. "Development of the Skull of *Dermophis mexicanus* (Amphibia: Gymnophiona) with Comments on Skull Kinesis and Amphibian Relationships." *Journal of Morphology* 173 (1982): 203–223.

Wake, Marvalee H., John C. Hafner, Mark S. Hafner, Lorrie L. Klosterman, and James L. Patton. "The Karyotype of *Typhlonectes compressicaudus* (Amphibia: Gymnophiona), with Comments on Chromosome Evolution in Caecilians." *Experientia* 36 (1980): 171–172.

Wilkinson, Mark, and Ronald A. Nussbaum. "Evolutionary Relationships of the Lungless Caecilian *Atretochoana eiselti* (Amphibia: Gymnophiona: Typhlonectidae)." *Biological Journal of the Linnean Society* 126 (1999): 192–223.

Marvalee H. Wake, PhD

For further reading

Alcala, A. C., and W. C. Brown. *Philippine Amphibians: An Illustrated Field Guide.* Makati City, Philippines: Bookmark, Inc.,1998.

AmphibiaWeb: Information on Amphibian Biology and Conservation. [Web site]. 2002 [accessed November 19, 2002]. <http://amphibiaweb.org>

Anajeva, Natalia, Leo J. Borkin, Ilya S. Darevsky, and N. L. Orlov. *Dictionary of Animal Names in Five Languages. Amphibians and Reptiles.* Moscow: Russky Yazk Publishers, 1988.

Anstis, M. *Tadpoles of South-eastern Australia: A Guide with Keys.* Sydney, Australia: Reed New Holland, 2002.

Arnold, E. N., and J. A. Burton. *Field Guide to the Reptiles and Amphibians of Britain and Europe,* 2nd ed. London: HarperCollins Publishers, 1998.

Ashton, Raymond E., and Patricia S. Ashton. *Handbook of Amphibian and Reptiles of Florida,* Parts 1–3. Miami: Windward Publishing, 1985–1988.

Barker, John, Gordon Grigg, and Michael J. Tyler. *A Field Guide to Australian Frogs.* Chipping Norton, Australia: Surrey Beatty and Sons, 1995.

Bekoff, M., and G. M. Burghardt, eds. *The Development of Behavior: Comparative and Evolutionary Aspects.* New York: Garland Press, 1978.

Birkhead, T. R. and A. P. Møller, eds. *Sperm Competition and Sexual Selection.* San Diego: Academic Press, 1998.

Bishop, Sherman C. *Handbook of Salamanders: The Salamanders of the United States, of Canada, and of Lower California.* Ithaca, NY: Comstock, 1994.

Bourret, René. *Les Batraciens de l'Indochine.* Vol. 6. Hanoi: Institut Océanographique de l'Indochine, 1942.

Bruce, R., L. D. Houck, and R. G. Jaeger. *The Biology of Plethodontid Salamanders.* New York: KluwerAcademic/Plenum Publishers, 2000.

Campbell, A., ed. *Declines and Disappearances of Australian Frogs.* Canberra, Australia: Environment Australia, 1999.

Campbell, Jonathan A. *Amphibians and Reptiles of Northern Guatemala, the Yuacatán, and Belize.* Norman: University of Oklahoma Press, 1998.

Carruthers, V. C. *Frogs and Frogging in Southern Africa.* Cape Town: Struik, 2001.

Catalogue of American Amphibians and Reptiles. New Haven, CT: Society for the Study of Amphibians and Reptiles, 1963–2002.

Cei, José M. *Batracios de Chile.* Santiago: Ediciones de la Universidad de Chile, 1962.

———. "The Amphibians of Argentina." *Monitore Zoologico Italiano (n.s.) Monogr* 2 (1980): 1–609.

Channing, A. *Amphibians of Central and Southern Africa.* Ithaca, NY: Cornell University Press, 2001.

Cogger, Harold G. *Reptiles and Amphibians of Australia.* 6th ed. Sydney, Australia: Reed New Holland, 2001.

Cogger, Harold G., and Richard G. Zweifel, eds. *Encyclopedia of Reptiles and Amphibians,* 2nd ed. San Diego: Academic Press, 1998.

Collins, Joseph T. *Amphibians and Reptiles in Kansas,* 3rd ed. Public Education Series No. 13. Lawrence: Museum of Natural History, University of Kansas, 1993.

Conant, Roger, and Joseph T. Collins. *A Field Guide to Reptiles and Amphibians, Eastern and Central North America.* New York: Houghton Mifflin, 1991.

Cook, Francis R. *Introduction to Canadian Amphibians and Reptiles.* Ottawa, Canada: National Museums of Canada, 1984.

Degenhardt, William G., Charles W. Painter, and Andrew H. Price. *Amphibians and Reptiles of New Mexico.* Albuquerque: University of New Mexico Press, 1996.

Duellman, William E. *Amphibian Species of the World: Additions and Corrections.* Special Publication 21. Lawrence: Museum of Natural History, University of Kansas, 1993.

———, ed. *Patterns of Distribution of Amphibians: A Global Perspective.* Baltimore: Johns Hopkins University Press, 1999.

———. *Hylid Frogs of Middle America.* Ithaca, NY: Society for the Study of Amphibians and Reptiles, 2001.

Duellman, William E., and Linda Trueb. *Biology of Amphibians.* New York: McGraw-Hill Book Co., 1986.

Dundee, Harold A., and Douglas A. Rossman. *The Amphibians and Reptiles of Louisiana.* Baton Rouge: Louisiana State University Press, 1989.

Dutta, Suchil K., and K. Manamendra-Arachchi. *The Amphibian Fauna of Sri Lanka.* Colombo, Sri Lanka: Wildlife Heritage Trust of Sri Lanka, 1996.

Estes, Richard. *Handbuch der Paläoherpetologie,* Vol. 2. Stuttgart: Gustav Fischer Verlag, 1988.

Exbrayat, Jean-Marie. *Les Gymnophiones Ces curieux Amphibiens.* Paris: Societe Nouvelle des Editions Boubee, 2000.

Feder, Martin E., and Warren W. Burggren, eds. *Environmental Physiology of the Amphibians.* Chicago: University of Chicago Press, 1992.

Frank, Norman, and Erica Ramus. *A Complete Guide to Scientific Names of Reptiles and Amphibians of the World.* Pottsville, PA: N. G. Publishing, Inc., 1995.

Frost, Darrel R. *Amphibian Species of the World: A Taxonomic and Geographical Reference.* Lawrence, KS: Allen Press and Association of Systematics Collections, 1985.

———. Amphibian Species of the World: An Online Reference. [Web site]. American Museum of Natural History. July 15, 2002 [accessed November 19, 2002]. <http://research.amnh.org/herpetology/amphibia/>.

Garcia Paris, Mario. *Los Anfibios de España.* Madrid: Ministerio de Agricultura, Pesca y Alimentación, 1985.

Gasc, Jean-Pierre, A. Cabela, J. Crnobrnja-Isailovic, et al., eds. *Atlas of Amphibians and Reptiles in Europe.* Paris: Societas Europaea Herpetologica and Muséum National d'Histoire Naturelle, 1997.

Gibson, A. C., ed. *Neotropical Biodiversity and Conservation.* Los Angeles: Mildred E. Mathias Botanical Garden Miscellaneous Publications, 1996.

Glaw, F., and M. Vences. *A Fieldguide to the Amphibians and Reptiles of Madagascar.* Frankfurt, Germany: Edition Chimaira, 1999.

Grassé, P.-P., and M. Delsol, eds. *Traité de Zoologie: Batraciens.* Paris: Masson, 1986.

Green, David, ed. *Amphibians in Decline: Canadian Studies of a Global Problem.* Vol. 1, *Herpetological Conservation.* St. Louis: Society for the Study of Amphibians and Reptiles Publications, 1997.

Grigg, G., R. Shine, and H. Ehmann, eds. *The Biology of Australasian Frogs and Reptiles.* Chipping Norton, Australia: Surrey Beatty and Sons, 1985.

Griffiths, Richard A. *Newts and Salamanders of Europe.* San Diego: Academic Press, 1996.

Halliday, Tim, and Kraig Adler, eds. *The New Encyclopedia of Reptiles and Amphibians.* Oxford, UK: Oxford University Press, 2002.

Hall, Brian K., and Marvalee H. Wake, eds. *The Origin and Evolution of Larval Forms.* San Diego: Academic Press, 1999.

Heatwole, Harold, and George T. Barthalmus, eds. *Amphibian Biology.* Vol. 1, *The Integument.* Chipping Norton, Australia: Surrey Beatty & Sons, 1994.

Heatwole, Harold, and Brian K. Sullivan, eds. *Amphibian Biology.* Vol. 2, *Social Behavior.* Chipping Norton, Australia: Surrey Beatty & Sons, 1995.

Heatwole, Harold, and Ellen M. Dawley, eds. *Amphibian Biology.* Vol. 3, *Sensory Biology.* Chipping Norton, Australia: Surrey Beatty & Sons, 1998.

Heatwole, Harold, and Robert L. Carroll, eds. *Amphibian Biology.* Vol. 4, *Palaeontology. The Evolutionary History of Amphibians.* Chipping Norton, Australia: Surrey Beatty & Sons, 2000.

Herrmann, Hans-Joachim. *Terrarien Atlas.* Vol. 1, *Kulturgeschichte, Biologie, und Terrarienhaltung von Amphibien, Schleichenlurche, Schwanzlurche, Froschlurche.* Melle, Germany: Mergus Verlag, 2001.

Himstedt, W. *Die Blindwuehlen.* Marburg: Die Neue Brehm-Buecherei Bd. 630, Westart Wissenschaften and Heidelberg: Spektrum Akademischer Verlag, 1996.

Hofrichter, Robert, ed. *Amphibians: The World of Frogs, Toads, Salamanders and Newts.* Buffalo, NY: Firefly Books, 2000.

Inger, R. F., and R. B. Stuebing. *A Field Guide to the Frogs of Borneo.* Kota, Indonesia: Natural History Publications, 1997.

Joglar, Rafael L. *Los Coquíes de Puerto Rico.* San Juan: Editorial de la Universidad de Puerto Rico, 1998.

Johnson, Tom R. *Amphibians and Reptiles of Missouri,* 2nd ed. Jefferson City, Missouri: Missouri Department of Conservation, 2000.

Kuzmin, S. *Amphibians of the Former Soviet Union.* Sofia, Bulgaria, and Moscow, Russia: Pensoft, 1999.

Lamar, William W. *The World's Most Spectacular Reptiles and Amphibians.* Tampa, FL: World Publications, 1997.

Lannoo, M. J., ed. *Status and Conservation of U.S. Amphibians.* Berkeley: University of California Press, 2003.

Lee, J. C. *Amphibians and Reptiles of the Yucatán Peninsula.* Ithaca, NY: Cornell University Press, 1996.

Lescure, Jean, and Christian Marty. *Atlas des Amphibiens de Guyane.* Paris: Muséum National d'Histoire Naturelle, 2000.

Maeda, N. and M. Matsui. *Frogs and Toads of Japan.* Tokyo: Bun-Ichi Sogo Shuppan Co., 1990.

McCranie, James R., and Larry David Wilson. *The Amphibians of Honduras.* Ithaca, NY: Society of the Study of Amphibians and Reptiles, 2002.

McDiarmid, Roy W., and Ronald Altig, eds. *Tadpoles: The Biology of Anuran Larvae.* Chicago: University of Chicago Press, 1999.

Menzies, James I. *Handbook of Common New Guinea Frogs*. Wau, Papua New Guinea: Wau Ecology Institute, 1976.

Meyer, John R., and Carol Farneti Foster. *A Guide to the Frogs and Toads of Belize*. Malabar, FL: Krieger Publishing Company, 1996.

Murphy, John C. *Amphibians and Reptiles of Trinidad and Tobago*. Malabar, FL: Krieger Publishing Company, 1997.

Noble, G. K. *The Biology of the Amphibia*. New York: Dover Publications, 1954.

Nussbaum, R. A., E. D. Brodie, and R. M. Storm. *Amphibians and Reptiles of the Pacific Northwest*. Moscow: University of Idaho Press, 1983.

Parker, H. W. *A Monograph of the Frogs of the Family Microhylidae*. London: British Museum, 1934.

Passmore, Neville, and Vincent Carruthers. *South African Frogs: A Complete Guide*, rev. ed. Halfway House, South Africa: Southern Book Publishers; Johannesburg: Witwatersrand University Press, 1995.

Petranka, J. W. *Salamanders of the United States and Canada*. Washington, DC: Smithsonian Institution Press, 1998.

Pleguezuelos, Juan M. ed. *Distribución y Biogeogrfía de los Anfibis y Reptiles en España y Portugal*. Granada, Spain: Universidad de Granada, 1997.

Pough, F. Henry, Robin M. Andrews, John E. Cadle, Mart L. Crump, Alan H. Savitzky, and Kentwood D. Wells. *Herpetology*, 2nd ed. Upper Saddle River, NJ: Prentice Hall, 2001.

Powell, Robert, and Robert C. Henderson. *Contributions to West Indian Herpetology*. Ithaca, NY: Society for the Study of Amphibians and Reptiles, 1996.

Rivero, Juan A. *Los Anfibios y Reptiles de Puerto Rico*, 2nd ed. San Juan: Universidad de Puerto Rico, 1998.

Robb, Joan. *New Zealand Amphibians and Reptiles in Colour*. Auckland: Collins Publishers, 1980.

Rödel, M. O. *Herpetofauna of West Africa*. I, *Amphibians of the West African Savanna*. Frankfurt, Germany: Edition Chimaira, 2000.

Rodriguez, Lily O., and William E. Duellman. *Guide to the Frogs of the Iquitos Region, Amazonian Peru*. Special Publication 22. Lawrence: Natural History Museum, University of Kansas, 1994.

Ryan, M. J., ed. *Anuran Communication*. Washington, DC: Smithsonian Institution Press, 2001.

Sanchíz, Borja. *Encyclopedia of Paleoherpetology*. Part 4, *Salientia*. Munich: Verlag Dr. Friedrich Pfeil, 1998.

Savage, Jay M. *The Amphibians and Reptiles of Costa Rica*. Chicago: University of Chicago Press, 2002.

Schiøtz, A. *Treefrogs of Africa*. Frankfurt, Germany: Edition Chimaira, 1999.

Schulte, Rainer. *Pfeilgiftfrösche, "Arteneil Peru."* INIBICO, Waiblingen, 1999.

Schultze, Hans-Peter, and Linda Trueb, eds. *Origins of the Higher Groups of Tetrapods. Controversy and Consensus*. Ithaca, NY: Comstock Publishing Associates, Cornell University Press, 1991.

Schwartz, Albert, and Robert C. Henderson. *Amphibians and Reptiles of the West Indies: Descriptions, Distributions, and Natural History*. Gainesville: University of Florida Press, 1991.

Schwenk, K., ed. *Feeding*. San Diego: Academic Press, 2000.

Stebbins, Robert C. *A Field Guide to Western Reptiles and Amphibians*, 2nd ed. Boston: Houghton Mifflin, 1985.

Stebbins, Robert C., and N. W. Cohen. *Natural History of Amphibians*. Princeton, NJ: Princeton University Press, 1995.

Taylor, Douglas H., and Sheldon I. Guttman, eds. *The Reproductive Biology of Amphibians*. New York: Plenum Press, 1977.

Taylor, Edward Harrison. *Caecilians of the World: A Taxonomic Review*. Lawrence: University of Kansas Press, 1968.

Thorn, R., and J. Raffaelli. *Les Salamandres de l'Ancien Monde*. Paris: Societe Nouvelle des Editions Boubee, 2001.

Tyler, Michael J. *Frogs*. Sydney, Australia: Collins, 1976.

———, ed. *The Gastric Brooding Frog*. London: Biddles, 1983.

Vial, James L., ed. *Evolutionary Biology of Anurans*. Columbia: University of Missouri Press, 1973.

Walls, Jerry G. *Poison Frogs of the Family Dendrobatidae: Jewels of the Rainforest*. Neptune City, NJ: TFH Publications, 1994.

Zhao, Er-Mi, and Kraig Adler. *Herpetology of China*. Ithaca, NY: Society for the Study of Amphibians and Reptiles, 1993.

Zhao, Er-mi, Qixiong Hu, Yaoming Jing, and Yuhua Yang. *Studies on Chinese Salamanders*. Oxford, OH: Society for the Study of Amphibians and Reptiles, 1988.

Zug, George R., Laurie J. Vitt, and Janalee P. Caldwell. *Herpetology: An Introductory Biology of Amphibians and Reptiles*, 2nd ed. San Diego: Academic Press, 2001.

Organizations

American Society of Ichthyologists and Herpetologists
donnelly@fiu.edu
Phone: (305)919-5651
<http://199.245.200.110/>

American Zoo and Aquarium Association
8403 Colesville Road, Suite 710
Silver Spring, MD 20910
<http://www.aza.org>

Asociación Herpetológica Española
Apartado de Correos 191
28911 Leganés
Madrid
Spain
<http://elebo.fbiolo.uv.es/zoologia/AHE/>

Australian Society of Herpetologists
c/o CSIRO Wildlife and Ecology
PO Box 84
Lyneham, ACT 2602
Australia
<http://www.gu.edu.au/school/asc/ppages/academic/jmhero/ash/frameintro.html>

AZA Amphibian Advisory Group
<http://www.amphibiantag.homestead.com/>

Declining Amphibian Populations Task Force
Phone: 44 (0)1908 653831
<http://www.open.ac.uk/daptf/>

Environment Australia
GPO Box 787
Canberra, ACT 2601
Australia
Phone: 61-2-6274-1111
<http://www.ea.gov.au>

Herpetologists' League
<http://www.inhs.uiuc.edu/cbd/HL/HL.html>

IUCN: The World Conservation Union
Rue Mauverney 28
1196
Gland
Switzerland
Phone: 41-22-999-0000
<http://www.iucn.org>

National Amphibian Conservation Center
Detroit Zoological Park
8450 W. Ten Mile Rd.
Royal Oak, MI 48067
<http://www.detroitzoo.org/critters2.html>

North American Amphibian Monitoring Program
<http://www.mp2-pwrc.usgs.gov/naamp/>

Societas Europaea Herpetologica
Natural Resources Institute, University of Greenwich
Central Avenue, Chatham Maritime
Kent
ME4 4TB
United Kingdom
<http://www.gli.cas.cz/SEH/>

Society for Research on Amphibians and Reptiles in New Zealand (SRARNZ)
SBS, Victoria University of Wellington, PO Box 600
Wellington
New Zealand

Society for the Study of Amphibians and Reptiles
<http://www.ukans.edu/~ssar/>

• • • • •

Contributors to the first edition

The following individuals contributed chapters to the original edition of Grzimek's Animal Life Encyclopedia, *which was edited by Dr. Bernhard Grzimek, Professor, Justus Liebig University of Giessen, Germany; Director, Frankfurt Zoological Garden, Germany; and Trustee, Tanzanian National Parks, Tanzania.*

Dr. Michael Abs
Curator, Ruhr University
Bochum, Germany

Dr. Salim Ali
Bombay Natural History Society
Bombay, India

Dr. Rudolph Altevogt
Professor, Zoological Institute, University of Münster
Münster, Germany

Dr. Renate Angermann
Curator, Institute of Zoology, Humboldt University
Berlin, Germany

Edward A. Armstrong
Cambridge University
Cambridge, England

Dr. Peter Ax
Professor, Second Zoological Institute and Museum, University of Göttingen
Göttingen, Germany

Dr. Franz Bachmaier
Zoological Collection of the State of Bavaria
Munich, Germany

Dr. Pedru Banarescu
Academy of the Roumanian Socialist Republic, Trajan Savulescu Institute of Biology
Bucharest, Romania

Dr. A. G. Bannikow
Professor, Institute of Veterinary Medicine
Moscow, Russia

Dr. Hilde Baumgärtner
Zoological Collection of the State of Bavaria
Munich, Germany

C. W. Benson
Department of Zoology, Cambridge University
Cambridge, England

Dr. Andrew Berger
Chairman, Department of Zoology, University of Hawaii
Honolulu, Hawaii, U.S.A.

Dr. J. Berlioz
National Museum of Natural History
Paris, France

Dr. Rudolf Berndt
Director, Institute for Population Ecology, Hiligoland Ornithological Station
Braunschweig, Germany

Dieter Blume
Instructor of Biology, Freiherr-vom-Stein School
Gladenbach, Germany

Dr. Maximilian Boecker
Zoological Research Institute and A. Koenig Museum
Bonn, Germany

Dr. Carl-Heinz Brandes
Curator and Director, The Aquarium, Overseas Museum
Bremen, Germany

Dr. Donald G. Broadley
Curator, Umtali Museum
Mutare, Zimbabwe

Dr. Heinz Brüll
Director; Game, Forest, and Fields Research Station
Hartenholm, Germany

Dr. Herbert Bruns
Director, Institute of Zoology and the Protection of Life
Schlangenbad, Germany

Hans Bub
Heligoland Ornithological Station
Wilhelmshaven, Germany

A. H. Chrisholm
Sydney, Australia

Herbert Thomas Condon
Curator of Birds, South Australian Museum
Adelaide, Australia

Dr. Eberhard Curio
Director, Laboratory of Ethology, Ruhr University
Bochum, Germany

Dr. Serge Daan
Laboratory of Animal Physiology, University of Amsterdam
Amsterdam, The Netherlands

Dr. Heinrich Dathe
Professor and Director, Animal Park and Zoological Research Station, German Academy of Sciences
Berlin, Germany

Dr. Wolfgang Dierl
Zoological Collection of the State of Bavaria
Munich, Germany

Dr. Fritz Dieterlen
Zoological Research Institute, A. Koenig Museum
Bonn, Germany

Dr. Rolf Dircksen
Professor, Pedagogical Institute
Bielefeld, Germany

Josef Donner
Instructor of Biology
Katzelsdorf, Austria

Dr. Jean Dorst
Professor, National Museum of Natural History
Paris, France

Dr. Gerti DÜcker
Professor and Chief Curator, Zoological Institute, University of Münster
Münster, Germany

Dr. Michael Dzwillo
Zoological Institute and Museum, University of Hamburg
Hamburg, Germany

Dr. Irenäus Eibl-Eibesfeldt
Professor and Director, Institute of Human Ethology, Max Planck Institute for Behavioral Physiology
Percha/Starnberg, Germany

Dr. Martin Eisentraut
Professor and Director, Zoological Research Institute and A. Koenig Museum
Bonn, Germany

Dr. Eberhard Ernst
Swiss Tropical Institute
Basel, Switzerland

R. D. Etchecopar
Director, National Museum of Natural History
Paris, France

Dr. R. A. Falla
Director, Dominion Museum
Wellington, New Zealand

Dr. Hubert Fechter
Curator, Lower Animals, Zoological Collection of the State of Bavaria
Munich, Germany

Dr. Walter Fiedler
Docent, University of Vienna, and Director, Schönbrunn Zoo
Vienna, Austria

Wolfgang Fischer
Inspector of Animals, Animal Park
Berlin, Germany

Dr. C. A. Fleming
Geological Survey Department of Scientific and Industrial Research
Lower Hutt, New Zealand

Dr. Hans Frädrich
Zoological Garden
Berlin, Germany

Dr. Hans-Albrecht Freye
Professor and Director, Biological Institute of the Medical School
Halle a.d.S., Germany

Günther E. Freytag
Former Director, Reptile and Amphibian Collection, Museum of Cultural History in Magdeburg
Berlin, Germany

Dr. Herbert Friedmann
Director, Los Angeles County Museum of Natural History
Los Angeles, California, U.S.A.

Dr. H. Friedrich
Professor, Overseas Museum
Bremen, Germany

Dr. Jan Frijlink
Zoological Laboratory, University of Amsterdam
Amsterdam, The Netherlands

Dr. Dr. H.C. Karl Von Frisch
Professor Emeritus and former Director, Zoological Institute, University of Munich
Munich, Germany

Dr. H. J. Frith
C.S.I.R.O. Research Institute
Canberra, Australia

Dr. Ion E. Fuhn
Academy of the Roumanian Socialist Republic, Trajan Savulescu Institute of Biology
Bucharest, Romania

Dr. Carl Gans
Professor, Department of Biology, State University of New York at Buffalo
Buffalo, New York, U.S.A.

Dr. Rudolf Geigy
Professor and Director, Swiss Tropical Institute
Basel, Switzerland

Dr. Jacques Gery
St. Genies, France

Dr. Wolfgang Gewalt
Director, Animal Park
Duisburg, Germany

Dr. Dr. H.C. Dr. H.C. Viktor Goerttler
Professor Emeritus, University of Jena
Jena, Germany

Dr. Friedrich Goethe
Director, Institute of Ornithology, Heligoland Ornithological Station
Wilhelmshaven, Germany

Dr. Ulrich F. Gruber
Herpetological Section, Zoological Research Institute and A. Koenig Museum
Bonn, Germany

Dr. H. R. Haefelfinger
Museum of Natural History
Basel, Switzerland

Dr. Theodor Haltenorth
Director, Mammalology, Zoological Collection of the State of Bavaria
Munich, Germany

Barbara Harrisson
Sarawak Museum, Kuching, Borneo
Ithaca, New York, U.S.A.

Dr. Francois Haverschmidt
President, High Court (retired)
Paramaribo, Suriname

Dr. Heinz Heck
Director, Catskill Game Farm
Catskill, New York, U.S.A.

Dr. Lutz Heck
Professor (retired), and Director, Zoological Garden, Berlin
Wiesbaden, Germany

Dr. Dr. H.C. Heini Hediger
Director, Zoological Garder
Zurich, Switzerland

Dr. Dietrich Heinemann
Director, Zoological Garden, Münster
Dörnigheim, Germany

Dr. Helmut Hemmer
Institute for Physiological Zoology, University of Mainz
Mainz, Germany

Dr. W. G. Heptner
Professor, Zoological Museum, University of Moscow
Moscow, Russia

Dr. Konrad Herter
Professor Emeritus and Director (retired), Zoological Institute, Free University of Berlin
Berlin, Germany

Dr. Hans Rudolf Heusser
Zoological Museum, University of Zurich
Zurich, Switzerland

Dr. Emil Otto Höhn
Associate Professor of Physiology,
University of Alberta
Edmonton, Canada

Dr. W. Hohorst
Professor and Director, Parasitological
Institute, Farbwerke Hoechst A.G.
Frankfurt-Höchst, Germany

Dr. Folkhart Hückinghaus
Director, Senckenbergische Anatomy,
University of Frankfurt a.M.
Frankfurt a.M., Germany

Francois Hüe
National Museum of Natural History
Paris, France

Dr. K. Immelmann
Professor, Zoological Institute, Tech-
nical University of Braunschweig
Braunschweig, Germany

Dr. Junichiro Itani
Kyoto University
Kyoto, Japan

Dr. Richard F. Johnston
Professor of Zoology, University of
Kansas
Lawrence, Kansas, U.S.A.

Otto Jost
Oberstudienrat, Freiherr-vom-Stein
Gymnasium
Fulda, Germany

Dr. Paul Kähsbauer
Curator, Fishes, Museum of Natural
History
Vienna, Austria

Dr. Ludwig Karbe
Zoological State Institute and Mu-
seum
Hamburg, Germany

Dr. N. N. Kartaschew
Docent, Department of Biology,
Lomonossow State University
Moscow, Russia

Dr. Werner Kästle
Oberstudienrat, Gisela Gymnasium
Munich, Germany

Dr. Reinhard Kaufmann
Field Station of the Tropical Institute,
Justus Liebig University, Giessen,
Germany
Santa Marta, Colombia

Dr. Masao Kawai
Primate Research Institute, Kyoto
University
Kyoto, Japan

Dr. Ernst F. Kilian
Professor, Giessen University and
Catedratico Universidad Austral, Val-
divia-Chile
Giessen, Germany

Dr. Ragnar Kinzelbach
Institute for General Zoology, Univer-
sity of Mainz
Mainz, Germany

Dr. Heinrich Kirchner
Landwirtschaftsrat (retired)
Bad Oldesloe, Germany

Dr. Rosl Kirchshofer
Zoological Garden, University of
Frankfort a.M.
Frankfurt a.M., Germany

Dr. Wolfgang Klausewitz
Curator, Senckenberg Nature Mu-
seum and Research Institute
Frankfurt a.M., Germany

Dr. Konrad Klemmer
Curator, Senckenberg Nature Mu-
seum and Research Institute
Frankfurt a.M., Germany

Dr. Erich Klinghammer
Laboratory of Ethology, Purdue Uni-
versity
Lafayette, Indiana, U.S.A.

Dr. Heinz-Georg Klös
Professor and Director, Zoological
Garden
Berlin, Germany

Ursula Klös
Zoological Garden
Berlin, Germany

Dr. Otto Koehler
Professor Emeritus, Zoological Insti-
tute, University of Freiburg
Freiburg i. BR., Germany

Dr. Kurt Kolar
Institute of Ethology, Austrian Acad-
emy of Sciences
Vienna, Austria

Dr. Claus König
State Ornithological Station of Baden-
Württemberg
Ludwigsburg, Germany

Dr. Adriaan Kortlandt
Zoological Laboratory, University of
Amsterdam
Amsterdam, The Netherlands

Dr. Helmut Kraft
Professor and Scientific Councillor,
Medical Animal Clinic, University of
Munich
Munich, Germany

Dr. Helmut Kramer
Zoological Research Institute and A.
Koenig Museum
Bonn, Germany

Dr. Franz Krapp
Zoological Institute, University of
Freiburg
Freiburg, Switzerland

Dr. Otto Kraus
Professor, University of Hamburg,
and Director, Zoological Institute and
Museum
Hamburg, Germany

Dr. Dr. Hans Krieg
Professor and First Director (retired),
Scientific Collections of the State of
Bavaria
Munich, Germany

Dr. Heinrich Kühl
Federal Research Institute for Fish-
eries, Cuxhaven Laboratory
Cuxhaven, Germany

Dr. Oskar Kuhn
Professor, formerly University
Halle/Saale
Munich, Germany

Dr. Hans Kumerloeve
First Director (retired), State Scien-
tific Museum, Vienna
Munich, Germany

Dr. Nagamichi Kuroda
Yamashina Ornithological Institute,
Shibuya-Ku
Tokyo, Japan

Dr. Fred Kurt
Zoological Museum of Zurich Univer-
sity, Smithsonian Elephant Survey
Colombo, Ceylon

Dr. Werner Ladiges
Professor and Chief Curator, Zoological Institute and Museum, University
of Hamburg
Hamburg, Germany

Leslie Laidlaw
Department of Animal Sciences, Purdue University
Lafayette, Indiana, U.S.A.

Dr. Ernst M. Lang
Director, Zoological Garden
Basel, Switzerland

Dr. Alfredo Langguth
Department of Zoology, Faculty of Humanities and Sciences, University of the Republic
Montevideo, Uruguay

Leo Lehtonen
Science Writer
Helsinki, Finland

Bernd Leisler
Second Zoological Institute, University of Vienna
Vienna, Austria

Dr. Kurt Lillelund
Professor and Director, Institute for Hydrobiology and Fishery Sciences, University of Hamburg
Hamburg, Germany

R. Liversidge
Alexander MacGregor Memorial Museum
Kimberley, South Africa

Dr. Dr. Konrad Lorenz
Professor and Director, Max Planck Institute for Behavioral Physiology
Seewiesen/Obb., Germany

Dr. Dr. Martin Lühmann
Federal Research Institute for the Breeding of Small Animals
Celle, Germany

Dr. Johannes Lüttschwager
Oberstudienrat (retired)
Heidelberg, Germany

Dr. Wolfgang Makatsch
Bautzen, Germany

Dr. Hubert Markl
Professor and Director, Zoological Institute, Technical University of Darmstadt
Darmstadt, Germany

Basil J. Marlow , B.SC. (Hons)
Curator, Australian Museum
Sydney, Australia

Dr. Theodor Mebs
Instructor of Biology
Weissenhaus/Ostsee, Germany

Dr. Gerlof Fokko Mees
Curator of Birds, Rijks Museum of Natural History
Leiden, The Netherlands

Hermann Meinken
Director, Fish Identification Institute, V.D.A.
Bremen, Germany

Dr. Wilhelm Meise
Chief Curator, Zoological Institute and Museum, University of Hamburg
Hamburg, Germany

Dr. Joachim Messtorff
Field Station of the Federal Fisheries Research Institute
Bremerhaven, Germany

Dr. Marian Mlynarski
Professor, Polish Academy of Sciences, Institute for Systematic and Experimental Zoology
Cracow, Poland

Dr. Walburga Moeller
Nature Museum
Hamburg, Germany

Dr. H.C. Erna Mohr
Curator (retired), Zoological State Institute and Museum
Hamburg, Germany

Dr. Karl-Heinz Moll

Waren/Müritz, Germany

Dr. Detlev Müller-Using
Professor, Institute for Game Management, University of Göttingen
Hannoversch-Münden, Germany

Werner Münster
Instructor of Biology
Ebersbach, Germany

Dr. Joachim Münzing
Altona Museum
Hamburg, Germany

Dr. Wilbert Neugebauer
Wilhelma Zoo
Stuttgart-Bad Cannstatt, Germany

Dr. Ian Newton
Senior Scientific Officer, The Nature Conservancy
Edinburgh, Scotland

Dr. Jürgen Nicolai
Max Planck Institute for Behavioral Physiology
Seewiesen/Obb., Germany

Dr. Günther Niethammer
Professor, Zoological Research Institute and A. Koenig Museum
Bonn, Germany

Dr. Bernhard Nievergelt
Zoological Museum, University of Zurich
Zurich, Switzerland

Dr. C. C. Olrog
Institut Miguel Lillo San Miguel de Tucuman
Tucuman, Argentina

Alwin Pedersen
Mammal Research and aRctic Explorer
Holte, Denmark

Dr. Dieter Stefan Peters
Nature Museum and Senckenberg Research Institute
Frankfurt a.M., Germany

Dr. Nicolaus Peters
Scientific Councillor and Docent, Institute of Hydrobiology and Fisheries, University of Hamburg
Hamburg, Germany

Dr. Hans-Günter Petzold
Assistant Director, Zoological Garden
Berlin, Germany

Dr. Rudolf Piechocki
Docent, Zoological Institute, University of Halle
Halle a.d.S., Germany

Dr. Ivo Poglayen-Neuwall
Director, Zoological Garden
Louisville, Kentucky, U.S.A.

Dr. Egon Popp
Zoological Collection of the State of Bavaria
Munich, Germany

Dr. Dr. H.C. Adolf Portmann
Professor Emeritus, Zoological Institute, University of Basel
Basel, Switzerland

Hans Psenner
Professor and Director, Alpine Zoo
Innsbruck, Austria

Dr. Heinz-Siburd Raethel
Oberveterinärrat
Berlin, Germany

Dr. Urs H. Rahm
Professor, Museum of Natural History
Basel, Switzerland

Dr. Werner Rathmayer
Biology Institute, University of Konstanz
Konstanz, Germany

Walter Reinhard
Biologist
Baden-Baden, Germany

Dr. H. H. Reinsch
Federal Fisheries Research Institute
Bremerhaven, Germany

Dr. Bernhard Rensch
Professor Emeritus, Zoological Institute, University of Münster
Münster, Germany

Dr. Vernon Reynolds
Docent, Department of Sociology,
University of Bristol
Bristol, England

Dr. Rupert Riedl
Professor, Department of Zoology,
University of North Carolina
Chapel Hill, North Carolina, U.S.A.

Dr. Peter Rietschel
Professor (retired), Zoological Institute, University of Frankfurt a.M.
Frankfurt a.M., Germany

Dr. Siegfried Rietschel
Docent, University of Frankfurt; Curator, Nature Museum and Research
Institute Senckenberg
Frankfurt a.M., Germany

Herbert Ringleben
Institute of Ornithology, Heligoland
Ornithological Station
Wilhelmshaven, Germany

Dr. K. Rohde
Institute for General Zoology, Ruhr
University
Bochum, Germany

Dr. Peter Röben
Academic Councillor, Zoological Institute, Heidelberg University
Heidelberg, Germany

Dr. Anton E. M. De Roo
Royal Museum of Central Africa
Tervuren, South Africa

Dr. Hubert Saint Girons
Research Director, Center for National Scientific Research
Brunoy (Essonne), France

Dr. Luitfried Von Salvini-Plawen
First Zoological Institute, University
of Vienna
Vienna, Austria

Dr. Kurt Sanft
Oberstudienrat, Diesterweg-Gymnasium
Berlin, Germany

Dr. E. G. Franz Sauer
Professor, Zoological Research Institute and A. Koenig Museum, University of Bonn
Bonn, Germany

Dr. Eleonore M. Sauer
Zoological Research Institute and A.
Koenig Museum, University of Bonn
Bonn, Germany

Dr. Ernst Schäfer
Curator, State Museum of Lower Saxony
Hannover, Germany

Dr. Friedrich Schaller
Professor and Chairman, First Zoological Institute, University of Vienna
Vienna, Austria

Dr. George B. Schaller
Serengeti Research Institute, Michael
Grzimek Laboratory
Seronera, Tanzania

Dr. Georg Scheer
Chief Curator and Director, Zoological Institute, State Museum of Hesse
Darmstadt, Germany

Dr. Christoph Scherpner
Zoological Garden
Frankfurt a.M., Germany

Dr. Herbert Schifter
Bird Collection, Museum of Natural
History
Vienna, Austria

Dr. Marco Schnitter
Zoological Museum, Zurich University
Zurich, Switzerland

Dr. Kurt Schubert
Federal Fisheries Research Institute
Hamburg, Germany

Eugen Schuhmacher
Director, Animals Films, I.U.C.N.
Munich, Germany

Dr. Thomas Schultze-Westrum
Zoological Institute, University of
Munich
Munich, Germany

Dr. Ernst Schüt
Professor and Director (retired), State
Museum of Natural History
Stuttgart, Germany

Dr. Lester L. Short , Jr.
Associate Curator, American Museum
of Natural History
New York, New York, U.S.A.

Dr. Helmut Sick
National Museum
Rio de Janeiro, Brazil

Dr. Alexander F. Skutch
Professor of Ornithology, University
of Costa Rica
San Isidro del General, Costa Rica

Dr. Everhard J. Slijper
Professor, Zoological Laboratory,
University of Amsterdam
Amsterdam, The Netherlands

Bertram E. Smythies
Curator (retired), Division of Forestry
Management, Sarawak-Malaysia
Estepona, Spain

Dr. Kenneth E. Stager
Chief Curator, Los Angeles County
Museum of Natural History
Los Angeles, California, U.S.A.

Dr. H.C. Georg H.W. Stein
Professor, Curator of Mammals, Institute of Zoology and Zoological Museum, Humboldt University
Berlin, Germany

Dr. Joachim Steinbacher
Curator, Nature Museum and Senckenberg Research Institute
Frankfurt a.M., Germany

Dr. Bernard Stonehouse
Canterbury University
Christchurch, New Zealand

Dr. Richard Zur Strassen
Curator, Nature Museum and Senckenberg Research Institute
Frandfurt a.M., Germany

Dr. Adelheid Studer-Thiersch
Zoological Garden
Basel, Switzerland

Dr. Ernst Sutter
Museum of Natural History
Basel, Switzerland

Dr. Fritz Terofal
Director, Fish Collection, Zoological
Collection of the State of Bavaria
Munich, Germany

Dr. G. F. Van Tets
Wildlife Research
Canberra, Australia

Ellen Thaler-Kottek
Institute of Zoology, University of
Innsbruck
Innsbruck, Austria

Dr. Erich Thenius
Professor and Director, Institute of
Paleontolgy, University of Vienna
Vienna, Austria

Dr. Niko Tinbergen
Professor of Animal Behavior, Depart-
ment of Zoology, Oxford University
Oxford, England

Alexander Tsurikov
Lecturer, University of Munich
Munich, Germany

Dr. Wolfgang Villwock
Zoological Institute and Museum,
University of Hamburg
Hamburg, Germany

Zdenek Vogel
Director, Suchdol Herpetological Sta-
tion
Prague, Czechoslovakia

Dieter Vogt
Schorndorf, Germany

Dr. Jiri Volf
Zoological Garden
Prague, Czechoslovakia

Otto Wadewitz
Leipzig, Germany

Dr. Helmut O. Wagner
Director (retired), Overseas Museum,
Bremen
Mexico City, Mexico

Dr. Fritz Walther
Professor, Texas A & M University
College Station, Texas, U.S.A.

John Warham
Zoology Department, Canterbury
University
Christchurch, New Zealand

Dr. Sherwood L. Washburn
University of California at Berkeley
Berkeley, California, U.S.A.

Eberhard Wawra
First Zoological Institute, University
of Vienna
Vienna, Austria

Dr. Ingrid Weigel
Zoological Collection of the State of
Bavaria
Munich, Germany

Dr. B. Weischer
Institute of Nematode Research, Fed-
eral Biological Institute
Münster/Westfalen, Germany

Herbert Wendt
Author, Natural History
Baden-Baden, Germany

Dr. Heinz Wermuth
Chief Curator, State Nature Museum,
Stuttgart
Ludwigsburg, Germany

Dr. Wolfgang Von Westernhagen
Preetz/Holstein, Germany

Dr. Alexander Wetmore
United States National Museum,
Smithsonian Institution
Washington, D.C., U.S.A.

Dr. Dietrich E. Wilcke
Röttgen, Germany

Dr. Helmut Wilkens
Professor and Director, Institute of
Anatomy, School of Veterinary Medi-
cine
Hannover, Germany

Dr. Michael L. Wolfe
Utah, U.S.A.

Hans Edmund Wolters
Zoological Research Institute and A.
Koenig Museum
Bonn, Germany

Dr. Arnfrid Wünschmann
Research Associate, Zoological Garden
Berlin, Germany

Dr. Walter Wüst
Instructor, Wilhelms Gymnasium
Munich, Germany

Dr. Heinz Wundt
Zoological Collection of the State of
Bavaria
Munich, Germany

Dr. Claus-Dieter Zander
Zoological Institute and Museum,
University of Hamburg
Hamburg, Germany

Dr. Dr. Fritz Zumpt
Director, Entomology and Parasitol-
ogy, South African Institute for Medical Research
Johannesburg, South Africa

Dr. Richard L. Zusi
Curator of Birds, United States National Museum, Smithsonian Institution
Washington, D.C., U.S.A.

Glossary

Advertisement call—Sound produced by male anurans during the mating season to attract females.

Agonistic behavior—Fighting behavior between members of the same species.

Amplexus—The copulatory embrace of frogs and toads, during which the male fertilizes the eggs that are released by the female.

Aposematic—The conspicuously recognizable markings of an animal, such as in poison frogs, that serve to warn off potential predators. Also called warning coloring.

Arciferal—A condition in which the two halves of the pectoral (shoulder) girdle are not fused ventrally.

Axillary—Of, relating to, or located near the axilla; the cavity beneath the junction of a forelimb and the body.

Barbels—Whisker-like appendages found on both sides of the mouth.

Basal—Arising from the base of a stem; of or relating to, or being essential for, maintaining the fundamental vital activities of an organism.

Cartilaginous—Consisting of cartilage, a tough, elastic skeletal tissue consisting mostly of collagen fibers.

Chromosome—Thread-like structure consisting mostly of genetic material (DNA) in the nucleus of cells.

Clade—An evolutionary lineage of organisms that includes the most recent common ancestor of all those organisms and all the descendants of that common ancestor.

Cladograms—Graphic, tree-like representations that show the evolutionary relationships of organisms.

Clavicle—Paired bony elements of the pectoral girdle.

Cloaca—The common chamber into which the urinary, digestive, and reproductive systems discharge their contents, and which opens to the exterior.

Cloud forest—Moist forest at mid- to high elevations on mountains in the tropics.

Clutch—Eggs deposited by a single female in one breeding.

Cocoon—A tough protective covering.

Conspecific—Of or belonging to the same species.

Continuous breeder—An animal that may breed throughout the year.

Coracoid—Paired bony elements of the pectoral girdle.

Courtship—Behavioral interactions between males and females that precede and accompany mating.

Cranial—Of or pertaining to the cranium (skull).

Crest—An elevated ridge-like structure.

Crypsis—Involves resemblance or imitation of some feature (background or object) in an organism's environment, including its form, color, and pattern; camouflage. It may also be referred to as cryptic coloration.

Cutaneous—Of or pertaining to the skin.

Dermis—The layer of skin immediately below the epidermis.

Desiccation—The process of drying out.

Diapause—A period of physiologically enforced dormancy between periods of activity.

Dimorphism—The existence of two different forms (color, size, sex) of a species in the same population.

Direct development—Transition from the egg to the adult form without passing through a free-living larval stage.

Diurnal—Active by day.

Dorsal—Pertaining to the back or upper surface or one of its parts.

Dorsolateral—Pertaining to the interface of the back and the sides.

Ectotherm—An animal whose body temperature is controlled by the environment.

Eft—The juvenile, terrestrial phase of newts.

Electrophoretic analysis—Running electric currents through tissues in a chemical medium or gel to cause various components to separate. Used in genetic analysis.

Embryo—The young before hatching from the egg.

Epicoracoid—Paired cartilaginous elements of the pectoral girdle.

Epidermis—The outermost layer of skin.

Estivation—A state of dormancy or torpor during prolonged hot or dry periods.

Exostosis—A proliferation of bone usually resulting in sculpturing on the surface of a bone.

Explosive breeder—A species in which the breeding season is very short, usually at the time of the first heavy rains of the rainy season.

External fertilization—The joining of sperm and eggs (fertilization) outside of the female's body.

Femoral gland—A gland on the thigh.

Fertilization—The penetration of an egg by sperm.

Fetus—The unborn young of a viviparous animal.

Firmisternal—A condition in which the two halves of the pectoral (shoulder) girdle are fused ventrally.

Fossorial—Living underground.

Frontoparietal—Paired bones forming most of the roof of the skull.

Gill—A respiratory structure in aquatic animals through which gas exchange occurs.

Girdle—The group of connected bones that provide support for a pair of limbs.

Gland—An organ that produces chemical compounds (secretions).

Gravid—Female carrying young or eggs.

Hatchling—A young animal that has just emerged from an egg.

Heterospecific—Members of a different species.

Home range—The area in which an individual lives, except for migrations.

Hybrid—Individual resulting from mating of parents that belong to different species.

Hyoid—The group of cartilages and bones in the throat.

Ilium (pl. ilia)—Dorsal or anterior part of the pelvic (hip) girdle.

Inguinal—Pertaining to the groin.

Internal fertilization—Penetration of eggs by sperm inside the female's body.

Intromittent organ—A male copulatory organ.

Juvenile—Young, not sexually mature.

Keratinous—Epidermal structures composed of tough, fibrous protein (e.g., claws).

Labial tooth-row formula—The LTRF is written as a fraction designating the location and number of labial tooth rows. The most common LTRF is 2/3. The numerator indicates the number of rows on A, the anterior labium, while the denominator indicates the number of rows on P, the posterior labium. A LTRF only tells how many tooth rows there are on each labium and which ones have medial gaps; the lengths or positions of the rows are not designated.

Larva—The early stage of development after hatchling and before metamorphosis.

Larynx—A sound-producing structure at the anterior end of the trachea (windpipe) in the throat, containing the vocal cords.

Lateral line organ—A sense organ embedded in the skin that responds to water-borne vibrations.

Live-bearing—Giving birth to young that have developed beyond the egg stage.

Mandible—The skeletal elements (bones) that make up the lower jaw.

Maxillary—The skeletal elements (bones) that make up most of the upper jaw.

Metamorphosis—The transformation from one stage to another in the life cycle (e.g., from larva to adult).

Monophyly—The monophyletic taxon, also called a clade, includes the most recent common ancestor of all those organisms and all the descendants of that common ancestor.

Nares—The paired openings of the nasal capsule.

Nasal—Paired bones forming the anterior roof of the skull.

Newt—Salamanders of the genera *Notophthalmus*, *Taricha*, and *Triturus* that are characteristically aquatic.

Nocturnal—Active at night.

Ovary—The female reproductive organ that produces eggs, or ova.

Oviduct—The duct in females through which eggs pass from the ovary to the cloaca.

Oviparous—Producing eggs that develop and hatch outside the mother's body.

Ovoviviparous—Producing eggs that develop within the mother's body and hatch within or immediately after extrusion from the parent.

Ovum (pl. ova)—The female gamete, an egg.

Paedomorphic—Retention of juvenile (or larval) characters in the adult stage.

Papilla—A small, nipple-like projection.

Parotoid gland—One of a pair of large glands situated behind the eye.

Pectoral girdle—The group of bones that support the forelimbs.

Pelvic girdle—The group of bones that support the hind limbs.

Penultimate—Next to the last.

Phalange—One of the digits in the hand or foot.

Phylogenetic—Pertaining to evolutionary history.

Premaxillary—Paired bones forming the anterior margin of the upper jaw.

Scansorial—Adapted to or specialized for climbing.

Seasonal breeder—A species that breeds at a specific time of the year.

Sexual dimorphism—Difference of physical form (shape, size, or coloration) between the sexes; any consistent difference between males and females beyond the basic functional portions of the sex organs.

Spermatheca—Organ in the female that receives and stores sperm from the male pending fertilization of the eggs.

Spermatogenesis—Synonymous to spermiogenesis, the initial stage of sperm formation.

Spermatophore—A gelatinous structure capped with sperm produced by most male salamanders.

Spiracle—A slit that opens the throat to the outside.

Sternum—A median element in the pectoral girdle.

Tubercle—A small knob-like projection.

Tympanic annulus—A cartilaginous ring surrounding the tympanum.

Tympanum—The membranous eardrum.

Unken reflex—A defensive posture in which the body is arched and the head and tail are lifted upward.

Ventral—Pertaining to the lower surfaces of the body or one of its parts.

Viviparous—Giving birth to live young that develop within and are nourished by the mother.

Amphibians species list

Anura [Order]
Leiopelmatidae [Family]
  Leiopelma [Genus]
    L. archeyi [Species]
    L. hamiltoni
    L. hochstetteri
    L. pakeka

Ascaphidae [Family]
  Ascaphus [Genus]
    A. montanus [Species]
    A. truei

Bombinatoridae [Family]
  Barbourula [Genus]
    B. busuangensis [Species]
    B. kalimantanensis
    B. bombina
    B. fortinuptialis
    B. lichuanensis
    B. maxima
    B. microdeladigitora
    B. orientalis
    B. pachypus
    B. variegata

Discoglossidae [Family]
  Alytes [Genus]
    A. cisternasii [Species]
    A. dickhilleni
    A. muletensis
    A. obstetricans
  Discoglossus [Genus]
    D. hispanicus [Species]
    D. jeanneae
    D. montalenti
    D. nigriventer
    D. pictus
    D. sardus

Rhinophrynidae [Family]
  Rhinophrynus [Genus]
    R. dorsalis [Species]

Pipidae [Family]
  Hymenochirus [Genus]
    H. boettgeri [Species]

H. boulengeri
H. curtipes
H. feae
Pipa [Genus]
  P. arrabali [Species]
  P. aspera
  P. carvalhoi
  P. myersi
  P. parva
  P. pipa
  P. snethlageae
Pseudhymenochirus [Genus]
  P. merlini [Species]
Silurana [Genus]
  S. epitropicalis [Species]
  S. petersii
  S. tropicalis
Xenopus [Genus]
  X. amieti [Species]
  X. andrei
  X. borealis
  X. boumbaensis
  X. clivii
  X. fraseri
  X. gilli
  X. laevis
  X. largeni
  X. longipes
  X. muelleri
  X. pygmaeus
  X. ruwenzoriensis
  X. vestitus
  X. wittei

Megophryidae [Family]
  Atympanophrys [Genus]
    A. gigantica [Species]
    A. shapingensis
  Brachytarsophrys [Genus]
    B. carinensis [Species]
    B. feae
    B. gigantica
    B. intermedia
    B. platyparietus

Leptobrachella [Genus]
  L. baluensis [Species]
  L. brevicrus
  L. mjobergi
  L. natunae
  L. palmata
  L. parva
  L. serasanae
Leptobrachium [Genus]
  L. abbotti [Species]
  L. banae
  L. chapaense
  L. gunungense
  L. hainanense
  L. hasseltii
  L. hendricksoni
  L. montanum
  L. nigrops
  L. pullum
  L. smithi
  L. xanthospilum
Leptolalax [Genus]
  L. alpinus [Species]
  L. arayai
  L. bourreti
  L. dringi
  L. gracilis
  L. hamidi
  L. heteropus
  L. liui
  L. maurus
  L. nahangensis
  L. pelodytoides
  L. pictus
  L. sungi
  L. tuberosus
  L. ventripunctatus
Megophrys [Genus]
  M. aceras [Species]
  M. baluensis
  M. boettgeri
  M. brachykolos
  M. dringi
  M. edwardinae

M. glandulosa
M. kobayashii
M. kuatunensis
M. lateralis
M. longipes
M. mangshanensis
M. minor
M. montana
M. nankiangensis
M. omeimontis
M. pachyproctus
M. palpebralespinosa
M. parva
M. robusta
M. spinata
Ophryophryne [Genus]
  O. microstoma [Species]
  O. pachyproctus
  O. poilani
Oreolalax [Genus]
  O. chuanbeiensis [Species]
  O. granulosus
  O. jingdongensis
  O. liangbeiensis
  O. lichuanensis
  O. major
  O. multipunctatus
  O. nanjiangensis
  O. omeimontis
  O. pingii
  O. popei
  O. puxiongensis
  O. rhodostigmatus
  O. rugosus
  O. schmidti
  O. weigoldi
  O. xiangchengensis
Scutiger [Genus]
  S. adungensis [Species]
  S. bhutanensis
  S. boulengeri
  S. brevipes
  S. chintingensis
  S. chuanbeiensis
  S. glandulatus
  S. gongshanensis
  S. granulosus
  S. jingdongensis
  S. jiulongensis
  S. lichuanensis
  S. liupanensis
  S. maculatus
  S. major
  S. mammatus
  S. muliensis
  S. multipunctatus
  S. nepalensis
  S. ningshanensis
  S. nyingchiensis
  S. occidentalis
  S. omeimontis

S. pingii
S. pingwuensis
S. popei
S. rhodostigmatus
S. ruginosus
S. rugosus
S. schmidti
S. sikimmensis
S. tuberculatus
S. weigoldi
S. xiangchengensis
Vibrissaphora [Genus]
  V. ailaonica [Species]
  V. boringii
  V. echinata
  V. leishanensis
  V. liui
Xenophrys [Genus]
  X. daweimontis [Species]
  X. jingdongensis
  X. medogensis
  X. wuliangshanensis
  X. wushanensis
  X. zhangi

Pelobatidae [Family]
  Pelobates [Genus]
    P. cultripes [Species]
    P. fuscus
    P. syriacus
    P. varaldii
  Scaphiopus [Genus]
    S. couchii [Species]
    S. holbrookii
    S. hurterii
  Spea [Genus]
    S. bombifrons [Species]
    S. hammondii
    S. intermontana
    S. multiplicata

Pelodytidae [Family]
  Pelodytes [Genus]
    P. caucasicus [Species]
    P. ibericus
    P. punctatus

Heleophrynidae [Family]
  Heleophryne [Genus]
    H. hewitti [Species]
    H. natalensis
    H. orientalis
    H. purcelli
    H. regis
    H. rosei

Sooglossidae [Family]
  Nesomantis [Genus]
    N. thomasseti [Species]
  Sooglossus [Genus]
    S. gardineri [Species]
    S. sechellensis

Limnodynastidae [Family]
  Adelotus [Genus]
    A. brevis [Species]
  Heleioporus [Genus]
    H. albopunctatus [Species]
    H. australiacus
    H. barycragus
    H. eyrei
    H. inornatus
    H. psammophilus
  Kyarranus [Genus]
    K. kundagungan [Species]
    K. loveridgei
    K. sphagnicolus
  Lechriodus [Genus]
    L. aganoposis [Species]
    L. fletcheri
    L. intergerivus
    L. melanopyga
    L. platyceps
  Limnodynastes [Genus]
    L. convexiusculus [Species]
    L. depressus
    L. dorsalis
    L. dumerilii
    L. fletcheri
    L. interioris
    L. ornatus
    L. peronii
    L. salmini
    L. spenceri
    L. tasmaniensis
    L. terraereginae
  Megistolotis [Genus]
    M. lignarius [Species]
  Mixophyes [Genus]
    M. balbus [Species]
    M. fasciolatus
    M. fleayi
    M. hihihorlo
    M. iteratus
    M. schevilli
  Neobatrachus [Genus]
    N. albipes [Species]
    N. aquilonius
    N. centralis
    N. fulvus
    N. kunapalari
    N. pelobatoides
    N. pictus
    N. sudelli
    N. sutor
    N. wilsmorei
  Notaden [Genus]
    N. bennettii [Species]
    N. melanoscaphus
    N. nichollsi
    N. weigeli
  Philoria [Genus]
    P. frosti [Species]

Myobatrachidae [Family]
  Arenophryne [Genus]
    A. rotunda [Species]
  Assa [Genus]
    A. darlingtoni [Species]
  Bryobatrachus [Genus]
    B. nimbus [Species]
  Crinia [Genus]
    C. bilingua [Species]
    C. deserticola
    C. georgiana
    C. glauerti
    C. insignifera
    C. parinsignifera
    C. pseudinsignifera
    C. remota
    C. riparia
    C. signifera
    C. sloanei
    C. subinsignifera
    C. tasmaniensis
    C. tinnula
  Geocrinia [Genus]
    G. alba [Species]
    G. laevis
    G. leai
    G. lutea
    G. rosea
    G. victoriana
    G. vitellina
  Metacrinia [Genus]
    M. nichollsi [Species]
  Myobatrachus [Genus]
    M. gouldii [Species]
  Paracrinia [Genus]
    P. haswelli [Species]
  Pseudophryne [Genus]
    P. australis [Species]
    P. bibronii
    P. coriacea
    P. corroboree
    P. covacevichae
    P. dendyi
    P. douglasi
    P. guentheri
    P. major
    P. occidentalis
    P. pengilleyi
    P. raveni
    P. semimarmorata
  Rheobatrachus [Genus]
    R. silus [Species]
    R. vitellinus
  Spicospina [Genus]
    S. flammocaerulea [Species]
  Taudactylus [Genus]
    T. acutirostris [Species]
    T. diurnus
    T. eungellensis
    T. liemi

    T. pleione
    T. rheophilus
  Uperoleia [Genus]
    U. altissima [Species]
    U. arenicola
    U. aspera
    U. borealis
    U. capitulata
    U. crassa
    U. fusca
    U. glandulosa
    U. inundata
    U. laevigata
    U. lithomoda
    U. littlejohni
    U. marmorata
    U. martini
    U. micromeles
    U. mimula
    U. minima
    U. mjobergi
    U. orientalis
    U. rugosa
    U. russelli
    U. talpa
    U. trachyderma
    U. tyleri

Leptodactylidae [Family]
  Adelophryne [Genus]
    A. adiastola [Species]
    A. baturitensis
    A. gutturosa
    A. maranguapensis
    A. pachydactyla
  Adenomera [Genus]
    A. andreae [Species]
    A. bokermanni
    A. H.edactyla
    A. lutzi
    A. marmorata
    A. martinezi
  Alsodes [Genus]
    A. australis [Species]
    A. barrioi
    A. gargola
    A. hugoi
    A. kaweshkari
    A. laevis
    A. montanus
    A. monticola
    A. nodosus
    A. pehuenche
    A. tumultuosus
    A. vanzolinii
    A. verrucosus
    A. vittatus
  Atelognathus [Genus]
    A. ceii [Species]
    A. grandisonae
    A. nitoi

    A. patagonicus
    A. praebasalticus
    A. reverberii
    A. salai
    A. solitarius
  Atopophrynus [Genus]
    A. syntomopus [Species]
  Barycholos [Genus]
    B. pulcher [Species]
    B. ternetzi
  Batrachophrynus [Genus]
    B. brachydactylus [Species]
    B. macrostomus
  Batrachyla [Genus]
    B. antartandica [Species]
    B. fitzroya
    B. leptopus
    B. nibaldoi
    B. taeniata
  Caudiverbera [Genus]
    C. caudiverbera [Species]
  Ceratophrys [Genus]
    C. aurita [Species]
    C. calcarata
    C. cornuta
    C. cranwelli
    C. joazeirensis
    C. ornata
    C. stolzmanni
    C. testudo
  Chacophrys [Genus]
    C. pierottii [Species]
  Crossodactylodes [Genus]
    C. bokermanni [Species]
    C. izecksohni
    C. pintoi
  Crossodactylus [Genus]
    C. aeneus [Species]
    C. bokermanni
    C. caramaschii
    C. dantei
    C. dispar
    C. gaudichaudii
    C. grandis
    C. lutzorum
    C. schmidti
    C. trachystomus
  Cycloramphus [Genus]
    C. asper [Species]
    C. bandeirensis
    C. bolitoglossus
    C. boraceiensis
    C. brasiliensis
    C. carvalhoi
    C. catarinensis
    C. cedrensis
    C. diringshofeni
    C. dubius
    C. duseni
    C. eleutherodactylus

C. fulginosus
C. granulosus
C. izecksohni
C. jordanensis
C. juimirim
C. lutzorum
C. migueli
C. mirandaribeiroi
C. ohausi
C. rhyakonastes
C. semipalmatus
C. stejnegeri
C. valae
Dischidodactylus [Genus]
  D. colonnelloi [Species]
  D. duidensis
Edalorhina [Genus]
  E. nasuta [Species]
  E. perezi
Eleutherodactylus [Genus]
  E. aaptus [Species]
  E. abbotti
  E. acatallelus
  E. acerus
  E. achatinus
  E. acmonis
  E. actinolaimus
  E. actites
  E. acuminatus
  E. acutirostris
  E. adamastus
  E. aemulatus
  E. affinis
  E. alalocophus
  E. alberchi
  E. albericoi
  E. albipes
  E. alcoae
  E. alfredi
  E. altae
  E. altamazonicus
  E. alticola
  E. amadeus
  E. amniscola
  E. amplinympha
  E. anatipes
  E. anciano
  E. andi
  E. andicola
  E. andrewsi
  E. anemerus
  E. angelicus
  E. angustidigitorum
  E. anolirex
  E. anomalus
  E. anonymus
  E. anotis
  E. anthrax
  E. antillensis
  E. aphanus

E. apiculatus
E. apostates
E. appendiculatus
E. araiodactylus
E. ardalonychus
E. armstrongi
E. ashkapara
E. atkinsi
E. atrabracus
E. atratus
E. audanti
E. augusti
E. aurantiguttatus
E. auriculatoides
E. auriculatus
E. aurilegulus
E. avicuporum
E. avius
E. azueroensis
E. babax
E. bacchus
E. baiotis
E. bakeri
E. balionotus
E. barlagnei
E. bartonsmithi
E. baryecuus
E. batrachylus
E. bearsei
E. bellona
E. berkenbuschii
E. bernali
E. bicolor
E. bicumulus
E. bilineatus
E. binotatus
E. biporcatus
E. blairhedgesi
E. bockermanni
E. boconoensis
E. bocourti
E. bogotensis
E. bolbodactylus
E. boulengeri
E. bransfordii
E. bresslerae
E. brevifrons
E. brevirostris
E. briceni
E. brittoni
E. brocchi
E. bromeliaceus
E. buccinator
E. buckleyi
E. bufoniformis
E. cabrerai
E. cacao
E. cadenai
E. cajamarcensis
E. calcaratus

E. calcarulatus
E. caliginosus
E. cantitans
E. capitonis
E. caprifer
E. caribe
E. carmelitae
E. carranguerorum
E. carvalhoi
E. caryophyllaceus
E. casparii
E. catalinae
E. cavernibardus
E. cavernicola
E. celator
E. cerasinus
E. cerastes
E. ceuthospilus
E. chac
E. chalceus
E. charadra
E. charlottevillensis
E. cheiroplethus
E. chiastonotus
E. chloronotus
E. chlorophenax
E. chlorosoma
E. chrysops
E. chrysozetetes
E. citriogaster
E. cochranae
E. coffeus
E. colodactylus
E. colomai
E. colostichos
E. condor
E. conspicillatus
E. cooki
E. coqui
E. cornutus
E. corona
E. cosnipatae
E. counouspeus
E. crassidigitus
E. cremnobates
E. crenunguis
E. crepitans
E. cristinae
E. croceoinguinis
E. crucifer
E. cruentus
E. cruralis
E. cruzi
E. cryophilius
E. cryptomelas
E. cuaquero
E. cubanus
E. cundalli
E. cuneatus
E. cuneirostris

E. curtipes
E. cystignathoides
E. danae
E. darlingtoni
E. daryi
E. decoratus
E. degener
E. deinops
E. delicatus
E. delius
E. dennisi
E. devillei
E. diadematus
E. diaphonus
E. diastema
E. dilatus
E. dimidiatus
E. diogenes
E. discoidalis
E. dissimulatus
E. dixoni
E. dolomedes
E. dolops
E. dorsopictus
E. douglasi
E. duellmani
E. duende
E. dundeei
E. eileenae
E. elassodiscus
E. elegans
E. emcelae
E. emiliae
E. emleni
E. eneidae
E. epipedus
E. epochthidius
E. eremitus
E. eriphus
E. ernesti
E. erythromerus
E. erythropleura
E. escoces
E. etheridgei
E. eugeniae
E. eunaster
E. euphronides
E. eurydactylus
E. exoristus
E. factiosus
E. fallax
E. fecundus
E. fenestratus
E. fetosus
E. fitzingeri
E. flavescens
E. fleischmanni
E. floridus
E. fowleri
E. frater

E. fraudator
E. furcyensis
E. fuscus
E. gaigeae
E. galdi
E. ganonotus
E. ginesi
E. gladiator
E. glamyrus
E. glandulifer
E. glanduliferoides
E. glandulosus
E. glaphycompus
E. glaucoreius
E. glaucus
E. goini
E. gollmeri
E. gossei
E. grabhami
E. gracilis
E. grahami
E. grandiceps
E. grandis
E. grandoculis
E. granulosus
E. greggi
E. greyi
E. griphus
E. gryllus
E. gualteri
E. guanahacabibes
E. guantanamera
E. guentheri
E. guerreroensis
E. gularis
E. gulosus
E. gundlachi
E. guttilatus
E. gutturalis
E. haitianus
E. hamiotae
E. hectus
E. hedricki
E. helonotus
E. helvolus
E. heminota
E. hernandezi
E. heterodactylus
E. hobartsmithi
E. hoehnei
E. holti
E. hybotragus
E. hylaeformis
E. hypostenor
E. iberia
E. ibischi
E. ignicolor
E. illotus
E. ilojsintuta
E. imitatrix

E. inachus
E. incanus
E. incertus
E. incomptus
E. infraguttatus
E. ingeri
E. inguinalis
E. inoptatus
E. insignitus
E. intermedius
E. interorbitalis
E. inusitatus
E. ionthus
E. izecksohni
E. jaimei
E. jamaicensis
E. jasperi
E. jaumei
E. johannesdei
E. johnstonei
E. jorgevelosai
E. jota
E. juanchoi
E. jugans
E. juipoca
E. junori
E. karcharias
E. karlschmidti
E. katopteroides
E. kelephas
E. kirklandi
E. klinikowskii
E. labiosus
E. lacrimosus
E. lacteus
E. lamprotes
E. lancinii
E. lanthanites
E. lasallorum
E. latens
E. laticeps
E. laticlavius
E. laticorpus
E. latidiscus
E. lauraster
E. leberi
E. lemur
E. lentiginosus
E. lentus
E. leoncei
E. leoni
E. leprus
E. leptolophus
E. leucopus
E. librarius
E. lichenoides
E. limbatus
E. lindae
E. lineatus
E. lirellus

E. lividus
E. llojsintuta
E. locustus
E. longipes
E. longirostris
E. loustes
E. lucioi
E. luscombei
E. luteolateralis
E. luteolus
E. lutitus
E. lymani
E. lynchi
E. lythrodes
E. macdougalli
E. maculosus
E. malkini
E. manezinho
E. mantipus
E. mariposa
E. marmoratus
E. marnockii
E. mars
E. martiae
E. martinicensis
E. matudai
E. maurus
E. medemi
E. megacephalus
E. megalops
E. megalotympanum
E. melacara
E. melanoproctus
E. melanostictus
E. memorans
E. mendax
E. mercedesae
E. merendonensis
E. merostictus
E. metabates
E. mexicanus
E. milesi
E. mimus
E. minutus
E. miyatai
E. mnionaetes
E. modestus
E. modipeplus
E. molybrignus
E. mondolfii
E. monensis
E. monnichorum
E. montanus
E. moro
E. muricatus
E. museosus
E. myersi
E. myllomyllon
E. myops
E. nasutus

E. nebulosus
E. necerus
E. necopinus
E. neodreptus
E. nephophilus
E. nervicus
E. nicefori
E. nigriventris
E. nigrogriseus
E. nigrovittatus
E. nitidus
E. nivicolimae
E. noblei
E. nortoni
E. nubicola
E. nyctophylax
E. obesus
E. obmutescens
E. occidentalis
E. ocellatus
E. ockendeni
E. ocreatus
E. octavioi
E. oeus
E. olanchano
E. olivaceus
E. omiltemanus
E. omoaensis
E. opimus
E. orcesi
E. orcutti
E. orestes
E. orientalis
E. ornatissimus
E. orpacobates
E. orphnolaimus
E. oxyrhyncus
E. paisa
E. palenque
E. pallidus
E. palmeri
E. pantoni
E. parabates
E. paramerus
E. paranaensis
E. parapelates
E. pardalis
E. parectatus
E. parvillus
E. parvus
E. pastazensis
E. pataikos
E. patriciae
E. paulodutrai
E. paulsoni
E. paululus
E. pechorum
E. pecki
E. pelorus
E. penelopus

E. pentasyringos
E. peraticus
E. percnopterus
E. percultus
E. permixtus
E. peruvianus
E. petersorum
E. petrobardus
E. pezopetrus
E. phalarus
E. phasma
E. philipi
E. phoxocephalus
E. phragmipleuron
E. piceus
E. pictissimus
E. pinarensis
E. pinchoni
E. pinguis
E. pipilans
E. pituinus
E. planirostris
E. platychilus
E. platydactylus
E. pleurostriatus
E. plicifer
E. pluvicanorus
E. podiciferus
E. polychrus
E. polymniae
E. poolei
E. portoricensis
E. pozo
E. principalis
E. probolaeus
E. prolatus
E. prolixodiscus
E. proserpens
E. pruinatus
E. psephosypharus
E. pseudoacuminatus
E. pteridophilus
E. ptochus
E. pugnax
E. pulidoi
E. pulvinatus
E. punctariolus
E. pusillus
E. pycnodermis
E. pygmaeus
E. pyrrhomerus
E. quantus
E. quaquaversus
E. quidditus
E. quinquagesimus
E. racemus
E. ramagii
E. randorum
E. raniformis
E. rayo

E. repens
E. restrepoi
E. reticulatus
E. rhabdolaemus
E. rhodesi
E. rhodopis
E. rhodoplichus
E. rhodostichus
E. rhyacobatrachus
E. richmondi
E. ricordii
E. ridens
E. riparius
E. riveroi
E. riveti
E. rivulus
E. ronaldi
E. rosadoi
E. roseus
E. rostralis
E. rozei
E. rubicundus
E. rubrimaculatus
E. ruedai
E. rufescens
E. rufifemoralis
E. rufioculis
E. rugosus
E. rugulosus
E. ruidus
E. ruizi
E. rupinius
E. ruthae
E. ruthveni
E. sabrinus
E. salaputium
E. saltator
E. saltuarius
E. samaipatae
E. sanctaemartae
E. sanguineus
E. sartori
E. satagius
E. savagei
E. saxatilis
E. schmidti
E. schultei
E. schwartzi
E. sciagraphus
E. scitulus
E. scoloblepharus
E. scolodiscus
E. scopaeus
E. semipalmatus
E. serendipitus
E. sernai
E. shrevei
E. sierramaestrae
E. signifer
E. silverstonei

E. silvicola
E. simonbolivari
E. simoteriscus
E. simoterus
E. siopelus
E. sisyphodemus
E. skydmainos
E. sobetes
E. spanios
E. spatulatus
E. spilogaster
E. spinosus
E. stadelmani
E. stejnegerianus
E. stenodiscus
E. sternothylax
E. stuarti
E. subsigillatus
E. suetus
E. sulcatus
E. sulculus
E. supernatis
E. surdus
E. symingtoni
E. syristes
E. taeniatus
E. talamancae
E. tamsitti
E. tarahumaraensis
E. taurus
E. taylori
E. tayrona
E. tenebrionis
E. teretistes
E. terraebolivaris
E. tetajulia
E. thectopternus
E. thomasi
E. thorectes
E. thymalopsoides
E. thymelensis
E. tigrillo
E. tinker
E. toa
E. toftae
E. tonyi
E. torrenticola
E. trachyblepharis
E. trachydermus
E. trepidotus
E. tribulosus
E. truebae
E. tubernasus
E. turquinensis
E. turumiquirensis
E. unicolor
E. unistrigatus
E. uno
E. uranobates
E. urichi

E. vanadise
E. variabilis
E. varians
E. varleyi
E. veletis
E. venancioi
E. ventrilineatus
E. ventrimarmoratus
E. verecundus
E. verrucipes
E. verruculatus
E. versicolor
E. vertebralis
E. vicarius
E. vidua
E. viejas
E. vilarsi
E. vinhai
E. viridicans
E. viridis
E. vocalis
E. vocator
E. w-nigrum
E. walkeri
E. warreni
E. weinlandi
E. wetmorei
E. wiensi
E. wightmanae
E. xeniolum
E. xestus
E. xucanebi
E. xylochobates
E. yaviensis
E. yucatanensis
E. zeuctotylus
E. zeus
E. zimmermanae
E. zongoensis
E. zophus
E. zugi
E. zygodactylus
Euparkerella [Genus]
    E. brasiliensis [Species]
    E. cochranae
    E. robusta
    E. tridactyla
Eupsophus [Genus]
    E. calcaratus [Species]
    E. contulmoensis
    E. emiliopugini
    E. insularis
    E. migueli
    E. nahuelbutensis
    E. roseus
    E. vertebralis
Geobatrachus [Genus]
    G. walkeri [Species]
Holoaden [Genus]
    H. bradei [Species]

H. luederwaldti
Hydrolaetare [Genus]
    H. schmidti [Species]
Hylodes [Genus]
    H. amnicola [Species]
    H. asper
    H. babax
    H. charadranaetes
    H. glaber
    H. heyeri
    H. lateristrigatus
    H. magalhaesi
    H. meridionalis
    H. mertensi
    H. nasus
    H. ornatus
    H. otavioi
    H. perplicatus
    H. phyllodes
    H. regius
    H. sazima
    H. uai
    H. vanzolinii
Hylorina [Genus]
    H. sylvatica [Species]
Insuetophrynus [Genus]
    I. acarpicus [Species]
Ischnocnema [Genus]
    I. quixensis [Species]
    I. sanctaecrucis
    I. saxatilis
    I. simmonsi
    I. verrucosa
Lepidobatrachus [Genus]
    L. asper [Species]
    L. laevis
    L. llanensis
Leptodactylus [Genus]
    L. albilabris [Species]
    L. bolivianus
    L. bufonius
    L. camaquara
    L. chaquensis
    L. colombiensis
    L. cunicularius
    L. dantasi
    L. didymus
    L. diedrus
    L. elenae
    L. fallax
    L. flavopictus
    L. furnarius
    L. fuscus
    L. geminus
    L. gracilis
    L. griseigularis
    L. hallowelli
    L. insularum
    L. jolyi
    L. knudseni

L. labialis
L. labrosus
L. labyrinthicus
L. laticeps
L. latinasus
L. leptodactyloides
L. lithonaetes
L. longirostris
L. macrosternum
L. magistris
L. marambaiae
L. melanonotus
L. myersi
L. mystaceus
L. mystacinus
L. nesiotes
L. notoaktites
L. ocellatus
L. pallidirostris
L. pascoensis
L. pentadactylus
L. petersii
L. plaumanni
L. podicipinus
L. poecilochilus
L. pustulatus
L. rhodomystax
L. rhodonotus
L. rhodostima
L. riveroi
L. rugosus
L. sabanensis
L. silvanimbus
L. spixii
L. stenodema
L. syphax
L. tapiti
L. troglodytes
L. validus
L. ventrimaculatus
L. viridis
L. wagneri
Limnomedusa [Genus]
    L. macroglossa [Species]
Lithodytes [Genus]
    L. lineatus [Species]
Macrogenioglottus [Genus]
    M. alipioi [Species]
Megaelosia [Genus]
    M. bocainensis [Species]
    M. boticariana
    M. goeldii
    M. lutzae
    M. massarti
Odontophrynus [Genus]
    O. achalensis [Species]
    O. americanus
    O. barrioi
    O. carvalhoi
    O. cultripes

O. lavillai
O. moratoi
O. occidentalis
O. salvatori
Paratelmatobius [Genus]
    P. cardosoi [Species]
    P. gaigeae
    P. lutzii
    P. mantiqueira
    P. pictiventris
    P. poecilogaster
Phrynopus [Genus]
    P. adenobrachius [Species]
    P. adenopleurus
    P. bagrecitoi
    P. bracki
    P. brunneus
    P. columbianus
    P. cophites
    P. dagmarae
    P. fallaciosus
    P. flavomaculatus
    P. heimorum
    P. horstpauli
    P. iatamasi
    P. juninensis
    P. kauneorum
    P. kempffi
    P. laplacai
    P. lucida
    P. montium
    P. nanus
    P. nebulanastes
    P. parkeri
    P. peraccai
    P. pereger
    P. peruanus
    P. peruvianus
    P. pinguis
    P. simonsii
    P. spectabilis
    P. thomsoni
    P. wettsteini
Phyllonastes [Genus]
    P. carrascoicola [Species]
    P. heyeri
    P. lochites
    P. lynchi
    P. myrmecoides
    P. ritarasquinae
Physalaemus [Genus]
    P. aguirrei [Species]
    P. albifrons
    P. albonotatus
    P. barrioi
    P. biligonigerus
    P. bokermanni
    P. caete
    P. centralis
    P. cicada

P. coloradorum
P. crombiei
P. cuqui
P. cuvieri
P. deimaticus
P. enesefae
P. ephippifer
P. evangelistai
P. fernandezae
P. fischeri
P. fuscomaculatus
P. gracilis
P. henselii
P. jordanensis
P. kroyeri
P. lisei
P. maculiventris
P. maximus
P. moreirae
P. nanus
P. nattereri
P. obtectus
P. olfersii
P. petersi
P. pustulatus
P. pustulosus
P. riograndensis
P. rupestris
P. santafecinus
P. signifer
P. soaresi
P. spinigerus
Phyzelaphryne [Genus]
   P. miriamae [Species]
Pleurodema [Genus]
   P. bibroni [Species]
   P. borellii
   P. brachyops
   P. bufonina
   P. cinerea
   P. diplolistris
   P. guayapae
   P. kriegi
   P. marmorata
   P. nebulosa
   P. thaul
   P. tucumana
Proceratophrys [Genus]
   P. appendiculata [Species]
   P. avelinoi
   P. bigibbosa
   P. boiei
   P. cristiceps
   P. cristinae
   P. cururu
   P. fryi
   P. goyana
   P. laticeps
   P. melanopogon
   P. moehringi

P. palustris
P. schirchi
Pseudopaludicola [Genus]
   P. boliviana [Species]
   P. ceratophryes
   P. falcipes
   P. llanera
   P. mineira
   P. mirandae
   P. mystacalis
   P. pusilla
   P. riopiedadensis
   P. saltica
   P. ternetzi
Rupirana [Genus]
   R. cardosoi [Species]
Scythrophrys [Genus]
   S. sawayae [Species]
Somuncuria [Genus]
   S. somuncurensis [Species]
Syncope [Genus]
   S. antenori [Species]
   S. carvalhoi
   S. tridactyla
Telmatobius [Genus]
   T. albiventris [Species]
   T. arequipensis
   T. atacamensis
   T. atahualpai
   T. brevipes
   T. brevirostris
   T. carillae
   T. ceiorum
   T. cirrhacelis
   T. colanensis
   T. contrerasi
   T. crawfordi
   T. culeus
   T. dankoi
   T. edaphonastes
   T. fronteriensis
   T. gigas
   T. halli
   T. hauthali
   T. hockingi
   T. hypselocephalus
   T. ifornoi
   T. ignavus
   T. intermedius
   T. jahuira
   T. jelskii
   T. laticeps
   T. latirostris
   T. marmoratus
   T. mayoloi
   T. necopinus
   T. niger
   T. oxycephalus
   T. pefauri
   T. peruvianus

T. pinguiculus
T. platycephalus
T. rimac
T. schreiteri
T. scrocchii
T. simonsi
T. stephani
T. thompsoni
T. truebae
T. vellardi
T. yuracare
T. zapahuirensis
Telmatobufo [Genus]
   T. australis [Species]
   T. bullocki
   T. venustus
Thoropa [Genus]
   T. lutzi [Species]
   T. megatympanum
   T. miliaris
   T. petropolitana
   T. saxatilis
Vanzolinius [Genus]
   V. discodactylus [Species]
Zachaenus [Genus]
   Z. carvalhoi [Species]
   Z. parvulus
   Z. roseus

Rhinodermatidae [Family]
   Rhinoderma [Genus]
      R. darwinii [Species]
      R. rufum

Brachycephalidae [Family]
   Brachycephalus [Genus]
      B. didactyla [Species]
      B. ephippium
      B. hermogenesi
      B. nodoterga
      B. pernix
      B. vertebralis

Bufonidae [Family]
   Adenomus [Genus]
      A. dasi [Species]
      A. kandianus
      A. kelaartii
   Altiphrynoides [Genus]
      A. malcolmi [Species]
   Andinophryne [Genus]
      A. atelopoides [Species]
      A. colomai
      A. olallai
   Ansonia [Genus]
      A. albomaculata [Species]
      A. anotis
      A. fuliginea
      A. guibei
      A. hanitschi
      A. inthanon

A. latidisca
A. leptopus
A. longidigita
A. malayana
A. mcgregori
A. minuta
A. muelleri
A. ornata
A. penangensis
A. platysoma
A. rubrigina
A. siamensis
A. spinulifer
A. tiomanica
A. torrentis
Atelophryniscus [Genus]
   A. chrysophorus [Species]
Atelopus [Genus]
   A. andinus [Species]
   A. angelito
   A. arsyecue
   A. arthuri
   A. balios
   A. bomolochos
   A. boulengeri
   A. carauta
   A. carbonerensis
   A. carrikeri
   A. certus
   A. chiriquiensis
   A. chocoensis
   A. chrysocorallus
   A. coynei
   A. cruciger
   A. ebenoides
   A. elegans
   A. erythropus
   A. eusebianus
   A. exiguus
   A. famelicus
   A. farci
   A. flavescens
   A. franciscus
   A. galactogaster
   A. glyphus
   A. guanujo
   A. guitarraensis
   A. halihelos
   A. ignescens
   A. laetissimus
   A. leoperezii
   A. limosus
   A. longibrachius
   A. longirostris
   A. lozanoi
   A. lynchi
   A. mandingues
   A. mindoensis
   A. minutulus
   A. mucubajiensis

A. muisca
A. nahumae
A. nanay
A. nepiozomus
A. nicefori
A. oxyrhynchus
A. pachydermus
A. palmatus
A. pedimarmoratus
A. peruensis
A. petriruizi
A. pictiventris
A. pinangoi
A. planispina
A. quimbaya
A. sanjosei
A. seminiferus
A. senex
A. sernai
A. simulatus
A. siranus
A. sonsonensis
A. sorianoi
A. spumarius
A. spurrelli
A. subornatus
A. tamaensis
A. tricolor
A. varius
A. walkeri
A. willimani
A. zeteki
Bufo [Genus]
   B. abatus [Species]
   B. achalensis
   B. acutirostris
   B. ailaoanus
   B. alvarius
   B. amatolicus
   B. amboroensis
   B. americanus
   B. anderssoni
   B. andrewsi
   B. angusticeps
   B. arabicus
   B. arborescandens
   B. arenarum
   B. arequipensis
   B. arunco
   B. asmarae
   B. asper
   B. aspinius
   B. atacamensis
   B. atukoralei
   B. bankorensis
   B. baxteri
   B. beddomii
   B. beebei
   B. beiranus
   B. bergi

B. biporcatus
B. blanfordii
B. blombergi
B. bocourti
B. boreas
B. brauni
B. brevirostris
B. brongersmai
B. buchneri
B. bufo
B. caeruleostictus
B. calamita
B. californicus
B. camerunensis
B. campbelli
B. canaliferus
B. canorus
B. castaneoticus
B. cataulaciceps
B. cavifrons
B. celebensis
B. ceratophrys
B. chavin
B. chlorogaster
B. chudeaui
B. claviger
B. coccifer
B. cognatus
B. compactilis
B. coniferus
B. cophotis
B. corynetes
B. cristatus
B. cristiglans
B. crucifer
B. cryptotympanicus
B. cycladen
B. cyphosus
B. damaranus
B. danatensis
B. danielae
B. dapsilis
B. debilis
B. dhufarensis
B. diptychus
B. divergens
B. djohongensis
B. dodsoni
B. dombensis
B. dorbignyi
B. empusus
B. exsul
B. fastidiosus
B. fenoulheti
B. fergusonii
B. fernandezae
B. fissipes
B. flavolineatus
B. fowleri
B. fuliginatus

B. funereus
B. gabbi
B. galeatus
B. gallardoi
B. gargarizans
B. gariepensis
B. garmani
B. gemmifer
B. glaberrimus
B. gnustae
B. gracilipes
B. grandisonae
B. granulosus
B. guttatus
B. gutturalis
B. hadramautinus
B. haematiticus
B. hemiophrys
B. himalayanus
B. hoeschi
B. holdridgei
B. hololius
B. houstonensis
B. hypomelas
B. ibarrai
B. ictericus
B. inca
B. intermedius
B. inyangae
B. iserni
B. japonicus
B. jimi
B. jordani
B. justinianoi
B. juxtasper
B. kabischi
B. kassasii
B. kavangensis
B. kavirensis
B. kelloggi
B. kerinyagae
B. kisoloensis
B. kotagamai
B. koynayensis
B. langanoensis
B. latastii
B. latifrons
B. lemairii
B. limensis
B. lindneri
B. lonnbergi
B. luetkenii
B. lughenisis
B. lughensis
B. luristanicus
B. macrocristatus
B. macrotis
B. maculatus
B. margaritifer
B. marinus

B. marmoreus
B. mauritanicus
B. mazatlanensis
B. melanochlorus
B. melanogaster
B. melanopleura
B. melanostictus
B. mexicanus
B. microscaphus
B. microtympanum
B. minshanicus
B. mocquardi
B. nasicus
B. nesiotes
B. noellerti
B. nouettei
B. nyikae
B. oblongus
B. occidentalis
B. ocellatus
B. olivaceus
B. pageoti
B. pantherinus
B. paracnemis
B. pardalis
B. parietalis
B. parkeri
B. parvus
B. pentoni
B. periglenes
B. peripatetes
B. perplexus
B. perreti
B. pewzowi
B. poeppigii
B. poweri
B. proboscideus
B. pseudoraddei
B. punctatus
B. pygmaeus
B. quadriporcatus
B. quechua
B. quercicus
B. raddei
B. rangeri
B. reesi
B. regularis
B. retiformis
B. robinsoni
B. roqueanus
B. rubropunctatus
B. rufus
B. rumbolli
B. schmidti
B. schneideri
B. sclerocephalus
B. scorteccii
B. shaartusiensis
B. silentvalleyensis
B. simus

B. speciosus
B. spiculatus
B. spinulosus
B. steindachneri
B. stejnegeri
B. sternosignatus
B. stomaticus
B. stuarti
B. sumatranus
B. superciliaris
B. surdus
B. tacanensis
B. taiensis
B. taitanus
B. terrestris
B. tibetanus
B. tienhoensis
B. tihamicus
B. togoensis
B. torrenticola
B. trifolium
B. tuberculatus
B. tuberosus
B. turkanae
B. tutelarius
B. urunguensis
B. uzunguensis
B. valhallae
B. valliceps
B. variegatus
B. vellardi
B. veraguensis
B. verrucosissimus
B. vertebralis
B. villiersi
B. viridis
B. vittatus
B. wolongensis
B. woodhousii
B. xeros
Bufoides [Genus]
B. meghalayanus [Species]
Capensibufo [Genus]
C. rosei [Species]
C. tradouwi
Crepidophryne [Genus]
C. epiotica [Species]
Dendrophryniscus [Genus]
D. berthalutzae [Species]
D. bokermanni
D. brevipollicatus
D. carvalhoi
D. leucomystax
D. minutus
D. stawiarskyi
Didynamipus [Genus]
D. sjostedti [Species]
Frostius [Genus]
F. pernambucensis [Species]
Laurentophryne [Genus]

L. parkeri [Species]
Leptophryne [Genus]
  L. borbonica [Species]
  L. cruentata
Melanophryniscus [Genus]
  M. atroluteus [Species]
  M. cambaraensis
  M. devincenzii
  M. macrogranulosus
  M. montevidensis
  M. moreirae
  M. orejasmirandai
  M. rubriventris
  M. sanmartini
  M. stelzneri
  M. tumifrons
Mertensophryne [Genus]
  M. micranotis [Species]
Metaphryniscus [Genus]
  M. sosai [Species]
Nectophryne [Genus]
  N. afra [Species]
  N. batesii
Nectophrynoides [Genus]
  N. asperginis [Species]
  N. cryptus
  N. minutus
  N. tornieri
  N. viviparus
  N. wendyae
  N. liberiensis
  N. occidentalis
Oreophrynella [Genus]
  O. cryptica [Species]
  O. hubneri
  O. macconnelli
  O. nigra
  O. quelchii
  O. vasquezi
Osornophryne [Genus]
  O. antisana [Species]
  O. bufoniformis
  O. guacamayo
  O. percrassa
  O. sumacoensis
  O. talipes
Pedostibes [Genus]
  P. everetti [Species]
  P. hosii
  P. kempi
  P. maculatus
  P. rugosus
  P. tuberculosus
Pelophryne [Genus]
  P. albotaeniata [Species]
  P. api
  P. brevipes
  P. guentheri
  P. lighti
  P. macrotis

P. misera
P. rhopophilius
P. scalptus
P. fluviatica
P. guentheri
P. gundlachi
P. lemur
P. longinasus
P. peltocephala
P. taladai
Pseudobufo [Genus]
  P. subasper [Species]
Rhamphophryne [Genus]
  R. acrolopha [Species]
  R. festae
  R. lindae
  R. macrorhina
  R. nicefori
  R. proboscidea
  R. rostrata
  R. tenrec
  R. truebae
Schismaderma [Genus]
  S. carens [Species]
Spinophrynoides [Genus]
  S. osgoodi [Species]
Stephopaedes [Genus]
  S. anotis [Species]
  S. loveridgei
Torrentophryne [Genus]
  T. aspinia [Species]
  T. burmana
Truebella [Genus]
  T. skoptes [Species]
  T. tothastes
Werneria [Genus]
  W. bambutensis [Species]
  W. mertensiana
  W. preussi
  W. tandyi
Wolterstorffina [Genus]
  W. chirioi [Species]
  W. mirei
  W. parvipalmata

Dendrobatidae [Family]
  Aromobates [Genus]
    A. nocturnus [Species]
  Colostethus [Genus]
    C. abditaurantius [Species]
    C. agilis
    C. alacris
    C. alagoanus
    C. alessandroi
    C. anthracinus
    C. argyrogaster
    C. atopoglossus
    C. awa
    C. ayarzaguenai
    C. beebei
    C. betancuri

C. bocagei
C. brachistriatus
C. breviquartus
C. bromelicola
C. brunneus
C. caeruleodactylus
C. capixaba
C. capurinensis
C. carioca
C. cevallosi
C. chalcopis
C. chocoensis
C. degranvillei
C. delatorreae
C. dunni
C. edwardsi
C. elachyhistus
C. exasperatus
C. faciopunctulatus
C. fallax
C. fascianiger
C. flotator
C. fraterdanieli
C. fugax
C. fuliginosus
C. furviventris
C. goianus
C. guanayensis
C. humilis
C. idiomelas
C. imbricolus
C. infraguttatus
C. inguinalis
C. jacobuspetersi
C. juanii
C. kingsburyi
C. lacrimosus
C. latinasus
C. lehmanni
C. littoralis
C. lynchi
C. machalilla
C. mandelorum
C. maquipucuna
C. marchesianus
C. marmoreoventris
C. mcdiarmidi
C. melanolaemus
C. mertensi
C. mittermeieri
C. murisipanensis
C. mystax
C. nexipus
C. nubicola
C. olfersioides
C. palmatus
C. parimae
C. parkerae
C. peculiaris
C. peruvianus

C. pinguis
C. poecilonotus
C. praderioi
C. pratti
C. pulchellus
C. pumilus
C. ramosi
C. ranoides
C. roraima
C. ruizi
C. ruthveni
C. saltuarius
C. saltuensis
C. sanmartini
C. sauli
C. shrevei
C. shuar
C. stepheni
C. subpunctatus
C. sylvaticus
C. talamancae
C. tamacuarensis
C. tepuyensis
C. thorntoni
C. toachi
C. trilineatus
C. utcubambensis
C. vergeli
C. vertebralis
C. whymperi
C. yaguara
Cryptophyllobates [Genus]
C. azureiventris [Species]
Dendrobates [Genus]
D. amazonicus [Species]
D. arboreus
D. auratus
D. azureus
D. biolat
D. captivus
D. castaneoticus
D. claudiae
D. duellmani
D. fantasticus
D. flavovittatus
D. galactonotus
D. granuliferus
D. histrionicus
D. imitator
D. labialis
D. lamasi
D. lehmanni
D. leucomelas
D. mysteriosus
D. occultator
D. pumilio
D. quinquevittatus
D. reticulatus
D. rubrocephalus
D. rufulus

D. speciosus
D. sylvaticus
D. tinctorius
D. truncatus
D. vanzolinii
D. ventrimaculatus
D. vicentei
Epipedobates [Genus]
E. andinus [Species]
E. bassleri
E. bilinguis
E. bolivianus
E. boulengeri
E. braccatus
E. cainarachi
E. erythromos
E. espinosai
E. femoralis
E. flavopictus
E. hahneli
E. ingeri
E. macero
E. maculatus
E. myersi
E. parvulus
E. petersi
E. pictus
E. planipaleae
E. pongoensis
E. pulchripectus
E. rubriventris
E. silverstonei
E. simulans
E. smaragdinus
E. tricolor
E. trivittatus
E. zaparo
Mannophryne [Genus]
M. caquetio [Species]
M. collaris
M. cordilleriana
M. herminae
M. lamarcai
M. neblina
M. oblitterata
M. olmonae
M. riveroi
M. trinitatis
M. yustizi
Minyobates [Genus]
M. abditus [Species]
M. altobueyensis
M. bombetes
M. fulguritus
M. minutus
M. opisthomelas
M. steyermarki
M. viridis
M. virolensis
Nephelobates [Genus]

N. alboguttatus [Species]
N. duranti
N. haydeeae
N. leopardalis
N. mayorgai
N. meridensis
N. molinarii
N. orostoma
N. serranus
Phyllobates [Genus]
P. aurotaenia [Species]
P. bicolor
P. lugubris
P. terribilis
P. vittatus

Allophrynidae [Family]
Allophryne [Genus]
A. ruthveni [Species]

Centrolenidae [Family]
Centrolene [Genus]
C. acanthidiocephalum [Species]
C. altitudinale
C. andinum
C. antioquiense
C. audax
C. azulae
C. bacatum
C. ballux
C. buckleyi
C. fernandoi
C. geckoideum
C. gemmatum
C. gorzulai
C. grandisonae
C. guanacarum
C. heloderma
C. hesperium
C. huilense
C. hybrida
C. ilex
C. lemniscatum
C. litoralis
C. lynchi
C. mariae
C. medemi
C. muelleri
C. notostictum
C. paezorum
C. papillahallicum
C. peristictum
C. petrophilum
C. pipilatum
C. prosoblepon
C. puyoense
C. quindianum
C. robledoi
C. sanchezi
C. scirtetes

C. tayrona
C. venezuelense
Cochranella [Genus]
C. adiazeta [Species]
C. albomaculata
C. ametarsia
C. anomala
C. armata
C. auyantepuiana
C. balionota
C. bejaranoi
C. cariticommata
C. castroviejoi
C. chami
C. chancas
C. cochranae
C. cristinae
C. croceopodes
C. daidalea
C. duidaeana
C. euhystrix
C. euknemos
C. flavopunctata
C. garciae
C. geijskesi
C. granulosa
C. griffithsi
C. helenae
C. ignota
C. luminosa
C. luteopunctata
C. megacheira
C. megistra
C. midas
C. nephelophila
C. nola
C. ocellata
C. ocellifera
C. orejuela
C. oreonympha
C. oyampiensis
C. phenax
C. pluvialis
C. posadae
C. prasina
C. punctulata
C. ramirezi
C. resplendens
C. ritae
C. riveroi
C. rosada
C. ruizi
C. savagei
C. saxiscandens
C. siren
C. solitaria
C. spiculata
C. spilota
C. spinosa
C. susatamai

C. tangarana
C. truebae
C. vozmedianoi
C. xanthocheridia
Hyalinobatrachium [Genus]
H. antisthenesi [Species]
H. aureoguttatum
H. bergeri
H. cardiacalyptum
H. chirripoi
H. colymbiphyllum
H. crurifasciatum
H. crybetes
H. duranti
H. esmeralda
H. eurygnathum
H. flavidigitatum
H. fleischmanni
H. fragile
H. iaspidiense
H. ibama
H. lemur
H. loreocarinatum
H. mondolfii
H. munozorum
H. nouraguensis
H. orientale
H. ostracodermoides
H. pallidum
H. parvulum
H. pellucidum
H. pleurolineatum
H. pulveratum
H. revocatum
H. ruedai
H. talamancae
H. taylori
H. uranoscopum
H. valerioi
H. vireovittatum

Hylidae [Family]
Acris [Genus]
A. crepitans [Species]
A. gryllus
Agalychnis [Genus]
A. annae [Species]
A. calcarifer
A. callidryas
A. craspedopus
A. litodryas
A. moreletii
A. saltator
A. spurrelli
Anotheca [Genus]
A. spinosa [Species]
Aparasphenodon [Genus]
A. bokermanni [Species]
A. brunoi
A. venezolanus

Aplastodiscus [Genus]
A. perviridis [Species]
Argenteohyla [Genus]
A. siemersi [Species]
Calyptahyla [Genus]
C. crucialis [Species]
Corythomantis [Genus]
C. greeningi [Species]
Cryptobatrachus [Genus]
C. boulengeri [Species]
C. fuhrmanni
C. nicefori
Cyclorana [Genus]
C. alboguttata [Species]
C. australis
C. brevipes
C. cryptotis
C. cultripes
C. longipes
C. maculosa
C. maini
C. manya
C. novaehollandiae
C. platycephala
C. vagitus
C. verrucosa
Duellmanohyla [Genus]
D. chamulae [Species]
D. ignicolor
D. lythrodes
D. rufioculis
D. salvavida
D. schmidtorum
D. soralia
D. uranochroa
Flectonotus [Genus]
F. fissilis [Species]
F. fitzgeraldi
F. goeldii
F. ohausi
F. pygmaeus
Gastrotheca [Genus]
G. abdita [Species]
G. albolineata
G. andaquiensis
G. angustifrons
G. antomia
G. argenteovirens
G. aureomaculata
G. bufona
G. christiani
G. chrysosticta
G. cornuta
G. dendronastes
G. dunni
G. espeletia
G. excubitor
G. fissipes
G. galeata
G. gracilis

G. griswoldi
G. guentheri
G. helenae
G. lateonota
G. lauzurica
G. litonedis
G. longipes
G. marsupiata
G. microdiscus
G. monticola
G. nicefori
G. ochoai
G. orophylax
G. ovifera
G. pacchamama
G. peruana
G. plumbea
G. pseustes
G. psychrophila
G. rebeccae
G. riobambae
G. ruizi
G. splendens
G. testudinea
G. trachyceps
G. walkeri
G. weinlandii
G. williamsoni
Hemiphractus [Genus]
H. bubalus [Species]
H. fasciatus
H. helioi
H. johnsoni
H. proboscideus
H. scutatus
Hyla [Genus]
H. acreana [Species]
H. albofrenata
H. alboguttata
H. albomarginata
H. albonigra
H. albopunctata
H. albopunctulata
H. albosignata
H. albovittata
H. alemani
H. allenorum
H. altipotens
H. alvarengai
H. alytolylax
H. ameibothalame
H. americana
H. amicorum
H. anataliasiasi
H. anceps
H. andersonii
H. andina
H. angustilineata
H. annectans
H. aperomea

H. araguaya
H. arborea
H. arborescandens
H. arenicolor
H. arildae
H. armata
H. aromatica
H. astartea
H. atlantica
H. auraria
H. avivoca
H. baileyi
H. balzani
H. battersbyi
H. benitezi
H. berthalutzae
H. bifurca
H. biobeba
H. bipunctata
H. bischoffi
H. bistincta
H. boans
H. bocourti
H. bogerti
H. bogotensis
H. bokermanni
H. branneri
H. brevifrons
H. bromeliacia
H. buriti
H. cachimbo
H. cadaverina
H. caingua
H. calcarata
H. callipeza
H. callipygia
H. calthula
H. calvicollina
H. calypsa
H. carnifex
H. carvalhoi
H. catracha
H. caucana
H. cavicola
H. celata
H. cembra
H. cerradensis
H. chaneque
H. charadricola
H. charazani
H. chimalapa
H. chinensis
H. chlorostea
H. chryses
H. chrysoscelis
H. cinerea
H. cipoensis
H. circumdata
H. claresignata
H. clepsydra

H. columbiana
H. colymba
H. crassa
H. crepitans
H. cruzi
H. cyanomma
H. cymbalum
H. debilis
H. decipiens
H. delarivai
H. dendrophasma
H. dendroscarta
H. dentei
H. denticulenta
H. dolloi
H. dutrai
H. ebraccata
H. echinata
H. ehrhardti
H. elegans
H. elianeae
H. euphorbiacea
H. eximia
H. faber
H. fasciata
H. femoralis
H. fernandoi
H. fimbrimembra
H. fluminea
H. fuentei
H. fusca
H. garagoensis
H. gaucheri
H. geographica
H. giesleri
H. godmani
H. goiana
H. gouveai
H. graceae
H. grandisonae
H. granosa
H. gratiosa
H. gryllata
H. guentheri
H. haddadi
H. hadroceps
H. hallowellii
H. haraldschultzi
H. hazelae
H. heilprini
H. helenae
H. hobbsi
H. hutchinsi
H. hylax
H. hypselops
H. ibitiguara
H. ibitipoca
H. imitator
H. inframaculata
H. inparquesi

H. insolitus
H. intermedia
H. izecksohni
H. jahni
H. japonica
H. jimi
H. joannae
H. juanitae
H. kanaima
H. karenanneae
H. koechlini
H. labedactyla
H. labialis
H. lancasteri
H. lanciformis
H. langei
H. larinopygion
H. lascinia
H. leali
H. lemai
H. leptolineata
H. leucophyllata
H. leucopygia
H. limai
H. lindae
H. loquax
H. loveridgei
H. luctuosa
H. luteoocellata
H. lynchi
H. marginata
H. marianae
H. marianitae
H. marmorata
H. martinsi
H. mathiassoni
H. melanargyrea
H. melanomma
H. melanopleura
H. melanorhabdota
H. meridensis
H. meridiana
H. meridionalis
H. microcephala
H. microderma
H. microps
H. miliaria
H. minima
H. minuscula
H. minuta
H. miotympanum
H. mixe
H. mixomaculata
H. miyatai
H. molitor
H. multifasciata
H. musica
H. mykter
H. nahdereri
H. nana

H. nanuzae
H. nephila
H. novaisi
H. nubicola
H. ocapia
H. oliveirai
H. ornatissima
H. pacha
H. pachyderma
H. padreluna
H. palaestes
H. palliata
H. palmeri
H. pantosticta
H. pardalis
H. parviceps
H. pauiniensis
H. pelidna
H. pellita
H. pellucens
H. pentheter
H. perkinsi
H. phlebodes
H. phyllognatha
H. picadoi
H. piceigularis
H. picta
H. pictipes
H. picturata
H. pinima
H. pinorum
H. platydactyla
H. plicata
H. polytaenia
H. praestans
H. prasina
H. psarolaima
H. pseudopseudis
H. pseudopuma
H. ptychodactyla
H. pugnax
H. pulchella
H. pulchrilineata
H. punctata
H. quadrilineata
H. raniceps
H. regilla
H. rhea
H. rhodopepla
H. riveroi
H. rivularis
H. robertmertensi
H. robertsorum
H. roeschmanni
H. roraima
H. rosenbergi
H. rossalleni
H. rubicundula
H. rubracyla
H. rufitela

H. ruschii
H. sabrina
H. salvaje
H. sanborni
H. sanchiangensis
H. sarampiona
H. sarayacuensis
H. sarda
H. sartori
H. savignyi
H. saxicola
H. sazimai
H. schubarti
H. secedens
H. semiguttata
H. senicula
H. sibleszi
H. simmonsi
H. simplex
H. siopela
H. smaragdina
H. smithii
H. soaresi
H. squirella
H. staufferorum
H. stenocephala
H. stingi
H. subocularis
H. sumichrasti
H. surinamensis
H. suweonensis
H. taeniopus
H. thorectes
H. thysanota
H. tica
H. timbeba
H. tintinnabulum
H. torrenticola
H. triangulum
H. tritaeniata
H. trux
H. tsinlingensis
H. tuberculosa
H. uruguaya
H. valancifer
H. varelae
H. vasta
H. versicolor
H. vigilans
H. virolinensis
H. walkeri
H. warreni
H. wavrini
H. werneri
H. weygoldti
H. wilderi
H. xanthosticta
H. xapuriensis
H. xera
H. yaracuyana

H. zeteki
H. zhaopingensis
Hylomantis [Genus]
    H. aspera [Species]
    H. granulosa
Litoria [Genus]
    L. adelaidensis [Species]
    L. albolabris
    L. amboinensis
    L. andiirrmalin
    L. angiana
    L. arfakiana
    L. aruensis
    L. aurea
    L. becki
    L. bicolor
    L. booroolongensis
    L. brevipalmata
    L. brongersmai
    L. bulmeri
    L. burrowsae
    L. caerulea
    L. capitula
    L. castanea
    L. cavernicola
    L. chloris
    L. chloronota
    L. citropa
    L. congenita
    L. contrastens
    L. cooloolensis
    L. coplandi
    L. cyclorhynchus
    L. dahlii
    L. darlingtoni
    L. dentata
    L. dorsalis
    L. dorsivena
    L. electrica
    L. eucnemis
    L. everetti
    L. ewingii
    L. exophthalmia
    L. fallax
    L. freycineti
    L. genimaculata
    L. gilleni
    L. gracilenta
    L. graminea
    L. havina
    L. impura
    L. inermis
    L. infrafrenata
    L. iris
    L. jervisiensis
    L. jeudii
    L. latopalmata
    L. lesueuri
    L. leucova
    L. littlejohni

L. longicrus
L. longirostris
L. lorica
L. louisiadensis
L. lutea
L. majikthise
L. meiriana
L. microbelos
L. micromembrana
L. modica
L. moorei
L. mucro
L. multiplica
L. mystax
L. nannotis
L. napaea
L. nasuta
L. nigrofrenata
L. nigropunctata
L. nyakalensis
L. obtusirostris
L. oenicolen
L. ollauro
L. olongburensis
L. pallida
L. paraewingi
L. pearsoniana
L. peronii
L. personata
L. phyllochroa
L. piperata
L. pratti
L. pronimia
L. prora
L. pygmaea
L. quadrilineata
L. raniformis
L. revelata
L. rheocola
L. rothii
L. rubella
L. sanguinolenta
L. spenceri
L. spinifera
L. subglandulosa
L. thesaurensis
L. timida
L. tornieri
L. tyleri
L. umbonata
L. vagabunda
L. verreauxii
L. vocivincens
L. wapogaensis
L. watjulumensis
L. wisselensis
L. wollastoni
L. xanthomera
Lysapsus [Genus]
    L. caraya [Species]

L. laevis
L. limellus
Nyctimantis [Genus]
    N. rugi [Species]
Nyctimystes [Genus]
    N. avocalis [Species]
    N. cheesmani
    N. dayi
    N. daymani
    N. disruptus
    N. fluviatilis
    N. foricula
    N. granti
    N. gularis
    N. humeralis
    N. kubori
    N. montanus
    N. narinosus
    N. obsoletus
    N. oktediensis
    N. papua
    N. perimetri
    N. persimilis
    N. pulcher
    N. rueppelli
    N. semipalmatus
    N. trachydermis
    N. tyleri
    N. zweifeli
Osteocephalus [Genus]
    O. ayarzaguenai [Species]
    O. buckleyi
    O. cabrerai
    O. deridens
    O. elkejungingerae
    O. exophthalmus
    O. fuscifacies
    O. langsdorffii
    O. leprieurii
    O. mutabor
    O. oophagus
    O. pearsoni
    O. planiceps
    O. subtilis
    O. taurinus
    O. verruciger
    O. yasuni
Osteopilus [Genus]
    O. brunneus [Species]
    O. dominicensis
    O. septentrionalis
Pachymedusa [Genus]
    P. dacnicolor [Species]
Pelodryas [Genus]
    P. splendida [Species]
Phasmahyla [Genus]
    P. cochranae [Species]
    P. exilis
    P. guttata
    P. jandaia
Phrynohyas [Genus]

P. coriacea [Species]
P. imitatrix
P. mesophaea
P. resinifictrix
P. venulosa
Phrynomedusa [Genus]
P. appendiculata [Species]
P. bokermanni
P. fimbriata
P. marginata
P. vanzolinii
Phyllodytes [Genus]
P. acuminatus [Species]
P. auratus
P. brevirostris
P. kautskyi
P. luteolus
P. melanomystax
P. tuberculosus
Phyllomedusa [Genus]
P. atelopoides [Species]
P. ayeaye
P. baltea
P. bicolor
P. boliviana
P. buckleyi
P. burmeisteri
P. centralis
P. coelestis
P. danieli
P. distincta
P. duellmani
P. ecuatoriana
P. hulli
P. hypochondrialis
P. iheringii
P. lemur
P. medinai
P. palliata
P. perinesos
P. psilopygion
P. rohdei
P. sauvagii
P. tarsius
P. tetraploidea
P. tomopterna
P. trinitatis
P. vaillantii
P. venusta
Plectrohyla [Genus]
P. acanthodes [Species]
P. avia
P. chrysopleura
P. dasypus
P. exquisita
P. glandulosa
P. guatemalensis
P. hartwegi
P. ixil
P. lacertosa

P. matudai
P. pokomchi
P. psiloderma
P. pycnochila
P. quecchi
P. sagorum
P. tecunumani
P. teuchestes
Pseudacris [Genus]
P. brachyphona [Species]
P. brimleyi
P. clarkii
P. crucifer
P. feriarum
P. illinoensis
P. maculata
P. nigrita
P. ocularis
P. ornata
P. streckeri
P. triseriata
Pseudis [Genus]
P. bolbodactyla [Species]
P. fusca
P. minuta
P. paradoxa
P. tocantins
Pternohyla [Genus]
P. dentata [Species]
P. fodiens
Ptychohyla [Genus]
P. erythromma [Species]
P. euthysanota
P. hypomykter
P. legleri
P. leonhardschultzei
P. macrotympanum
P. panchoi
P. salvadorensis
P. sanctaecrucis
P. spinipollex
Scarthyla [Genus]
S. ostinodactyla [Species]
Scinax [Genus]
S. acuminata [Species]
S. agilis
S. albicans
S. alcatraz
S. alleni
S. altera
S. angrensis
S. argyreornata
S. ariadne
S. atrata
S. aurata
S. baumgardneri
S. berthae
S. blairi
S. boesemani
S. boulengeri

S. brieni
S. caldarum
S. canastrensis
S. cardosoi
S. carnevallii
S. castroviejoi
S. catharinae
S. centralis
S. chiquitana
S. crospedospila
S. cruentomma
S. cuspidatus
S. cynocephala
S. danae
S. duartei
S. ehrhardti
S. elaeochroa
S. eurydice
S. exigua
S. flavoguttata
S. funerea
S. fuscomarginata
S. fuscovaria
S. garbei
S. goinorum
S. granulata
S. hayii
S. heyeri
S. hiemalis
S. humilis
S. icterica
S. jolyi
S. jureia
S. kautskyi
S. kennedyi
S. lindsayi
S. littoralis
S. littorea
S. longilinea
S. luizotavioi
S. machadoi
S. maracaya
S. megapodia
S. melloi
S. nasica
S. nebulosa
S. obtriangulata
S. opalinus
S. oreites
S. pachycrus
S. parkeri
S. pedromedinae
S. perereca
S. perpusilla
S. proboscidea
S. quinquefasciata
S. ranki
S. rizibilis
S. rostrata
S. rubra

S. similis
S. squalirostris
S. staufferi
S. strigilata
S. sugillata
S. trachythorax
S. trapicheiroi
S. trilineata
S. v-signata
S. wandae
S. x-signata
Smilisca [Genus]
S. baudinii [Species]
S. cyanosticta
S. phaeota
S. puma
S. sila
S. sordida
Sphaenorhynchus [Genus]
S.bromelicola [Species]
S.carneus
S.dorisae
S.lacteus
S.orophilus
S.palustris
S.pauloalvini
S.planicola
S.platycephalus
S.prasinus
S.surdus
Stefania [Genus]
S. ackawaio [Species]
S. ayangannae
S. coxi
S. evansi
S. ginesi
S. goini
S. marahuaquensis
S. oculosa
S. percristata
S. riae
S. riveroi
S. roraimae
S. satelles
S. scalae
S. schuberti
S. tamacuarina
S. woodleyi
Tepuihyla [Genus]
T. aecii [Species]
T. celsae
T. edelcae
T. galani
T. luteolabris
T. rimarum
T. rodriguezi
T. talbergae
Trachycephalus [Genus]
T. atlas [Species]
T. jordani
T. nigromaculatus

Triprion [Genus]
T. petasatus [Species]
T. spatulatus
Xenohyla [Genus]
X. eugenioi [Species]
X. truncata

Ranidae [Family]
Amolops [Genus]
A. chakratensis [Species]
A. chapaensis
A. chunganensis
A. cremnobatus
A. formosus
A. gerbillus
A. granulosus
A. hainanensis
A. himalayanus
A. hongkongensis
A. jaunsari
A. jinjiangensis
A. kangtingensis
A. kaulbacki
A. larutensis
A. liangshanensis
A. lifanensis
A. loloensis
A. longimanus
A. mantzorum
A. marmoratus
A. monticola
A. nepalicus
A. ricketti
A. spinapectoralis
A. taiwanianus
A. tormotus
A. torrentis
A. tuberodepressus
A. viridimaculatus
A. wuyiensis
Anhydrophryne [Genus]
A. rattrayi [Species]
Arthroleptella [Genus]
A. bicolor [Species]
A. drewesii
A. hewitti
A. landdrosia
A. lightfooti
A. ngongoniensis
A. villiersi
Arthroleptides [Genus]
A. dutoiti [Species]
A. martiensseni
Aubria [Genus]
A. masako [Species]
A. subsigillata
Batrachylodes [Genus]
B. elegans [Species]
B. gigas
B. mediodiscus

B. minutus
B. montanus
B. trossulus
B. vertebralis
B. wolfi
Cacosternum [Genus]
C. boettgeri [Species]
C. capense
C. karooicum
C. leleupi
C. namaquense
C. nanum
C. platys
C. poyntoni
C. striatum
Ceratobatrachus [Genus]
C. guentheri [Species]
Chaparana [Genus]
C. aenea [Species]
C. delacouri
C. fansipani
C. quadranus
C. sikimensis
C. unculuanus
Conraua [Genus]
C. alleni [Species]
C. beccarii
C. crassipes
C. derooi
C. goliath
C. robusta
Dimorphognathus [Genus]
D. africanus [Species]
Discodeles [Genus]
D. bufoniformis [Species]
D. guppyi
D. malukuna
D. opisthodon
D. vogti
Elachyglossa [Genus]
E. gyldenstolpei [Species]
Ericabatrachus [Genus]
E. baleensis [Species]
Euphlyctis [Genus]
E. cornii [Species]
E. cyanophlyctis
E. ehrenbergii
E. hexadactylus
Fejervarya [Genus]
F. iskandari [Species]
Hildebrandtia [Genus]
H. macrotympanum [Species]
H. ornata
H. ornatissima
Hoplobatrachus [Genus]
H. crassus [Species]
H. demarchii
H. occipitalis
H. rugulosus
H. tigerinus

H. verruculosus
Huia [Genus]
  H. cavitympanum [Species]
  H. javana
  H. nasica
  H. sumatrana
Indirana [Genus]
  I. beddomii [Species]
  I. brachytarsus
  I. diplosticta
  I. gundia
  I. leithii
  I. leptodactyla
  I. phrynoderma
  I. semipalmata
  I. tenuilingua
Ingerana [Genus]
  I. alpina [Species]
  I. baluensis
  I. liui
  I. mariae
  I. medogensis
  I. reticulata
  I. sariba
  I. tasanae
  I. tenasserimensis
  I. xizangensis
Lankanectes [Genus]
  L. corrugatus [Species]
Lanzarana [Genus]
  L. largeni [Species]
Limnonectes [Genus]
  L. acanthi [Species]
  L. andamanensis
  L. arathooni
  L. asperatus
  L. blythii
  L. brevipalmatus
  L. cancrivorus
  L. dabanus
  L. dammermani
  L. diuatus
  L. doriae
  L. finchi
  L. fragilis
  L. fujianensis
  L. greenii
  L. grunniens
  L. heinrichi
  L. ibanorum
  L. ingeri
  L. kadarsani
  L. kenepaiensis
  L. keralensis
  L. khammonensis
  L. khasiensis
  L. kirtisinghe
  L. kohchangae
  L. kuhlii
  L. laticeps

L. leporinus
L. leytensis
L. limnocharis
L. macrocephalus
L. macrodon
L. macrognathus
L. magnus
L. malesianus
L. mawlyndipi
L. mawphlangensis
L. micrixalus
L. microdiscus
L. microtympanum
L. modestus
L. murthii
L. mysorensis
L. namiyei
L. nepalensis
L. nilagiricus
L. nitidus
L. orissaensis
L. palavanensis
L. paramacrodon
L. parambikulamana
L. parvus
L. pierrei
L. pileatus
L. plicatellus
L. raja
L. rhacoda
L. rufescens
L. sauriceps
L. shompenorum
L. syhadrensis
L. teraiensis
L. timorensis
L. toumanoffi
L. tweediei
L. visayanus
L. vittiger
L. woodworthi
Meristogenys [Genus]
  M. amoropalamus [Species]
  M. jerboa
  M. kinabaluensis
  M. macrophthalmus
  M. orphnocnemis
  M. phaeomerus
  M. poecilus
  M. whiteheadi
Micrixalus [Genus]
  M. fuscus [Species]
  M. gadgili
  M. nudis
  M. phyllophilus
  M. saxicola
  M. silvaticus
  M. thampii
Microbatrachella [Genus]
  M. capensis [Species]

Minervarya [Genus]
  M. sahyadris [Species]
Nannophrys [Genus]
  N. ceylonensis [Species]
  N. guentheri
  N. marmorata
Nanorana [Genus]
  N. parkeri [Species]
  N. pleskei
  N. ventripunctata
Natalobatrachus [Genus]
  N. bonebergi [Species]
Nothophryne [Genus]
  N. broadleyi [Species]
Nyctibatrachus [Genus]
  N. aliciae [Species]
  N. beddomii
  N. deccanensis
  N. humayuni
  N. hussaini
  N. kempholeyensis
  N. major
  N. minor
  N. modestus
  N. sanctipalustris
  N. sylvaticus
  N. vasanthi
Occidozyga [Genus]
  O. baluensis [Species]
  O. borealis
  O. celebensis
  O. diminutivus
  O. floresianus
  O. laevis
  O. lima
  O. magnapustulosus
  O. martensii
  O. semipalmatus
  O. vittatus
Odorrana [Genus]
  O. exiliversabilis [Species]
  O. hainanensis
  O. jingdongensis
  O. nasuta
Paa [Genus]
  P. annandalii [Species]
  P. arnoldi
  P. blanfordii
  P. boulengeri
  P. bourreti
  P. chayuensis
  P. conaensis
  P. ercepeae
  P. exilispinosa
  P. fasciculispina
  P. feae
  P. hazarensis
  P. jiulongensis
  P. liebigii
  P. liui

P. maculosa
P. minica
P. mokokchungensis
P. polunini
P. rarica
P. rostandi
P. shini
P. spinosa
P. sternosignata
P. verrucospinosa
P. vicina
P. yunnanensis
Palmatorappia [Genus]
  P. solomonis [Species]
Petropedetes [Genus]
  P. cameronensis [Species]
  P. johnstoni
  P. natator
  P. newtoni
  P. palmipes
  P. parkeri
  P. perreti
Phrynobatrachus [Genus]
  P. accraensis [Species]
  P. acridoides
  P. acutirostris
  P. albolabris
  P. albomarginatus
  P. alleni
  P. alticola
  P. annulatus
  P. anotis
  P. asper
  P. auritus
  P. batesii
  P. bequaerti
  P. bottegi
  P. brevipalmatus
  P. calcaratus
  P. congicus
  P. cornutus
  P. cricogaster
  P. cryptotis
  P. dalcqi
  P. dendrobates
  P. dispar
  P. elberti
  P. fraterculus
  P. gastoni
  P. ghanensis
  P. giorgii
  P. graueri
  P. guineensis
  P. gutturosus
  P. hylaios
  P. irangi
  P. keniensis
  P. kinangopensis
  P. krefftii
  P. latifrons
  P. liberiensis

P. mababiensis
P. manengoubensis
P. minutus
P. nanus
P. natalensis
P. ogoensis
P. pakenhami
P. parkeri
P. parvulus
P. perpalmatus
P. petropedetoides
P. plicatus
P. pygmaeus
P. rouxi
P. rungwensis
P. scapularis
P. sciangallarum
P. steindachneri
P. sternfeldi
P. stewartae
P. sulfureogularis
P. taiensis
P. tellinii
P. tokba
P. ukingensis
P. uzungwensis
P. versicolor
P. villiersi
P. vogti
P. werneri
P. zavattarii
Phrynodon [Genus]
  P. sandersoni [Species]
Platymantis [Genus]
  P. acrochordus [Species]
  P. aculeodactylus
  P. akarithymus
  P. banahao
  P. batantae
  P. boulengeri
  P. browni
  P. cheesmanae
  P. cornutus
  P. corrugatus
  P. dorsalis
  P. gilliardi
  P. guentheri
  P. guppyi
  P. hazelae
  P. ingeri
  P. insulatus
  P. isarog
  P. lawtoni
  P. levigatus
  P. luzonensis
  P. macrops
  P. macrosceles
  P. magnus
  P. meyeri
  P. mimicus
  P. mimulus

P. montanus
P. myersi
P. naomii
P. neckeri
P. negrosensis
P. nexipus
P. panayensis
P. papuensis
P. parkeri
P. pelewensis
P. polillensis
P. punctatus
P. pygmaeus
P. rabori
P. rhipiphalcus
P. schmidti
P. sierramadrensis
P. solomonis
P. spelaeus
P. subterrestris
P. vitiana
P. vitiensis
P. weberi
Poyntonia [Genus]
  P. paludicola [Species]
Ptychadena [Genus]
  P. aequiplicata [Species]
  P. anchietae
  P. ansorgii
  P. arnei
  P. bibroni
  P. broadleyi
  P. bunoderma
  P. christyi
  P. chrysogaster
  P. cooperi
  P. erlangeri
  P. filwoha
  P. grandisonae
  P. guibei
  P. harenna
  P. ingeri
  P. keilingi
  P. largeni
  P. longirostris
  P. mahnerti
  P. mapacha
  P. mascareniensis
  P. mossambica
  P. nana
  P. neumanni
  P. newtoni
  P. obscura
  P. oxyrhynchus
  P. perplicata
  P. perreti
  P. porosissima
  P. pujoli
  P. pumilio
  P. retropunctata
  P. schillukorum

P. schubotzi
P. stenocephala
P. straeleni
P. submascareniensis
P. subpunctata
P. superciliaris
P. taenioscelis
P. tournieri
P. trinodis
P. upembae
P. uzungwensis
Pyxicephalus [Genus]
P. adspersus [Species]
P. edulis
P. obbianus
Rana [Genus]
R. adenopleura [Species]
R. albolabris
R. altaica
R. alticola
R. amamiensis
R. amieti
R. amnicola
R. amurensis
R. andersonii
R. angolensis
R. anlungensis
R. aragonensis
R. archotaphus
R. areolata
R. arfaki
R. arvalis
R. asiatica
R. asperrima
R. attigua
R. aurantiaca
R. aurora
R. bannanica
R. baramica
R. bedriagae
R. bergeri
R. berlandieri
R. blairi
R. bonaespei
R. boylii
R. brownorum
R. bwana
R. caldwelli
R. camerani
R. cascadae
R. catesbeiana
R. celebensis
R. cerigensis
R. chalconota
R. chaochiaoensis
R. chapaensis
R. charlesdarwini
R. chensinensis
R. chevronta
R. chichicuahutla
R. chiricahuensis

R. chitwanensis
R. chosenica
R. clamitans
R. cordofana
R. crassiovis
R. cretensis
R. cubitalis
R. curtipes
R. daemeli
R. dalmatina
R. danieli
R. darlingi
R. daunchina
R. debussyi
R. demarchii
R. desaegeri
R. dracomontana
R. draytonii
R. dunni
R. dybowskii
R. elberti
R. emelijanovi
R. epeirotica
R. erythraea
R. esculenta
R. everetti
R. fasciata
R. fasciatus
R. florensis
R. forreri
R. fuelleborni
R. fukienensis
R. fuscigula
R. galamensis
R. garoensis
R. garritor
R. ghoshi
R. glandulosa
R. gracilis
R. graeca
R. grahami
R. grandocula
R. grayii
R. grisea
R. grylio
R. guentheri
R. heckscheri
R. hejiangensis
R. hispanica
R. holsti
R. holtzi
R. honnorati
R. hosii
R. huanrenensis
R. hubeiensis
R. humeralis
R. hymenopus
R. iberica
R. igorota
R. inyangae
R. ishikawae

R. italica
R. japonica
R. jimiensis
R. johni
R. johnsi
R. johnstoni
R. juliani
R. kampeni
R. khare
R. kreffti
R. kuangwuensis
R. kunyuensis
R. kurtmuelleri
R. latastei
R. lateralis
R. latouchii
R. lemairei
R. lemairii
R. leptoglossa
R. lepus
R. lessonae
R. livida
R. longicrus
R. longipes
R. luctuosa
R. lungshengensis
R. luteiventris
R. macrocnemis
R. macrodactyla
R. macroglossa
R. macrops
R. maculata
R. magnaocularis
R. malabarica
R. mangyanum
R. maosonensis
R. margaretae
R. margariana
R. maritima
R. megapoda
R. melanomenta
R. miadis
R. milleti
R. minima
R. miopus
R. moellendorffi
R. moluccana
R. montezumae
R. montivaga
R. multidenticulata
R. muscosa
R. narina
R. neovolcanica
R. nicobariensis
R. nigrolineata
R. nigromaculata
R. nigrotympanica
R. nigrovittata
R. novaeguineae
R. oatesii

R. occidentalis
R. okaloossae
R. okinavana
R. omeimontis
R. omiltemana
R. onca
R. ornativentris
R. palmipes
R. palustris
R. papua
R. parkeriana
R. perezi
R. persimilis
R. picturata
R. pipiens
R. pirica
R. plancyi
R. pleuraden
R. porosa
R. pretiosa
R. psaltes
R. pueblae
R. pustulosa
R. pyrenaica
R. raniceps
R. rhodesianus
R. ridibunda
R. rugosa
R. ruwenzorica
R. saharica
R. sakuraii
R. sanguinea
R. sangzhiensis
R. sauteri
R. schmackeri
R. scutigera
R. senchalensis
R. septentrionalis
R. sevosa
R. shqiperica
R. shuchinae
R. siberu
R. sierramadrensis
R. signata
R. similis
R. spectabilis
R. sphenocephala
R. spinidactyla
R. spinulosa
R. springbokensis
R. subaquavocalis
R. subaspera
R. supragrisea
R. supranarina
R. swinhoana
R. sylvatica
R. tagoi
R. taipehensis
R. tarahumarae
R. taylori

R. temporalis
R. temporaria
R. tenggerensis
R. terentievi
R. tiannanensis
R. tientaiensis
R. tipanan
R. tlaloci
R. tsushimensis
R. utsunomiyaorum
R. vaillanti
R. vandijki
R. varians
R. versabilis
R. vertebralis
R. vibicaria
R. virgatipes
R. wageri
R. warszewitschii
R. weiningensis
R. wittei
R. wuchuanensis
R. yavapaiensis
R. zhengi
R. zhenhaiensis
R. zweifeli
Sphaerotheca [Genus]
S. breviceps [Species]
S. dobsonii
S. leucorhynchus
S. maskeyi
S. rolandae
S. strachani
S. swani
Staurois [Genus]
S. latopalmatus [Species]
S. natator
S. tuberilinguis
Taylorana [Genus]
T. hascheana [Species]
T. limborgi
Tomopterna [Genus]
T. cryptotis [Species]
T. delalandii
T. dobsoni
T. krugerensis
T. marmorata
T. natalensis
T. tandyi
T. tuberculosa

Arthroleptidae [Family]
Arthroleptis [Genus]
A. adelphus [Species]
A. adolfifriederici
A. affinis
A. bivittatus
A. brevipes
A. carquejai
A. francei
A. reichei

A. stenodactylus
A. tanneri
A. tuberosus
A. variabilis
A. wahlbergii
Astylosternus [Genus]
A. batesi [Species]
A. corrugatus
A. diadematus
A. fallax
A. laurenti
A. montanus
A. nganhanus
A. occidentalis
A. perreti
A. ranoides
A. rheophilus
A. schioetzi
Cardioglossa [Genus]
C. aureoli [Species]
C. cyaneospila
C. dorsalis
C. elegans
C. escalerae
C. gracilis
C. gratiosa
C. leucomystax
C. liberiensis
C. melanogaster
C. nigromaculata
C. oreas
C. pulchra
C. schioetzi
C. trifasciata
C. venusta
Leptodactylodon [Genus]
L. albiventris [Species]
L. axillaris
L. bicolor
L. blanci
L. boulengeri
L. erythrogaster
L. mertensi
L. ornatus
L. ovatus
L. perreti
L. polyacanthus
L. ventrimarmoratus
Schoutedenella [Genus]
S. crusculum [Species]
S. discodactyla
S. hematogaster
S. lameerei
S. loveridgei
S. milletihorsini
S. mossoensis
S. nimbaensis
S. phrynoides
S. poecilonotus
S. pyrrhoscelis

S. schubotzi
S. spinalis
S. sylvatica
S. taeniata
S. troglodytes
S. vercammeni
S. xenochirus
S. xenodactyla
S. xenodactyloides
S. zimmeri
S. gabonicus
Trichobatrachus [Genus]
T. robustus [Species]

Hemisotidae [Family]
Hemisus [Genus]
H. barotseensis [Species]
H. brachydactylus
H. guineensis
H. guttatus
H. marmoratus
H. microscaphus
H. olivaceus
H. perreti
H. sudanensis
H. wittei

Hyperoliidae [Family]
Acanthixalus [Genus]
A. spinosus [Species]
Afrixalus [Genus]
A. aureus [Species]
A. brachycnemis
A. clarkei
A. crotalus
A. delicatus
A. dorsalis
A. enseticola
A. equatorialis
A. fornasini
A. fornasinii
A. fulvovittatus
A. knysnae
A. lacteus
A. laevis
A. leucostictus
A. lindholmi
A. morerei
A. nigeriensis
A. orophilus
A. osorioi
A. paradorsalis
A. schneideri
A. septentrionalis
A. spinifrons
A. stuhlmanni
A. sylvaticus
A. uluguruensis
A. upembae
A. vibekensis
A. vittiger

A. weidholzi
A. wittei
Alexteroon [Genus]
A. hypsiphonus [Species]
A. jynx
A. obstetricans
Arlequinus [Genus]
A. krebs [Species]
Callixalus [Genus]
C. pictus [Species]
Chlorolius [Genus]
C. koehleri [Species]
Chrysobatrachus [Genus]
C. cupreonitens [Species]
Cryptothylax [Genus]
C. greshoffii [Species]
C. minutus
Heterixalus [Genus]
H. alboguttatus [Species]
H. andrakata
H. betsileo
H. boettgeri
H. luteostriatus
H. madagascariensis
H. mocquardi
H. punctatus
II. rutenbergi
H. tricolor
H. variabilis
Hyperolius [Genus]
H. acutirostris [Species]
H. adametzi
H. albofrenatus
H. alticola
H. angolensis
H. argus
H. atrigularis
H. balfouri
H. baumanni
H. benguellensis
H. bicolor
H. bobirensis
H. bocagei
H. bolifambae
H. bopeleti
H. brachiofasciatus
H. castaneus
H. chabanaudi
H. chlorosteus
H. chrysogaster
H. cinereus
H. cinnamomeoventris
H. concolor
H. cystocandicans
H. destefanii
H. diaphanus
H. discodactylus
H. endjami
H. fasciatus
H. ferreirai

H. ferrugineus
H. fimbriolatus
H. frontalis
H. fuscigula
H. fusciventris
H. ghesquieri
H. gularis
H. guttulatus
H. horstockii
H. houyi
H. hutsebauti
H. inornatus
H. kachalolae
H. kibarae
H. kihangensis
H. kivuensis
H. kuligae
H. lamottei
H. langi
H. lateralis
H. laticeps
H. laurenti
H. leleupi
H. leucotaenius
H. lucani
H. maestus
H. major
H. marmoratus
H. minutissimus
H. mitchelli
H. montanus
H. mosaicus
H. nasutus
H. nienokouensis
H. nimbae
H. obscurus
H. occidentalis
H. ocellatus
H. orkarkarri
H. pardalis
H. parkeri
H. phantasticus
H. pickersgilli
H. picturatus
H. pictus
H. platyceps
H. polli
H. polystictus
H. poweri
H. protchei
H. pseudargus
H. puncticulatus
H. punctulatus
H. pusillus
H. pustulifer
H. quadratomaculatus
H. quinquevittatus
H. raveni
H. reesi
H. rhizophilus

H. riggenbachi
H. robustus
H. rubrovermiculatus
H. sankuruensis
H. schoutedeni
H. seabrai
H. semidiscus
H. sheldricki
H. soror
H. spinigularis
H. steindachneri
H. stenodactylus
H. sylvaticus
H. tannerorum
H. thoracotuberculatus
H. tornieri
H. torrentis
H. tuberculatus
H. tuberilinguis
H. vilhenai
H. viridiflavus
H. viridigulosus
H. viridis
H. wermuthi
H. xenorhinus
H. zavattarii
H. zonatus
Kassina [Genus]
K. arboricola [Species]
K. cassinoides
K. cochranae
K. decorata
K. fusca
K. kuvangensis
K. lamottei
K. maculata
K. maculifer
K. maculosa
K. mertensi
K. parkeri
K. schioetzi
K. senegalensis
K. somalica
Kassinula [Genus]
K. wittei [Species]
Leptopelis [Genus]
L. anchietae [Species]
L. argenteus
L. aubryi
L. barbouri
L. bequaerti
L. bocagei
L. bocagii
L. boulengeri
L. brevipes
L. brevirostris
L. broadleyi
L. bufonides
L. calcaratus
L. christyi
L. concolor

L. cynnamomeus
L. fenestratus
L. fiziensis
L. flavomaculatus
L. gramineus
L. hyloides
L. jordani
L. karissimbensis
L. kivuensis
L. lebeaui
L. macrotis
L. marginatus
L. millsoni
L. modestus
L. mossambicus
L. natalensis
L. nordequatorialis
L. notatus
L. occidentalis
L. ocellatus
L. omissus
L. oryi
L. palmatus
L. parbocagii
L. parkeri
L. parvus
L. ragazzii
L. rufus
L. susanae
L. uluguruensis
L. vannutellii
L. vermiculatus
L. viridis
L. xenodactylus
L. yaldeni
L. zebra
Nesionixalus [Genus]
N. molleri [Species]
N. thomensis
Opisthothylax [Genus]
O. immaculatus [Species]
Paracassina [Genus]
P. kounhiensis [Species]
P. obscura
Phlyctimantis [Genus]
P. boulengeri [Species]
P. keithae
P. leonardi
P. verrucosus
Semnodactylus [Genus]
S. wealii [Species]
Tachycnemis [Genus]
T. seychellensis [Species]

Rhacophoridae [Family]
Aglyptodactylus [Genus]
A. laticeps [Species]
A. madagascariensis
A. securifer
Boophis [Genus]
B. albilabris [Species]

B. albipunctatus
B. andohahela
B. andreonei
B. anjanaharibeensis
B. ankaratra
B. blommersae
B. boehmei
B. brachychir
B. burgeri
B. difficilis
B. elenae
B. englaenderi
B. erythrodactylus
B. feonnyala
B. goudotii
B. granulosus
B. haematopus
B. hillenii
B. idae
B. jaegeri
B. laurenti
B. lichenoides
B. luteus
B. madagascariensis
B. majori
B. mandraka
B. marojezensis
B. microtis
B. microtympanum
B. miniatus
B. occidentalis
B. opisthodon
B. pauliani
B. perigetes
B. picturatus
B. pyrrhus
B. rappiodes
B. reticulatus
B. rhodoscelis
B. rufioculis
B. sibilans
B. tephraeomystax
B. viridis
B. vittatus
B. williamsi
B. xerophilus
Buergeria [Genus]
B. buergeri [Species]
B. japonica
B. pollicaris
B. robusta
Chirixalus [Genus]
C. doriae [Species]
C. dudhwaensis
C. eiffingeri
C. hansenae
C. idiootocus
C. laevis
C. nongkhorensis
C. palpebralis
C. simus

C. vittatus
Chiromantis [Genus]
   C. petersii [Species]
   C. rufescens
   C. xerampelina
Laliostoma [Genus]
   L. labrosum [Species]
Mantella [Genus]
   M. aurantiaca [Species]
   M. baroni
   M. bernhardi
   M. betsileo
   M. cowanii
   M. crocea
   M. expectata
   M. haraldmeieri
   M. laevigata
   M. madagascariensis
   M. nigricans
   M. pulchra
   M. viridis
Mantidactylus [Genus]
   M. acuticeps [Species]
   M. aerumnalis
   M. aglavei
   M. albofrenatus
   M. albolineatus
   M. alutus
   M. ambohimitombi
   M. ambohitra
   M. ambreensis
   M. argenteus
   M. asper
   M. bertini
   M. betsileanus
   M. bicalcaratus
   M. biporus
   M. blommersae
   M. boulengeri
   M. brevipalmatus
   M. cornutus
   M. corvus
   M. curtus
   M. decaryi
   M. depressiceps
   M. domerguei
   M. eiselti
   M. elegans
   M. femoralis
   M. fimbriatus
   M. flavobrunneus
   M. grandidieri
   M. grandisonae
   M. granulatus
   M. guibei
   M. guttulatus
   M. horridus
   M. kelyi
   M. klemmeri
   M. leucomaculatus

M. liber
M. lugubris
M. luteus
M. madecassus
M. majori
M. malagasius
M. massi
M. microtympanum
M. mocquardi
M. moseri
M. opiparis
M. peraccae
M. phantasticus
M. plicifer
M. pseudoasper
M. pulcher
M. punctatus
M. redimitus
M. rivicola
M. sculpturatus
M. silvanus
M. spinifer
M. spiniferus
M. striatus
M. thelenae
M. tornieri
M. ulcerosus
M. ventrimaculatus
M. webbi
M. wittei
Nyctixalus [Genus]
   N. moloch [Species]
   N. pictus
   N. spinosus
Philautus [Genus]
   P. abditus [Species]
   P. acutirostris
   P. acutus
   P. adspersus
   P. albopunctatus
   P. alticola
   P. amoenus
   P. andersoni
   P. annandalii
   P. aurantium
   P. aurifasciatus
   P. banaensis
   P. beddomii
   P. bombayensis
   P. bunitus
   P. carinensis
   P. chalazodes
   P. charius
   P. cherrapunjiae
   P. cornutus
   P. crnri
   P. disgregus
   P. dubius
   P. elegans
   P. emembranatus

P. eximius
P. femoralis
P. flaviventris
P. garo
P. glandulosus
P. gracilipes
P. gryllus
P. hassanensis
P. hosii
P. hypomelas
P. ingeri
P. jacobsoni
P. jerdonii
P. jinxiuensis
P. kempiae
P. kerangae
P. kottigeharensis
P. leitensis
P. leucorhinus
P. longchuanensis
P. longicrus
P. maosonensis
P. medogensis
P. melanensis
P. menglaensis
P. microdiscus
P. microtympanum
P. mjobergi
P. namdaphaensis
P. narainensis
P. nasutus
P. noblei
P. ocellatus
P. odontotarsus
P. pallidipes
P. parkeri
P. parvulus
P. petersi
P. pleurostictus
P. poecilius
P. pulcherrimus
P. refugii
P. rhododiscus
P. romeri
P. sanctisilvaticus
P. saueri
P. schmackeri
P. shillongensis
P. shyamrupus
P. signatus
P. similis
P. stictomerus
P. surdus
P. surrufus
P. swamianus
P. tectus
P. temporalis
P. terebrans
P. travancoricus
P. tytthus

P. umbra
P. variabilis
P. vermiculatus
P. vittiger
P. worcesteri
Polypedates [Genus]
  P. chenfui [Species]
  P. colletti
  P. cruciger
  P. dennysii
  P. dorsoviridis
  P. dugritei
  P. eques
  P. feae
  P. hecticus
  P. hungfuensis
  P. insularis
  P. leucomystax
  P. longinasus
  P. macrotis
  P. maculatus
  P. megacephalus
  P. mutus
  P. naso
  P. omeimontis
  P. otilophus
  P. pingbianensis
  P. prasinatus
  P. pseudocruciger
  P. puerensis
  P. taeniatus
  P. yaoshanensis
  P. zed
  P. zhaojuensis
Rhacophorus [Genus]
  R. achantharrhena [Species]
  R. angulirostris
  R. annamensis
  R. appendiculatus
  R. arboreus
  R. arvalis
  R. aurantiventris
  R. baliogaster
  R. baluensis
  R. barisani
  R. bimaculatus
  R. bipunctatus
  R. bisacculus
  R. calcadensis
  R. calcaneus
  R. catamitus
  R. cyanopunctatus
  R. depressus
  R. dulitensis
  R. edentulus
  R. everetti
  R. exechopygus
  R. fasciatus
  R. fergusonianus

R. gauni
R. georgii
R. gongshanensis
R. harrissoni
R. hoanglienensis
R. javanus
R. kajau
R. macropus
R. malabaricus
R. margaritifer
R. maximus
R. modestus
R. moltrechti
R. monticola
R. namdaphaensis
R. nigropalmatus
R. nigropunctatus
R. notater
R. orlovi
R. owstoni
R. oxycephalus
R. pardalis
R. poecilonotus
R. prominanus
R. prominanus
R. reinwardtii
R. reticulatus
R. rhodopus
R. robinsonii
R. rufipes
R. schlegelii
R. taipeianus
R. taroensis
R. translineatus
R. tuberculatus
R. turpes
R. verrucopus
R. verrucosus
R. viridis
Theloderma [Genus]
  T. asperum [Species]
  T. bicolor
  T. corticale
  T. gordoni
  T. horridum
  T. kwangsiense
  T. leporosum
  T. phrynoderma
  T. schmarda
  T. stellatum

Microhylidae [Family]
  Adelastes [Genus]
    A. hylonomus [Species]
  Albericus [Genus]
    A. brunhildae [Species]
    A. darlingtoni
    A. fafniri
    A. gudrunae
    A. gunnari

A. laurini
A. rhenaurum
A. siegfriedi
A. swanhildae
A. tuberculus
A. valkuriarum
A. variegatus
Altigius [Genus]
  A. alios [Species]
Anodonthyla [Genus]
  A. boulengerii [Species]
  A. montana
  A. nigrigularis
  A. rouxae
Aphantophryne [Genus]
  A. minuta [Species]
  A. pansa
  A. sabini
Arcovomer [Genus]
  A. passarellii [Species]
Asterophrys [Genus]
  A. leucopus [Species]
  A. turpicola
Balebreviceps [Genus]
  B. hillmani [Species]
Barygenys [Genus]
  B. atra [Species]
  B. cheesmanae
  B. exsul
  B. flavigularis
  B. maculata
  B. nana
  B. parvula
Breviceps [Genus]
  B. acutirostris [Species]
  B. adspersus
  B. fuscus
  B. gibbosus
  B. macrops
  B. maculatus
  B. montanus
  B. mossambicus
  B. namaquensis
  B. poweri
  B. rosei
  B. sylvestris
  B. verrucosus
Calluella [Genus]
  C. brooksii [Species]
  C. flava
  C. guttulata
  C. smithi
  C. volzi
  C. yunnanensis
Callulina [Genus]
  C. kreffti [Species]
Callulops [Genus]
  C. boettgeri [Species]
  C. comptus
  C. doriae

C. dubius
C. eurydactylus
C. fuscus
C. glandulosus
C. humicola
C. kopsteini
C. personatus
C. robustus
C. sagittatus
C. slateri
C. stictogaster
C. wilhelmanus
Chaperina [Genus]
C. fusca [Species]
Chiasmocleis [Genus]
C. alagoanus [Species]
C. albopunctata
C. anatipes
C. atlantica
C. bassleri
C. bicegoi
C. capixaba
C. carvalhoi
C. centralis
C. hudsoni
C. leucosticta
C. mehelyi
C. panamensis
C. schubarti
C. shudikarensis
C. urbanae
C. ventrimaculata
Choerophryne [Genus]
C. longirostris [Species]
C. rostellifer
Cophixalus [Genus]
C. ateles [Species]
C. biroi
C. bombiens
C. cheesmanae
C. concinnus
C. crepitans
C. cryptotympanum
C. daymani
C. exiguus
C. hosmeri
C. infacetus
C. kaindiensis
C. mcdonaldi
C. montanus
C. monticola
C. neglectus
C. nubicola
C. ornatus
C. parkeri
C. peninsularis
C. pipilans
C. riparius
C. saxatilis

C. shellyi
C. sphagnicola
C. tagulensis
C. verecundus
C. verrucosus
C. zweifeli
Cophyla [Genus]
C. phyllodactyla [Species]
Copiula [Genus]
C. fistulans [Species]
C. minor
C. oxyrhina
C. pipiens
C. tyleri
Ctenophryne [Genus]
C. geayi [Species]
C. minor
Dasypops [Genus]
D. schirchi [Species]
Dermatonotus [Genus]
D. muelleri [Species]
Dyscophus [Genus]
D. antongilii [Species]
D. guineti
D. insularis
Elachistocleis [Genus]
E. bicolor [Species]
E. erythrogaster
E. ovalis
E. piauiensis
E. surinamensis
Gastrophryne [Genus]
G. carolinensis [Species]
G. elegans
G. olivacea
G. pictiventris
G. usta
Gastrophrynoides [Genus]
G. borneensis [Species]
Genyophryne [Genus]
G. thomsoni [Species]
Glyphoglossus [Genus]
G. molossus [Species]
Hamptophryne [Genus]
H. boliviana [Species]
Hoplophryne [Genus]
H. rogersi [Species]
H. uluguruensis
Hylophorbus [Genus]
H. nigrinus [Species]
H. picoides
H. richardi
H. rufescens
H. sextus
H. tetraphonus
H. wondiwoi
Hyophryne [Genus]
H. histrio [Species]
Hypopachus [Genus]
H. barberi [Species]

H. variolosus
Kalophrynus [Genus]
K. baluensis [Species]
K. bunguranus
K. heterochirus
K. interlineatus
K. intermedius
K. menglienicus
K. nubicola
K. orangensis
K. palmatissimus
K. pleurostigma
K. punctatus
K. robinsoni
K. subterrestris
Kaloula [Genus]
K. baleata [Species]
K. borealis
K. conjuncta
K. kokacii
K. mediolineata
K. picta
K. pulchra
K. rigida
K. rugifera
K. taprobanica
K. verrucosa
Madecassophryne [Genus]
M. truebae [Species]
Mantophryne [Genus]
M. infulata [Species]
M. lateralis
M. louisiadensis
Melanobatrachus [Genus]
M. indicus [Species]
Metaphrynella [Genus]
M. pollicaris [Species]
M. sundana
Microhyla [Genus]
M. achatina [Species]
M. annamensis
M. annectens
M. berdmorei
M. borneensis
M. butleri
M. chakrapanii
M. erythropoda
M. fowleri
M. fusca
M. heymonsi
M. karunaratnei
M. maculifera
M. mixtura
M. okinavensis
M. ornata
M. palmipes
M. perparva
M. petrigena
M. picta

M. pulchra
M. rubra
M. superciliaris
M. zeylanica
Micryletta [Genus]
  M. inornata [Species]
  M. steinegeri
Myersiella [Genus]
  M. microps [Species]
Nelsonophryne [Genus]
  N. aequatorialis [Species]
  N. aterrimus
Oreophryne [Genus]
  O. albopunctata [Species]
  O. anthonyi
  O. anulata
  O. atrigularis
  O. biroi
  O. brachypus
  O. brevicrus
  O. celebensis
  O. crucifer
  O. flava
  O. frontifasciata
  O. geislerorum
  O. idenburgensis
  O. inornata
  O. insulana
  O. jeffersoniana
  O. kampeni
  O. moluccensis
  O. monticola
  O. nana
  O. parkeri
  O. rookmaakeri
  O. variabilis
  O. wapoga
  O. wolterstorffi
  O. zimmeri
Otophryne [Genus]
  O. pyburni [Species]
  O. robusta
  O. steyermarki
Parhoplophryne [Genus]
  P. usambarica [Species]
Pherohapsis [Genus]
  P. menziesi [Species]
Phrynella [Genus]
  P. pulchra [Species]
Phrynomantis [Genus]
  P. affinis [Species]
  P. annectens
  P. bifasciatus
  P. microps
  P. somalicus
Platypelis [Genus]
  P. alticola [Species]
  P. barbouri
  P. cowanii
  P. grandis
  P. milloti

P. occultans
P. pollicaris
P. tsaratananaensis
P. tuberifera
Plethodontohyla [Genus]
  P. alluaudi [Species]
  P. angulifera
  P. bipunctata
  P. brevipes
  P. coudreaui
  P. guentherpetersi
  P. inguinalis
  P. laevipes
  P. minuta
  P. notosticta
  P. ocellata
  P. serratopalpebrosa
  P. tuberata
Probreviceps [Genus]
  P. macrodactylus [Species]
  P. rhodesianus
  P. uluguruensis
Ramanella [Genus]
  R. anamalaiensis [Species]
  R. minor
  R. montana
  R. mormorata
  R. obscura
  R. palmata
  R. triangularis
  R. variegata
Relictivomer [Genus]
  R. pearsei [Species]
Rhombophryne [Genus]
  R. testudo [Species]
Spelaeophryne [Genus]
  S. methneri [Species]
Sphenophryne [Genus]
  S. adelphe [Species]
  S. brevicrus
  S. brevipes
  S. cornuta
  S. crassa
  S. dentata
  S. fryi
  S. gracilipes
  S. hooglandi
  S. macrorhyncha
  S. mehelyi
  S. palmipes
  S. pluvialis
  S. polysticta
  S. pusilla
  S. rhododactyla
  S. robusta
  S. schlaginhaufeni
Stereocyclops [Genus]
  S. incrassatus [Species]
Stumpffia [Genus]
  S. gimmeli [Species]

S. grandis
S. psologlossa
S. pygmaea
S. tetradactyla
S. tridactyla
Synapturanus [Genus]
  S. mirandaribeiroi [Species]
  S. rabus
  S. salseri
Syncope [Genus]
  S. antenori [Species]
  S. carvalhoi
  S. tridactyla
Uperodon [Genus]
  U. globulosus [Species]
  U. systoma
Xenobatrachus [Genus]
  X. anorbis [Species]
  X. arfakianus
  X. bidens
  X. fuscigula
  X. giganteus
  X. huon
  X. macrops
  X. mehelyi
  X. multisica
  X. obesus
  X. ocellatus
  X. ophiodon
  X. rostratus
  X. scheepstrai
  X. schiefenhoeveli
  X. subcroceus
  X. tumulus
  X. zweifeli
Xenorhina [Genus]
  X. arboricola [Species]
  X. bouwensi
  X. eiponis
  X. minima
  X. oxycephala
  X. parkerorum
  X. similis

Scaphiophryninae [Family]
  Paradoxophyla [Genus]
    P. palmata [Species]
  Scaphiophryne [Genus]
    S. brevis [Species]
    S. calcarata
    S. gottlebei
    S. madagascariensis
    S. marmorata
    S. obscura
    S. pustulosa
    S. verrucosa

Caudata [Order]

Sirenidae [Family]
  Pseudobranchus [Genus]

P. axanthus [Species]
P. striatus
Siren [Genus]
  S. intermedia [Species]
  S. lacertina

Hynobiidae [Family]
  Batrachuperus [Genus]
    B. gorganensis [Species]
    B. karlschmidti
    B. londongensis
    B. mustersi
    B. persicus
    B. pinchonii
    B. taibaiensis
    B. tibetanus
    B. yenyuanensis
  Hynobius [Genus]
    H. abei [Species]
    H. amjiensis
    H. arisanensis
    H. boulengeri
    H. chinensis
    H. dunni
    H. formosanus
    H. hidamontanus
    H. kimurae
    H. leechii
    H. lichenatus
    H. mantschuriensis
    H. naevius
    H. nebulosus
    H. nigrescens
    H. okiensis
    H. retardatus
    H. sonani
    H. stejnegeri
    H. takedai
    H. tenuis
    H. tokyoensis
    H. tsuensis
    H. turkestanicus
    H. yiwuensis
    H. yunanicus
  Liua [Genus]
    L. shihi [Species]
  Onychodactylus [Genus]
    O. fischeri [Species]
    O. japonicus
  Pachyhynobius [Genus]
    P. shangchengensis [Species]
  Protohynobius [Genus]
    P. puxiongensis [Species]
  Pseudohynobius [Genus]
    P. tsinpaensis [Species]
  Ranodon [Genus]
    R. sibiricus [Species]
  Salamandrella [Genus]
    S. keyserlingii [Species]

Cryptobranchidae [Family]
  Andrias [Genus]

A. davidianus [Species]
A. japonicus
Cryptobranchus [Genus]
  C. alleganiensis [Species]

Dicamptodontidae [Family]
  Dicamptodon [Genus]
    D. aterrimus [Species]
    D. copei
    D. ensatus
    D. tenebrosus

Ambystomatidae [Family]
  Ambystoma [Genus]
    A. altamirani [Species]
    A. amblycephalum
    A. andersoni
    A. annulatum
    A. barbouri
    A. bombypellum
    A. californiense
    A. cingulatum
    A. dumerilii
    A. flavipiperatum
    A. gracile
    A. granulosum
    A. jeffersonianum
    A. laterale
    A. leorae
    A. lermaense
    A. mabeei
    A. macrodactylum
    A. maculatum
    A. mavortium
    A. mexicanum
    A. opacum
    A. ordinarium
    A. rivulare
    A. rosaceum
    A. talpoideum
    A. taylori
    A. texanum
    A. tigrinum
    A. velasci

Salamandridae [Family]
  Chioglossa [Genus]
    C. lusitanica [Species]
  Cynops [Genus]
    C. chenggongensis [Species]
    C. cyanurus
    C. ensicauda
    C. orientalis
    C. orphicus
    C. pyrrhogaster
    C. wolterstorffi
  Echinotriton [Genus]
    E. andersoni [Species]
    E. chinhaiensis
  Euproctus [Genus]
    E. asper [Species]

E. montanus
E. platycephalus
Mertensiella [Genus]
  M. caucasica [Species]
Neurergus [Genus]
  N. crocatus [Species]
  N. kaiseri
  N. microspilotus
  N. strauchii
Notophthalmus [Genus]
  N. meridionalis [Species]
  N. perstriatus
  N. viridescens
Pachytriton [Genus]
  P. brevipes [Species]
  P. labiatus
Paramesotriton [Genus]
  P. caudopunctatus [Species]
  P. chinensis
  P. deloustali
  P. fuzhongensis
  P. guanxiensis
  P. hongkongensis
  P. laoensis
Pleurodeles [Genus]
  P. poireti [Species]
  P. waltl
Salamandra [Genus]
  S. algira [Species]
  S. atra
  S. corsica
  S. infraimmaculata
  S. lanzai
  S. luschani
  S. salamandra
Salamandrina [Genus]
  S. terdigitata [Species]
Taricha [Genus]
  T. granulosa [Species]
  T. rivularis
  T. torosa
Triturus [Genus]
  T. alpestris [Species]
  T. boscai
  T. carnifex
  T. cristatus
  T. dobrogicus
  T. helveticus
  T. italicus
  T. karelinii
  T. marmoratus
  T. montandoni
  T. vittatus
  T. vulgaris
Tylototriton [Genus]
  T. asperrimus [Species]
  T. hainanensis
  T. kweichowensis
  T. shanjing
  T. taliangensis

T. verrucosus
T. wenxianensis

Proteidae [Family]
  Necturus [Genus]
    N. alabamensis [Species]
    N. beyeri
    N. lewisi
    N. maculosus
    N. punctatus
  Proteus [Genus]
    P. anguinus [Species]

Rhyacotritonidae [Family]
  Rhyacotriton [Genus]
    R. cascadae [Species]
    R. kezeri
    R. olympicus
    R. variegatus

Plethodontidae [Family]
  Aneides [Genus]
    A. aeneus [Species]
    A. ferreus
    A. flavipunctatus
    A. hardii
    A. lugubris
    A. vagrans
  Batrachoseps [Genus]
    B. aridus [Species]
    B. attenuatus
    B. campi
    B. diabolicus
    B. gabrieli
    B. gavilanensis
    B. gregarius
    B. incognitus
    B. kawia
    B. luciae
    B. major
    B. minor
    B. nigriventris
    B. pacificus
    B. regius
    B. relictus
    B. simatus
    B. stebbinsi
    B. wrighti
  Bolitoglossa [Genus]
    B. adspersa [Species]
    B. altamazonica
    B. alvaradoi
    B. anthracina
    B. arborescandens
    B. biseriata
    B. borburata
    B. capitana
    B. carri
    B. celaque
    B. cerroensis
    B. chica

B. colonnea
B. compacta
B. conanti
B. cuchumatana
B. cuna
B. decora
B. diaphora
B. digitigrada
B. diminuta
B. dofleini
B. dunni
B. engelhardti
B. epimela
B. equatoriana
B. flavimembris
B. flaviventris
B. franklini
B. gracilis
B. guaramacalensis
B. hartwegi
B. helmrichi
B. hermosa
B. hiemalis
B. hypacra
B. jacksoni
B. lignicolor
B. lincolni
B. longissima
B. lozanoi
B. macrinii
B. marmorea
B. medemi
B. meliana
B. mexicana
B. minutula
B. mombachoensis
B. morio
B. mulleri
B. nicefori
B. nigrescens
B. oaxacensis
B. occidentalis
B. odonnelli
B. orestes
B. palmata
B. pandi
B. peruviana
B. pesrubra
B. phalarosoma
B. platydactyla
B. porrasorum
B. ramosi
B. riletti
B. robusta
B. rostrata
B. rufescens
B. salvinii
B. savagei
B. schizodactyla
B. silverstonei

B. sima
B. sooyorum
B. spongai
B. striatula
B. stuarti
B. subpalmata
B. synoria
B. taylori
B. vallecula
B. veracrucis
B. walkeri
B. yucatana
B. zapoteca
Bradytriton [Genus]
  B. silus [Species]
Chiropterotriton [Genus]
  C. arboreus [Species]
  C. chiropterus
  C. chondrostega
  C. cracens
  C. dimidiatus
  C. lavae
  C. magnipes
  C. mosaueri
  C. multidentatus
  C. orculus
  C. priscus
  C. terrestris
Cryptotriton [Genus]
  C. adelos [Species]
  C. alvarezdeltoroi
  C. monzoni
  C. nasalis
  C. veraepacis
  C. wakei
Dendrotriton [Genus]
  D. bromeliacius [Species]
  D. cuchumatanus
  D. megarhinus
  D. rabbi
  D. sanctibarbarus
  D. xolocalcae
Desmognathus [Genus]
  D. aeneus [Species]
  D. apalachicolae
  D. auriculatus
  D. brimleyorum
  D. carolinensis
  D. conanti
  D. folkertsi
  D. fuscus
  D. imitator
  D. marmoratus
  D. monticola
  D. ochrophaeus
  D. ocoee
  D. orestes
  D. quadramaculatus
  D. santeetlah
  D. welteri

D. wrighti
Ensatina [Genus]
  E. eschscholtzii [Species]
Eurycea [Genus]
  E. aquatica [Species]
  E. bislineata
  E. chisholmensis
  E. cirrigera
  E. guttolineata
  E. junaluska
  E. latitans
  E. longicauda
  E. lucifuga
  E. multiplicata
  E. nana
  E. naufragia
  E. neotenes
  E. pterophila
  E. quadridigitata
  E. rathbuni
  E. robusta
  E. sosorum
  E. tonkawae
  E. tridentifera
  E. troglodytes
  E. tynerensis
  E. waterlooensis
  E. wilderae
Gyrinophilus [Genus]
  G. gulolineatus [Species]
  G. palleucus
  G. porphyriticus
  G. subterraneus
Haideotriton [Genus]
  H. wallacei [Species]
Hemidactylium [Genus]
  H. scutatum [Species]
Hydromantes [Genus]
  H. ambrosii [Species]
  H. brunus
  H. flavus
  H. genei
  H. imperialis
  H. italicus
  H. platycephalus
  H. shastae
  H. strinatii
  H. supramontis
Ixalotriton [Genus]
  I. niger [Species]
  I. parvus
Lineatriton [Genus]
  L. lineolus [Species]
  L. orchileucos
  L. orchimelas
Nototriton [Genus]
  N. abscondens [Species]
  N. barbouri
  N. brodiei
  N. gamezi
  N. guanacaste

N. lignicola
N. limnospectator
N. major
N. picadoi
N. richardi
N. saslaya
N. stuarti
N. tapanti
Nyctanolis [Genus]
  N. pernix [Species]
Oedipina [Genus]
  O. alfaroi [Species]
  O. alleni
  O. altura
  O. carablanca
  O. collaris
  O. complex
  O. cyclocauda
  O. elongata
  O. gephyra
  O. gracilis
  O. grandis
  O. ignea
  O. maritima
  O. pacificensis
  O. parvipes
  O. paucidentata
  O. poelzi
  O. pseudouniformis
  O. savagei
  O. stenopodia
  O. stuarti
  O. taylori
  O. uniformis
Parvimolge [Genus]
  P. townsendi [Species]
Phaeognathus [Genus]
  P. hubrichti [Species]
Plethodon [Genus]
  P. ainsworthi [Species]
  P. albagula
  P. amplus
  P. angusticlavius
  P. aureolus
  P. caddoensis
  P. chattahoochee
  P. cheoah
  P. chlorobryonis
  P. cinereus
  P. cylindraceus
  P. dorsalis
  P. dunni
  P. electromorphus
  P. elongatus
  P. fourchensis
  P. glutinosus
  P. grobmani
  P. hoffmani
  P. hubrichti
  P. idahoensis
  P. jordani

P. kentucki
P. kiamichi
P. kisatchie
P. larselli
P. meridianus
P. metcalfi
P. mississippi
P. montanus
P. neomexicanus
P. nettingi
P. ocmulgee
P. oconaluftee
P. ouachitae
P. petraeus
P. punctatus
P. richmondi
P. savannah
P. sequoyah
P. serratus
P. shenandoah
P. shermani
P. stormi
P. teyahalee
P. vandykei
P. variolatus
P. vehiculum
P. ventralis
P. virginia
P. websteri
P. wehrlei
P. welleri
P. yonahlossee
Pseudoeurycea [Genus]
  P. ahuitzotl [Species]
  P. altamontana
  P. amuzga
  P. anitae
  P. aquatica
  P. bellii
  P. brunnata
  P. cephalica
  P. cochranae
  P. conanti
  P. exspectata
  P. firscheini
  P. gadovii
  P. galeanae
  P. gigantea
  P. goebeli
  P. juarezi
  P. leprosa
  P. longicauda
  P. lynchi
  P. melanomolga
  P. mixcoatl
  P. mystax
  P. naucampatepetl
  P. nigromaculata
  P. praecellens
  P. rex
  P. robertsi

P. saltator
P. scandens
P. smithi
P. tenchalli
P. teotepec
P. tlahcuiloh
P. unguidentis
P. werleri
Pseudotriton [Genus]
  P. montanus [Species]
  P. ruber
Stereochilus [Genus]
  S. marginatus [Species]
Thorius [Genus]
  T. arboreus [Species]
  T. aureus
  T. boreas
  T. dubitus
  T. grandis
  T. infernalis
  T. insperatus
  T. lunaris
  T. macdougalli
  T. magnipes
  T. maxillabrochus
  T. minutissimus
  T. minydemus
  T. munificus
  T. narismagnus
  T. narisovalis
  T. omiltemi
  T. papaloae
  T. pennatulus
  T. pulmonaris
  T. schmidti
  T. smithi
  T. spilogaster
  T. troglodytes
Typhlotriton [Genus]
  T. spelaeus [Species]

Amphiumidae [Family]
Amphiuma [Genus]
  A. means [Species]
  A. pholeter
  A. tridactylum

Gymnophiona [Order]

Rhinatrematidae [Family]
Epicrionops [Genus]
  E. bicolor [Species]
  E. columbianus
  E. lativittatus
  E. marmoratus
  E. niger
  E. parkeri
  E. peruvianus
  E. petersi
Rhinatrema [Genus]

R. bivittatum [Species]

Ichthyophiidae [Family]
Caudacaecilia [Genus]
  C. asplenia [Species]
  C. larutensis
  C. nigroflava
  C. paucidentula
  C. weberi
Ichthyophis [Genus]
  I. acuminatus [Species]
  I. atricollaris
  I. bannanicus
  I. beddomei
  I. bernisi
  I. biangularis
  I. billitonensis
  I. bombayensis
  I. dulitensis
  I. elongatus
  I. glandulosus
  I. glutinosus
  I. humphreyi
  I. husaini
  I. hypocyaneus
  I. javanicus
  I. kohtaoensis
  I. laosensis
  I. longicephalus
  I. malabarensis
  I. mindanaoensis
  I. monochrous
  I. orthoplicatus
  I. paucisulcus
  I. peninsularis
  I. pseudangularis
  I. sikkimensis
  I. singaporensis
  I. subterrestris
  I. sumatranus
  I. supachaii
  I. tricolor
  I. youngorum

Uraeotyphlidae [Family]
Uraeotyphlus [Genus]
  U. interruptus [Species]
  U. malabaricus
  U. menoni
  U. narayani
  U. oxyurus

Scolecomorphidae [Family]
Crotaphatrema [Genus]
  C. bornmuelleri [Species]
  C. lamottei
  C. tchabalmbaboensis
Scolecomorphus [Genus]
  S. kirkii [Species]
  S. uluguruensis
  S. vittatus

Caeciliidae [Family]
Atretochoana [Genus]
  A. eiselti [Species]
Boulengerula [Genus]
  B. boulengeri [Species]
  B. changamwensis
  B. fischeri
  B. taitana
  B. uluguruensis
Brasilotyphlus [Genus]
  B. braziliensis [Species]
Caecilia [Genus]
  C. abitaguae [Species]
  C. albiventris
  C. antioquiaensis
  C. armata
  C. attenuata
  C. bokermanni
  C. caribea
  C. corpulenta
  C. crassisquama
  C. degenerata
  C. disossea
  C. dunni
  C. flavopunctata
  C. gracilis
  C. guntheri
  C. inca
  C. isthmica
  C. leucocephala
  C. marcusi
  C. mertensi
  C. nigricans
  C. occidentalis
  C. orientalis
  C. pachynema
  C. perdita
  C. pressula
  C. subdermalis
  C. subnigricans
  C. subterminalis
  C. tentaculata
  C. tenuissima
  C. thompsoni
  C. volcani
Chthonerpeton [Genus]
  C. arii [Species]
  C. braestrupi
  C. exile
  C. indistinctum
  C. onorei
  C. perissodus
  C. viviparum
Dermophis [Genus]
  D. costaricense [Species]
  D. glandulosus
  D. gracilior
  D. mexicanus
  D. oaxacae

D. occidentalis
D. parviceps
Gegeneophis [Genus]
  G. carnosus [Species]
  G. fulleri
  G. ramaswamii
Gegenophis [Genus]
  G.s krishni [Species]
Geotrypetes [Genus]
  G. angeli [Species]
  G. pseudoangeli
  G. seraphini
Grandisonia [Genus]
  G. alternans [Species]
  G. brevis
  G. diminutiva
  G. larvata
  G. sechellensis
Gymnopis [Genus]
  G. multiplicata [Species]
  G. syntremus
Herpele [Genus]
  H. multiplicata [Species]
  H. squalostoma
Hypogeophis [Genus]
  H. rostratus [Species]

Idiocranium [Genus]
  I. russeli [Species]
Indotyphlus [Genus]
  I. battersbyi [Species]
Luetkenotyphlus [Genus]
  L. brasiliensis [Species]
Microcaecilia [Genus]
  M. albiceps [Species]
  M. rabei
  M. supernumeraria
  M. taylori
  M. unicolor
Mimosiphonops [Genus]
  M. reinhardti [Species]
  M. vermiculatus
Nectocaecilia [Genus]
  N. petersii [Species]
Oscaecilia [Genus]
  O. bassleri [Species]
  O. elongata
  O. equatorialis
  O. hypereumeces
  O. koepckeorum
  O. ochrocephala
  O. osae
  O. polyzona

O. zweifeli
Parvicaecilia [Genus]
  P. nicefori [Species]
  P. pricei
Potomotyphlus [Genus]
  P. kaupii [Species]
Praslinia [Genus]
  P. cooperi [Species]
Schistometopum [Genus]
  S. garzonheydti [Species]
  S. gregorii
  S. thomense
Siphonops [Genus]
  S. annulatus [Species]
  S. hardyi
  S. insulanus
  S. leucoderus
  S. paulensis
Sylvacaecilia [Genus]
  S. grandisonae [Species]
Typhlonectes [Genus]
  T. compressicauda [Species]
  T. cunhai
  T. natans

A brief geologic history of animal life

A note about geologic time scales: A cursory look will reveal that the timing of various geological periods differs among textbooks. Is one right and the others wrong? Not necessarily. Scientists use different methods to estimate geological time—methods with a precision sometimes measured in tens of millions of years. There is, however, a general agreement on the magnitude and relative timing associated with modern time scales. The closer in geological time one comes to the present, the more accurate science can be—and sometimes the more disagreement there seems to be. The following account was compiled using the more widely accepted boundaries from a diverse selection of reputable scientific resources.

Geologic time scale

Era	Period	Epoch	Dates	Life forms
Proterozoic			2,500-544 mya*	First single-celled organisms, simple plants, and invertebrates (such as algae, amoebas, and jellyfish)
Paleozoic	Cambrian		544-490 mya	First crustaceans, mollusks, sponges, nautiloids, and annelids (worms)
	Ordovician		490-438 mya	Trilobites dominant. Also first fungi, jawless vertebrates, starfish, sea scorpions, and urchins
	Silurian		438-408 mya	First terrestrial plants, sharks, and bony fish
	Devonian		408-360 mya	First insects, arachnids (scorpions), and tetrapods
	Carboniferous	Mississippian	360-325 mya	Amphibians abundant. Also first spiders, land snails
		Pennsylvanian	325-286 mya	First reptiles and synapsids
	Permian		286-248 mya	Reptiles abundant. Extinction of trilobytes
Mesozoic	Triassic		248-205 mya	Diversification of reptiles: turtles, crocodiles, therapsids (mammal-like reptiles), first dinosaurs
	Jurassic		205-145 mya	Insects abundant, dinosaurs dominant in later stage. First mammals, lizards, frogs, and birds
	Cretaceous		145-65 mya	First snakes and modern fish. Extinction of dinosaurs, rise and fall of toothed birds
Cenozoic	Tertiary	Paleocene	65-55.5 mya	Diversification of mammals
		Eocene	55.5-33.7 mya	First horses, whales, and monkeys
		Oligocene	33.7-23.8 mya	Diversification of birds. First anthropoids (higher primates)
		Miocene	23.8-5.6 mya	First hominids
		Pliocene	5.6-1.8 mya	First australopithecines
	Quaternary	Pleistocene	1.8 mya-8,000 ya	Mammoths, mastodons, and Neanderthals
		Holocene	8,000 ya-present	First modern humans

*Millions of years ago (mya)

Index

Bold page numbers indicate the primary discussion of a topic; page numbers in italics indicate illustrations.

A

Acanthixalus spp., 6:279, 6:283
Acanthixalus spinosus. See African wart frogs
Acanthostega spp., 6:7
Acid rain, 6:57
Acris spp., 6:48, 6:66, 6:225, 6:228–229
Acris crepitans. See Northern cricket frogs
Actinopterygians, 6:7
Adelogyrinids, 6:10, 6:11
Adelophryne spp., 6:156
Adelotus spp., 6:35, 6:141
Adelotus brevis. See Tusked frogs
Adenomera spp., 6:32, 6:34, 6:156, 6:158
Adenomus spp., 6:184
Adrenal glands, lissamphibian, 6:20–21
Afrana spp., 6:248
African bullfrogs, 6:247, 6:249, 6:254, 6:261
African clawed frogs. *See* Common plantanna
African gray treefrogs. *See* Gray treefrogs
African treefrogs, 6:3–4, 6:6, 6:**279–290**, 6:284
African wart frogs, 6:284, 6:285
Afrixalus spp., 6:32, 6:279, 6:280, 6:281, 6:282–283
Afrixalus brachycnemis, 6:283
Afrixalus delicatus, 6:283
Afrixalus fornasinii. See Greater leaf-folding frogs
Agalychnis spp., 6:49, 6:227
Agalychnis callidryas. See Red-eyed treefrogs
Agalychnis craspedopus, 6:227
Agalychnis moreletii, 6:228–229
Aggressive vocalizations, 6:47–48
Aglyptodactylus spp., 6:281, 6:291, 6:292, 6:293
Ailao moustache toads, 6:112, 6:114–115
Ailao spiny toads. *See* Ailao moustache toads
Aïstopods, 6:10
Alajuela toads. *See* Golden toads
Alatau salamanders. *See* Semirechensk salamanders
Albanerpetontidae, 6:13
Alexteroon spp., 6:279, 6:283
Alexteroon obstetricans, 6:35
Alkaloids, 6:197, 6:198, 6:199, 6:200, 6:208
Allophryne ruthveni. See Ruthven's frogs
Allophrynidae. *See* Ruthven's frogs
Alpine salamanders, 6:38, 6:367
Alpine toads, 6:110, 6:111
Alsodes spp., 6:157
Altiphrynoides spp., 6:184
Altiphrynoides malcomi. See Malcolm's Ethiopian toads
Alytes spp. *See* Midwife toads

Alytes cisternasii. See Iberian midwife toads
Alytes dickhilleni, 6:90
Alytes muletensis. See Majorca midwife toads
Alytes obstetricans. See Midwife toads
Amazonian poison frogs, 6:202, 6:206, 6:208
Amazonian skittering frogs, 6:229, 6:231, 6:239
Ambystoma spp., 6:39, 6:42, 6:49, 6:355, 6:358
Ambystoma californiense, 6:356, 6:359, 6:360
Ambystoma cingulatum. See Flatwoods salamanders
Ambystoma gracile. See Northwestern salamanders
Ambystoma lermaense, 6:356
Ambystoma macrodactylum, 6:356
Ambystoma mavortium. See Tiger salamanders
Ambystoma mexicanum. See Mexican axolotl
Ambystoma opacum, 6:31, 6:35, 6:358
Ambystoma talpoideum, 6:356
Ambystoma tigrinum. See Tiger salamanders
Ambystomatidae. *See* Mole salamanders
Amerana spp., 6:248
American tailed caecilians, 6:5, 6:39, 6:411, 6:412, 6:**415–418**, 6:417
American toads, 6:45, 6:184, 6:190
Amero-Australian treefrogs, 6:**225–243**, 6:231–232
  behavior, 6:228–229, 6:233–242
  conservation status, 6:230, 6:233–242
  defense mechanisms, 6:66
  distribution, 6:5, 6:6, 6:225, 6:228, 6:233–242
  evolution, 6:4, 6:225–226
  feeding ecology, 6:229, 6:233–242
  habitats, 6:228, 6:233–242
  humans and, 6:230, 6:233–242
  physical characteristics, 6:215, 6:226–228, 6:233–242, 6:281
  reproduction, 6:229–230, 6:233–242
  tadpoles, 6:227–228
  taxonomy, 6:225–226, 6:233–242
Amietia spp., 6:248
Amietia vertebralis. See Wide-mouthed frogs
Amo spp., 6:247, 6:248
Amolopini, 6:247
Amolops spp., 6:247, 6:248, 6:249, 6:250
Amolops formosus. See Beautiful torrent frogs
Amphibians
  art and, 6:52, 6:54
  as food, 6:54, 6:252, 6:256
  as introduced species, 6:54, 6:58, 6:191, 6:251, 6:262
  as pets, 6:54–55, 6:58
  behavior, 6:**44–50**

biogeography, 6:4–5
communication, 6:44–48
conservation status, 6:6, 6:**56–60**
definition and description, 6:**3–6**
deformities in, 6:56–57, 6:59
diseases of, 6:57, 6:60
distribution, 6:4, 6:5–6
egg attendance, 6:34–35, 6:216, 6:217
egg deposition, 6:31–32, 6:39, 6:230, 6:251, 6:364
egg development and hatching, 6:33–34
egg fertilization, 6:32–33
egg transportation, 6:35–37
evolution, 6:3, 6:4–5, 6:**7–14**, 6:28
feeding ecology, 6:6, 6:54
habitats, 6:6, 6:7
humans and, 6:**51–55**
larvae, 6:28, 6:36, 6:**39–43**
literature and, 6:52, 6:53
medicinal uses of, 6:53
metamorphosis, 6:28, 6:39, 6:42–43
migrating, 6:59, 6:356
physical characteristics, 6:3, 6:**15–27**, 6:16
population decline of, 6:56–59
protection of, 6:59–60
reproduction, 6:3, 6:18, 6:**28–38**
taxonomy, 6:3–4, 6:11
See also specific topics and types of amphibians
Amphiuma spp., 6:35, 6:405, 6:407
Amphiuma means. See Two-toed amphiumas
Amphiuma pholeter. See One-toed amphiumas
Amphiuma tridactylum. See Three-toed amphiumas
Amphiumas, 6:5, 6:13, 6:323, 6:325, 6:**405–410**, 6:408
Amphiumidae. *See* Amphiumas
Amplectic positions. *See* Amplexus
Amplexus, 6:65, 6:68, 6:304, 6:365–366
Anatomy. *See* Physical characteristics
Andinophryne spp., 6:184
Andrias spp., 6:34, 6:49
Andrias davidianus. See Chinese giant salamanders
Andrias japonicus. See Japanese giant salamanders
Aneides spp., 6:391
Aneides ferreus, 6:391
Aneides lugubris. See Arboreal salamanders
Aneides vagrans, 6:391
Anhydrophryne spp., 6:245, 6:251
Annam broad-headed toads, 6:111, 6:112, 6:115
Annam spadefoot toads. *See* Annam broad-headed toads

INDEX

mythology and, 6:51–52
physical characteristics, 6:15, *6:18*, 6:23–25,
6:24, 6:323–324
predators and, 6:325
reproduction, 6:28–*29*, 6:*31*–35, 6:37–38,
6:325, 6:326
taxonomy, 6:3, 6:4, 6:323–*324*
torrent, 6:3, 6:5, 6:323, 6:325, 6:349,
6:**385–388**, *6:387*
toxins, 6:325
See also specific types of salamanders
Salamandra atra. See Alpine salamanders
Salamandra japonica. See Japanese clawed
salamanders
Salamandra lanzai. See Laza's alpine
salamanders
Salamandra salamandra. See European fire
salamanders
Salamandra tigrina. See Tiger salamanders
Salamandrella spp., 6:335
Salamandrella keyserlingii. See Siberian
salamanders
Salamandridae, 6:**363–375**, *6:368–369*
behavior, 6:*365*–366, 6:370–375
communication, 6:44
conservation status, 6:367, 6:370–375
courtship behavior, 6:49, 6:365–366
distribution, 6:*363*, 6:364, 6:*370–375*
evolution, 6:13, 6:363
feeding ecology, 6:366–367, 6:370–375
habitats, 6:364–365, 6:370–375
humans and, 6:367, 6:370–375
physical characteristics, 6:17, 6:24, 6:325,
6:363, 6:370–375
reproduction, 6:30–31, 6:326, 6:367,
6:370–375
species of, 6:370–375
taxonomy, 6:323, 6:363, 6:370–375
See also Newts
Salamandrina spp., 6:17, 6:24
Salamandrina terdigitata. See Spectacled
salamanders
Salamandroidea, 6:13, 6:323
Salientia, 6:11
Saltenia spp., 6:12
Saltenia ibanezi, 6:99
Sambava tomato frogs, *6:304*
San Marcos salamanders. *See* Texas blind
salamanders
Sanderson's hook frogs, 6:247, *6:254*, 6:260
Sandhill frogs, *6:148*, 6:149, 6:150, *6:151*
Sarcopterygians, 6:7
Savanna squeaking frogs. *See* Common
squeakers
Scapherpetontids, 6:13
Scaphiophryne spp., 6:317, 6:318
Scaphiophryne brevis, 6:318
Scaphiophryne calcarata. See Mocquard's rain
frogs
Scaphiophryne gottlebei. See Red rain frogs
Scaphiophryne madagascariensis, 6:318
Scaphiophryne marmorata, 6:317, 6:318
Scaphiophrynidae. *See* Madagascan toadlets
Scaphiopus spp., 6:119
Scaphiopus couchii. See Couch's spadefoot toads
Scaphiopus holbrookii. See Eastern spadefoot
toads
Scarthyla goinorum. See Amazonian skittering
frogs
Schismaderma spp., 6:184

Schistometopum thomense, 6:411
Schmidt's lazy toads, *6:112*, 6:114
Schoutedenella poecilonotus. See Ugandan
squeakers
Scinax spp., 6:227
Scinax rizibilis, 6:230
Scolecomorphidae. *See* Buried-eye caecilians
Scolecomorphus spp., 6:431, 6:432, 6:437
Scolecomorphus kirkii. See Kirk's caecilians
Scolecomorphus vittatus. See Banded caecilians
Scotiophryne spp., 6:89
Scramble competition, 6:48–49
Scutiger spp., 6:110, 6:111
Scutiger nyingchiensis. See Nyingchi lazy toads
Scutiger schmidti. See Schmidt's lazy toads
Scythrodes spp., 6:156
Sedge frogs. *See* Painted reed frogs
Semirechensk salamanders, 6:53, 6:335, 6:337,
6:*338*, 6:341
Semnodactylus spp., 6:281
Sensory system, lissamphibians, 6:15, 6:21–23
Seychelles frogs, 6:6, 6:36, 6:68, 6:**135–138**,
6:137
Seychelles treefrogs, *6:282*, *6:283*, *6:284*,
6:289
Seymouriamorphs, 6:10
Shakespeare, William, 6:52
Sharp-nosed caecilians. *See* Frigate Island
caecilians
Sharp-nosed grass frogs, *6:254*, 6:260
Sharp-nosed reed frogs, 6:281, *6:284*, *6:286*
Shelania laurenti, 6:99
Shelania pascuali, 6:99
Shomronella spp., 6:12
Shovel-footed squeakers. *See* Common
squeakers
Shovel-nosed frogs, 6:6, 6:**273–278**, *6:274*,
6:276
Siberian salamanders, 6:51, 6:325, 6:336,
6:337, 6:*338*, 6:341–342
Sierra Nevada ensatina. *See Ensatina* spp.
Silurana spp., 6:102
Silurana tropicalis. See Tropical clawed frogs
Sinerpeton spp., 6:13
Siphonops annulatus, 6:411, 6:436
Siphonops mexicanus. See Mexican caecilians
Siredon gracilis. See Northwestern salamanders
Siredon mexicanum. See Mexican axolotl
Siren intermedia. See Lesser sirens
Siren lacertina. See Greater sirens
Siren striata. See Northern dwarf sirens
Sirena maculosa. See Mudpuppies
Sirenidae. *See* Sirens
Sirenoidea. *See* Sirens
Sirens, 6:17, 6:**327–333**, *6:330*
distribution, 6:5, 6:*327*, 6:328, 6:*331–332*
dwarf, 6:**327–333**, *6:330*
evolution, 6:13, 6:327
physical characteristics, 6:323, 6:325,
6:327–328, 6:331–332
reproduction, 6:32, 6:329, 6:331–332
taxonomy, 6:323, 6:327, 6:331–332, 6:335
Skin. *See* Integumentary system
Slender frogs, 6:110
Slender mud frogs, 6:111, *6:112*, *6:113*–114
Smooth newts, 6:*365*, 6:*366*, 6:374–375
Somuncuria spp., 6:157, 6:158
Sonoran tiger salamanders, 6:360
Sooglossidae. *See* Seychelles frogs
Sooglossus gardineri. See Gardiner's frogs

Sooglossus sechellensis. See Seychelles frogs
South American bullfrogs, 6:*157*, 6:*160*,
6:166–167
Southeast Asian broad-skulled toads. *See*
Common Sunda toads
Southern foam nest treefrogs. *See* Gray
treefrogs
Southern gastric brooding frogs, 6:149, 6:*150*,
6:153
Southern ground-hornbills, 6:68
Southern platypus frogs. *See* Southern gastric
brooding frogs
Southern three-toed toadlets, 6:*182*
Southern waterdogs. *See* Dwarf waterdogs
Spadefoot toads, 6:4–5, 6:64, 6:**119–125**,
6:*120–121*, 6:*123*
See also Burmese spadefoot toads
Spanish sharp-ribbed newts, 6:*372*
Spea spp., 6:119
Spea bombifrons. See Plains spadefoot toads
Spea intermontana, 6:122
Spectacled salamanders, 6:*365*
Speleomantes spp., 6:391
Spencer's burrowing frogs, 6:140, 6:*140*, 6:*141*
Spermatophores, 6:33, 6:.365–366, 6:371
Sphaenorhynchus lacteus, 6:229
Sphaerotheca spp., 6:246, 6:248
Sphenophryne cornuta. See Horned land frogs
Spicospina flammocaerulea, 6:148
Spinophrynoides spp., 6:184
Spiny-armed frogs, 6:*253*, 6:258
Spiny-headed treefrogs, 6:37, 6:226,
6:228–229, 6:230, 6:*232*, 6:234–*235*
Spiracles, 6:41
Spix's saddleback toads. *See* Pumpkin toadlets
Spondylophryne, 6:89
Spotted burrowing frogs. *See* Spotted snout-
burrowers
Spotted Cochran frogs, 6:*218*, 6:221
Spotted salamanders, 6:49
Spotted shovel-nosed frogs. *See* Spotted
snout-burrowers
Spotted snout-burrowers, 6:*274*, 6:*276*, 6:277
Spring lizards. *See* Salamanders
Spring peepers, 6:47, 6:228–229
Spring salamanders, 6:390, 6:*391*
Squeakers, 6:**265–271**, 6:267
Staurois spp., 6:248
Staurois latopalmatus, 6:46
Stefania spp., 6:37, 6:230
Stegocephalians, 6:7–11
Stem-caecilians, 6:13
Stem-salamandroids, 6:13
Stem-tetrapods, 6:8, 6:9, 6:10
Stephen's rocket frogs, 6:*201*, 6:204
Stephopaedes spp., 6:184
Stephopaedes anotis. See Chirinda toads
Stereochilus spp., 6:390
Stereochilus marginatus, 6:390
Strawberry poison frogs, 6:45, 6:198–199,
6:200, 6:*202*, 6:*205*, 6:207
Striped newts, 6:372
Strongylopus spp., 6:245
Sumaco horned treefrogs, 6:*232*, 6:233–234
Sung's slender frogs, 6:111
Surinam horned frogs, 6:*158*, 6:161, 6:*162*
Surinam toads, 6:36–37, 6:**99–107**, 6:*100*,
6:*101*, 6:*106*
Synapturanus salseri. See Timbo disc frogs
Syncope spp., 6:304

T